AI and Emerging Technologies for Emergency Response and Smart Cities

Froilan D. Mobo
Philippine Merchant Marine Academy, Philippines

IGI Global
Scientific Publishing
Publishing Tomorrow's Research Today

Vice President of Editorial	Melissa Wagner
Director of Acquisitions	Mikaela Felty
Director of Book Development	Jocelynn Hessler
Production Manager	Mike Brehm
Cover Design	Phillip Shickler

Published in the United States of America by
IGI Global Scientific Publishing
701 East Chocolate Avenue
Hershey, PA, 17033, USA
Tel: 717-533-8845
Fax: 717-533-7115
Website: https://www.igi-global.com E-mail: cust@igi-global.com

Library of Congress Cataloging-in-Publication Data

Names: Mobo, Froilan Delute, 1979- editor
Title: AI and emerging technologies for emergency response and smart cities / [edited by] Froilan Mobo.
Description: Hershey, PA : IGI Global Scientific Publishing, [2025] | Includes bibliographical references and index. | Summary: "In emergency response, AI-powered systems have demonstrated their ability to process vast amounts of data in real time, enabling quicker and more efficient decision-making. From predictive analytics that forecast natural disasters to autonomous drones that aid in search and rescue operations, emerging technologies are revolutionizing the field. This book delves into the tools and frameworks that harness these technologies to mitigate risks, minimize damage, and save lives in times of crisis"-- Provided by publisher.
Identifiers: LCCN 2025004045 (print) | LCCN 2025004046 (ebook) | ISBN 9798369397701 hardcover | ISBN 9798369397718 paperback | ISBN 9798369397725 ebook
Subjects: LCSH: Emergency management--Technological innovations | Smart cities | Artificial intelligence
Classification: LCC HV551.2 .A37 2026 (print) | LCC HV551.2 (ebook) | DDC 363.34/80285--dc23/eng/20250425
LC record available at https://lccn.loc.gov/2025004045
LC ebook record available at https://lccn.loc.gov/2025004046

British Cataloguing in Publication Data
A Cataloguing in Publication record for this book is available from the British Library.

Table of Contents

Detailed Table of Contents

Artificial intelligence has emerged as a key technology for addressing the challenges of urban resilience against disasters. This paper explores how AI can be pivotal in strengthening the resilience of smart cities, enabling them to withstand and adapt to the increasingly frequent and intense natural disasters that threaten urban centers worldwide. The research has delved into the networks that make a smart city resilient to disasters and their interplay. The paper also highlights the complementarity between smart and resilient cities frameworks. The integration of AI within the framework of smart city represents a significant leap toward building disaster-resilient urban environments. It emphasizes the synergistic relationship between smart and resilient city frameworks, showcasing that the incorporation of AI is a crucial advancement toward creating disaster-resilient urban environments.

A smart city is an urban environment that gathers data through diverse electrical means and sensors. Smart cities utilize information and communication technologies (ICT) to enhance service quality through the management of public resources, emphasizing comfort and preserving environmental sustainability. The fifth generation (5G) of wireless mobile communication facilitates a novel form of communication network to link all individuals and entities. 5G will have a significant impact on economies and civilizations by providing the essential communication infrastructure required by diverse smart devices. One of the many applications in smart cities is the Intelligent Transport System (ITS). 5G technology has the potential to transform ITS. The paper intends to examine the effects and significance of 5G on ITS from multiple perspectives.

Disaster management systems. This introduction explores the role of disaster management technologies, focusing on their evolution, current applications, and future potential, particularly in the context of AI-based solutions.Historically, disaster management relied heavily on manual processes and rudimentary tools for assessment and response. This often resulted in delayed reactions and inadequate resource allocation during emergencies. Geographic Information Systems (GIS), remote sensing, and telecommunications have all played crucial roles in improving situational awareness and enabling faster decision-making.ICT facilitates real-time data sharing among stakeholders, including government agencies, NGOs, and the public. This connectivity enhances coordination and ensures that critical information reaches affected populations promptly. For instance, mobile technology has empowered communities to report incidents and receive alerts, thereby improving early warning systems.Moreover, data analytics has emerged as a key component in disaster management.

The revolutionary potential of AI-driven personalized learning to improve teacher effectiveness and student engagement is examined in this chapter. By customizing content to each learners unique learning needs and preferences AI technologies like intelligent tutoring platforms and adaptive learning systems are revolutionizing educational experiences. By letting students advance at their own speed and addressing their weaknesses and strengths personalized learning creates a more stimulating and productive learning environment. The importance of artificial intelligence (AI) in boosting student motivation is examined in this chapter using interactive tools such as gamification augmented and virtual reality apps and real-time feedback systems. The chapter discusses how AI affects student learning as well as how it can help teachers become more efficient by automating repetitive administrative duties like attendance and grading. Teachers can concentrate more on professional development teamwork and instruction by freeing up valuable time.

This chapter explores the transformative role of Artificial Intelligence (AI) in advancing sustainable urban public transportation. As cities grow, challenges like congestion, emissions, and efficiency demand innovative solutions. AI addresses these through data-driven methods in route optimization, traffic

This chapter explores the potential of mobile applications in enhancing earthquake preparedness and response in Taiwan, a region with high seismic activity. It investigates the design and development of a comprehensive mobile application integrating real-time earthquake information, interactive communication tools, and personalized safety instructions using Geographic Information Systems (GIS) technology. The study examines the effectiveness of such an application in enhancing user preparedness and response, focusing on the role of real-time data delivery, personalized safety guidance, communication capabilities, and GIS integration. By analyzing user perception and behavior changes, the research assesses the impact of the mobile application on individual and community preparedness and response during earthquake events. This investigation provides valuable insights into leveraging mobile technology, GIS, and real-time communication to improve disaster management strategies in Taiwan.

Attendance management in various settings, such as classrooms and workplaces, presents challenges due to its time-consuming nature and susceptibility to proxy attendance. Conventional methods of attendance marking, including fingerprint and radio frequency identification (RFID) systems, lack reliability and efficiency. The system utilizes models such as Haar cascade for face detection, OpenCV-Python for image processing, a Caffe model for deep learning-based face detection, and Support Vector Machines (SVM) for facial recognition. Instead of traditional methods, this system aims to streamline attendance recording by utilizing facial recognition technology. The primary objective is to develop a system that is efficient, time-saving, and user- friendly. Compared to traditional methods, this system significantly reduces the workload associated with attendance management. The proposed system will be implemented in four phases: Image Capture, Group Image Segmentation and Face Detection, Face Comparison and Recognition, and Attendance Database Updating.

With mobile devices in widespread use, accessible public charging has become essential. However, traditional charging kiosks often lack security, increasing the risk of theft. This study introduces a new anti-theft mobile charging station designed to maximize security and user convenience. Key features include a keyless lock, an alarm system, a coin-operated payment mechanism, and solar power, along with a fingerprint sensor for user authentication and an infrared alarm to prevent unauthorized access. The design process involved detailed diagrams and flowcharts, leading to the station's systematic construction with an electronic box, secure locker, and robust frame. User testing showed high acceptability, confirming the system's feasibility as a secure, reliable, and user-friendly solution for public places. This extensible system offers a practical way for users to charge devices securely and conveniently in any public environment.

The rapid growth of urban populations has led to increased waste production, challenging conventional waste management systems. This paper presents an IoT-based Smart Waste Bin System designed to optimize waste collection and enhance urban cleanliness. Equipped with sensors, the system monitors waste levels in real-time and transmits data via LoRaWAN or NB-IoT to a cloud platform. It triggers alerts for improper waste sorting and guides users on correct disposal. A compression mechanism activates at 70% capacity to reduce waste volume, and when the bin reaches 90%, the system notifies municipal authorities with GPS data. Instant data analysis facilitates efficient waste collection scheduling and route optimization, lowering operational costs and enhancing efficiency. The system supports sustainability by minimizing unnecessary collection trips and fuel consumption. Initial tests show its potential to improve traditional practices, contributing to smarter, cleaner cities and aligning with broader smart city strategies.

This chapter explores the critical role of cross-border collaborations in advancing AI-enabled smart cities. As urban areas grapple with challenges in resource management, sustainability, and quality of life, AI offers transformative solutions, spanning areas like transportation, energy, public safety, and healthcare. However, the development of smart cities transcends national boundaries, necessitating robust international partnerships. This chapter also examines successful models of cross-border collaborations, such as international consortia and public-private partnerships, and presents case studies that highlight the benefits of knowledge-sharing, resource pooling, and harmonized regulatory frameworks. Ultimately, this chapter advocates for collaborative efforts that build resilient, AI-driven urban environments equipped to meet the demands of an increasingly urbanized future, fostering sustainability, equity, and quality of life for all citizens.

The current research focuses on investigating how the smart homes' use of IoT uplifts energy efficiency and sustainability. Though IoT is a means of satisfying growing energy demands and combating ecological problems, there is a problem of incorporating devices into an IoT system. Our study responds to these challenges and suggests that better positioned antennas, enhanced RF link quality, and better cooling systems can greatly lower energy consumption. This contributes to the literature on how IoT may be used to support positive sustainability initiatives and expands general knowledge of its use in energy management and preservation.

With accuracy in customer insights, personalization at its best is made possible, and all of this is automated within the creation of content itself. AI helps brands sift through huge datasets, foretell consumer preferences, and even adjust campaigns in real time to increase the importance and impact of marketing initiatives. Brands use tools such as chatbots and recommendation engines to connect with the customers through targeted content for higher customer satisfaction and loyalty and eventually influence consumer decisions in purchasing. Another area is programmatic advertising and DCO, in which AI maximizes ad efficiency and reach. Ethical challenges in data privacy, bias, and transparency form the essence of these aspects and require brands to strike a balance between innovation and responsible AI use in maintaining consumer trust.

Preface

The increasing integration of Artificial Intelligence (AI) and emerging technologies into our daily lives is transforming the way we respond to emergencies and build smart, sustainable cities. As urbanization accelerates and global challenges such as climate change, pandemics, and natural disasters become more prevalent, the need for innovative solutions to enhance resilience and adaptability in urban environments has never been more urgent. This book, *AI and Emerging Technologies for Emergency Response and Smart Cities*, brings together cutting-edge research, practical applications, and visionary insights from experts worldwide to address these critical challenges.

In emergency response, AI-powered systems have demonstrated their ability to process vast amounts of data in real time, enabling quicker and more efficient decision-making. From predictive analytics that forecast natural disasters to autonomous drones that aid in search and rescue operations, emerging technologies are revolutionizing the field. This book delves into the tools and frameworks that harness these technologies to mitigate risks, minimize damage, and save lives in times of crisis.

Similarly, smart cities rely on AI and advanced technologies to create sustainable, efficient, and citizen-centered urban environments. Smart transportation, energy management, waste disposal systems, and enhanced public safety mechanisms are some of the innovations reshaping urban spaces. The chapters in this book explore how AI-driven solutions enable cities to optimize resources, improve infrastructure, and deliver superior services while addressing the environmental and social challenges of modern urbanization.

This edited volume aims to serve as a resource for academics, policymakers, technology developers, and practitioners who seek to understand and leverage the transformative potential of AI and emerging technologies. Each chapter offers unique perspectives, case studies, and actionable insights, highlighting the collaborative and interdisciplinary efforts required to tackle complex problems.

We extend our heartfelt gratitude to the contributors whose expertise and commitment have made this book a valuable contribution to the growing body of knowledge in the fields of emergency response and smart cities. We also express our appreciation to the readers who recognize the importance of advancing these fields for a more secure, sustainable, and inclusive future.

It is our hope that this book inspires innovative thinking and fosters meaningful collaboration in harnessing AI and emerging technologies to build safer and smarter communities worldwide.

Prof. Dr. Froilan Delute Mobo

January 2025

Chapter 1
AI Integration to Strengthen Disaster Resilience in Smart Cities

Ravinder Singh
https://orcid.org/0009-0007-7161-6709
SR University, India

Geetha Manoharan
https://orcid.org/0000-0002-8644-8871
SR University, India

ABSTRACT

Artificial intelligence has emerged as a key technology for addressing the challenges of urban resilience against disasters. This paper explores how AI can be pivotal in strengthening the resilience of smart cities, enabling them to withstand and adapt to the increasingly frequent and intense natural disasters that threaten urban centers worldwide. The research has delved into the networks that make a smart city resilient to disasters and their interplay. The paper also highlights the complementarity between smart and resilient cities frameworks. The integration of AI within the framework of smart city represents a significant leap toward building disaster-resilient urban environments. It emphasizes the synergistic relationship between smart and resilient city frameworks, showcasing that the incorporation of AI is a crucial advancement toward creating disaster-resilient urban environments.

1. INTRODUCTION

In the twenty-first century, cities around the world have become engines of economic growth, innovation, and social development. Driven by rapid advancements in technology and fueled by the promise of enhanced quality of life, the concept of "smart cities" has gained traction globally. Smart cities are urban areas that integrate information and communication technology (ICT) and Internet of Things (IoT) technologies to manage resources more efficiently, provide better public services, and improve overall urban sustainability. Governments and private entities alike are investing substantial financial resources to develop these cities, embedding them with technology that can respond dynamically to

DOI: 10.4018/979-8-3693-9770-1.ch001

urban challenges and improve citizens' lives. The global smart cities market is projected to expand from USD 410.8 billion in 2020 to USD 820.7 billion by 2025 (Markets, 2020), with projections indicating sustained growth in the coming years. However, as urban populations swell due to enhanced economic opportunities, cities face significant vulnerabilities. These risks, exacerbated by climate change, include increased susceptibility to natural disasters such as floods, earthquakes, and hurricanes.

Urbanization, while offering economic opportunities, has created sprawling metropolises with critical infrastructure—transport, healthcare, and utilities—that are highly susceptible to disruption (Srivastava, 2020). Today's smart cities, equipped with sensors, data analytics, and autonomous systems, provide an excellent platform for integrating disaster resilience (Sharifi et al., 2021; Jung et al., 2020). By leveraging artificial intelligence (AI), cities can make substantial strides in disaster prediction, preparedness, response, and recovery. This chapter explores how AI can be pivotal in strengthening the resilience of smart cities, enabling them to withstand and adapt to the increasingly frequent and intense natural disasters that threaten urban centers worldwide.

1.1. The Smart City Landscape: Opportunities and Challenges

The shift toward smart cities reflects a broader global movement toward urban centers that are not only more livable but also more responsive to the needs of their residents. From intelligent traffic systems that reduce congestion to energy grids that optimize power distribution, smart cities embody the ideals of connectivity, convenience, and efficiency. Examples include Singapore's Smart Nation initiative, which prioritizes the use of technology to enhance mobility, sustainability, and public safety while also promoting efficient governance for the intelligent application of technology (Chang & Das, 2020), and Barcelona's Smart City program, renowned for its environmental monitoring systems and resource-efficient utilities (Bibri & Krogstie, 2020).

In regions experiencing high urbanization rates, where better job prospects, healthcare, education, and social services drive rural-to-urban migration, the allure of smart cities is especially potent. Estimates suggest that nearly 68% of the global population will live in urban areas by 2050 (Gu et al., 2021), accentuate the critical need for resilient, sustainable city planning. However, as cities expand and concentrate on economic activities, they inadvertently increase their exposure to hazards (Singh & Manoharan, 2024). Urban centers with densely packed infrastructure and large populations become magnets for climate-related and other environmental risks. Floods, hurricanes, and even pandemics can ripple through a city's interconnected systems, disrupting lives and livelihoods.

1.2. Rising Vulnerabilities in Urban Centers

We expect an increase in the frequency, intensity, and unpredictability of disasters as climate change accelerates. Natural disasters worldwide caused losses exceeding $275 billion in 2022, with urban centers bearing a significant portion of these losses (Yuryeva et al., 2023). Coastal cities, for instance, face heightened risks of flooding due to rising sea levels, while inland cities might grapple with extreme heat or water scarcity. These scenarios necessitate resilient infrastructure capable of withstanding shocks and stresses. Reaction must give way to proactive resilience, where cities not only prepare for disasters but also design to absorb, adapt, and recover from them with minimal impact on lives and resources. A multi-faceted strategy, leveraging the very technologies that are the strength of smart cities, is necessary for this shift. Here, AI stands out as a transformative tool, capable of analyzing vast amounts of data in real time, identifying patterns, and generating insights that would be impossible for humans to achieve unaided (Manoharan et al., 2024).

1.3. The Transformative Role of AI in Building Resilient Cities

The integration of AI into the disaster resilience framework of smart cities represents a significant leap toward safer, more adaptable urban environments. As cities worldwide continue to invest in technology-driven solutions, AI offers a unique opportunity to shift from reactive disaster management to proactive resilience-building (Alahi et al., 2023). By predicting potential risks, enabling real-time monitoring, automating emergency response, and ensuring the resilience of essential infrastructure, AI can help cities navigate the challenges posed by climate change and rapid urbanization. To achieve this vision, it is imperative that cities prioritize the integration of AI into their smart city strategies, fostering collaboration among public and private entities, academia, and local communities. In doing so, urban centers can transform into resilient hubs that not only survive disasters but continue to thrive in the face of adversity. As we move forward, the synergy between AI and smart city infrastructure will become a cornerstone in the global effort to create disaster-resilient urban spaces, paving the way for a safer, more sustainable future for generations to come.

2. SMART CITIES AND THEIR EVOLUTION

2.1. The Evolution of Urban Growth: From Garden Cities to Smart Cities

The transformation of rural areas into urban centers is driving the rapid expansion of cities. Improved employment prospects, enhanced social benefits, modern infrastructure, and natural population growth largely motivate this shift (Eremia et al., 2017). Urban architects and planners face the challenge of designing smart systems that promote resource efficiency and foster sustainable development. Sir Ebenezer Howard, in his 1898 book Garden Cities of Tomorrow, envisioned a utopian city where residents coexisted harmoniously with nature, laying the groundwork for sustainable city planning. The Smart City concept emerged in the 1990s, aiming to tackle key urban challenges: establishing a robust economic foundation, building efficient infrastructure, enhancing quality of life and place, fostering social inclusion, conserving natural resources, and ensuring effective governance (Anthopoulos LG, 2017).

In 2001, the California Institute for Smart Communities introduced the term “smart community,” emphasizing the crucial role of community involvement in enhancing urban life and work environments. This marked a shift toward a more inclusive and participatory approach (Moura & de Abreu e Silva, 2021). By 2007, Giffinger et al. further expanded this notion by integrating governance and citizen-centric dimensions into the smart city model. This shift focused on the importance of a government’s role in shaping a city that prioritizes its residents’ needs and promotes engagement, transparency, and accountability. In 2010, Chen introduced the concept of “sensorizing” cities—embedding sensors throughout urban environments to collect and analyze real-time data. The early 2010s saw multiple scholars refine the smart city definition. For example, Lazaroiu and Roscia (2012) envisioned a smart city as a “community of average technology size” focused on interconnectivity, sustainability, and security. Barrionuevo et al. (2012) emphasized using technology and resources intelligently and in coordination to develop integrated, livable, and sustainable urban centers. Kourtit and Nijkamp (2012) added that smart cities leverage knowledge-intensive strategies to boost their socioeconomic, ecological, and competitive performance. They highlighted a "promising mix" of human, infrastructural, social, and entrepreneurial capital as the backbone of a truly smart city. Lombardi et al. (2012) further supported this approach by advocating for the role of ICT in enhancing human capital, social relations, and environmental initiatives. Cretu (2012) introduced two primary schools of thought regarding smart cities: one focused on leveraging new governance and economic paradigms, and the other on creating a network of sensors, devices, and ICT integration to enhance every facet of urban life. In 2013, Bakıcı et al. described smart cities as "high-tech-intensive" environments connecting people, data, and infrastructure, aiming to create a greener, more competitive, and sustainable city that enhances quality of life. The idea of a smart city as a defined geographical area powered by high technologies like ICT and energy logistics was solidified by Dameri (2013). Dameri highlighted the benefits of smart city development, from well-being and environmental quality to intelligent urban growth. Zygiaris (2013) further refined this view by categorizing smart cities as "green," "interconnected," and "intelligent," underscoring their ability to generate value through real-time data from sensors and innovative, knowledge-driven human capital. In 2015, Marsal-Llacuna et al. encapsulated the modern vision of smart cities, focusing on using data and IT to optimize urban services, improve infrastructure, foster collaboration, and encourage innovation within both the private and public sectors.

2.2. Definitions of Smart Cities

Smart cities are defined as the integration of physical, digital, and human systems to create sustainable, inclusive urban environments. They utilize IoT sensors embedded in infrastructure, vehicles, and devices to collect real-time data on systems like energy, transportation, and law enforcement. This data is integrated into ICT platforms, enabling city managers to optimize operations, enhance resilience, and engage stakeholders. Evolving definitions emphasize aspects like participatory governance, energy efficiency, environmental sustainability, and talent attraction. Early definitions highlighted community involvement, while later perspectives stressed technological, creative, and entrepreneurial dimensions, shaping smart cities as hubs of innovation and sustainable urban development (Moura et al., 2021).

2.3. The Role of AI, IoT, and Data-Driven Solutions in Advancing Smart Cities.

2.3.1. Defining Internet of Things

The IoT is a transformative concept in information technology, though it lacks a universally agreed definition. Broadly, IoT refers to a framework where devices equipped with sensors and embedded technology collect, transmit, and interact with data autonomously over the internet. These "Things" can engage with both their internal states and external environments, aiding decision-making processes. The IEEE defines IoT as enabling new applications and services that bridge physical and virtual worlds, with Machine-to-Machine (M2M) communication as its foundation, facilitating interactions between devices and cloud-based applications (IEEE Communications Magazine, IEEE Communications Society, 2024). Oxford Dictionaries simplifies IoT as the interconnection of everyday objects via the internet to exchange data. IoT connects smaller systems, such as smart devices and homes, into broader ecosystems like smart cities, creating an inclusive and complex network. This interconnected system enhances functionality and efficiency but also represents a multi-faceted and evolving technological domain (Mouha, 2021).

2.3.2. IoT-Driven Smart Cities: Enhancing Urban Living and Sustainability

Smart cities aim to elevate the quality of life by providing residents with enhanced healthcare, safety, convenience, and access to knowledge. Central to this vision, IoT emerges as a transformative technology, fostering significant advancements and reshaping urban landscapes worldwide. By connecting devices and systems, IoT enables smarter, more responsive cities, driving efficiency, sustainability, and accessibility. This integration marks a global shift toward interconnected urban environments that prioritize the well-being and needs of their citizens (Talebkhah et al., 2020). We have made extensive efforts to advance smart homes, intelligent transportation, traffic management, waste disposal, energy systems, healthcare, and numerous other facilities, all working together to build a smart city. These integrated developments aim to create interconnected, efficient urban environments that enhance daily living and foster sustainable, responsive infrastructure. Together, they support the vision of cities that are more adaptable, resourceful, and centered on improving residents' quality of life (Talebkhah et al., 2021). The smart city model has gained greater popularity compared to other urban development approaches (such as the information city, digital city, and telicity) and serves as a practical application of IoT, thereby incorporating its core operational systems. Talebkhah (2021) has outlined IoT's role in developing smart cities, providing essential components for their structure, as illustrated in Figure 1. Bibri & Krogstie (2020) analyzed the development and practical implementation of data-driven smart cities, specifically focusing on sustainable and innovative solutions. The research employed a case study approach, highlighting the realization of this new urban model by comparing London and Barcelona, two of Europe's top data-driven smart cities. The findings indicated that London excels in ICT infrastructure and diverse data sources, making it a leader in technical capabilities. Barcelona, on the other hand, excelled in data-oriented skills, showcasing horizontal information platforms, operations centers, dashboards, training and education programs, innovation labs, research centers, and strategic planning initiatives.

Figure 1. IoT's role in developing smart cities

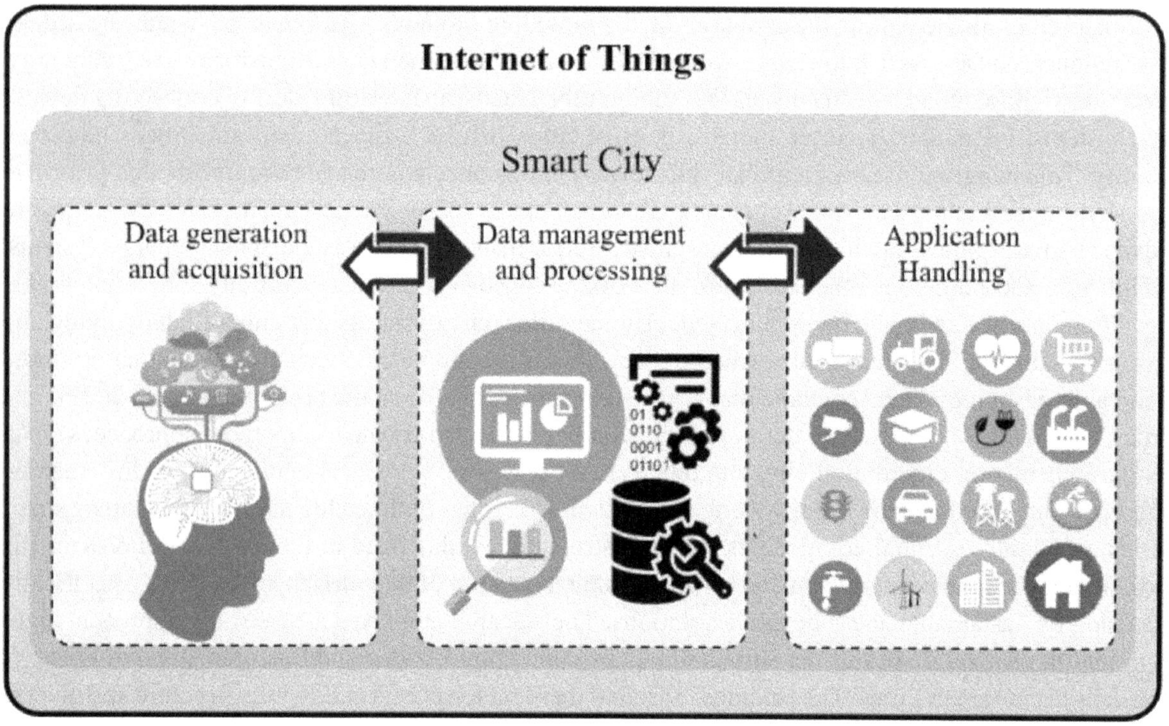

(Talebkhah et al., 2021)

2.3.3. Transforming Smart Cities through Digital Innovation

Radu's (2020) research highlights the integration of IoT, AI, big data, and blockchain in smart cities, transforming daily life for citizens. The study identifies that disruptive technologies can drive positive change by promoting sustainable economic and social practices. IoT, AI, and big data automate decision-making and problem-solving, contributing to smarter cities, while blockchain enhances trust by ensuring secure communication and optimizing existing infrastructure. These technologies, particularly AI, big data, IoT, and blockchain, are interdependent and complement other innovations like wearable tech, social robotics, VR/AR, 3D printing, and digital twins. The research also emphasizes that these technologies address key areas such as mobility, environmental sustainability, health, security, business, energy, and education in smart cities.

3. THE IMPORTANCE OF INTEGRATING DISASTER RESILIENCE IN SMART CITY FRAMEWORKS

3.1. Building Urban Resilience: Strategies for Disaster Risk Management in Complex Systems

Resilience has become a key focus in urban planning and research, offering valuable insights into managing disaster challenges within complex socio-ecological systems. Urban resilience refers to "the ability of a city to recover from devastation," achieved through three mechanisms: persistence (the capacity of systems to resist disruptions), transition, and transformation. Enhancing resilience is viewed as essential for reducing immediate casualties, injuries, and economic impacts from disasters through four primary strategies: (1) conducting systematic risk assessments and monitoring to increase public and government understanding of disaster risks; (2) fostering a culture and establishing incentives that hold stakeholders accountable for disaster preparedness, response, and recovery; (3) prioritizing long-term planning through targeted investments and the use of existing frameworks; and (4) fostering international collaboration through support, research, and ongoing evaluation (Hofmann, 2021).

3.2. Leveraging Smart City Technologies for Sustainable Urban Development

With the rise of urbanization and population growth, cities face increasingly visible challenges like energy shortages, air pollution, and waste management issues (Zhu et al., 2019). In response, the concept of the smart city aims to leverage information and communication technologies along with other advanced technologies to tackle these issues and enhance urban functionality. Technology and people are at the core of the smart city framework. Researchers are investigating and adapting innovative technologies like the IoT, big data, and fog computing for effective implementation in smart city initiatives. People-focused elements like education, learning, and knowledge are also central to smart cities, as they contribute to productivity and improved quality of life. Continued research underscores that technology alone cannot solve all urban problems, highlighting the importance of green and sustainable development factors, such as environmental protection and public safety, which are essential for creating healthier, more satisfying urban environments (Yigitcanlar et al., 2019; Zhu et al., 2020).

3.3. A Global Imperative for Disaster Preparedness and Risk Management

In the past two decades, the frequency of devastating disaster events worldwide has led to an increased global focus on the resilience of communities to disasters. Rapid urbanization and poor development planning have heightened the vulnerability of communities to hazards, often creating new risks or worsening existing ones, resulting in a sharp rise in disaster-related losses. The primary reason for the escalation of hazards into disasters is the inability of communities to effectively manage emerging risks. To minimize damage and recover swiftly, communities must take proactive measures to mitigate risks and build resilience (Meduri et al., 2024). However, recovery speed and effectiveness vary across communities, influenced by factors such as socioeconomic status, external support, and prior disaster experience. As a result, understanding the unique characteristics of resilient communities is crucial for policymakers and researchers to help improve disaster preparedness and recovery. Resilience has thus become a central concept in managing risks related to climate change, environmental disasters, critical infrastructure, security, and terrorism. Global discussions now embed resilience, forming the foundation of key international frameworks such as the UN's Sendai Framework for Disaster Risk Reduction, the Paris Agreement on Climate Change, and the Sustainable Development Goals (Parker, 2020). Slovakia only carries out disaster preparedness tasks in accordance with legal obligations and regulations, according to a study on the importance of disaster preparedness factors for sustainable disaster risk management. There are no specific initiatives, policies, or strategies aimed at enhancing societal resilience to disasters, nor are there any efforts focused on the strategic development of the population's preparedness in this regard (Titko & Ristvej, 2020).

3.4. Smart City vs. Resilient City: Balancing Urban Innovation and Sustainability

Increasing concerns over urban challenges and environmental stress have sparked global interest in the concepts of the Smart Cities and Resilient Cities. Research shows that smart cities development is advancing more quickly than resilient cities, with smart cities studies primarily rooted in engineering, telecommunications, and computer science, whereas resilient cities research focuses on environmental and urban studies. These distinctions suggest that both smart cities and resilient cities are crucial for urban progress and warrant balanced consideration in decision-making. Given rising threats from natural hazards (such as floods, earthquakes, and hurricanes) and social issues (such as terrorism, COVID-19, and poverty), Resilient city frameworks are seen as capable of addressing a wide range of crises (Zhu et al., 2020; Bertilsson et al., 2018). Forecasts indicate that urban populations will surpass 66% by 2050, positioning resilient cities as a proactive model to equip cities for a variety of urban crises. Unlike Resilient cities, smart cities offer both advantages and challenges, particularly due to rapid technological advancements and a limited focus on environmental and social concerns. Smart cities faces issues such as network vulnerability, cybersecurity risks, data privacy, and social inequality, with sustainability emerging as a significant challenge (Zhu et al., 2020; Yigitcanlar et al., 2019). Therefore, ensuring the sustainable development of smart cities is essential for the future.

3.5. Integrating Disaster Resilience into Smart Cities for Sustainable Futures

Integrating disaster resilience into smart city frameworks is essential for fostering sustainable, adaptable urban environments that can withstand various crises. As urban populations are on the increase, cities face heightened vulnerabilities from natural hazards like floods, earthquakes, and storms, as well as human-made threats such as cybersecurity breaches and social unrest. Resilient infrastructure and proactive disaster management strategies in a smart city can mitigate the impact of such events, safeguarding both its citizens and technological assets. Disaster resilience within smart cities involves embedding risk reduction, preparedness, and response capabilities into urban planning, digital infrastructure, and public services. Leveraging data from IoT devices, big data analytics, and AI, smart cities can monitor real-time environmental conditions, predict hazards, and facilitate rapid response strategies. For instance, flood-prone areas can utilize sensors to monitor water levels and provide early warnings to residents, reducing casualties and damages. Additionally, smart transportation systems equipped with AI can help reroute traffic and prioritize emergency response vehicles during crises, significantly improving response efficiency. Moreover, disaster resilience is crucial for safeguarding the continuity of smart city services. Cyber threats, for example, pose unique risks in smart cities, where interconnected systems control critical infrastructure. Resilience measures, such as data redundancy and robust cybersecurity protocols, ensure these systems remain operational during disruptions, maintaining public safety and trust. Integrating resilience also supports sustainable development, a core element in frameworks like the Sendai Framework for Disaster Risk Reduction and the Sustainable Development Goals. By proactively addressing potential risks, smart cities can foster safer, more equitable communities and encourage sustainable growth. Ultimately, disaster resilience is not merely an add-on but a fundamental component of a robust smart city, ensuring cities are not only technologically advanced but also prepared for the complexities of the modern world.

4. SYSTEMS FOR INTEGRATING DISASTER RESILIENCE IN SMART CITIES

4.1. Integrating Resilience into Urban Systems for Smart Cities

Integrating disaster resilience into urban systems is essential for building robust and adaptable smart cities that can withstand natural and human-made hazards. Resilience-focused systems create a framework for cities to anticipate, absorb, and recover from disruptions efficiently. These systems include critical response and support networks, community and societal resilience networks, and sustainable and adaptive infrastructure networks, all of which are essential for ensuring continued operations and minimizing the impact of disasters on residents (Meduri et al., 2024). By embedding resilience into these systems, cities can better prepare for disasters, enabling rapid response and recovery efforts. For example, resilient transportation systems facilitate timely evacuations, while healthcare networks ensure continuous medical support for affected populations (Elvas et al., 2021). Similarly, community networks play a vital role in building resilient systems by fostering local engagement, coordination, and resource-sharing, which enhances both preparedness and response capabilities during disasters. They create a foundation of mutual support and inclusivity, enabling communities to mobilize quickly, address diverse needs, and accelerate recovery efforts effectively (Ramirez Lopez & Grijalba Castro, 2020). Furthermore, integrating resilience enhances adaptive capacity, allowing systems to evolve based on past incidents and new data. For smart cities, leveraging AI, IoT, and data analytics within these systems can improve situational awareness, monitor vulnerabilities, and predict potential risks. Overall, resilience-focused systems foster a culture of preparedness, mitigate economic and social losses, and promote sustainable urban development. Prioritizing resilience in these interconnected systems is vital for safeguarding cities, improving their quality of life, and enabling a rapid, well-coordinated response to future crises.

4.2. Challenges in Integrating Resilience Systems in Smart Cities

Smart cities often face challenges in fully integrating resilience systems due to several factors, including limited resources, complexity, and competing priorities. One key issue is the lack of sufficient funding and political will to prioritize resilience within urban planning. Investments in resilience systems, like infrastructure upgrades and risk mitigation measures, may appear less urgent or secondary to immediate development goals, despite the focus on technology and innovation. Moreover, integrating resilience requires collaboration across various sectors, such as transportation, healthcare, and environmental management, which can be difficult due to fragmented governance structures and lack of coordination between different agencies. The complexity of resilience itself, which encompasses social, environmental, economic, and infrastructural aspects, makes it harder to create holistic systems that address all dimensions of risk and vulnerability. Additionally, there is often a lack of long-term vision in smart city planning. Many urban development projects focus on short-term gains and technological advancements, neglecting the need for systems that can withstand or recover from disasters. Furthermore, the lack of engagement from urban populations in resilience planning can result in gaps in community-based preparedness and response mechanisms. As a result, while smart cities invest in digital and technological advancements, they often overlook the necessity of integrating comprehensive resilience strategies into their urban systems.

4.3. Comparison of Smart City Components with Disaster Resilience

Meduri et al. (2024) conducted a comprehensive review comparing smart city frameworks with disaster mitigation and resilience strategies. Their literature review identified specific smart city components critical to disaster mitigation and resilience, organized under key areas such as architecture, governance, planning, data management, facilities, services, people, and the environment. For example, Information and Communication Technology (ICT) tools are essential for smart city architecture, aiding not only in efficient urban operations but also in disaster response coordination. Similarly, components like smart mobility in transport and water distribution play dual roles in enhancing daily functionality and providing resilience in emergencies. Key elements of governance in smart cities, such as social inclusion and efficient logistics, strengthen community bonds and streamline resource allocation during crises. Planning and management involve ICT tools, while data and knowledge rely on integrated database management, remote sensing, and educational initiatives, all of which contribute to informed decision-making in disaster scenarios. Social, economic, institutional, infrastructural, and community components further strengthen disaster resilience. Social resilience emphasizes educational equity and healthcare access, while economic resilience focuses on income equity and labor force participation. Institutional resilience underscores governance and planning, and infrastructural resilience highlights shelter availability, healthcare facilities, and road access. Community resilience fosters relationships, citizen participation, and social innovation. Together, these interconnected components enhance a city's overall adaptability and robustness against disasters. Table 1 in their study lists the critical components that bridge smart city initiatives with disaster resilience strategies, providing a structured overview of areas such as transportation, public safety, healthcare systems, environmental monitoring, and emergency response.

Table 1. Critical Components of Smart Cities and Disaster Resilience

Smart City Components	Disaster Resilience Components
Architecture • Information and Communication Technology (ICT) Tools • Transport (Smart Mobility) • Water Distribution • Health Care	Economic resilience • Equitable income • Female labor force participation • Diversified economic base • % of employment • % of house ownership
Governance • Social Inclusion • Logistics & supply chain	Social resilience • Educational equality • Health security • Transport access
Planning & Management • ICT Tools	Institutional resilience • Planning organizational systems • Strengthening local governance • % of population covered in mitigation
Data & Knowledge • Integrated Database Management • Remote sensing data • Knowledge driver → education	Infrastructural resilience • Community response & recovery capacity • % of shelters • Health care facilities • Frequency of vulnerability assessment • Road access • Shelters for pre-disaster evacuation • Waste management
Environment • Water Management • Pollution control • Housing • Water resource management	Community • Relationships between communities • Sense of community • Place attachment • Citizen participation • % of knowledge workers • Prospective social innovation
Services • Health • Water	
People • Employment • Knowledge • Quality of life	
Facilities • Energy → generation, distribution, consumption	

(Source Meduri et al., 2024)

4.4. Core Networks in Smart Cities Mission: Enhancing Disaster Resilience

The study by Meduri et al. (2024) demonstrates the interconnectedness between smart cities and disaster resilience. The Smart Cities Mission (SCM) fosters disaster resilience through an integrated focus on advanced urban planning, data-driven governance, and community-centered services. The robust and adaptable design of smart city infrastructure inherently supports disaster preparedness and response by ensuring the continuity of critical services and mitigating economic and social disruptions during crises. Central to this interconnectedness between SCM and disaster resilience are the core networks-transportation, healthcare, inclusion, education, employment, environment, community, and infrastructure. Each network, when integrated into smart cities, contributes distinct resilience elements that enhance the city's ability to withstand and recover from disasters. For instance, resilient transportation and healthcare networks ensure that movement and medical services remain accessible, even in emergencies. Inclusion and educational networks strengthen social cohesion and preparedness, enabling communities to be more self-reliant in crises. Additionally, the SCM's emphasis on environmental sustainability plays a critical role in disaster resilience. Green infrastructure, such as parks and wetlands, not only improves urban quality of life but also mitigates the impacts of hazards like floods and heatwaves by enhancing water absorption and cooling. In essence, each element of the SCM works symbiotically to create urban environments that can anticipate, respond to, and recover from disruptions, making disaster resilience a fundamental outcome of well-implemented smart city initiatives. Fig. 2 is a representation of the essential networks needed to build disaster-resilient smart cities (Meduri et al., 2024).

Figure 2. Key networks to create disaster resilient smart cities

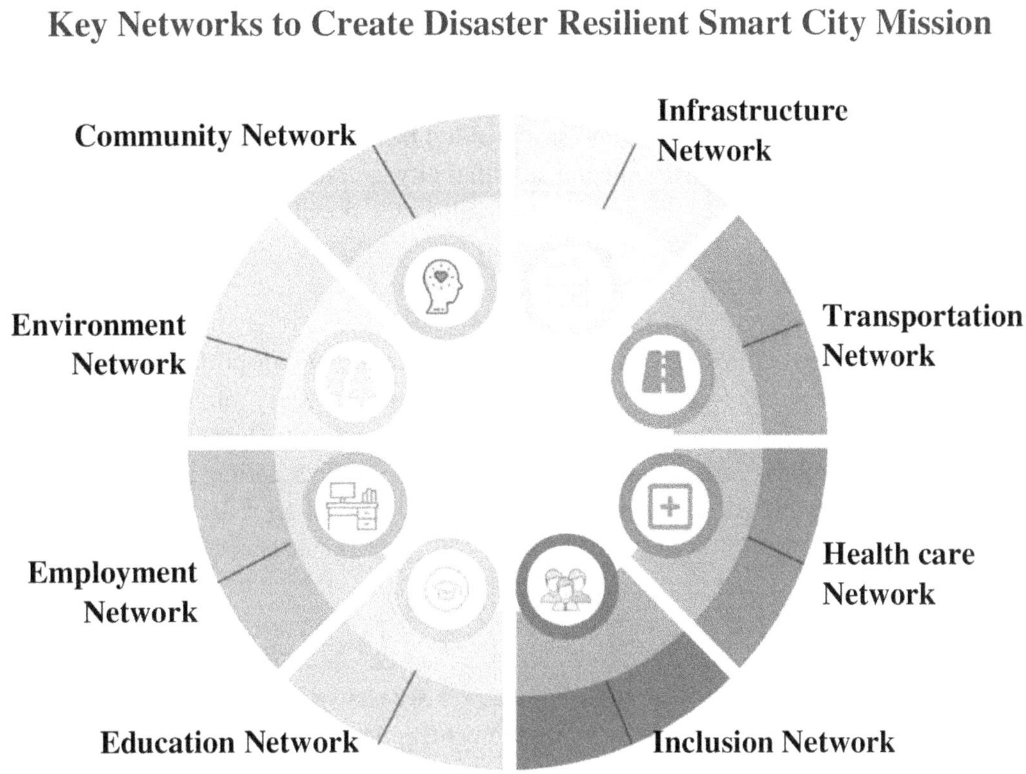

Resilient infrastructure, material stipulation, building codes

- **Transportation Network**

Smart mobility, road mapping, regulatory measures for human action, transport access

- **Health care Network**

Hospital infrastructure, health security, affordable healthcare, ICT in healthcare, emergency medicine

- **Inclusion Network**

Social inclusion, cultural diversity, percentage of female workforce, percentage of knowledge workforce

- **Education Network**

Education equality, knowledge-driven, ICT know-how, emergency preparedness

- **Employment Network**

Percentage of knowledge workforce, percentage of employment, entrepreneurship, diversified income base

- **Environment Network**

Natural habitats, land and air quality, hazard zoning, pollution control, waste management, green technology, green energy

- **Community Network**

Quality of life, sense of community, place attachment, citizen participation, social innovation

(Source Meduri et al., 2024)

4.5. Key Networks for Disaster Resilience in Smart Cities

Key networks in smart cities—transportation, healthcare, inclusion, education, employment, environment, community, and infrastructure—are vital for enhancing disaster resilience. These networks ensure cities are prepared to withstand, respond to, and recover from disasters by providing robust systems for evacuation, healthcare, social support, and sustainable practices. Integrated networks foster social cohesion, economic stability, and environmental sustainability, creating adaptable urban environments that minimize risk, ensure quick recovery, and protect vulnerable populations during emergencies.

4.5.1. Transportation Network

India designed the Smart Cities Mission (SCM) to enhance urban transport infrastructure through the "walk-to-work" concept, incorporating smart mobility to facilitate easy intra-city commuting. This network includes road, rail, air, and sea routes, essential for disaster preparedness, evacuation, and emergency transportation. A robust transportation system enables swift evacuation and the delivery of aid post-disaster, while road networks help shape land use to mitigate disaster risks by ensuring resilience, safety, and repairability.

4.5.2. Healthcare Network

The National Health Policy of 2017 in India emphasizes accessible healthcare at all levels. Disaster-resilient healthcare focuses on human resource management, patient transfer, environmental health, and inter-agency coordination. Disaster-ready healthcare systems should adapt to mass casualty scenarios, enhance capacity, and ensure mental health support, helping communities recover quickly.

4.5.3. Inclusion Network

SCM promotes inclusive development, prioritizing the needs of vulnerable populations—women, children, the elderly, and people with disabilities—by ensuring their access to education, healthcare, housing, and protection. This network fosters social and economic resilience, emphasizing cultural awareness, equal opportunity, and participation, which collectively support disaster resilience by empowering and protecting marginalized groups.

4.5.4. Education Network

Education in smart cities fosters resilience through awareness and preparedness training, covering topics such as emergency response and first aid. Schools and community centers can serve as hubs for disaster education, teaching safety practices, and resilience strategies, such as sustainable urban planning. Integrating disaster management into education builds a community skilled in crisis response.

4.5.5. Employment Network

SCM supports economic growth and job creation, focusing on knowledge-driven industries. This network boosts disaster resilience by promoting skill development in technology, data analytics, and infrastructure management, which are critical in DRR. Employment in disaster management roles contributes to economic stability, helping communities to sustain and rebuild after disasters.

4.5.6. Environment Network

SCM's environmental network promotes sustainable practices, like green infrastructure, renewable energy, and pollution control, to enhance resilience. By preserving natural habitats, implementing hazard zoning, and reducing carbon emissions, cities can reduce their vulnerability to climate change and natural disasters. Waste management, clean energy, and early warning systems contribute to a healthier, more resilient urban environment.

4.5.7. Community Network

SCM encourages community engagement to strengthen disaster resilience by fostering social bonds and community-led disaster response. By integrating traditional knowledge, promoting social innovation, and supporting local volunteer groups, the network enables communities to actively participate in planning, which enhances disaster preparedness and creates inclusive, resilient neighborhoods.

4.5.8. Infrastructure/Built Network

SCM prioritizes resilient urban infrastructure in areas such as water, electricity, sanitation, transportation, and housing. Disaster-resilient infrastructure requires rigorous risk assessments and adaptive planning to withstand hazards. This network enhances resilience through a combination of pre-disaster preparation, infrastructure adaptation, and post-disaster recovery, ensuring the city's infrastructure can continue to serve its residents in times of crisis.

4.6. Enhancing Disaster Resilience in Smart Cities through AI Integration

The integration of AI plays a key role in strengthening disaster resilience across various components of smart cities. In the transportation network, AI can optimize traffic management by predicting traffic patterns, adjusting signals, reducing traffic jams, and ensuring the fastest routes, thereby optimizing transportation systems (Chavhan & Sambare, 2024). It also supports predictive maintenance, ensuring that critical infrastructure, such as roads and vehicles, remains operational during crises (Karballaeezadeh et al., 2020). Similarly, in the healthcare network, AI enhances disaster preparedness by prioritizing patient care through automated triage systems and improving the efficiency of medical operations (Weisberg et al., 2020). AI-driven data analytics can anticipate healthcare needs during disasters, ensuring better resource allocation and more effective responses to mass casualty scenarios (Li et al., 2021).

In the inclusion network, AI ensures equitable access to disaster response services for vulnerable populations. AI systems can track and cater to the specific needs of marginalized groups, such as the elderly and disabled, ensuring they receive appropriate support during disasters (Moon, 2023). AI also enhances educational networks by integrating disaster preparedness into learning platforms, offering personalized disaster management training and simulations. Moreover, AI can monitor the operational status of educational institutions, ensuring continuity of learning during emergencies through online platforms (Chen et al., 2020).

AI's impact extends to the employment network, where it helps identify job opportunities in critical sectors and supports skill development for disaster management roles. AI systems can also predict workforce needs, ensuring a resilient labor force that can aid in recovery (Abdeldayem & Aldulaimi, 2020). In the environmental network, AI helps optimize green infrastructure, improve water management, and reduce climate-induced risks by predicting and managing environmental hazards. AI-powered early warning systems for natural disasters further bolster cities' preparedness and response capabilities (Habib et al., 2024; Dharmarathne et al., 2024).

Finally, AI enhances community resilience by facilitating communication, resource mobilization, and community-led disaster response (Chakravarty & Gattupalli, 2024). Social media analysis using AI can identify emerging risks and enable faster reactions (Sufi & Khalil, 2022). In the infrastructure network, AI improves the management and maintenance of critical systems like water, electricity, and sanitation. It also assists in assessing infrastructure damage post-disaster, accelerating recovery efforts (Lozano & Tien, 2023; Rane et al., 2024). AI's ability to analyze data, predict outcomes, and automate responses significantly strengthens disaster resilience, making smart cities more adaptable and capable of withstanding future challenges.

5. AI'S ROLE IN BRIDGING THE DISASTER RESILIENCE GAP

5.1. AI-Driven Resilience: Learning and Improving Disaster Management

AI-driven resilience evolves over time by learning from past disaster responses. By analyzing previous successes and failures, AI systems refine their predictive models and response strategies, increasing their accuracy with each event. This adaptability enhances community resilience by improving preparedness, response, and recovery efforts. AI's capabilities in predictive analysis, real-time monitoring, and resource optimization make it a vital tool for disaster resilience, helping communities not only withstand disasters but also recover more quickly, contributing to a safer, more resilient world.

5.2. AI for Disaster Resilience: Predictive Insights and Early Warnings

The unique capabilities of AI in data analysis, real-time processing, and predictive modeling are gaining recognition as a transformative tool for disaster resilience. Resilience against disasters depends on accurate, timely information, as well as the ability to respond dynamically to rapidly changing situations (Rane et al., 2024). AI excels at these tasks, making it highly suitable for anticipating, managing, and mitigating the impact of disasters. One of the key strengths of AI in disaster resilience is its ability to process massive volumes of data from diverse sources. AI algorithms can analyze information from weather forecasts, seismic data, satellite imagery, and historical patterns to make highly accurate predictions about potential disasters, such as earthquakes, floods, and hurricanes (Venkadesh et al., 2024). By detecting patterns and anomalies in this data, AI models can provide early warnings that give communities crucial time to prepare, potentially saving lives and reducing damage.

5.3. Real-time Monitoring

Real-time monitoring is another area where AI plays a crucial role. Through IoT sensors, CCTV, and social media feeds, AI systems continuously monitor environmental and urban conditions, providing instant situational awareness (Panduman et al., 2024). When a disaster strikes, these AI systems can detect changes in the environment, analyze social media posts for distress signals, and monitor traffic patterns to inform evacuation routes. This enables emergency responders to act swiftly and make informed decisions, enhancing the safety and efficiency of their operations.

5.4. Resource Management for Optimizing Allocation and Recovery

Resource management is another critical aspect where AI significantly enhances disaster resilience. During an emergency, resources like food, water, medical supplies, and shelter are essential. AI-driven algorithms optimize the allocation and distribution of these resources by analyzing demand, supply, and logistical constraints. This helps ensure that the right resources reach the affected areas promptly and are allocated efficiently, minimizing waste and improving response times. AI also aids in post-disaster recovery by assessing structural damage, prioritizing repairs, and assisting with rehabilitation plans. Through machine learning, AI can analyze images and data from damaged sites to evaluate the extent of destruction. This helps authorities prioritize repair efforts and allocate resources to areas most in need, accelerating recovery and reducing downtime (Manoharan et al., 2024).

5.5. Emergency Decision-Making for Disaster Resilience

Emergency decision-making is critical to building resilience against such disasters, and AI and machine learning (ML) are emerging as transformative tools in this domain. These technologies provide data-driven insights, enhance predictive capabilities, optimize resource allocation, and empower communities to withstand crises. By rapidly analyzing vast and diverse datasets during emergencies, they can address the chaos and data deluge characteristic of disaster scenarios. Advanced tools for real-time data analysis and predictive modeling guide disaster management decision-makers through the intricate dynamics of crisis response. This integration of technology and resilience showcases the adaptability and efficiency in mitigating rising disaster risks (Singh & Manoharan, 2024).

5.6. Revolutionizing Disaster Management in Smart Cities through AI & Drones

The integration of AI and unmanned aerial vehicles (UAVs), or drones, into disaster management systems within smart cities has significantly enhanced response capabilities. These technologies, when combined with the IoT, offer a transformative approach to managing disasters. AI-powered drones provide real-time surveillance, assess damage, and facilitate rapid search and rescue operations, even in challenging urban environments. Additionally, the IoT connectivity of drones enables seamless coordination among disaster management teams, ensuring timely and accurate decision-making. By leveraging AI and drones, smart cities can improve community resilience and adapt to the dynamic challenges of disaster scenarios, demonstrating the vital role of technology in building safer and more responsive urban ecosystems (Singh et al., 2024).

5.7. ML Disaster Resilience in Smart Cities

ML is a transformative tool for enhancing disaster resilience in smart cities, offering robust capabilities for predictive modeling, hazard detection, and decision-making. By analyzing vast datasets, ML algorithms, such as decision trees, support vector machines (SVM), artificial neural networks (ANN), and advanced hybrid and ensemble methods, provide valuable insights for disaster preparedness and response. Hybrid and ensemble models stand out for their high accuracy and cost-effectiveness, while deep learning techniques, although requiring more computational resources, achieve superior predictive performance (Band et al., 2022). These models enable early warning systems, real-time disaster monitoring, and damage assessment, allowing for faster and more informed decision-making during emergencies (Linardor et al., 2022). In addition to disaster-specific applications, ML contributes to healthcare resilience, aiding in health monitoring, treatment decisions for critically ill patients, and providing remote care. This ensures continuity of healthcare services during crises, further strengthening urban resilience (Ghazal et al., 2021). The adaptability of ML extends to various disaster stages, from early warnings to post-disaster recovery, enabling efficient resource allocation and strategic planning. By incorporating ML into smart city frameworks, urban areas can better anticipate, withstand, and recover from disasters, enhancing the safety and well-being of communities. This integration is essential for the critical role of advanced technologies in building resilient, future-ready cities.

5.8. Natural Language Processing for Disaster Resilience in Smart Cities

Natural Language Processing (NLP) is a transformative technology for strengthening disaster resilience in smart cities by enabling efficient analysis and interpretation of vast textual and spoken data generated during crises. It leverages advanced algorithms to extract meaningful insights from sources such as social media, news reports, emergency calls, and sensor-generated alerts, providing decision-makers with real-time, actionable information. One of NLP's critical contributions lies in its ability to detect emerging threats through early warning systems. By analyzing trends, keywords, and anomalies across social media platforms and other communication channels, NLP can identify signs of impending disasters, such as floods, fires, or earthquakes, enabling faster response times (Sit et al., 2020; Sufi & Khalil, 2022). This proactive approach is particularly valuable in smart cities, where rapid interventions can minimize loss of life and property. NLP aids in voice-to-text transcription for emergency calls, speeding up the process of dispatching assistance. NLP-powered chatbots and virtual assistants provide real-time guidance to residents, answering queries and offering safety instructions during emergencies (Chaurasia et al., 2023). Post-disaster, NLP supports recovery efforts by analyzing feedback from affected communities to identify gaps in resource allocation and improve the efficiency of aid distribution. Sentiment analysis tools assess the emotional and psychological impacts of disasters, helping governments and NGOs prioritize mental health interventions (Sufi & Khalil, 2022). In the interconnected systems of smart cities, NLP seamlessly integrates with other technologies such as the IoT and ML to enhance disaster preparedness and management. By automating data analysis, prioritizing critical information, and fostering effective communication, NLP plays a vital role in building disaster-resilient smart cities, ensuring safety, inclusivity, and adaptive recovery for urban communities.

5.9. Computer Vision for Enhancing Disaster Resilience

Computer vision (CV) is a groundbreaking technology that significantly enhances disaster resilience in smart cities by enabling machines to interpret and analyze visual data from various sources, such as cameras, drones, and satellites. By leveraging advanced algorithms and deep learning, CV processes real-time images and videos, providing critical insights that aid in disaster preparedness, response, and recovery (Ramos et al., 2024). In disaster monitoring, CV-powered systems can detect anomalies such as rising water levels, structural damages, or landslides through real-time surveillance. For instance, during floods, CV algorithms analyze satellite imagery to map affected areas, predict the flood's progression, and identify safe zones for evacuation. Similarly, CV-equipped drones can assess the extent of damage in inaccessible or hazardous regions, enabling rescue teams to strategize their operations effectively (Muhadi et al., 2021; Gomez, 2023; Wang & Nishio, 2024). CV also plays a crucial role in early warning systems by identifying visual cues of impending disasters, such as smoke plumes for wildfires or cracks in infrastructure signaling potential collapses (Ramos et al., 2024). By automating these detections, CV reduces the reliance on manual observation, ensuring quicker response times. In the aftermath of disasters, CV assists in damage assessment by analyzing imagery to estimate the extent of destruction and categorizing it for prioritizing recovery efforts. It also supports urban planning by providing insights into the resilience of infrastructure, helping cities rebuild more robustly (Wang, 2021). Integrated with smart city technologies, CV enhances situational awareness by processing data from interconnected systems like traffic cameras and drones. This enables the efficient coordination of emergency services, optimizing resource allocation and minimizing response delays (Bosch et al., 2020). Overall, computer vision empowers smart cities to predict, monitor, and respond to disasters with unprecedented precision and efficiency. Its ability to transform raw visual data into actionable intelligence strengthens urban resilience, safeguarding lives and infrastructure against the growing challenges of natural and human-made disasters.

5.10. Neural Networks and Deep Learning for Disaster Resilience

Neural networks and deep learning (DL) are transformative technologies that play a critical role in enhancing disaster resilience in smart cities. These advanced AI techniques process vast and complex datasets to extract valuable insights, enabling cities to better prepare for, respond to, and recover from disasters. Deep learning, with its multi-layered neural networks, excels in identifying patterns and anomalies in data, making it indispensable for early disaster prediction and detection. For instance, convolutional neural networks (CNNs) analyze satellite imagery to forecast weather patterns, detect floods, or monitor wildfire progression. Similarly, recurrent neural networks (RNNs) process time-series data, such as seismic readings, to predict earthquakes or tsunamis, providing crucial early warnings. During disasters, neural networks aid in real-time decision-making by integrating data from diverse sources like drones, IoT sensors, and social media. These models can identify affected areas, assess damage severity, and optimize resource allocation for rescue and relief operations. For example, deep learning-powered drones can analyze live video feeds to locate stranded individuals or navigate challenging terrains, speeding up search and rescue efforts. Post-disaster, neural networks support recovery and reconstruction by assessing infrastructure damage and predicting the long-term impacts of disasters on urban systems. They can model recovery scenarios to guide efficient rebuilding strategies that prioritize resilience. Furthermore, these technologies enhance disaster communication by enabling natural language processing for multilingual alert systems and sentiment analysis, ensuring inclusive and effective communication with diverse urban populations. In the context of smart cities, neural networks and deep learning integrate seamlessly with other AI tools to create adaptive, responsive systems that strengthen disaster resilience. Their ability to process large-scale, unstructured data and provide actionable insights makes them invaluable for mitigating the impacts of disasters and building safer, smarter urban environments.

5.11. Enhancing Disaster Management and Urban Resilience with Fuzzy Logic Approaches

Fuzzy logic is a computational approach that handles uncertainties and imprecise information, making it highly effective in disaster management scenarios where data is often incomplete or ambiguous. Urban areas in the 21st century face significant challenges, raising concerns about the future of urban living. Evaluating smart cities requires addressing numerous conflicting criteria due to their increasing complexities. Multi-criteria evaluation techniques are essential for assessing the performance of multiple decision-making entities, but these methods often involve inherent uncertainties and vagueness. Fuzzy-based multi-criteria models effectively handle such complexities, offering solutions for real-world problems. A study by Kutty et al. (2023) introduces a novel fuzzy expert-driven decision support model, integrating the Analytical Hierarchy Process (AHP) with the Evaluation based on Distance from Average Solution (EDAS) approach in a spherical fuzzy environment to create a composite performance index. Assessing urban resilience is a critical step in improving resilience performance, as it helps guide action plans, identify deficiencies, and enable spatial and temporal comparisons. Gaber et al. (2021) created the Resilience Performance Index (RPI) by establishing a conceptual framework, selecting relevant resilience indicators, and modeling resilience performance using a fuzzy logic approach. This approach integrates resilience analysis with AI tools and dynamic modeling techniques, offering a comprehensive method for evaluating and enhancing resilience. Emergency vehicles (EVs) must reach their destinations swiftly to improve the chances of survival for those in critical situations. An efficient vehicle routing solution can help navigate challenges such as damaged roads, collapsed bridges, and accidents. Roy et al. (2024) suggested a simulation of an IoT network for emergency vehicle routing. It used a linguistic interval type 2 fuzzy logic system (LIT2FLS) to combine data and make decisions better in these situations. Jatobá et al. (2023) aimed to develop and evaluate a framework for assessing the resilience potential of public health systems. Using fuzzy logic, the study created a mathematical model to score four key resilience abilities—monitoring, anticipation, learning, and response—and to calculate an aggregated coefficient of resilience potential in healthcare. Findings revealed that cities most impacted by COVID-19 had lower resilience potential before the pandemic. While some local health systems demonstrated adequate response capabilities, deficiencies in other resilience abilities hindered effective management of the virus's spread.

6. ETHICAL CONSIDERATIONS

Although AI holds immense promise for disaster response, it must overcome several challenges. It is crucial to safeguard sensitive information from cyber threats by protecting data privacy and security in AI systems. Furthermore, clear ethical frameworks are necessary to guide AI use in disaster response, ensuring that decisions made by AI are fair and unbiased. Responsible development and deployment of

AI technologies are essential to maximize their potential benefits while minimizing risks. This involves tackling issues such as data privacy, algorithmic bias, and the digital divide. By focusing on ethical principles and involving a broad range of stakeholders in the development process, AI can be leveraged to promote social justice and equity, thereby strengthening community resilience. Collaborative efforts between governments, organizations, and technology providers are key to overcoming these challenges and unlocking the full potential of AI in disaster management (Rane et al., 2024). Maintaining human control over AI systems and informing individuals impacted by these technologies about associated risks are essential for establishing "trustworthy AI". Ethical frameworks proposed by organizations like UNESCO and legislative efforts by the European Union focus on data rights, accountability, and transparency. Singapore's Model AI Governance Framework offers practical guidance for AI deployment. The rise of Explainable Artificial Intelligence (XAI) is pivotal in fostering trust, especially in disaster management, by improving transparency and accountability in AI systems. XAI models, like SHAP, help detect and predict disasters, reducing biases and enhancing decision-making processes, thereby strengthening the reliability of AI systems (Albahri et al., 2024).

Ethical considerations in AI for disaster prediction include informed consent, equity, transparency, human-centric approaches, and continual evaluation. Informed consent ensures individuals are aware of data use, promoting transparency and accountability. AI must be accessible to all, preventing inequality and ensuring vulnerable populations benefit. Transparency in AI models builds trust, while a human-centric approach ensures ethical decision-making. Continuous evaluation of AI models is vital to mitigate biases, ensuring fair outcomes and preventing discriminatory impacts. These considerations are crucial for creating equitable, transparent, and effective AI-driven disaster management systems that respect human rights and promote societal well-being (Varsha et al., 2024).

The use of AI in emergency management faces numerous ethical and operational challenges, as illustrated by real-world examples. Bias in AI systems, such as Amazon's 2018 recruiting tool, can lead to unequal treatment, while lack of transparency, as seen with a 2019 criminal risk algorithm, raises accountability concerns. Incidents such as the 2017 Equifax breach underscore privacy and security risks, highlighting the potential misuse of sensitive data during emergencies. AI's reliance on high-quality data, evidenced during the 2020 COVID-19 pandemic, can falter with limited or unreliable inputs, compromising decision-making. Ethical dilemmas, such as resource allocation, mirror the challenges faced in cases like Uber's 2018 autonomous vehicle fatality. Public trust, safety, and regulation remain critical, as resistance or cybersecurity threats could disrupt emergency responses. Accountability issues, as with Tesla's vehicle crashes, further complicate AI deployment. Navigating these challenges is vital for responsible AI integration (Visave, 2024).

7. DISCUSSION

The integration of AI into smart city frameworks marks a key milestone in the evolution of addressing the challenges of urban resilience against disasters. This paper explores how AI-driven technologies significantly enhance the disaster resilience of smart cities by enabling proactive risk management, real-time monitoring, and optimized resource allocation. While the concept of smart cities revolves around technological advancement and urban efficiency, it is crucial to embed disaster resilience as a core component to ensure sustainability.

AI plays a transformative role in smart cities through its predictive analytics, which allow urban planners to anticipate disasters by analyzing data from diverse sources, such as IoT devices and weather systems. This capability is crucial in mitigating the impacts of natural disasters like floods and hurricanes, especially in highly urbanized areas prone to climate risks. Additionally, AI enhances urban systems such as healthcare, transportation, and community networks by enabling tailored disaster preparedness and response strategies. For example, AI-driven healthcare systems prioritize patient care during emergencies through automated triage systems, while smart transportation networks optimize evacuation routes, thereby reducing casualties and enhancing efficiency.

Despite these advancements, integrating disaster resilience within smart cities remains challenging. Limited funding, fragmented governance, and competing urban priorities often hinder the adoption of holistic resilience strategies. Furthermore, the complexity of resilience, which encompasses infrastructural, social, and environmental aspects, requires the need for multidisciplinary collaboration and long-term planning. AI can mitigate these challenges by fostering data-driven governance and enabling better coordination among stakeholders.

Ethical considerations, including data privacy, algorithmic bias, and transparency, remain critical in AI integration. Trustworthy AI models that prioritize explainability, such as Explainable AI (XAI), are necessary to address these concerns, ensuring equitable and reliable disaster responses. Furthermore, involving local communities in resilience planning is vital for fostering inclusivity and ensuring that technological interventions align with the needs of all urban residents, particularly vulnerable populations.

The paper also highlights the complementarity between smart and resilient city frameworks. While smart cities focus on leveraging technology for efficiency and innovation, resilience frameworks emphasize sustainability and adaptive capacity. Bridging these paradigms through AI can create urban environments that are not only technologically advanced but also capable of withstanding and recovering from diverse crises.

8. CONCLUSION

In conclusion, the integration of AI within the framework of smart cities represents a significant leap toward building disaster-resilient urban environments. As cities face increasing vulnerability due to rapid urbanization and the escalating impacts of climate change, AI provides critical tools for disaster prediction, preparedness, response, and recovery. By leveraging AI technologies such as predictive analytics, real-time monitoring, and data-driven decision-making, smart cities can enhance their capacity to anticipate and mitigate disaster risks, ultimately safeguarding lives and minimizing economic losses.

However, achieving disaster resilience through AI integration requires overcoming several challenges, including ethical concerns around data privacy, algorithmic bias, and transparency. Moreover, the successful implementation of AI-driven resilience strategies demands collaborative efforts between public and private sectors, academia, and local communities. Ensuring that AI systems are inclusive and equitable is essential for achieving comprehensive resilience.

Ultimately, the future of smart cities hinges on the harmonious integration of technological advancements with sustainable urban planning and resilient infrastructure. As urban populations continue to grow, cities that effectively combine AI with resilience strategies will not only survive the challenges posed by natural disasters but will also thrive, creating safer, more sustainable, and adaptable urban spaccs for future generations.

REFERENCES

Abdeldayem, M. M., & Aldulaimi, S. H. (2020). Trends and opportunities of artificial intelligence in human resource management: Aspirations for public sector in Bahrain. *International journal of scientific and technology research, 9*(1), 3867-3871.

Alahi, M. E. E., Sukkuea, A., Tina, F. W., Nag, A., Kurdthongmee, W., Suwannarat, K., & Mukhopadhyay, S. C. (2023). Integration of IoT-Enabled Technologies and Artificial Intelligence (AI) for Smart City Scenario: Recent Advancements and Future Trends. *Sensors (Basel)*, 23(11), 5206. DOI: 10.3390/s23115206 PMID: 37299934

Albahri, A. S., Khaleel, Y. L., Habeeb, M. A., Ismael, R. D., Hameed, Q. A., Deveci, M., Homod, R. Z., Albahri, O. S., Alamoodi, A. H., & Alzubaidi, L. (2024). A systematic review of trustworthy artificial intelligence applications in natural disasters. *Computers & Electrical Engineering*, 118, 109409. DOI: 10.1016/j.compeleceng.2024.109409

Anthopoulos, L. G. (2017). The rise of the smart city. In *Understanding smart cities: a tool for smart government or an industrial trick?* (pp. 5–45). Springer., DOI: 10.1007/978-3-319-57015-0_2

Bakıcı, T., Almirall, E., & Wareham, J. (2013). A smart city initiative: The case of Barcelona. *Journal of the Knowledge Economy*, 4(2), 135–148. DOI: 10.1007/s13132-012-0084-9

Band, S. S., Ardabili, S., Sookhak, M., Chronopoulos, A. T., Elnaffar, S., Moslehpour, M., Csaba, M., Torok, B., Pai, H.-T., & Mosavi, A. (2022). When smart cities get smarter via machine learning: An in-depth literature review. *IEEE Access : Practical Innovations, Open Solutions*, 10, 60985–61015. DOI: 10.1109/ACCESS.2022.3181718

Barrionuevo, J. M., Berrone, P., & Ricart, J. E. (2012). Smart cities, sustainable progress. *IESE insight, 14*(14), 50-57.

Bertilsson, L., Wiklund, K., de Moura Tebaldi, I., Rezende, O. M., Veról, A. P., & Miguez, M. G. (2019). Urban flood resilience–A multi-criteria index to integrate flood resilience into urban planning. *Journal of Hydrology (Amsterdam)*, 573, 970–982. DOI: 10.1016/j.jhydrol.2018.06.052

Bibri, S. E., & Krogstie, J. (2020). The emerging data–driven Smart City and its innovative applied solutions for sustainability: The cases of London and Barcelona. *Energy Informatics*, 3(1), 5. DOI: 10.1186/s42162-020-00108-6

Bosch, M., Conroy, C., Ortiz, B., & Bogden, P. (2020, September). Improving emergency response during hurricane season using computer vision. In *Earth Resources and Environmental Remote Sensing/GIS Applications XI* (Vol. 11534, pp. 72–79). SPIE., DOI: 10.1117/12.2574639

Chakravarty, P., & Gattupalli, S. (2024). Integration of Indigenous Traditional Knowledge and AI in Hurricane Resilience and Adaptation. In *Advances in Hurricane Risk in a Changing Climate* (pp. 125–158). Springer Nature Switzerland., DOI: 10.1007/978-3-031-63186-3_6

Chang, F., & Das, D. (2020). Smart Nation Singapore: Developing Policies for a Citizen-Oriented Smart City Initiative. Developing National Urban Policies. *doi, 10*, 978-981. https://doi.org/DOI: 10.1007/978-981-15-3738-7_18

Chaurasia, S. H. U. B. H. A. M., Jain, S. H. U. B. H. A., Vishwkarma, H. O., & Singh, N. I. S. H. A. N. T. (2023). Conversational AI Unleashed: A Comprehensive Review of NLP-Powered Chatbot Platforms. *Iconic Research and Engineering Journals*, 7(3), 1–8.

Chavhan, R. D., & Sambare, G. B. (2024). AI-Driven Traffic Management Systems In Smart Cities: A Review. *Educational Administration: Theory and Practice*, 30(5), 105–116. DOI: 10.53555/kuey.v30i5.2780

Chen, L., Chen, P., & Lin, Z. (2020). Artificial intelligence in education: A review. *IEEE Access : Practical Innovations, Open Solutions*, 8, 75264–75278. DOI: 10.1109/ACCESS.2020.2988510

Chen, T. (2010). Smart grids, smart cities need better networks [Editor's note]. *IEEE Network*, 24(2), 2–3. DOI: 10.1109/MNET.2010.5430136

Cretu, L. G. (2012). Smart cities design using event-driven paradigm and semantic web. *Informações Econômicas*, 16(4), 57.

Dameri, R. P. (2017). *Smart city implementation. Progress in IS.* Springer., DOI: 10.1007/978-3-319-45766-6

Dharmarathne, G., Waduge, A. O., Bogahawaththa, M., Rathnayake, U., & Meddage, D. P. P. (2024). Adapting cities to the surge: A comprehensive review of climate-induced urban flooding. *Results in Engineering*, 102123, 102123. Advance online publication. DOI: 10.1016/j.rineng.2024.102123

Elvas, L. B., Mataloto, B. M., Martins, A. L., & Ferreira, J. C. (2021). Disaster management in smart cities. *Smart Cities*, 4(2), 819–839. DOI: 10.3390/smartcities4020042

Eremia, M., Toma, L., & Sanduleac, M. (2017). The smart city concept in the 21st century. *Procedia Engineering*, 181, 12–19. DOI: 10.1016/j.proeng.2017.02.357

Gaber, R. M., El-Kader, A., Hussein, M., & Okba, E. M. (2022). The Resilience Performance Index, a Fuzzy Logic Approach to Assess Urban Resilience. *International Journal of Sustainable Development and Planning*, 17(4), 1225–1235. Advance online publication. DOI: 10.18280/ijsdp.170421

Ghazal, T. M., Hasan, M. K., Alshurideh, M. T., Alzoubi, H. M., Ahmad, M., Akbar, S. S., Al Kurdi, B., & Akour, I. A. (2021). IoT for smart cities: Machine learning approaches in smart healthcare—A review. *Future Internet*, 13(8), 218. DOI: 10.3390/fi13080218

Giffinger, R., Fertner, C., Kramar, H., Kalasek, R., Pichler-Milanović, N., & Meijers, E. (2007). Smart cities ranking of European medium-sized cities. https://www.smart-cities.eu/download/smart_cities_final_report.pdf

Gomez Llerena, J. A. (2023). *Investigation of the Integrity of Tailings Dam Surfaces Using Aerial Imaging and Computer Vision* (Doctoral dissertation).

Gu, D., Andreev, K., & Dupre, M. E. (2021). Major Trends in Population Growth Around the World. *China CDC Weekly*, 3(28), 604–613. DOI: 10.46234/ccdcw2021.160 PMID: 34594946

Habib, M., Habib, A., Albzaie, M., & Farghal, A. (2024). Sustainability benefits of AI-based engineering solutions for infrastructure resilience in arid regions against extreme rainfall events. *Discover Sustainability*, 5(1), 278. DOI: 10.1007/s43621-024-00500-2

Hofmann, S. Z. (2021). 100 Resilient Cities program and the role of the Sendai framework and disaster risk reduction for resilient cities. *Progress in Disaster Science*, 11, 100189. DOI: 10.1016/j.pdisas.2021.100189

IEEE Communications Magazine | IEEE Communications Society. (2024, October 1). https://www.comsoc.org/publications/magazines/ieee-communications-magazine

Jatobá, A., de Castro Nunes, P., & de Carvalho, P. V. (2023). A framework to assess potential health system resilience using fuzzy logic. *Revista Panamericana de Salud Pública = Pan American Journal of Public Health*, 47, e73. DOI: 10.26633/RPSP.2023.73 PMID: 37123641

Jung, D., Tuan, V. T., Tran, D. Q., Park, M., & Park, S. (2020). Conceptual Framework of an Intelligent Decision Support System for Smart City Disaster Management. *Applied Sciences (Basel, Switzerland)*, 10(2), 666. DOI: 10.3390/app10020666

Karballaeezadeh, N., Zaremotekhases, F., Shamshirband, S., Mosavi, A., Nabipour, N., Csiba, P., & Várkonyi-Kóczy, A. R. (2020). Intelligent road inspection with advanced machine learning; hybrid prediction models for smart mobility and transportation maintenance systems. *Energies*, 13(7), 1718. DOI: 10.3390/en13071718

Kourtit, K., & Nijkamp, P. (2012). Smart cities in the innovation age. *Innovation (Abingdon)*, 25(2), 93–95. DOI: 10.1080/13511610.2012.660331

Kutty, A. A., Kucukvar, M., Onat, N. C., Ayvaz, B., & Abdella, G. M. (2023). Measuring sustainability, resilience and livability performance of European smart cities: A novel fuzzy expert-based multi-criteria decision support model. *Cities (London, England)*, 137, 104293. DOI: 10.1016/j.cities.2023.104293

Lazaroiu, G. C., & Roscia, M. (2012). Definition methodology for the smart cities model. *Energy*, 47(1), 326–332. DOI: 10.1016/j.energy.2012.09.028

Li, W., Chai, Y., Khan, F., Jan, S. R. U., Verma, S., Menon, V. G., Kavita, , & Li, X. (2021). A comprehensive survey on machine learning-based big data analytics for IoT-enabled smart healthcare system. *Mobile Networks and Applications*, 26(1), 234–252. DOI: 10.1007/s11036-020-01700-6

Linardos, V., Drakaki, M., Tzionas, P., & Karnavas, Y. L. (2022). Machine learning in disaster management: Recent developments in methods and applications. *Machine Learning and Knowledge Extraction*, 4(2), 446–473. Advance online publication. DOI: 10.3390/make4020020

Lombardi, P., Giordano, S., Farouh, H., & Yousef, W. (2012). Modelling the smart city performance. *Innovation (Abingdon)*, 25(2), 137–149. DOI: 10.1080/13511610.2012.660325

Lozano, J. M., & Tien, I. (2023). Data collection tools for post-disaster damage assessment of building and lifeline infrastructure systems. *International Journal of Disaster Risk Reduction*, 94, 103819. DOI: 10.1016/j.ijdrr.2023.103819

Manoharan, G., Razak, A., Rao, B. S., Singh, R., Ashtikar, S. P., & Nivedha, M. (2024). Navigating the Crescendo of Challenges in Harnessing Artificial Intelligence for Disaster Management. In *Predicting Natural Disasters With AI and Machine Learning* (pp. 64–94). IGI Global., DOI: 10.4018/979-8-3693-2280-2.ch003

Markets, R. A. (2020, October 5). Smart Cities Market Report 2020 - Global Forecast to 2025: Market Size is Expected to Grow from $410.8 Billion in 2020 to $820.7 Billion. *GlobeNewswire News Room*. https://www.globenewswire.com/news-release/2020/10/05/2103315/0/en/Smart-Cities-Market-Report-2020-Global-Forecast-to-2025-Market-Size-is-Expected-to-Grow-from-410-8-Billion-in-2020-to-820-7-Billion.html

Marsal-Llacuna, M. L., Colomer-Llinàs, J., & Meléndez-Frigola, J. (2015). Lessons in urban monitoring taken from sustainable and livable cities to better address the Smart Cities initiative. *Technological Forecasting and Social Change*, 90, 611–622. DOI: 10.1016/j.techfore.2014.01.012

Meduri, Y., Singh, R., & Manoharan, G. (2024). Key networks to create disaster resilient Smart Cities Mission: A case for remodeling India's Smart Cities Mission to include disaster resilience. *Resilient Cities and Structures*, 3(4), 52–62. DOI: 10.1016/j.rcns.2024.09.001

Moon, M. J. (2023). Searching for inclusive artificial intelligence for social good: Participatory governance and policy recommendations for making AI more inclusive and benign for society. *Public Administration Review*, 83(6), 1496–1505. DOI: 10.1111/puar.13648

Mouha, R. A. R. A. (2021). Internet of things (IoT). *Journal of Data Analysis and Information Processing*, 9(02), 77–101. DOI: 10.4236/jdaip.2021.92006

Moura, F., & de Abreu e Silva, J. (2021). Smart cities: definitions, evolution of the concept, and examples of initiatives. In Industry, innovation and infrastructure (pp. 989-997). Cham: Springer International Publishing.

Muhadi, N. A., Abdullah, A. F., Bejo, S. K., Mahadi, M. R., & Mijic, A. (2021). Deep learning semantic segmentation for water level estimation using surveillance camera. *Applied Sciences (Basel, Switzerland)*, 11(20), 9691. DOI: 10.3390/app11209691

Panduman, Y. Y. F., Funabiki, N., Fajrianti, E. D., Fang, S., & Sukaridhoto, S. (2024). A Survey of AI Techniques in IoT Applications with Use Case Investigations in the Smart Environmental Monitoring and Analytics in Real-Time IoT Platform. *Information (Basel)*, 15(3), 153. DOI: 10.3390/info15030153

Parker, D. J. (2020). Disaster resilience–a challenged science. *Environmental Hazards*, 19(1), 1–9. DOI: 10.1080/17477891.2019.1694857

Radu, L. D. (2020). Disruptive technologies in smart cities: A survey on current trends and challenges. *Smart Cities*, 3(3), 1022–1038. DOI: 10.3390/smartcities3030051

Ramirez Lopez, L. J., & Grijalba Castro, A. I. (2020). Sustainability and resilience in smart city planning: A review. *Sustainability (Basel)*, 13(1), 181. DOI: 10.3390/su13010181

Ramos, L., Casas, E., Bendek, E., Romero, C., & Rivas-Echeverría, F. (2024). Computer vision for wildfire detection: A critical brief review. *Multimedia Tools and Applications*, 83(35), 1–44. DOI: 10.1007/s11042-024-18685-z

Rane, N., Choudhary, S., & Rane, J. (2024). Artificial intelligence for enhancing resilience. *Journal of Applied Artificial Intelligence*, 5(2), 1–33. DOI: 10.48185/jaai.v5i2.1053

Roy, S., Jana, D. K., & Mishra, A. (2024). Linguistic interval type 2 fuzzy logic-based Exigency Vehicle routing: IoT system development for smart city applications with soft computing-based optimization. *Franklin Open*, 6, 100057. DOI: 10.1016/j.fraope.2023.100057

Sharifi, A., Khavarian-Garmsir, A. R., & Kummitha, R. K. R. (2021). Contributions of Smart City Solutions and Technologies to Resilience against the COVID-19 Pandemic: A Literature Review. *Sustainability (Basel)*, 13(14), 8018. DOI: 10.3390/su13148018

Singh, C. R., & Manoharan, G. (2024). Strengthening Resilience: AI and Machine Learning in Emergency Decision-Making for Natural Disasters. In *Internet of Things and AI for Natural Disaster Management and Prediction* (pp. 249-278). IGI Global. https://doi.org/DOI: 10.4018/979-8-3693-4284-8.ch012

Singh, R., Manoharan, G., & Singh, S. (2024). Strategic Deployment of AI and Drones Enhancing Disaster Management in Natural Disasters. *AI and IoT for Proactive Disaster Management*, 172-203. https://doi.org/DOI: 10.4018/979-8-3693-3896-4.ch009

Sit, M. A., Koylu, C., & Demir, I. (2020). Identifying disaster-related tweets and their semantic, spatial and temporal context using deep learning, natural language processing and spatial analysis: a case study of Hurricane Irma. In *Social Sensing and Big Data Computing for Disaster Management* (pp. 8–32). Routledge. DOI: 10.4324/9781003106494-2

Srivastava, R. K. (2020). Urbanization-Led Neo Risks and Vulnerabilities: A New Challenge. In *Disaster Studies and Management* (pp. 251–296). https://doi.org/DOI: 10.1007/978-981-15-2410-3_6

Sufi, F. K., & Khalil, I. (2022). Automated disaster monitoring from social media posts using AI-based location intelligence and sentiment analysis. *IEEE Transactions on Computational Social Systems*. Advance online publication. DOI: 10.1109/TCSS.2022.3157142

Talebkhah, M., Sali, A., Marjani, M., Gordan, M., Hashim, S. J., & Rokhani, F. Z. (2020, September). Edge computing: architecture, applications and future perspectives. In *2020 IEEE 2nd International Conference on Artificial Intelligence in Engineering and Technology (IICAIET)* (pp. 1-6). IEEE. https://doi.org/DOI: 10.1109/IICAIET49801.2020.9257824

Talebkhah, M., Sali, A., Marjani, M., Gordan, M., Hashim, S. J., & Rokhani, F. Z. (2021). IoT and big data applications in smart cities: Recent advances, challenges, and critical issues. *IEEE Access : Practical Innovations, Open Solutions*, 9, 55465–55484. DOI: 10.1109/ACCESS.2021.3070905

Titko, M., & Ristvej, J. (2020). Assessing importance of disaster preparedness factors for sustainable disaster risk management: The case of the Slovak Republic. *Sustainability (Basel)*, 12(21), 9121. DOI: 10.3390/su12219121

Varsha, V. R., Naganandini, S., & Hariharan, C. (2024). Utilizing AI and machine learning for natural disaster management: predicting natural disasters with AI and machine learning. In *Internet of Things and AI for Natural Disaster Management and Prediction* (pp. 279–304). IGI Global., DOI: 10.4018/979-8-3693-4284-8.ch013

Venkadesh, P., Divya, S. V., Marymariyal, P., & Keerthana, S. (2024). Predicting Natural Disasters With AI and Machine Learning. In *Utilizing AI and Machine Learning for Natural Disaster Management* (pp. 39–64). IGI Global., DOI: 10.4018/979-8-3693-3362-4.ch003

Visave, J. (2024). AI in Emergency Management: Ethical Considerations and Challenges. *Journal of Emergency Management and Disaster Communications*, 4(3), 1–19. DOI: 10.1142/S268998092450009X

Wang, M. (2021). *Active learning for improved damage detection and disaster response* (Doctoral dissertation).

Wang, S., & Nishio, M. (2024). Review for vision-based structural damage evaluation in disasters focusing on nonlinearity. *Smart Structures and Systems*, 33(4), 263. DOI: 10.12989/sss.2024.33.4.263

Weisberg, E. M., Chu, L. C., & Fishman, E. K. (2020). The first use of artificial intelligence (AI) in the ER: Triage not diagnosis. *Emergency Radiology*, 27(4), 361–366. DOI: 10.1007/s10140-020-01773-6 PMID: 32643069

Yigitcanlar, T., Kamruzzaman, M., Foth, M., Sabatini-Marques, J., Da Costa, E., & Ioppolo, G. (2019). Can cities become smart without being sustainable? A systematic review of the literature. *Sustainable Cities and Society*, 45, 348–365. DOI: 10.1016/j.scs.2018.11.033

Yuryeva, O., Kovaleva, N., & Shukhova, O. (2023). Increasing economic losses from natural disasters as a last decade trend. In *E3S Web of Conferences* (Vol. 458, p. 05005). EDP Sciences., DOI: 10.1051/e3sconf/202345805005

Zhu, S., Li, D., & Feng, H. (2019). Is smart city resilient? Evidence from China. *Sustainable Cities and Society*, 50, 101636. DOI: 10.1016/j.scs.2019.101636

Zhu, S., Li, D., Feng, H., Gu, T., Hewage, K., & Sadiq, R. (2020). Smart city and resilient city: Differences and connections. *Wiley Interdisciplinary Reviews. Data Mining and Knowledge Discovery*, 10(6), e1388. DOI: 10.1002/widm.1388

Zygiaris, S. (2013). Smart city reference model: Assisting planners to conceptualize the building of smart city innovation ecosystems. *Journal of the Knowledge Economy*, 4(2), 217–231. DOI: 10.1007/s13132-012-0089-4

Chapter 2
AI Mobility:
AI Transforming Smart Mobility With 5G, IoT, and mm-Wave to Enable and Enhance Public Safety in Future Smart Cities

J. Jesy Janet Kumari
https://orcid.org/0000-0002-3318-1632
Amrita School of Computing, Amrita Vishwa Vidhyapeetham, Bengaluru, India

S. Thangam
https://orcid.org/0000-0003-2251-3651
Amrita School of Computing, Amrita Vishwa Vidyapeetham, Bengaluru, India

ABSTRACT

A smart city is an urban environment that gathers data through diverse electrical means and sensors. Smart cities utilize information and communication technologies (ICT) to enhance service quality through the management of public resources, emphasizing comfort and preserving environmental sustainability. The fifth generation (5G) of wireless mobile communication facilitates a novel form of communication network to link all individuals and entities. 5G will have a significant impact on economies and civilizations by providing the essential communication infrastructure required by diverse smart devices. One of the many applications in smart cities is the Intelligent Transport System (ITS). 5G technology has the potential to transform ITS. The paper intends to examine the effects and significance of 5G on ITS from multiple perspectives.

1. INTRODUCTION

Cities consider complex systems with large numbers of interconnected populations, transportation, communication networks, a variety of services and enterprises, and utilities to improve urban residents' lifestyles. Large numbers of people are flocking to cities, and the municipal government is under pressure to provide the basic services essential for daily life. Rapid urbanization and overpopulation provide socio-economic, technical, and organizational challenges, as well as hazards to the sustainability of cities. Several modern cities have experienced significant growth while adhering to norms, leading to

DOI: 10.4018/979-8-3693-9770-1.ch002

issues such as pollution, traffic congestion, and socioeconomic disparity (Myeong & Shahzad, 2021). Many people have moved to cities in recent years, and projections show that by 2030, 60% of the world's population will live in cities (Gohar & Nencioni, 2021). As the population grows, smart applications are becoming more prevalent, promoting the development of smart cities (Gupta & Jha, 2015).

1.1 Motivation:

The smart city concept comprises the intelligent administration of valuable components such as transportation, medical services, utility services, housing, agriculture (Agiwal et al., 2016), and environmental building. Furthermore, in smart cities, various telecommunications or wireless infrastructures provide services effectively and connect millions of devices using a variety of technologies, including machine-to-machine (M2M) communication (Li et al., 2018), network virtualization, wireless sensor networks, and gateways (Shafique et al., 2020). Figure 1 depicts the evolution of the mobile communications.

Figure 1. The evolution of mobile communications

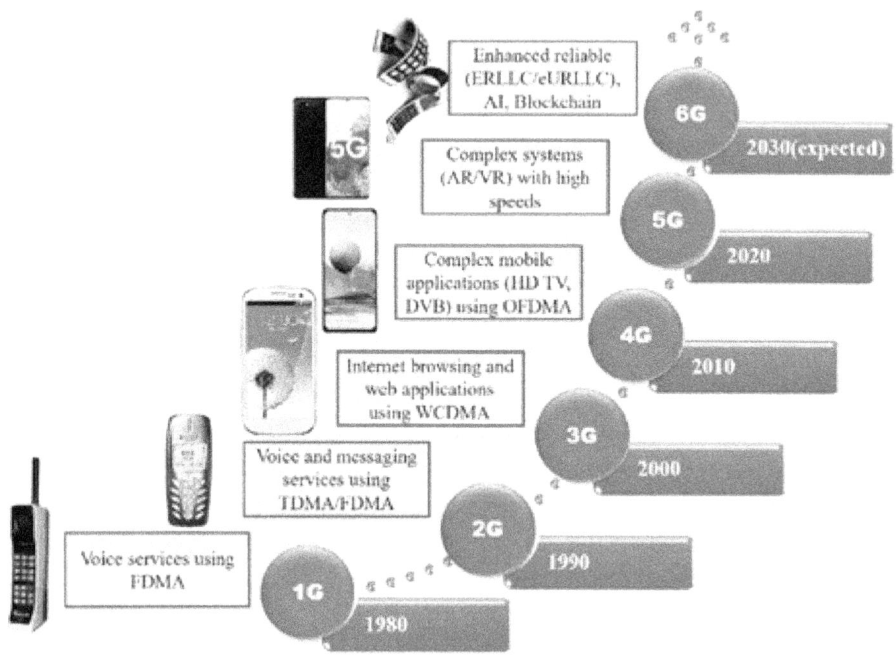

Microcontrollers, radio modules, and appropriate communication protocols are becoming standard in contemporary gadgets, with the Internet of Things (IoT) (Chettri & Bera, 2019) playing a crucial part in this change in thinking. IoT enables the operation of smart city infrastructure. Numerous public and commercial organizations, both nationally and internationally, employ the concept of IoT, which is fundamentally grounded in microcontrollers, radio modules, and appropriate communication protocols. The IoT enables the operation of smart city infrastructure. In the management of smart city initiatives, numerous public and private entities depend on ICT solutions that integrate the IoT concept. ICT solutions can efficiently govern the concept of a smart city, optimizing community resources, improving service quality, and minimizing operational and managerial costs. Enhancing service quality (QoS) while reducing operational and management expenses is a primary objective of the IoT framework in smart cities. The evolution of contemporary smart cities is significantly dependent on IoT technology. IoT technologies are essential for defining the future of smart cities and steering smart city standards toward substantial data volumes.

In 2022, Statista Research projected that the number of devices linked to the IoT will exceed twenty-nine billion by 2030, nearly doubling from 9.7 billion in 2020 (Muhammad & Safdar, 2021). The previously examined data indicates that the IoT is a highly promising contemporary technology, presenting intriguing new opportunities for services while simultaneously introducing recent challenges in the domain of intelligent application and service execution. IoT is pivotal in fostering sustainable growth and promoting innovative concepts, both of which are vital for the development and enhancement of diverse smart city applications. Experts have proposed a multitude of techniques, scenarios, technical solutions, and use cases to enhance the efficiency of smart city management. The integration of AI and IoT in smart cities is a nascent paradigm that is revolutionizing urban management and development. This integration involves utilizing AI algorithms to analyze the vast data generated by IoT sensors in smart cities. This integration improves development processes, offering unique opportunities and features while significantly reducing human involvement. A smart city is an urban area that collects data via various electronic methods and sensors. Smart cities employ ICT to improve service quality by managing public resources and prioritizing comfort and environmental sustainability. The fifth generation (5G) (Roger et al., 2020) of wireless mobile communication enables a new communication network to connect all individuals and entities. 5G will profoundly influence economies and societies by delivering the critical communication infrastructure for various smart gadgets. One of the many applications in smart cities is the Intelligent Transport System (ITS). 5G technology possesses the capacity to revolutionize ITS.

1.2 Contribution:

The research aims to analyze the impacts and importance of 5G on Intelligent Transportation Systems from many viewpoints. The work begins with an overview of the technological features of 5G. We will meticulously analyze the conditions and economic benefits of 5G, along with its influence on key vertical businesses in urban sectors, including energy, healthcare, manufacturing, entertainment, automotive, and public transit. Owing to a deficiency in critical smart technology, metropolitan areas often encounter difficulties in providing comfortable, safe, and sustainable living conditions as global population growth and urbanization escalate. The IoT has emerged as a solution to this challenge by connecting physical objects via electronics, sensors, software, and communication networks. This has transformed smart city infrastructures, incorporating many technologies that enhance sustainability, productivity, and comfort for urban inhabitants. Employing artificial intelligence (AI) to analyze the vast IoT data is generating new opportunities for the development and administration of sophisticated smart cities.

- This work introduces smart cities, outlining their features and analyzing the architecture of the IoT. We do a thorough analysis of various wireless communication technologies employed in smart city applications and perform an extensive investigation to determine the most appropriate communication technologies for particular use cases.

- It clarifies several AI algorithms and their suitability for smart city applications. It discusses the incorporation of IoT and AI in smart city contexts, highlighting the prospective advantages of 5G networks and AI in the advancement of modern urban environments.
- This work enhances the current literature by emphasizing the considerable potential of integrating IoT and AI, facilitating the creation of smart cities that improve urban residents' quality of life while fostering sustainability and productivity.
- It also analyzes the potential integration of IoT and AI, providing valuable insights into the future of smart cities and demonstrating how these technologies could improve urban environments and the well-being of their inhabitants.

Figure 2. Comparison of data rate, power consumption, cost, and a coverage range of available communication technologies.

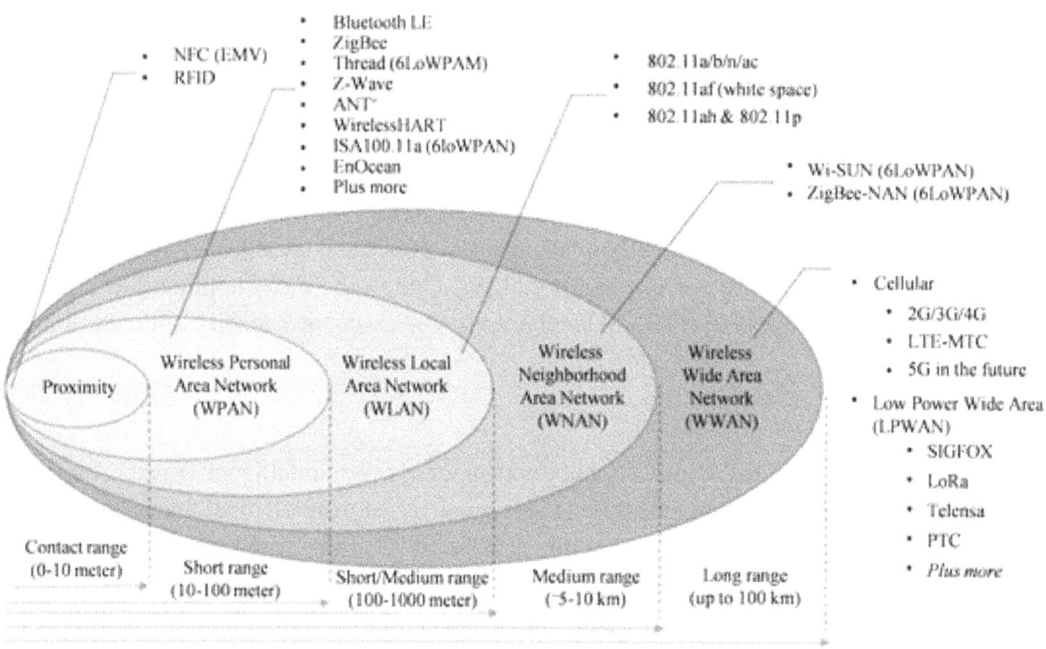

A variety of city infrastructure elements, including public spaces, bridges, buildings, and highways, incorporate IoT sensors as gadgets. Figure 2 shows the comparison of data rate, power consumption, cost, and a coverage range of available communication technologies. These sensors collect and transmit data on several parameters, including temperature, humidity, traffic flow, energy consumption, and air quality. The deployment of IoT devices generates significant volumes of data. These statistics therefore enhance urban governance and elevate the quality of life for residents. Nonetheless, the large volume of data generated by IoT sensors can be overwhelming for human operators to analyze and interpret. AI aids by utilizing machine learning (ML) algorithms to analyze large data sets and discern patterns and trends that may elude human perception. Assessing vast amounts of complex data might complicate the identification of the most accurate and efficient course of action. A crucial area where AI may enhance IoT in smart cities is predictive maintenance. By analyzing data from IoT sensors, AI algorithms can predict the need for maintenance of urban infrastructure, such as bridges and buildings, before any malfunctions occur. This can prevent costly and dangerous failures, ensuring the city's infrastructure remains in prime condition. Artificial intelligence has enabled several uses, such as intelligent water supply systems, energy management, waste management, and the reduction of pedestrian and traffic congestion, noise, and environmental pollution. Numerous smart city programs and technologies have focused on consolidating vast data and formulating solutions that tackle the complexities and dynamics of specific applications. AI-driven applications utilize vast data and knowledge to improve decision-making processes. About 30% of smart city applications are significantly integrating AI to enhance urban sustainability, resilience, social welfare, and vibrancy, especially within urban transportation networks. Experts anticipate that this trend will persist, with an expected rise in AI-driven smart city efforts by 2025. The rapid expansion of AI-driven smart city projects results from the steadfast endeavors of researchers, government officials, and urban inhabitants to explore innovative information and methodologies for smart city development. The escalating requirement for the incorporation of automation and artificial intelligence in smart city construction renders their expansion increasingly unavoidable. Despite producing substantial data relevant to diverse smart city applications, smart sensor nodes remain markedly underutilized. The existing ICT infrastructure can generate varied information, which is essential for consolidation. Artificial intelligence can improve public safety by analyzing data from surveillance cameras, microphones, and various sensors to detect and address criminal activity and other security issues. The integration of AI with IoT in smart cities has the potential to revolutionize urban planning and management; yet it also presents challenges around privacy, data security, and biases in AI systems. Therefore, smart cities must develop strong ethical and regulatory frameworks to ensure the appropriate and transparent use of these technologies.

We can examine IoT-based technologies linked with AI in a smart city from various perspectives, including smart sensors, communication technologies, and diverse applications. Key IoT technologies encompass the establishment of reliable and robust networking and communication frameworks to facilitate effective data and information exchange among the many elements of smart city services. Various scales dictate the development of IoT services for smart cities, contingent upon the application. Multiple networking and communication technologies may be requisite for their execution and functionality. Numerous review articles have examined big data, network security, and the Internet of Things in smart cities. Nevertheless, there are limited resources that thoroughly elucidate the amalgamation of IoT technology and AI's role in the smart city paradigm. This review article examines the concept of smart cities and the function of IoT in their advancement, illustrating how IoT facilitates communication among items, rendering them essential elements of smart cities. The integration of AI and IoT in smart cities can expedite development processes, generate new opportunities, and diminish human involvement. The essay examines the potential of AI to assess the vast quantities of data produced by IoT sensors in smart cities. It illustrates how this technology could enhance urban governance and, eventually, elevate the quality of life for residents.

Figure 3. Available communication technologies in various applications in the Internet of Things

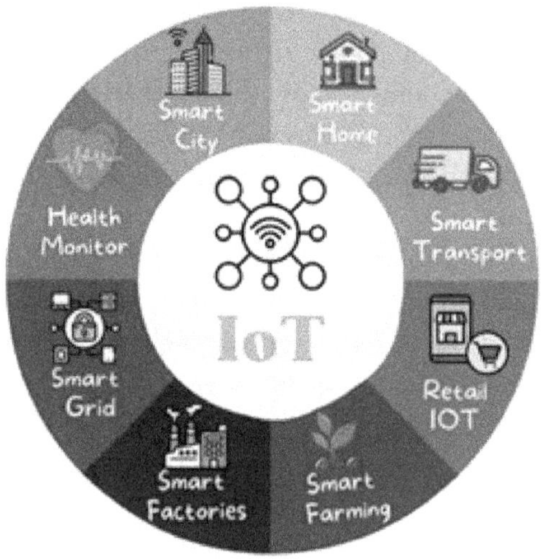

Multiple viewpoints, including intelligent sensors, communication technologies, and various applications, can analyze the integration of IoT technologies with AI in a smart city. Figure 3 highlights the available communication technologies in various applications in the Internet of Things. Crucial IoT technologies focus on creating reliable and robust networking and communication infrastructures to facilitate efficient data and information exchange across the many elements of smart city services. Various scales influence the advancement of IoT services for smart cities, contingent upon the application type. They may require diverse networking and communication technologies for their implementation and operation. Multiple review articles have previously examined various issues related to big data, network security, and the potential of IoT in smart cities. However, there are not many resources that look deeply into how IoT and AI work together in smart cities. This would allow for more in-depth study from many angles, such as smart sensors, communication technologies, and different applications. To facilitate efficient data and information exchange among the various elements of linked cities, the essential IoT technologies necessitate the establishment of reliable and robust networking and communication infrastructures. The size of IoT services in smart cities varies depending on the type of application. They may require diverse networking and communication technologies for their implementation and operation. The article expresses how AI can analyze the vast amounts of data generated by IoT sensors in smart cities. It highlights how these technologies could enhance local governance and improve inhabitants' quality of life.

1.3 5G Intelligent Ecosystem:

In the context of smart cities, artificial intelligence has an enormous impact on intelligent configurations. Artificial intelligence has the potential to enhance sustainability, reduce waste, and optimize energy consumption. Artificial intelligence algorithms can monitor and regulate energy consumption in buildings by autonomously adjusting the temperature and lighting based on occupancy patterns. This can lead to significant energy savings and improved energy efficiency. Artificial intelligence can facilitate trash management by forecasting garbage accumulation, determining the optimal time for waste collection, and reducing transportation expenses. We can implement artificial intelligence to monitor and preserve the environment. It has the potential to aid in the monitoring of air and water quality, the observation of vegetation changes, and the prediction of natural disasters. The incorporation of AI and IoT devices enables the creation of a more comprehensive and real-time comprehension of the environment, thereby bolstering environmental management and conservation initiatives. Artificial intelligence can revolutionize intelligent environmental solutions within a smart city context. Smart cities have the potential to improve the quality of life for their residents by leveraging AI to improve sustainability, efficiency, and resilience.

2. RELATED WORK

A smart environment is one that uses data-driven solutions and high-tech technologies to make built and natural ecosystems more sustainable, efficient, and resilient.

Halhoul Merabet et al. conducted a comprehensive study on AI-based approaches to control system development. The objective is to preserve comfortable interior conditions while simultaneously decreasing energy consumption. L. Li et al. investigated the obstacles encountered in the application of AI to enhance energy efficiency and comfort, as well as proposed potential avenues for future research in this field. Their research outlines the development and application of a multifaceted AI algorithm with numerous objectives, designed for practical environmental control. The algorithm aims to enhance the energy efficiency, thermal comfort, and indoor air quality (IAQ) of buildings in a prompt and accurate manner. The results indicated an elevated level of precision in the prediction of outcomes, surpassing 90% accuracy. Additionally, the system achieved average energy savings of 31%, 45%, and 35%, resulting in improved thermal comfort and reduced air pollution. A. I. Dounis examined the potential of AI in the design of building automation systems, with an emphasis on the optimization of energy consumption, the assurance of comfort, the promotion of health, and the improvement of productivity in residential environments. In response to sustainability and climate change concerns, L. Anthopoulos and V. Kazantzi investigated the increasing emphasis on energy efficiency, particularly in urban areas. It underscores the impact of emerging technologies such as blockchain and electricity and suggests an analysis of the impact of autonomous vehicles, intelligent building systems, artificial intelligence, and big data on urban energy systems. It also recommends a unified assessment framework for assessing the energy efficacy of big data and artificial intelligence.

S. Rubab et al. conducted a comparative analysis of various methodologies and proposed a comprehensive strategy for managing COVID-19 waste. This strategy considers the potential applications and advantages of waste, while also defining both immediate and long-term waste management objectives. A. Shreyas Madhav et al. proposed a mobile automaton capable of identifying and segregating common electronic waste from households. The robot utilizes a convolutional neural network-based identification algorithm to accurately classify e-waste with 96% accuracy.

To establish intelligent waste management practices in smart cities within existing waste management frameworks, S. Shukla and S. Hait examined the integration of technologies such as IoT, AI, intelligent transportation systems, and cloud computing.

Y. Himeur et al. investigated the application of artificial intelligence and data fusion techniques in the appraisal and surveillance of the environment, utilizing remote sensing data and satellite images. The discourse concludes with recommendations and proposals for future trajectories. AI-enabled mobile marine protected areas (MMPAs) are a novel approach to environmental monitoring and maritime governance that K. Bakker proposed. MMPAs employed machine learning algorithms to evaluate the data and promptly adjust to changes in environmental conditions and interruptions, and they employed digital devices to acquire information from a variety of sources. In a study by S. J. Soheli et al., they investigated whether using an adaptive neuro-fuzzy inference system (ANFIS) and the Internet of Things (IoT) in a smart, automated greenhouse could increase crop yield and efficiency. The system utilized sensors to acquire real-time data on meteorological variables, such as temperature, humidity, solar radiation, and soil moisture content. Technology, both economical and efficient, can significantly increase the greenhouse's crop yield.

D. Pamucar et al. investigated the potential impact of Metaverse technologies on the transportation system, underscored the ongoing efforts to integrate transportation into the Metaverse, and determined that these technologies have the potential to revolutionize the transportation system. The evaluation system aids decision-makers in prioritizing the implementation of sustainable transportation alternatives within the metaverse. I. Ahmed et al. proposed a theoretical model that is both environmentally sustainable and intelligent, utilizing cloud computing, IoT devices, and AI for data administration and acquisition. The configuration enables the analysis and archiving of digital data, resulting in decentralized cloud repositories that employ blockchain technology to facilitate a variety of applications.

E. K. Nti et al. emphasized the critical significance of AI in addressing environmental sustainability challenges, such as energy, water management, biodiversity, and transportation. The transportation sector has implemented artificial intelligence models, including computer vision and decision support systems. We must consistently oversee initiatives designed to enhance environmental sustainability. Table 1. shows the analysis of the influence of artificial intelligence on smart city technology.

Table 1. Analysis of the influence of artificial intelligence on smart city technology

Major Domains	Impact of AI
Smart Mobility	Improved traffic management Autonomous vehicles Predictive maintenance
Smart Governance	Improving decision-making Enhancing public services Increasing transparency and accountability
Smart Education	Personalized learning Intelligent tutoring Curriculum development:
Smart Economy	Increased productivity Improved decision-making Enhanced customer experience
Smart Healthcare	Medical diagnosis Predictive analytics Drug discovery
Smart Environment	Energy efficiency Waste management Environmental monitoring Sustainable transportation
Smart Living	Home automation Personalized recommendations Security and safety Health and wellness

Figure 4. Probable characteristics of IoT-enabled 5G networks that would be useful for smart cities.

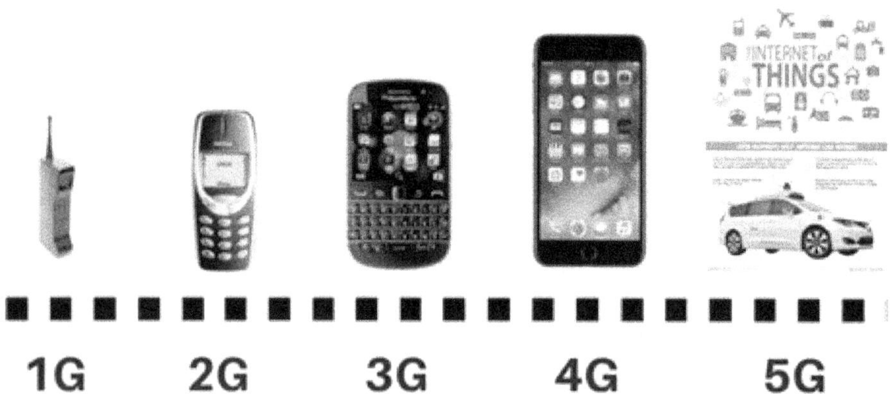

Each of the first four generations established a new level of connectivity. We expect 5G to carry on this trend, opening new possibilities for innovation and creativity. Figure 5 shows the probable characteristics of IoT-enabled 5G networks that would be useful for smart cities. 5G can connect more devices, is faster, more responsive, and uses less energy than previous generations of mobile communication. But to make rapid and substantial advancement, it must solve specific issues and shortcomings, just like any modern technology. Increasingly, as gadgets become more interconnected, consumers and businesses are worried about security and privacy. Technological maturity, costs, government restrictions, and international standards are the main obstacles. Many companies are facing problems with data security and privacy, a lack of resource criteria, and insufficient overall execution, says Ericsson. As 5G expands its use of millimeter waves, the limited range of these wavelengths will necessitate the incorporation of increasingly susceptible equipment. We require modern technology, new business models, and investments in skills and people to make the most of 5G networking. Most vertical industry workers must have the right gear. To reap the interdependent benefits of 5G deployment to the fullest, infrastructure must acquire 5G-capable devices, which calls for further investment in training and money. The consensus is that the main benefit of 5G will be an increase in operational efficiency in vertical industries. It is essential to integrate socially. The United Nations Sustainable Development Goals make use of socially beneficial vertical industry applications. Goals for the Future of Sustainability (SDGs). Informing the general population of the merits of fifth-generation technology is of the utmost importance. The common people's lack of technology knowledge is the root cause of this problem. This lack of knowledge creates a barrier to its implementation. The narrative establishes a link between 5G technology and the idea that resulted in the destruction of seventy-seven mobile towers in the UK.

The United Nations' Sustainable Development Goals serve as a guide for public education regarding the effects of 5G mobile technology. Eleven principal domains perceive the promotion of health and well-being as the most significant social benefit that communication technology contributes.

- In Section 2, we outline how 5G will facilitate and expedite innovation in a diverse array of smart city vertical industries, such as energy, health care, manufacturing, entertainment, and transportation. Additionally, in this section, we investigate the economic implications of 5G technology on each vertical business before proceeding to the next segment, which concentrates on a single vertical industry. Consequently, we concentrate on the influence of 5G technology on transportation.
- Section 3 concentrates on transportation as a vertical industry. In this section, we examine the necessity of an intelligent transportation system and the ways that 5G would facilitate this vertical industry. We examine the benefits of 5G use cases for transportation systems and the requirements that it can satisfy.
- In Section 4, we examine the implications of the intelligent transportation system in various dimensions. This section outlines the optimization of ITS to reduce the economic and environmental effects of transportation. It also addresses new challenges, including privacy and security, as well as ethical and social concerns that may impede its implementation.
- Section 5 provides the conclusion of the emerging work.

2.1 From the First generation to the fifth generation of communication system:

The early 1980s saw the launch of first-generation (1G) mobile communication, which used only analog radio waves for voice communication. 1G utilized the Frequency Division Multiple Access (FDMA) methodology, partitioning the frequency into channels and allocating each channel to an individual user exclusively at any given moment. The 1G mobile network exhibited inadequate roaming support, substandard call quality, and a lack of communication encryption. The maximum data transfer speed on the 1G network was 2.4 Kbps. Launched in the 1990s, the second generation (2G) featured digital radio signals, offering increased data rate capacity and improved speech quality. The Global System for Mobile Communications (GSM) was a digital standard for 2G that facilitated data services, such as Short Message Service (SMS) and Multimedia Message Service (MMS). 2G introduced two methodologies: Time Division Multiple Access (TDMA) and Code Division Multiple Access (CDMA). TDMA assigns a distinct code to each user, while CDMA divides signals into time segments.

Figure 5. Framework of 1G to 5G communication system

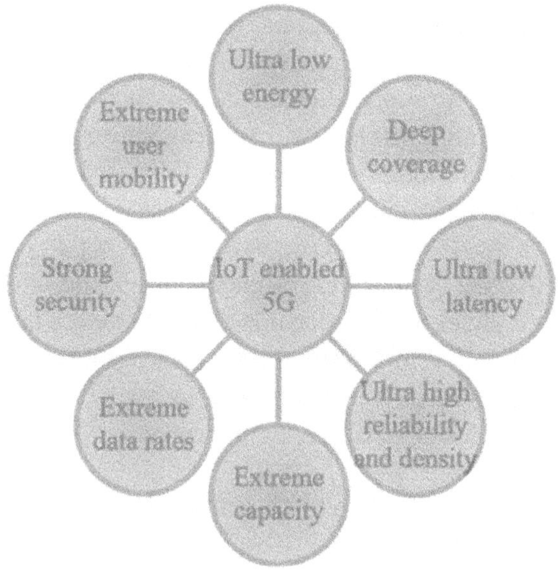

The data transfer rates of 2G were between 30 and 35 Kbps. Theoretically, Enhanced Data Rates for GSM Evolution (EDGE), a subsequent generation of 2G, was capable of transmitting data at frequencies of up to 1 Mbps.

The third generation (3G) was launched in the late 2000s. The International Telecommunication Union (ITU) defined 3G specifications as part of the International Mobile Telephone 2000 (IMT-2000) project, sometimes referred to as the Universal Mobile Telecommunications System (UMTS) because of its global roaming capabilities. The fundamental distinction between 2G and 3G is the use of packet switching instead of circuit switching. This enabled access to fundamental multimedia streaming and Internet apps. Initially, 3G technology permitted a data transfer rate of 200 Kbps; however, later versions may accommodate up to 3.1 Mbps. In 2008, the International Telecommunication Union (ITU) released the specifications for IMT-Advanced, commonly referred to as fourth-generation (4G) mobile communication. Following the advent of 4G in the early 2010s, Long-Term Evolution (LTE) emerged as the industry standard. Orthogonal Frequency Division Multiplexing (OFDM) and Multiple Input Multiple Output (MIMO) constituted the fundamental technologies of 4G. OFDM segments a digital signal into multiple narrowband channels with differing frequencies. Moreover, the fundamental difference between 3G and 4G was that 4G was exclusively dependent on the Internet Protocol (IP). 4G technology enables rapid Internet connectivity and high-definition (HD) video streaming, achieving data speeds of up to 100 Mbps.

2.2 Emerging trends in mobile networks

Mobile networks have emerged as the predominant and most widely utilized communication technology, boasting approximately 7.95 billion connections. Individuals and various devices, known as the IoT, employ mobile networks. In ten years, we anticipate the emergence of a new distributed ecosystem known as the Internet of Everything (IoE), which will include billions of globally interconnected IoTs.

The extensive interconnectedness of individuals and objects will lead to a substantial rise in data flow, placing strain on the current generation of wireless mobile communication. As a result, the extraordinary rise in connected devices, mobile data traffic, and the limitations of 4G technologies are prompting businesses and academics to focus on developing the requirements for the fifth generation (5G) of wireless mobile communication.

The aim of smart cities is to improve service quality through efficient management of public resources, emphasizing sustainability, maintenance, and comfort. The IoT is a communication framework that facilitates the interaction of ordinary devices via the Internet. Various sectors such as manufacturing, healthcare, energy, utility management, automotive, and public transportation employ the IoT paradigm.

Technological innovations propel the fifth generation (5G) network, transforming the core of mobile communication networks. 5G will enable a wide range of application possibilities through the introduction of a completely new network architecture. For the development of immersive solutions that are widespread and useful, modern technologies like Artificial Intelligence (AI) and Multi-access Edge Computing (MEC), along with network slicing and notarization technologies like Software-Defined Networking (SDN) and Network Function Virtualization (NFV), will be especially important. MEC is an essential facilitator for applications requiring near-real-time performance and decision-making. Multi-Access Edge Computing facilitates data processing at its source, thereby diminishing network latency. SDN is a networking paradigm that facilitates network programming, while NFV utilizes virtualization technology to deploy many types of hardware-based network devices as software instances.

Compared to 4G, 5G will allow data rates that are 10 to 100 times greater and accommodate a larger number of connected devices. Moreover, 5G will provide comprehensive geographical coverage and accessibility, along with improved security and privacy. Moreover, 5G will increase the lifespan of device batteries by ten times and decrease energy consumption by the same amount. The deployment of 5G technology focuses on essential technologies, such as new radio access, extensive Multiple Input Multiple Output (MIMO), heterogeneous ultra-densification, channel coding and decoding, and Millimetre Wave (mmWave). It does not encompass detailed information about the technology utilized by 5G. By 2035, 5G is projected to generate $12.3 trillion in products and services across various industries and create employment opportunities for up to twenty-two million people inside the 5G value chain.

5G enables the development of a productive industry, advanced technology, and creative communication techniques. The 5G communication network will enable the IoT infrastructure required by various smart city vertical businesses. 5G will support diverse use case requirements across many vertical industries, including energy, healthcare, manufacturing, media, entertainment, automotive, and public transit. We have developed network slicing to cater to the diverse service needs of vertical businesses. We consider the collection of network functions, to meet the communication service requirements of specific scenarios or business models.

2.3 Vehicle-to-Everything (V2X):

The energy crises and environmental issues have led to a surge in the adoption of electric vehicles. The emergence of smart grid concepts, integrating vehicle-to-everything (V2X) technology, has further amplified the importance of electric vehicles. Vehicle-to-everything (V2X) technology utilizes electric car batteries to generate additional income during periods of inactivity. Bi-directional charging allows battery assets to decrease energy usage in buildings and residences, earn income for the electric grid, or provide backup power to loads. This article presents a summary of the latest advancements in electric vehicle V2X functionality. We initially address many frameworks: vehicle-to-grid (V2G), vehicle-to-home (V2H), vehicle-to-building (V2B), vehicle-to-vehicle (V2V), vehicle-to-load (V2L), and vehicle-for-grid (V4G). We analyze the roles of aggregators and the IoT in the V2X infrastructure, including control modes, power flows, types of charging systems, and charging/discharging methodologies, among other aspects. We also explore V2X operational techniques and standards. This article also addresses the importance of power electronics in cyber-physical systems and vehicle-to-everything (V2X) operations. Our investigation also explores how V2X can aid in the adoption and spread of renewable energy. To maximize the potential of this technology for a sustainable future, we finish by identifying the challenges to V2X adoption and offering future insights and policy recommendations.

3. ROLE OF FUTURE SMART CITIES

The third generation (3G) was launched in the late 2000s. The International Telecommunication Union (ITU) defined 3G specifications as part of the International Mobile Telephone 2000 (IMT-2000) project, sometimes referred to as the Universal Mobile Telecommunications System (UMTS) because of its global roaming capabilities. The fundamental distinction between 2G and 3G is the use of packet switching instead of circuit switching. This enabled access to fundamental multimedia streaming and Internet apps. Initially, 3G technology permitted a data transfer rate of 200 Kbps; however, later versions may accommodate up to 3.1 Mbps. In 2008, the International Telecommunication Union (ITU) released the specifications for IMT-Advanced, commonly referred to as fourth-generation (4G) mobile communication. Following the advent of 4G in the early 2010s, Long-Term Evolution (LTE) emerged as the industry standard. Orthogonal Frequency Division Multiplexing (OFDM) and Multiple Input Multiple Output (MIMO) constituted the principal technologies of 4G. OFDM segments a digital signal into multiple narrowband channels with distinct frequencies. The principal difference between 3G and 4G was that 4G was dependent on the Internet Protocol (IP). 4G technology enables rapid Internet connectivity and high-definition (HD) video streaming, achieving data speeds of up to 100 Mbps.

Mobile networks have emerged as the predominant and most widely utilized communication technology, boasting approximately 7.95 billion connections. Individuals and various devices, known as the Internet of Things (IoT), employ mobile networks. In ten years, we anticipate the emergence of a new distributed ecosystem known as the Internet of Everything (IoE), which will include billions of globally interconnected IoTs. The extensive interconnectedness of individuals and objects will lead to a substantial rise in data flow, placing strain on the current generation of wireless mobile communication. As a result, the extraordinary surge in connected devices, mobile data traffic, and the limitations of 4G technologies are prompting both businesses and academics to focus on developing the requirements for the fifth generation (5G) of wireless mobile communication. Smart cities aim to improve service quality through the efficient use of public resources, emphasizing sustainability, maintenance, and comfort. The Internet of Things (IoT) is a communication framework that allows ordinary devices to interact with each other via the Internet. Various sectors such as manufacturing, healthcare, energy, utility management, automotive, and public transportation employ the IoT paradigm.

Technological innovations propel the fifth generation (5G) network, transforming the core of mobile communication networks. 5G will enable a wide range of application possibilities with the implementation of completely new network architecture. For the development of immersive solutions that are widespread and useful, innovative technologies like Artificial Intelligence (AI) and Multi-access Edge

Computing (MEC), along with network slicing and notarization technologies like Software-Defined Networking (SDN) and Network Function Virtualization (NFV), will be particularly important. MEC is an essential facilitator for applications requiring near-real-time performance and decision-making. MEC facilitates data processing at its source, thus minimizing network latency. SDN is a networking paradigm that facilitates network programming, while NFV utilizes virtualization technology to deploy many types of hardware-based network devices as software instances.

Compared to 4G, 5G will allow data rates that are 10 to 100 times greater and accommodate a larger number of connected devices. Moreover, 5G will provide comprehensive geographical coverage and accessibility, along with improved security and privacy. Moreover, 5G will enhance device battery longevity by a factor of ten and reduce energy consumption by a factor of ten. The deployment of 5G technology focuses on essential technologies, such as new radio access, extensive Multiple Input Multiple Output (MIMO), heterogeneous ultra-densification, channel coding and decoding, and Millimetre Wave (mmWave). This document does not provide detailed information about the technology that 5G utilizes. By 2035, 5G is projected to generate $12.3 trillion in products and services across various industries and create employment opportunities for up to twenty-two million people inside the 5G value chain.

5G enables the development of an efficient industry, intelligent technology, and novel communication techniques. The 5G communication network will enable the IoT infrastructure required by various smart city vertical businesses. 5G will support diverse use case requirements across many vertical industries, including energy, healthcare, manufacturing, media, and entertainment, automotive, and public transit. We have developed network slicing to cater to the diverse service needs of vertical businesses. Conceptually tailored network functions, known as slices, support the communication service requirements of specific scenarios or business models.

ICT frameworks connect sensors, mobile devices, communication gateways, data centers, power plants, hospitals, schools, libraries, water supply networks, waste management, transportation systems, law enforcement, and community services, facilitating the functioning of smart cities. As the Internet of Things (IoT) becomes a critical communication channel in smart cities, its implementations present numerous engineering and research challenges. To confront these obstacles, we require novel protocols, architectures, and services. 5G will transform smart cities by offering high-speed connectivity, low latency, extended battery life, and the ability to manage connections. This will enable the establishment of new services and the modification of existing ones.

The research investigates the role of IoT and 5G connectivity in a smart city, with a focus on mobility. Nevertheless, the investigation fails to demonstrate the economic implications of 5G technology on various industries or the requisite infrastructure for 5G-enabled transportation. The research delineates the technology and vertical sectors that are essential for the development of smart cities. Nevertheless, there is a dearth of information regarding the economic implications of 5G technology in numerous industries, as well as the role of mobile communication in smart cities. It offers a thorough assessment of 5G communication technologies for vehicle networks, with a particular emphasis on 5G-V2X communications. Consequently, it is the most pertinent study for this review. The authors address the 5G-V2X specification, architecture, use cases, and key enablers, such as edge computation and network slicing. The authors failed to consider the more extensive implications of 5G-enabled transportation, such as its influence on the economy and smart cities. The paper investigates the elements of smart cities, such as algorithms, networking technologies, and architectural design. Furthermore, it offers a comprehensive overview of the numerous applications of a smart city. However, it neglects to address the latest advancements in 5G technology, which have the potential to accelerate the development of smart cities and have significant economic implications. The authors provide a comprehensive description of the function of 5G New Radio (NR) for vehicle-to-everything communication. They introduced the technology components of 5G NR and illustrated how it can support a variety of V2X use cases for cooperative and autonomous transportation. The automotive sector experienced significant growth in the late 1920s, which had a significant impact on every industry. The affordability of vehicles consistently stimulates economic expansion in numerous industries. Technological advancements are enabling self-driving vehicles to revolutionize the transportation industry. The introduction of autonomous vehicles will result in substantial modifications. 5G facilitates and expedites the societal advantages of connected, automated, shared, and electrified vehicles. It will be instrumental in the advancement of the autonomous vehicle industry. The objective of this research is to investigate the implications of a 5G-enabled intelligent transportation system in various aspects, with a particular emphasis on the accounts and assurances of experts.

4. SMART MOBILITY WITH 5G, IOT & MM-WAVE TO ENABLE & ENHANCE PUBLIC SAFETY IN FUTURE SMART CITIES

A smart city can be characterized in several diverse ways. People frequently use terms like "intelligent city," "knowledgeable city," and "digital city" interchangeably with the term "smart city." This has resulted in a variety of speculative options for the concept of a smart city. Many contexts use the ambiguous term "smart city". There is no universally recognized definition or standardized blueprint for the establishment of a smart city. Smart city concepts emerged as a response to the challenges of urbanization and the need for sustainable urban development. The objective of the incorporation of a variety of technologies, such as IoT, AI, and big data analytics, is to improve the efficiency of urban services, reduce resource consumption, and increase living standards.

Technological advancement has made significant strides, allowing a modern smart city to collect data from a variety of applications, such as smart agriculture, smart industry, smart farming, smart healthcare, smart traffic, and smart pedestrian systems. We can then analyze and integrate this data to make decisions that enhance the quality of life. The rapid proliferation of IoT, cloud computing technologies, and AI may significantly improve the performance, quality, and interactivity of urban facilities, leading to a reduction in management expenses.

A variety of factors, including societal changes, economic challenges, and geriatric populations, are driving an emerging stage of generational transformation, which encompasses 5G and smart cities. This section discusses the key vertical businesses that 5G technologies are transforming into an IoT-based smart city, along with the role of 5G as the primary technology driving these changes. We can categorize the applications of 5G technology in a smart city into various dimensions, including distinct industrial verticals like energy, health care, manufacturing, media, and entertainment, automotive, and public transportation. Furthermore, each vertical industry may have a plethora of applications.

Although there are numerous methods for defining a smart city, two primary approaches frequently concentrate on the critical areas that it should prioritize. The initial trend in smart city definitions emphasizes a single urban characteristic, such as technology or environment, while disregarding other aspects of the city. The definitions do not address the fundamental objective of a smart city, which is to establish an innovative framework for urban administration and evaluation.

While the concept of a smart city can vary, it typically incorporates two primary methodologies that concentrate on critical areas. The initial trend in smart city definitions emphasizes a single urban characteristic, such as technology or environment, while failing to consider the remaining attributes of a city. The relevant definitions fail to address the fundamental objective of a smart city, which is to establish an innovative framework for the administration of urban areas and the evaluation of the interrelationships among diverse elements within an urban ecosystem. Concentrating exclusively on the advancement of one aspect of an urban system does not guarantee the resolution of the system's comprehensive issues. To confront the obstacles that contemporary cities encounter and to ensure a comprehensive and effective strategy, it is imperative to recognize the interdependence of all urban attributes.

The smart-city approach is emphasized as embodying a holistic approach that considers all urban characteristics as essential components of a system, as per an alternative set of definitions. This method emphasizes that the smart city is a new paradigm for urban organization and development, as it considers both technological and social factors to ensure the stability and sustainability of the urban system. These descriptions propose a comprehensive approach to urban challenges, leveraging emerging technology to reinvent the smart city paradigm and the interactions among its stakeholders.

The automotive industry is acutely aware of innovation in all areas, such as safety, manufacturing, and driving capabilities. As a result, all the world's most prominent manufacturers are striving to create automobiles that are more intelligent and connected, with a projected total of 250 million by 2020. 5G facilitates and accelerates the societal benefits of interconnected automation. We believe that shared and electrified vehicles will make a substantial contribution and will become fundamental to the development of autonomous vehicles.

4.1 Intelligent Transportation Systems

The implementation of 5G-enabled ITS can improve urban intelligence and automation. We are in the process of creating conveyance systems that are more efficient and secure than the current networks. We design the public transportation system to tackle the primary transportation challenges in metropolitan areas, such as traffic congestion, environmental pollution, and accidents. By enabling a genuinely intelligent transportation system, 5G can resolve these challenges. High-speed Internet connectivity equips the public conveyance system. A connected traffic cloud consolidates and analyzes real-time data from connected vehicles. Among other factors, the efficiency of petroleum and time resources matters. We have achieved a 20% outcome in preliminary trials of CITS. This has led to a 15% decrease in collisions, improved task efficiency, and reduced traffic congestion.

4.2 Development of Smart City

Infrastructural support, data-driven decision-making, citizen-centric strategies, enterprising spirit, creativity, teamwork, and adaptability are all crucial to building a smart city. Smart cities are urban ecosystems that are sophisticated and rely on ICT infrastructure to acquire and implement real-time data from a wide range of sensors and devices. This infrastructure, incorporating innovative technologies like IoT, cloud computing, and big data analytics, enables real-time monitoring and effective administration of urban systems. It also contributes to the realization of UN Sustainable Development Goal 1, which promotes inclusive, secure, resilient, and sustainable urban environments by improving urban governance, crisis management, and integrated planning. Technology integration, including artificial intelligence, the Internet of Things, big data analytics, and 5G/6G technologies, facilitates the digitization and sustainability of smart cities. As a result, smart city applications are expanding in a variety of sectors, such as transportation, energy, healthcare, public safety, and the environment. The integration of these technologies enables sustainable urban development and cultivates new commercial opportunities, thereby preserving the financial viability of the city within the virtual world. Metaverse can improve virtual cities by putting smart city features into a digital framework, creating immersive spaces for people to interact with each other, and planning to help people produce creative solutions to problems in cities. Institutional and social infrastructure are also essential components of smart cities. Institutional infrastructure is the integration of governance and collaboration among public, private, and civil entities to ensure the provision of transparent and efficient services. In contrast, social infrastructure is essential for augmenting the quality of life for residents by involving them and providing innovative solutions. To extract pertinent insights into urban planning, transportation, and environmental management, data-driven decision-making is essential for reality-based and temporally focused urban administration. This involves the utilization of extensive datasets from IoT devices and sensors. Additionally, smart cities implement integrated platforms, which encompass advanced transportation systems, energy management coordination, and facilities services. Additionally, the government's public service platform enhances overall efficiency and convenience by providing residents with convenient access to key services.

4.3 5G for Intelligent Transportation Systems in Smart Cities

Transportation accounts for 6 to 12 percent of the GDP in numerous developed countries. By 2030, we expect the quadrupling of registered vehicles to significantly impact on the transportation infrastructure. Additionally, we expect the fatality rate to increase to 1.8 million by 2030, resulting in an estimated annual loss of approximately 1.3 million lives due to road incidents. Addressing transportation challenges necessitates the concept of Intelligent Transportation Systems (ITS). It endeavors to provide innovative services that address traffic management and a variety of transportation modes. As a result, ITS has substantial potential to impact user productivity, energy consumption, carbon emissions, land utilization, congestion, safety, and mobility. 5G has the potential to significantly transform the ITS landscape. 5G and its diverse slicing types, including eMBB, mIoT, uRLLC, and V2X, can enhance and optimize the operation and performance of Intelligent Transportation Systems (ITS). To gain a comprehensive understanding of the concept of ITS, it is necessary to examine the advancement of vehicles and the technologies that facilitate it.

4.4 Vehicle Platooning in Smart Cities

Platooning enables vehicles to form a closely coordinated cluster, resulting in significantly reduced inter-vehicle spacing. This process improves road capacity, fuel efficiency, and productivity by enabling drivers to engage in alternative duties. Additionally, it reduces the number of accidents. All platoon vehicles engage in platoon operations and communicate with the main vehicle regularly. Each vehicle in a particular squadron is capable of autonomous operation. Vehicle-to-vehicle communications that are dependable are essential for platooning.

- Joining and leaving the platoon: The vehicle can join or leave a platoon at any time if it is in an active position, and it can use additional signals to complete the operation.

- Signaling and notification: This feature enables vehicles to recognize the platoon's formation and presence. This will facilitate the seamless incorporation of vehicles into or out of a platoon, enabling the vehicle that wishes to join to do so without disrupting the platoon.
- Management messaging: This feature enables the exchange of management messages to ensure the stable operational condition of vehicles within a platoon, including acceleration, deceleration, route selection, and platoon leader transitions, among other things. Vehicle-to-vehicle communication must enable reliable and secure message transmission to achieve effective and secure platooning, as the inter-vehicle distance is minimal while the vehicles are traveling at high velocities. The subsequent items outline the primary V2V communication prerequisites for platooning:

Cars with the lowest level of automation experience an end-to-end communication delay of 25 milliseconds, while those with the maximum level of automation experience a delay of 10 milliseconds.

- A relative longitudinal position accuracy of 0.5 meters is required for proximity platooning. In a batch of vehicles with minimal automation, message dependability is 90 percent, and in a cohort of vehicles with maximal automation, it is 99.99 percent.

The rate of message dissemination fluctuates between 10 and 30 messages per second.

- Remote driving is a cloud-based program that allows a human driver to remotely operate a vehicle via a vehicle-to-network connection. Remote driving may be implemented in numerous scenarios, including:

Edge cases can render a vehicle incapable of selecting a secure navigation route or method, potentially immobilizing it. In situations where the vehicle necessitates remote-control assistance from a remote driver, remote driving can be of assistance to automobiles. Edge cases include situations where the vehicle faces obstacles such as adverse weather, double-parked cars blocking the ingress/egress yellow lines, nearby accidents blocking lanes, and other unanticipated events that make it difficult to execute a safe maneuver or figure out what to do next.

Remote driving could provide benefits to the elderly, youth, and individuals who lack driving licenses or the ability to operate a vehicle. Moreover, fleet proprietors may require remote vehicle control in scenarios such as relocating trucks, delivering rental vehicles to clients, and providing remotely operated taxi services.

Services with established routes and terminals are more suitable for cloud-based public transportation. Remote driving may reduce the costs associated with completely autonomous driving for specific applications due to its reduced technical prerequisites, such as reduced computational demands for complex algorithms and reduced in-vehicle sensors.

Here are some potential V2X criteria for enabling remote driving:

In order to facilitate downlink data velocities of up to 1 Mbps and uplink data rates of up to 25 Mbps, two H.265/HEVC HD streams of up to 10 MB/s each are necessary. The automobile and the V2X application server communicate at a maximum speed of 250 km/h.

- The car and the V2X application server experienced a total delay of five ms.

5. CONCLUSION

Reliable and constant connectivity in smart cities is impossible to achieve without communication technology. However, there may be difficulties brought about by the proliferation of smart devices. Therefore, current studies explore the potential applications of IoT-enabled communication technologies in smart cities and their capacities. When compared to long-range technologies like LoRaWAN, short-range technologies such as Wi-Fi, ZigBee, and Z-Wave offer better data throughput. Every application has unique needs in terms of coverage area; thus, choosing the right communication technology is key.

Progress in several areas is crucial to the smart city's future: the environment, agriculture, economy, business, and government. Integrating modern 5G communication technology, Internet of Things devices, and artificial intelligence algorithms into urban infrastructure is crucial for a smart city to succeed.

Urban areas can become more efficient, sustainable, and livable with the help of this technology. This document outlines the prerequisites for integrating all sectors to facilitate the establishment of a smart city. Internet of Things (IoT) networks cannot use the current data transport standard. Thus, a concerted effort is needed to enable sensor nodes to communicate using different protocols while using little power. To keep operational costs down, researchers and stakeholders should focus on developing efficient storage approaches and designs for low-power devices. With 5G playing a pivotal role in future smart city paradigms, we must merge diverse networks designed for certain purposes into an all-encompassing smart city network. The prospective applications of AI are vast. We can enhance data fusion techniques to better access distinct types of data and use intelligent data reduction to remove "uninteresting" or unnecessary data from the AI development pipeline. In summary, the limited communication capabilities of current communication technologies, initially designed to serve a small number of devices, render them inadequate for ensuring constant connectivity in smart cities. Therefore, to meet the needs of future smart cities, it is critical to design advanced and standardized protocols, like the ones made possible by 5G on the Internet of Things.

5G networks have not yet realized their full potential since they remain in the research and development phase. Therefore, we anticipate the implementation of numerous 5G applications, enhancing current services, and introducing novel products in the upcoming years. There will be four main paradigms supported by 5G technologies: the Internet of Things (IoT), smart cities, ITS, and Industry 4.0. The advent of 5G, with its ultra-low latency and high bandwidth, will spur innovation and economic growth in every industry by encouraging the birth of new ones, reducing prices, and improving product and service quality. This study explores the impact of 5G technology applications and the innovative or transformational developments on several vertical companies within a smart city.

Additionally, we highlight the potential benefits of vehicular communication in a smart city's intelligent transportation system and how it could pave the way for its implementation. We take a multi-faceted look at the potential repercussions of an ITS, considering factors such as sustainability, ethics, economics, privacy, and security.

The advent of new industries, minimization of costs, and improvements in product and service quality will promote economic growth and innovation across all sectors, thanks to the revolutionary impact of 5G, which is marked by high bandwidth, integration of the Internet of Things, and ultra-low latency. Nevertheless, it will raise new technical, social, and ethical issues. Progress in several areas is crucial to the smart city's future: the environment, agriculture, economy, business, and government. Integrating modern 5G communication technology, Internet of Things devices, and artificial intelligence algorithms into urban infrastructure is crucial for a smart city to succeed. The synergistic integration of various technologies can achieve a more efficient, sustainable, and livable city. This document outlines the prerequisites for integrating all sectors to facilitate the establishment of a smart city. Internet of Things (IoT) networks cannot use the current data transport standard. Thus, a concerted effort is needed to enable sensor nodes to communicate using different protocols while using little power. To keep operational costs down, researchers and stakeholders should focus on developing efficient storage approaches and designs for low-power devices. With 5G playing a pivotal role in future smart city concepts, we should connect heterogeneous networks for individual applications to a comprehensive smart city network. The prospective applications of AI are vast. For instance, we might fine-tune data fusion techniques and employ intelligent data reduction to eliminate unnecessary or "uninteresting" data, thereby streamlining the AI development process and making more relevant data easily accessible. In summary, the limited communication capabilities of current communication technologies, initially designed to manage a small number of devices, cannot ensure continuous connectivity in smart cities. With the advent of 5G and the Internet of Things, there is a pressing need to standardize and develop advanced protocols to meet the needs of future smart cities.

REFERENCES:

Agiwal, M., Roy, A., & Saxena, N. (2016). Next generation 5G wireless networks: A comprehensive survey. *IEEE Communications Surveys and Tutorials*, 18(3), 1617–1655. DOI: 10.1109/COMST.2016.2532458

Angelidou, M. (2014). Smart city policies: A spatial approach. *Cities (London, England)*, 41, S3–S11. DOI: 10.1016/j.cities.2014.06.007

Anthopoulos, L. G. (2015). Understanding the smart city domain: A literature review. *Transforming city governments for successful smart cities*, 9-21.

Balzano, W., & Vitale, F. (2017, March). DiG-Park: a smart parking availability searching method using V2V/V2I and DGP-class problem. In *2017 31st International Conference on Advanced Information Networking and Applications Workshops (WAINA)* (pp. 698-703). IEEE.

Chettri, L., & Bera, R. (2019). A comprehensive survey on Internet of Things (IoT) toward 5G wireless systems. *IEEE Internet of Things Journal*, 7(1), 16–32. DOI: 10.1109/JIOT.2019.2948888

. El Zorkany, M., Yasser, A., & Galal, A. I. (2020). Vehicle to vehicle "V2V" communication: scope, importance, challenges, research directions and future. *The Open Transportation Journal, 14*(1).

Gohar, A., & Nencioni, G. (2021). The role of 5G technologies in a smart city: The case for intelligent transportation system. *Sustainability (Basel)*, 13(9), 5188. DOI: 10.3390/su13095188

Gupta, A., & Jha, R. K. (2015). A survey of 5G network: Architecture and emerging technologies. *IEEE Access : Practical Innovations, Open Solutions*, 3, 1206–1232. DOI: 10.1109/ACCESS.2015.2461602

Li, S., Da Xu, L., & Zhao, S. (2018). 5G Internet of Things: A survey. *Journal of Industrial Information Integration*, 10, 1–9. DOI: 10.1016/j.jii.2018.01.005

Muhammad, M., & Safdar, G. A. (2021). 5G-based V2V broadcast communications: A security perspective. *Array (New York, N.Y.)*, 11, 100084. DOI: 10.1016/j.array.2021.100084

Myeong, S., & Shahzad, K. (2021). Integrating data-based strategies and advanced technologies with efficient air pollution management in smart cities. *Sustainability (Basel)*, 13(13), 7168. DOI: 10.3390/su13137168

Parrado, N., & Donoso, Y. (2015). Congestion based mechanism for route discovery in a V2I-V2V system applying smart devices and IoT. *Sensors (Basel)*, 15(4), 7768–7806. DOI: 10.3390/s150407768 PMID: 25835185

Roger, S., Martín-Sacristán, D., Garcia-Roger, D., Monserrat, J. F., Kousaridas, A., Spapis, P., & Ayaz, S. (2020). 5G V2V communication with antenna selection based on context awareness: Signaling and performance study. *IEEE Transactions on Intelligent Transportation Systems*, 23(2), 1044–1057. DOI: 10.1109/TITS.2020.3019530

Shafique, K., Khawaja, B. A., Sabir, F., Qazi, S., & Mustaqim, M. (2020). Internet of things (IoT) for next-generation smart systems: A review of current challenges, future trends, and prospects for emerging 5G-IoT scenarios. *IEEE Access : Practical Innovations, Open Solutions*, 8, 23022–23040. DOI: 10.1109/ACCESS.2020.2970118

Zhao, D., Qin, H., Song, B., Zhang, Y., Du, X., & Guizani, M. (2020). A reinforcement learning method for joint mode selection and power adaptation in the V2V communication network in 5G. *IEEE Transactions on Cognitive Communications and Networking*, 6(2), 452–463. DOI: 10.1109/TCCN.2020.2983170

Chapter 3
AI-Based Disaster Prediction and Early Warning Systems

Ajay N. Upadhyaya
https://orcid.org/0000-0002-7583-6430
SAL Engineering and Technical Institute, India

Pranoy Debnath
https://orcid.org/0000-0001-6269-3867
Indian Institute of Technology, Bombay, India

Poonam Rani
https://orcid.org/0009-0004-4506-7871
Chandigarh Group of Colleges, India

P. Selvakumar
https://orcid.org/0000-0002-3650-4548
Department of Science and Humanities, Nehru Institute of Technology, Coimbatore, India

T. C. Manjunath
https://orcid.org/0000-0003-2545-9160
Rajarajeswari College of Engineering, India

Sumanta Bhattacharya
https://orcid.org/0000-0003-2563-2787
Maulana Abul Kalam Azad University of Technology, India

ABSTRACT

Disaster management systems. This introduction explores the role of disaster management technologies, focusing on their evolution, current applications, and future potential, particularly in the context of AI-based solutions.Historically, disaster management relied heavily on manual processes and rudimentary tools for assessment and response. This often resulted in delayed reactions and inadequate resource allocation during emergencies. Geographic Information Systems (GIS), remote sensing, and telecommunications have all played crucial roles in improving situational awareness and enabling faster decision-making. ICT facilitates real-time data sharing among stakeholders, including government agencies, NGOs, and the public. This connectivity enhances coordination and ensures that critical information reaches affected populations promptly. For instance, mobile technology has empowered communities to report incidents and receive alerts, thereby improving early warning systems.Moreover, data analytics has emerged as a key component in disaster management.

DOI: 10.4018/979-8-3693-9770-1.ch003

INTRODUCTION TO DISASTER MANAGEMENT TECHNOLOGIES

Disaster management systems. This introduction explores the role of disaster management technologies, focusing on their evolution, current applications, and future potential, particularly in the context of AI-based solutions.Historically, disaster management relied heavily on manual processes and rudimentary tools for assessment and response. This often resulted in delayed reactions and inadequate resource allocation during emergencies. Geographic Information Systems (GIS), remote sensing, and telecommunications have all played crucial roles in improving situational awareness and enabling faster decision-making.ICT facilitates real-time data sharing among stakeholders, including government agencies, NGOs, and the public. This connectivity enhances coordination and ensures that critical information reaches affected populations promptly. For instance, mobile technology has empowered communities to report incidents and receive alerts, thereby improving early warning systems.Moreover, data analytics has emerged as a key component in disaster management By leveraging big data, agencies can analyze historical patterns, predict future occurrences, and assess vulnerabilities within communities. Predictive analytics, in particular, has shown promise in forecasting disasters like floods, hurricanes, and earthquakes, allowing for preemptive actions to minimize risks. The use of land use that may indicate an increased risk of wildfires, enabling timely interventions. Similarly, AI-powered chatbots and virtual assistants can provide essential information to communities during crises, ensuring that individuals have access to guidance and resources when they need it most.The application of drones and unmanned aerial vehicles (UAVs) has also gained traction in disaster management. Drones can conduct aerial surveys of disaster-stricken areas, providing real-time imagery that helps assess damage and plan recovery efforts. They are particularly useful in reaching areas that are inaccessible to ground teams due to debris or unsafe conditions. Drones can also deliver supplies to remote locations quickly, improving response times and reducing reliance on traditional logistics.For example, flood sensors placed in vulnerable areas can provide immediate alerts when water levels rise, triggering early warning systems and allowing residents to evacuate before a disaster strikes. Training and capacity building are essential to empower individuals and organizations to leverage these technologies effectively.In conclusion, disaster management technologies have evolved significantly, particularly with the integration of AI and data analytics. These advancements offer immense potential to enhance disaster preparedness, response, and recovery efforts. However, realizing this potential requires

TYPES OF DISASTERS AND THEIR IMPACT

Historically, disaster management relied heavily on manual processes and rudimentary tools for assessment and response. This often resulted in delayed reactions and inadequate resource allocation during emergencies. Geographic Information Systems (GIS), remote sensing,(Alwindawi, A. F., et al.,2022) and telecommunications have all played crucial roles in improving situational awareness and enabling faster decision-making. This connectivity enhances coordination and ensures that critical information reaches affected populations promptly. For instance, mobile technology has empowered communities to report incidents and receive alerts, thereby improving early warning systems. Moreover, data analytics has emerged as a key component in disaster management. By leveraging big data, agencies can analyze historical patterns, predict future occurrences, and assess vulnerabilities within communities. Predictive analytics, in particular, has shown promise in forecasting disasters like floods, hurricanes, and earthquakes, allowing for preemptive actions to minimize risks.The application of drones and unmanned aerial vehicles (UAVs) has also gained traction in disaster management. Drones can conduct aerial surveys of disaster-stricken areas, providing real-time imagery that helps assess damage and plan recovery efforts. They are particularly useful in reaching areas that are inaccessible to ground teams due to debris or unsafe conditions. For example, flood sensors placed in vulnerable areas can provide immediate alerts when water levels rise, triggering early warning systems and allowing residents to evacuate before a disaster strikes. Despite the many benefits of technology in disaster management, challenges remain. Data privacy concerns, the digital divide(Al-Yahyai, S., et al.,2010), Furthermore, successful implementation of these technologies requires and communities. Building partnerships and fostering a culture of knowledge sharing can lead to more innovative solutions and improved disaster resilience. Training and capacity building are essential to empower individuals and organizations to leverage these technologies effectively.In conclusion, disaster management technologies have evolved significantly, particularly with the integration of AI and data analytics. These advancements offer immense potential to enhance disaster preparedness, response, and recovery efforts. However, realizing this potential requires addressing existing challenges and fostering collaboration among stakeholders. capabilities, ultimately leading to safer and more resilient communities in the face of inevitable disasters.

MACHINE LEARNING TECHNIQUES FOR PREDICTION

This overview delves into the primary machine learning techniques used for prediction, their methodologies, applications, and the implications of their use in real-world scenarios. Despite its simplicity, linear regression is effective for various applications, such as predicting Decision trees, on the other hand, are a more complex and versatile algorithm and easy to interpret, making it particularly useful in scenarios where transparency is essential, such as in healthcare or finance. However, decision trees can be prone to overfitting if not properly regulated, often necessitating techniques like pruning or the use of ensemble methods.Support vector machines (SVMs) offer another powerful technique. Unsupervised learning techniques, while less common in predictive analytics, are invaluable for exploratory data analysis and pattern discovery. In unsupervised learning, models are trained on datasets without labeled outcomes, allowing them to identify hidden structures and relationships. By simplifying datasets(Bhoi, A., et al.,2022), dimensionality reduction can enhance the performance of predictive models and improve computational efficiency. learn to take actions that maximize cumulative rewards based on feedback from their interactions with the environment. This technique has been particularly successful in robotics, game playing, and real-time decision-making pplications. Notable examples include AlphaGo, which defeated a world champion in the game(Brester, C., et al.,2023) In conclusion, machine learning techniques for prediction encompass a diverse array of methodologies that empower organizations to make informed decisions based on data-driven insights. From supervised learning approaches, such as regression and decision trees, to advanced techniques like deep learning and reinforcement learning, these methods provide the foundation for effective predictive analytics.

BUILDING EFFECTIVE EARLY WARNING SYSTEMS

This comprehensive overview explores the essential components, design principles, challenges, and best practices for building effective early warning systems involves understanding the hazards that a community faces and the vulnerabilities that exist within the population and infrastructure. This foundational knowledge is critical for designing an EWS tailored to specific risk.(Ramakrishnan T.,et al.,2022)s. Conducting comprehensive risk assessments, mapping hazard zones, and analyzing historical data on past disasters are essential steps in building a robust risk knowledge base. This information helps identify high-risk areas, vulnerable populations, and critical infrastructure that may be affected, allowing for more focused preparedness efforts.Monitoring and Warning Services such as meteorological sensors, seismographs, satellite imagery, and remote sensing. Advanced data analytics, including machine learning algorithms, can enhance the accuracy of predictions by analyzing real-time data and recognizing patterns that precede disasters. For example, in flood forecasting, hydrological models can predict water levels based on rainfall data, while seismic networks can detect and assess earthquake activity.Once a potential hazard is identified, the next crucial step in reaction to warnings. in preparedness activities. Local authorities should develop contingency plans and conduct regular reviews to ensure that they can respond effectively when warnings are issued. Community involvement in response planning fosters a sense of ownership and enhances the overall resilience of the population.Despite the critical importance of early warning systems, several challenges can hinder their effectiveness. One significant obstacle is the communication systems and explore alternative methods for disseminating warnings, such as community networks or local leaders who can relay information.Another challenge is the need for and workshops can foster collaboration and improve information-sharing among agencies.are also critical factors in the success of early warning systems. False alarms or inaccurate predictions can lead to public distrust and complacency, reducing the effectiveness of future warnings(Vijayakumar G, et al., 2024. Continuous monitoring and validation of data sources are essential to improve prediction models and ensure that the information disseminated is reliable. Engaging scientists and researchers in the development and maintenance of predictive models can enhance the scientific rigor behind the warnings issued.(Chakrabortty, R., et al.,2023)In recent years, advancements in technology have opened new avenues for enhancing early warning systems.

The integration of Moreover, the rise of offers exciting possibilities for disseminating early warnings. Mobile apps can provide real-time alerts, emergency information, and safety resources directly to users' smartphones. These apps can also allow for two-way communication, enabling communities to report local conditions and hazards, further enhancing situational awareness. is another essential aspect of building effective early warning systems. Involving local communities in the design and implementation of EWS fosters a sense of ownership and increases the likelihood of compliance with warning instructions. Participatory approaches, such as community workshops and focus groups, can help identify specific needs and preferences, ensuring that the system is tailored to local contexts. Training community Transparency in how warnings are generated, who is responsible for issuing them, and what actions should be taken can help establish confidence in the system. Regular feedback loops, where communities can provide input on the effectiveness of warnings and response measures, can further enhance trust and improve the system over time.In conclusion, building effective early warning systems requires a comprehensive approach that integrates risk knowledge, advanced monitoring technologies, clear communication strategies, and robust response capabilities. By addressing challenges such as the digital divide, ensuring inter-agency collaboration, and leveraging technological advancements, communities can enhance their resilience to disasters.

DATA SOURCES AND COLLECTION METHODS

In the realm of disaster management, the effective collection and utilization of data are paramount for enhancing preparedness, response, and recovery efforts. Data sources and collection methods itations, is essential for building robust disaster management systems. This overview explores the various types of data sources, methodologies for data collection, and best Both types of data are valuable in their own right and can be utilized to complement one another in the context of disaster management. Surveys are a commonly used method for gathering information from affected populations regarding their needs, vulnerabilities, and perceptions of risks. For instance, post-disaster surveys can provide critical insights into community impacts and recovery needs, informing future preparedness efforts.Interviews, both structured and unstructured, enable deeper insights into individual experiences and local knowledge regarding disaster risks. Engaging with community leaders, emergency responders, and affected individuals can uncover nuanced information that quantitative data may not fully capture. This qualitative data is invaluable for understanding the social dimensions of disasters, including how cultural factors influence risk perception and response behaviors.(Chamola, V., et al.,2021)Field observations involve direct monitoring of environmental conditions and community practices during disaster events. This method is particularly useful for assessing immediate impacts and collecting real-time data during emergencies. For example, during a flood, field teams can document water levels, infrastructure damage, and community responses, providing critical information for assessing the situation and planning interventions.is another powerful primary data collection method that utilizes satellite imagery and aerial surveys to gather information about the Earth's surface. This technology is instrumental in monitoring environmental changes, such as land use, vegetation cover, and the extent of disaster impacts. For example, remote sensing can provide accurate data on the extent of flooding, enabling authorities to assess affected areas quickly and allocate resources effectively.

Secondary data sources include historical records, governmental databases, academic research, and reports from non-governmental organizations (NGOs). Historical data is invaluable for understanding past disaster events, their frequency, and impacts on communities. This information can inform risk analysis and forecasting efforts.Academic research provides a wealth of information on disaster risks, response strategies, and recovery processes. Peer-reviewed studies can offer insights into best practices, innovative technologies, and lessons learned from past disasters. Collaborating with academic institutions can also facilitate access to specialized datasets and expertise. Platforms like Twitter, Facebook, and Instagram can provide immediate insights into public sentiment, emerging needs, and local conditions. enabling visual analysis and decision-making.Data analysis techniques vary based when collecting sensitive information from communities. Ensuring that data collection methods comply with ethical standards and legal requirements is paramount.Additionally, the (Chen, J.-Y et al.,2022) poses challenges in data collection efforts, particularly in low-resource settings where access to technology may be limited. Employing diverse data collection methods, including traditional approaches, can help bridge these gaps and ensure inclusive data gathering.Finally, data collection and analysis should be an ongoing process rather than a one-time effort. Continuous monitoring and periodic assessments allow for the adaptation of disaster management strategies based on evolving risks and community needs.

ConclusionIn summary, utilizing various collection methodologies, and integrating and analyzing data effectively, organizations can enhance their understanding of disaster risks and improve preparedness and response strategies. Overcoming challenges related to data privacy, access, and integration will be critical in ensuring that data-driven insights translate into actionable measures for safeguarding communities against the impacts of disasters.

REAL-TIME DATA ANALYSIS WITH AI

technologies. In an age where data is generated at unprecedented rates, the ability to analyze this data in real time is critical for organizations aiming to enhance operational efficiency, improve customer experiences, and respond swiftly to emerging challenges. This exploration of real-time data analysis with AI delves into its methodologies, applications, benefits, and challenges, highlighting its transformative potential across industries Real-time data analysis refers to the immediate processing and evaluation of data as it is generated or received. This contrasts with batch processing, where data is collected over time and analyzed at a later stage. Real-time analysis enables organizations to gain instantaneous insights, facilitating prompt decision-making and enabling proactive responses to dynamic situations (Daneshvar, M. H. et al.,2023) by automating the processing and interpretation of vast datasets. The first step in real-time data analysis involves gathering data from various sources, such as IoT devices, social media, transaction records, and customer interactions. Advanced data collection techniques, including streaming data from sensors and web scraping, enable organizations to obtain continuous streams of information. These methods ensure that data is available for analysis as soon as it is generated. Once data is collected, it needs to be processed in real time. devices real-time market Retailers use real-time data analysis to enhance customer experiences and optimize inventory management. In manufacturing, real-time data analysis enables predictive maintenance(Danish, M. 2022) by continuously monitoring machinery and equipment. Real-time data analysis is dynamic environments where conditions change rapidly.Automating data analysis with AI reduces the manual effort required to process and interpret data, allowing organizations to allocate resources more effectively.

This agility positions them favorably against competitors who may rely on slower, traditional analysis methods.Challenges of Real-Time Data AnalysisDespite its numerous advantages, real-time data analysis with AI also presents several challenges:As data volumes grow, scaling data processing systems to handle increasing amounts of real-time data can be challenging. data silos, and the need for specialized skills in AI and data analytics.The rapid evolution of AI and data analytics technologies has created a skills gap in many organizations. Training and retaining personnel with the necessary expertise to manage and analyze real-time data is essential for success. As the reliance on AI in real-time data analysis grows, ethical considerations surrounding data usage and algorithmic bias will come to the forefront. The development of collaborative AI systems, which integrate human insights with machine learning algorithms, will enhance the decision-making process. This hybrid approach will enable organizations to leverage the strengths of both human expertise and AI capabilities. Conclusion In summary, real-time data analysis with AI is revolutionizing the way organizations operate, enabling them to make informed decisions quickly and effectively. By leveraging diverse data sources, advanced processing techniques,-foster proactive problem-solving. While challenges related to data quality, scalability, and security remain(Esrafilian-Najafabadi, M., et al.,2022), the benefits of real-time data analysis far outweigh these obstacles. As technology continues to advance, organizations that embrace real-time data analysis with

AI will be well-positioned to thrive in an increasingly dynamic and competitive landscape. The future of real-time data analysis promises to unlock even greater potential, driving innovation and improving outcomes across various sectors.

CASE STUDIES: SUCCESS STORIES IN DISASTER PREDICTION

Disaster prediction has evolved significantly with the advent of advanced technologies and methodologies, leading to enhanced preparednes. This overview explores several notable success stories, illustrating how innovative approaches have transformed disaster management practices and saved lives.One of the most compelling success stories in disaster prediction occurred during the aftermath of the 2015 Nepal earthquake. This devastating event highlighted the necessity for improved early warning systems. In response, the United Nations and various NGOs collaborated to implement the "Nepal Earthquake Recovery and Reconstruction Plan," which integrated AI-driven data analytics for real-time monitoring and response.By utilizing satellite imagery and ground-based sensors(Fischer, J., et al.,2023), the team was able to assess damage to infrastructure and the displacement of populations. reports, and environmental sensors, allowing responders to identify the most affected areas quickly. The timely analysis facilitated the efficient distribution of humanitarian aid and resources, ensuring that assistance reached those in need without delay. This case exemplifies how leveraging technology can enhance situational awareness and improve disaster response efficacy. as another remarkable success story in disaster prediction. Established after the catastrophic tsunami of December 2004, which resulted in immense loss of life and property across several countries, the system utilizes a combination of seismic monitoring,

The integration of AI and machine learning into this system has significantlyiftly. trends that could signal impending to disaster prediction, particularly concerning earthquakes and tsunamis, has set a global standard. The Japan Meteorological Agency (JMA) employs a sophisticated network of seismometers and tsunami buoys that continuously monitorachievement in using satellite data for disaster prediction and management. By utilizing remote sensing technologies For instance, during the severe flooding in Europe in 2021, the Copernicus responders. The service utilized satellite imagery to assess the extent of flooding and identify areas at risk, facilitating targeted evacuation and resource allocation. The integration of AI to analyze satellite data in real-time further enhanced the program's ability to predict and respond to natural disasters effectively.Bangladesh, a country frequently affected . The Flood Forecasting and Warning Center (FFWC) employs a combination of hydrological models and satellite data to predict flood events accurately.By analyzing rainfall patterns and has also been instrumental in reaching remote populations, demonstrating the importance of technology in community engagement and disaster readiness.Australia's FireGuard project represents an innovative approach to predicting and managing bushfires, which are a significant risk in the region. By integrating The integration of community feedback and real-time data enhanced the system's accuracy and effectiveness. This case illustrates the vital role of predictive modeling and community involvement in managing disaster(Fowdur, T. P., et al.,2022) risks effectively.Conclusion These success stories in disaster prediction underscore the transformative impact of integrating advanced technologies, such as artificial intelligence and data analytics, into disaster management strategies. Each case highlights the importance of timely data collection, effective communication, and community engagement in enhancing disaster preparedness and response. As global challenges like climate change and urbanization increase the frequency and intensity of disasters, the continued evolution of predictive technologies will be crucial in safeguarding vulnerable populations and mitigating the impacts of natural disasters. The lessons learned from these case studies not only inform best practices but also inspire further innovations in disaster prediction, ultimately fostering resilience and sustainability in communities around the world.

COMMUNITY ENGAGEMENT IN DISASTER PREPAREDNESS

This comprehensive overview examines the importance of community engagement, successful strategies, challenges faced, and the broader implications for disaster management.k, leading to more targeted mitigation efforts. loss of life and property.Additionally, community engagement strengthens social networks and enhances collective efficacy.(Ghebleh Goydaragh, M., et al.,2021) Stronger community ties lead to improved communication and collaboration during emergencies. Research has shown that communities with robust social networks are more resilient, as neighbors look out for one another and share resources during crises. Effective engagement strategies can facilitate the establishment of these networks, ensuring that communities can mobilize quickly when disaster strikes.Successful Strategies for Community EngagementTo maximize the effectiveness of community engagement in disaster preparedness, several strategies have proven successful: Involving community members in the planning process allows them to contribute their insights and preferences. Workshops, focus groups, and public forums can be utilized to gather input on disaster preparedness plans. For example, in the Philippines, local government units have successfully implemented participatory risk assessment workshops, allowing residents to map hazards and identify vulnerabilities together.

These tools can be used to disseminate information about preparedness resources, emergency alerts, and training opportunities. For instance, apps like "ShakeAlert" provide real-time earthquake alerts and safety information, fostering proactive engagement among users.Collaborating with local organizations, schools, and businesses can amplify community engagement efforts. Partnerships can enhance resource sharing and provide additional expertise. For example, local businesses may support disaster preparedness initiatives by offering facilities for training sessions or donating supplies for emergency kits. and materials to meet diverse needs can enhance inclusivity. In many regions, organizations have successfully employed multilingual materials and accessible formats to ensure that critical information reaches everyone.Challenges in Community Engagement Despite its importance, community engagement in disaster preparedness faces several challenges:Limited Resources Many communities struggle with limited financial and human resources to implement engagement initiatives effectively. Organizations may lack funding to conduct training sessions or outreach programs, hindering their ability to engage residents meaningfully.Some community members may be skeptical of engagement efforts, viewing them as mere formalities without genuine intent. Building encourages community members to develop their emergency plans, fostering a culture of preparedness across(Ganguly, K. K.,et al.,2019) the nation. The Resilient Communities Project in Canada This project engages local communities in developing resilience plans to address climate change impacts. By facilitating workshops and collaborative planning sessions, community members identify local vulnerabilities and develop strategies to enhance resilience. This participatory approach has empowered communities to take ownership of their preparedness efforts, resulting in more robust resilience planning. Following devastating earthquakes and tsunamis, Indonesia implemented a community-led disaster risk reduction program. This initiative trained local volunteers to lead preparedness activities and educate their communities about risks and response strategies. By leveraging local knowledge and leadership, the program has strengthened community resilience and improved disaster preparedness in high-risk areas. The lessons learned from community engagement in disaster preparedness extend beyond individual communities. They underscore the importance of a collaborative approach to disaster management that prioritizes local knowledge and community involvement. By integrating community perspectives into disaster planning, policymakers can create more effective and relevant strategies that address specific needs and vulnerabilities.Furthermore, fostering a culture of preparedness within communities contributes to overall resilience. Communities that prioritize engagement are better equipped to respond to disasters, reducing the burden on emergency services and enhancing recovery efforts. By recognizing the value of community involvement, governments and organizations can build more robust disaster management frameworks that empower citizens to take an active role in safeguarding their futures.. While challenges remain, successful case studies demonstrate that with the right strategies,

ETHICAL IMPLICATIONS OF AI IN DISASTER MANAGEMENT

carefully considered. As AI systems are increasingly employed in high-stakes environments such as disaster prediction, response, and recovery, ethical challenges surrounding data privacy, bias, accountability, transparency, and the potential for dehumanization emerge. Addressing these concerns is crucial for ensuring that AI technologies contribute positively to disaster management efforts without compromising ethical standards or exacerbating existing inequalities.and surveillance technologies to monitor population movements during a disaster. While this information can enhance situational awareness and inform response strategies, it also poses risks to individual privacy. Unauthorized data collection or breaches can lead to sensitive information being exposed,(Han, S., et al.,2023) potentially resulting in harm to individuals and communities.Moreover, the use of surveillance technologies in disaster scenarios can create a culture of monitoring that undermines civil liberties. Communities may feel scrutinized, leading to distrust between residents and authorities. In disaster management, biased algorithms can lead to unequal resource distribution, inadequate responses to marginalized communities, or misinterpretation of data from diverse populations. For example, if an AI system disproportionately favors data from affluent neighborhoods when predicting disaster impacts, vulnerable communities may be overlooked, exacerbating existing disparities in disaster preparedness and response.

To mitigate these risks, developers and practitioners must prioritize fairness in AI design. This includes using s, and engaging community stakeholders in the development process. By prioritizing inclusivity, disaster management systems can become more equitable, ensuring that all communities receive the support they need during emergencies.As AI systems increasingly make decisions in disaster management—such as resource allocation or risk assessments—the question of accountability becomes paramount. If an AI system fails to predict a disaster accurately or mismanages resources, who is responsible for the consequences? The ambiguity surrounding accountability raises ethical concerns about the reliance o automated systems, particularly in high-stakes situations where lives are at risk.Clear lines of accountability must be established to address these concerns. Disaster management agencies should ensure that human oversight remains integral to AI-driven decision-making processes. Human experts should validate AI outputs, and decision-makers must be prepared to take responsibility for the outcomes of those decisions. By maintaining a human-centric approach, agencies can enhance accountability while leveraging AI's strengths. Transparency in AI algorithms is another ethical consideration crucial for fostering trust and understanding among stakeholders. Many AI systems operate as "black boxes," where the reasoning behind To promote trust, disaster management agencies must prioritize the explainability of AI systems. Stakeholders should understand how algorithms function, mechanisms, allowing communities to voice concerns and contribute to ongoing improvements in disaster management strategies To counteract the risk of dehumanization, disaster management agencies must prioritize human-centered approaches that emphasize empathy and community engagement. AI should serve as a tool to augment human decision-making rather than replace it. Integrating human perspectives into AI-driven processes can ensure that the unique needs and experiences of affected individuals and communities remain central to disaster response efforts.To navigate these ethical implications effectively, the establishment of robust rency, and the risk of dehumanization. By prioritizing ethical frameworks, fostering community engagement, and maintaining a human-centric approach, disaster management agencies can harness the benefits of AI while minimizing its risks.

REGULATORY FRAMEWORKS FOR DISASTER RESPONSE

Regulatory frameworks for disaster response are critical for ensuring effective preparedness, coordination, and recovery during and after disaster events. These frameworks encompass a wide range of policies, guidelines, and legal instruments that govern how governments, organizations, and communities respond to disasters. Given the increasing frequency and intensity of natural and human-made disasters, the need for robust regulatory frameworks has become more pressing(Haver, S. M., et al.,2018). This overview examines the essential components of disaster response regulations, the role of various stakeholders, the challenges faced in implementation, and emerging trends that shape the regulatory landscape. A clear legislative foundation is vital for establishing roles, responsibilities, and authority in disaster response. This includes laws that define the responsibilities of various government agencies at local, regional, and national levels, as well as the private sector and non-governmental organizations (NGOs). In many countries, specific disaster management acts outline these responsibilities, creating a legal basis for coordinated response efforts. Regulatory frameworks should mandate regular risk assessments to identify vulnerabilities and hazards within communities. These assessments inform planning and resource allocation, enabling stakeholders to prioritize mitigation efforts. Regulations may require local governments to develop and maintain hazard maps, vulnerability assessments, and preparedness plans, which is critical for ensuring a timely and effective response during emergencies. Regulations must address the allocation of resources for disaster preparedness and response.

This includes establishing funding mechanisms, such as emergency funds or disaster relief budgets, to ensure that necessary resources are available when disasters strike. Clarity in funding guidelines helps streamline financial support and promotes efficiency in response efforts.Regulatory frameworks should include provisions for effective public communication strategies. Informing communities about risks, response plans, and recovery efforts fosters public awareness and preparedness. After-action reviews and public reporting can enhance accountability and transparency in disaster management.A wide array of stakeholders implementation of regulatory frameworks for disaster response: National, state, and local governments are primary actors in disaster response regulation. They establish legal frameworks, allocate resources, and coordinate response efforts. Emergency management agencies often serve as the central authority responsible for disaster planning, response, and recovery.: NGOs are vital partners in disaster response, providing essential services and support during emergencies. Regulatory frameworks should facilitate collaboration between government agencies and NGOs, ensuring that their expertise and resources are integrated into response efforts. Businesses play an important role in disaster management, offering resources, logistics, and expertise. Regulations can encourage private sector participation by providing incentives for businesses to engage in disaster preparedness and response planning. Local community organizations and grassroots groups are essential for effective disaster response. Regulations should empower these organizations by recognizing their knowledge and resources, ensuring that community perspectives are incorporated into planning and response efforts. Collaborations between governments and academic institutions can lead to evidence-based policies and innovative solutions to enhance disaster management.(Jameel, S. M.,et al.,2022)Regulatory frameworks may suffer from fragmentation, with various agencies and organizations operating independently rather than collaboratively. This can lead to duplication of efforts, confusion, and inefficiencies in response.Limited financial and human resources often impede the effectivehinder swift action during emergencies, emphasizing the need for streamlined procedures that facilitate timely responses. disaster response regulations can limit community preparedness. Regulatory frameworks should prioritize outreach efforts to ensure that communities understand their roles and responsibilities during disasters. This includes addressing issues related to data privacy, cybersecurity, and the ethical use of technology in disaster response. Ensuring that regulations keep pace with technological advancements is essential for effective disaster management., regulatory frameworks must incorporate climate resilience into disaster planning and response. This includes recognizing the interconnectedness of climate risks and implementing strategies to mitigate their impacts on vulnerable communities.

There is a growing emphasis on community-based disaster management, recognizing the importance of local knowledge and participation. Regulatory frameworks are increasingly incorporating mechanisms for community engagement and empowerment, ensuring that local voices are heard in decision-making processes.Disasters often transcend national borders, necessitating collaboration among countries and international organizations. Regulatory frameworks are evolving to facilitate cross-border cooperation, information sharing, and coordinated response efforts during large-scale disasters.There is a heightened awareness of the need to address social inequities in disaster response. Regulatory frameworks are increasingly incorporating provisions to ensure that vulnerable and marginalized populations receive adequate support and protection during disasters.(Medhat, M., et al.,2022). Regulatory frameworks for disaster response are essential for establishing clear roles, responsibilities, and processes that guide effective action in the face of disasters. By incorporating key components such as legislative authority, risk assessment, emergency planning, and public communication, these frameworks can enhance preparedness, coordination, and recovery efforts. While challenges exist in implementation, the involvement of diverse stakeholders and the integration of emerging trends can strengthen disaster response frameworks. As the landscape of disaster risks evolves, and effective in safeguarding communities against the growing threats posed by disasters. Through ongoing collaboration and innovation, we can build resilient systems that protect lives and livelihoods in an increasingly unpredictable world.

CHALLENGES IN AI IMPLEMENTATION

AI implementation, exploring the complexities of data quality, algorithmic bias, integration issues, regulatory compliance, workforce readiness, and public acceptance.s face difficulties in obtaining the necessary data that is both comprehensive and representative. Issues such as incomplete datasets, inconsistencies, and inaccuracies can lead to suboptimal model performance and unreliable outcomes.

Moreover, silos within organizations often hinder the sharing and integration of information across departments or systems. This fragmentation can limit the ability to develop using diverse training datasets, and implementing fairness metrics to evaluate model performance. Engaging with stakeholders from various backgrounds can also provide valuable insights and perspectives, helping to create more equitable AI systems.Organizations often have legacy systems that are not designed to accommodate new AI solutions. Integrating AI requires substantial changes to infrastructure, data management practices, and workflows, which can be resource-intensive and disruptive.. Ensuring seamless integration involves not only technological upgrades but also change management strategies that foster buy-in from stakeholders at all levels. Organizations must invest in training and support to facilitate the transition and ensure that employees feel equipped to work alongside AI technologies. regulatory and ethical challenges. Governments and regulatory bodies are still developing frameworks to govern AI applications, leading to uncertainty for organizations seeking to implement these technologies. Navigating the legal landscape, particularly concerning data protection, privacy, and accountability, can be daunting.

This involves establishing clear policies for data handling, user consent, and transparency in AI decision-(Li, Z. et al.,2021) Organizations may struggle to recruit or develop talent capable of designing, implementing, and managing AI systems effectively. Engaging with the public through outreach, education, and stakeholder involvement can help demystify AI and address concerns. Building a reputation for responsible and ethical AI use is essential for garnering public support and acceptance.AI implementation often requires collaboration across various disciplines, including computer science, engineering, social sciences, and ethics. AI projects, ensuring that multiple viewpoints trust, interdisciplinary collaboration, and adaptability to emerging technologies. As organizations strive to harness the potential of AI, they must navigate these complexities with a commitment to ethical practices, transparency, and inclusivity. By proactively addressing these challenges, stakeholders can unlock the transformative power of AI while fostering a responsible and equitable future for all. The successful implementation of AI technologies will ultimately depend on collaboration, innovation, and a shared vision for leveraging AI to enhance human capabilities and societal well-being.

FUTURE TRENDS IN AI FOR DISASTER PREDICTION

into disaster prediction is poised to transform the landscape of emergency management and response, bringing forth a multitude of future trends that promise to enhance predictive capabilities, improve resilience, and ultimately save lives. As the frequency and severity of natural disasters continue to escalate due to climate change and other factors, the urgency for effective prediction methodologies becomes paramount. The evolution of AI technologies, coupled with advancements in data collection, processing, and analysis, is reshaping how organizations approach disaster prediction. This overview explores key future trends in AI for disaster prediction, highlighting advancements in machine learning, data integration, real-time analytics,, but the future will see a shift towards more sophisticated techniques, such as deep learning and ensemble methods. These advanced algorithms can analyze complex datasets with higher accuracy, recognizing patterns and correlations that may not be apparent to human analysts.social media platforms, and mobile applications generates vast amounts of real-time data that can significantly enhance predictive models. AI systems will increasingly harness this big data to refine predictions and improve situational awareness.For example, data from sensors insights into the likelihood of wildfires or floods. Social media data can also be analyzed to gauge public sentiment and identify emerging crises.

By integrating these varied data sources, AI systems can offer a more holistic view of potential disaster scenarios and facilitate timely interventions..(Moradi, H., et al.,2021) AI technologies will enable the continuous monitoring of multiple data streams, allowing organizations to detect early warning signs and predict the onset of disasters with greater precision.

This shift towards real-time analytics will empower emergency responders to act swiftly, allo on-the-ground conditions, helping to validate and refine AI predictions.For instance, during a flooding event, local residents can report real-time conditions through mobile applications, contributing to a more accurate understanding of the situation. By fostering community involvement, play a pivotal role in shaping its future. This includes ensuring fairness and equity in predictive models, particularly for vulnerable populations. Organizations will need to conduct regular audits to assess the impact of AI systems on different communities and make necessary adjustments to mitigate biases.The future of AI in disaster prediction will also be characterized (Mahadev Madgule, et al.,2023)by collaborative frameworks that bring together diverse stakeholders, including government agencies, NGOs, academia, and the private sector. Multi-stakeholder partnerships will be essential for pooling resources, expertise, and data to enhance predictive capabilities. For example, drones can be deployed to assess damage in disaster-affected areas, monitor environmental conditions, or deliver supplies to isolated communities. The integration of autonomous systems into disaster prediction. The future of AI in disaster prediction is marked by transformative trends that promise to enhance predictive capabilities, improve resilience, and empower communities. Advanced machine learning techniques, big data integration, real-time analytics, community engagement, ethical considerations, collaborative frameworks, climate data integration, and the development of autonomous systems all contribute to a more comprehensive approach to disaster prediction. and inclusivity in AI development and deployment.

CAPACITY BUILDING AND TRAINING INITIATIVES

challenges and capitalize on opportunities. By investing in capacity building, stakeholders—including governments, non-profits, and private sector entities—can create a skilled workforce that drives innovation, improves service delivery, and promotes resilience in the face of change.At the core of capacity building is the recognition that individuals and organizations possess untapped potential that can be harnessed through targeted training and support. Effective capacity building initiatives typically involve a comprehensive needs assessment to identify gaps in knowledge, skills, and resources. This assessment forms the foundation for developing tailored training (Salcedo, J. D., et al.,2023)programs that address specific requirements. One of the key aspects of successful capacity building is the emphasis on participatory approaches that engage stakeholders at all levels. By involving community members, employees, and local leaders in the design and implementation of training to more culturally sensitive and contextually appropriate training materials, which can significantly improve learning outcomes.

The impact of capacity building and training initiatives extends beyond individual skill enhancement; it can lead to systemic change within organizations and communities. When employees are equipped with new skills and knowledge, they are better positioned to drive innovation and improve processes. This can result in increased efficiency, higher productivity, (Sugumaran K. M, D.,et al.,2024)and ultimately, improved service delivery. Furthermore, organizations that prioritize capacity building are often more agile and adaptable, enabling them to navigate the complexities of modern challenges, such as technological advancements and shifting market demands.Moreover, capacity building initiatives play a critical role in addressing social and economic inequalities. For instance, training programs that focus on entrepreneurship can help individuals in disadvantaged areas to start their own businesses, creating jobs and stimulating local economies. Evaluating the effectiveness of capacity building and training initiatives is vital for ensuring continuous improvement. Monitoring and evaluation frameworks should be established to assess the impact of training programs to focus on remote work skills and crisis management, enabling them to maintain operations and support their employees during a time of uncertainty. Furthermore, technology plays an increasingly significant role in capacity building and training initiatives. Online courses, webinars, and virtual workshops can complement traditional training methods, reaching a broader, organizational effectiveness, and community resilience. By prioritizing the development of skills and knowledge, organizations can empower their workforce, promote social equity, and adapt to changing circumstances.

COLLABORATIVE APPROACHES TO DISASTER MANAGEMENT

Collaborative approaches to disaster management have emerged as essential strategies for effectively mitigating risks, responding to emergencies, and fostering resilience within communities. These approaches emphasize the importance of cooperation among various stakeholders—including government agencies, collaborative disaster management fosters a holistic response that enhances preparedness, recovery, and long-term resilience.At the heart of collaborative disaster management is the recognition that no single entity possesses all the necessary expertise, resources, or local knowledge to effectively manage disasters. This realization has prompted the development of multi-sector partnerships that integrate the capabilities of various stakeholders. For example, local governments may partner with NGOs to develop community-based disaster risk reduction strategies, while private sector organizations can provide technical expertise and resources. Collaborative frameworks facilitate the sharing of data and resources, ensuring that all stakeholders are on the same page regarding the status of the disaster and ongoing response efforts. (Naik, S. R., et al.,2021)Technology plays a pivotal role in this regard, with tools such as emergency communication systems, social media platforms, and data-sharing applications enabling real-time information exchange. engaging in simulated disaster scenarios, organizations can practice coordination, identify potential gaps, and develop effective response plans.

These collaborative training efforts foster trust valuable local knowledge that can inform risk assessments and response planning. Collaborative approaches that prioritize community engagement not only empower residents but also enhance the overall resilience of the community. For instance, local volunteers can be trained as first responders, providing immediate assistance during emergencies and facilitating recovery efforts., GIS can be used to map vulnerable populations and critical infrastructure, allowing stakeholders to prioritize resources and plan evacuation routes more effectively. Moreover, social media platforms can serve as tools for real-time communication and information dissemination, enabling communities to share resources, updates, and support during disasters. By fostering partnerships among various stakeholders, collaborative frameworks can ensure that recovery efforts are inclusive, equitable, and environmentally sustainable. This might involve integrating local economic recovery initiatives, mental health support services, and infrastructure rebuilding projects into a cohesive recovery plan. Engaging diverse stakeholders in these efforts can lead to innovative solutions that address the root causes of vulnerabilities while promoting long-term resilience.Another critical aspect of collaborative disaster management is the importance of developing and maintaining strong relationships before a

Additionally, establishing formal agreements and protocols among partners can clarify roles and responsibilities, streamline communication, and facilitate coordinated action during disasters.Indigenous and local communities often possess a wealth of knowledge regarding their environments and have developed effective strategies for managing risks over generations.Finally, the evaluation of collaborative disaster management efforts is crucial for continuous improvement. After-action reviews and identify lessons learned, and refine future strategies. Engaging all stakeholders in and leverage resources and expertise. Through clear communication, joint training, community engagement, and the integration of technology and traditional knowledge, collaborative disaster management can lead to more sustainable and equitable recovery efforts(Tiwari, S., et al.,2023).

EVALUATING THE IMPACT OF AI SYSTEMS

Evaluating including ethical considerations, social implications, economic effects, and technical performance. As artificial intelligence technologies become increasingly integrated into everyday life—ranging from healthcare and finance to education and transportation—the need for comprehensive evaluation frameworks has never been more pressing. These frameworks must assess not only the efficacy of AI systems but also their broader .might focus on user satisfaction(Pan, C. et al.,2023), trust, and the ethical implications of AI deployment. For example, in healthcare, evaluating an AI system designed for diagnostic purposes would involve assessing its accuracy in diagnosing conditions compared to human practitioners, as well as examining patient outcomes and experiences. A comprehensive evaluation should combine these metrics to provide a holistic view of the system's performance and impact.efficiency in hiring processes but could also inadvertently perpetuate biases if not carefully designed and monitored. Therefore, evaluations must consider contextual factors to ensure that AI systems are not only effective but also equitable and inclusive. This requires collaboration with stakeholders from diverse backgrounds . This involves not only technical audits of algorithms but also broader assessments provide valuable insights into public attitudes toward specific AI applications. Understanding the concerns of users—such as fears about job displacement, privacy violations, or loss of control—can inform strategies to enhance trust and acceptance. Engaging with communities early for ecosystem-level assessments. As AI technologies proliferate, understanding their cumulative effects on industries and society becomes increasingly important. This involves examining how multiple AI systems interact and influence one another, as well as their impact on labor markets, economic structures, and social dynamics. For example,is essential for robust evaluation processes. Multi-disciplinary teams can bring together diverse expertise to assess AI systems from multiple perspectives, fostering a more comprehensive understanding of their impacts. Collaborative initiatives, such as shared databases for benchmarking (Zhang, Y., et al.,2023)AI performance and ethical considerations, can enhance the quality and consistency of evaluations across different sectors.In conclusion, evaluating the impact of AI systems is a multifaceted endeavor that requires a systematic and comprehensive approach.

CONCLUSION AND FUTURE DIRECTIONS IN AI-BASED SYSTEMS

Driving innovation and enhancing efficiency while also raising important ethical, social, and technical considerations. As we conclude this exploration of AI's current landscape, it is essential to recognize the significant impact these technologies have already made and the vast potential they hold for the future. However, as we move forward, the challenges of ensuring ethical deployment, addressing biases, and maintaining transparency remain critical to realizing AI's full benefits.Looking ahead, one of the primary. As AI systems often reflect the data they are trained on, they can inadvertently perpetuate existing biases and inequalities. Future efforts must focus on creating comprehensive guidelines and best practices for data collection(Raghavendra, R., et al.,2023), algorithm development, and impact assessment. This includes promoting diversity in datasets and involving a wide range of stakeholders in the AI development process, ensuring that the systems we create are inclusive and fair. Collaborations among technologists, ethicists, sociologists, and policymakers will be crucial in establishing a framework that prioritizes ethical considerations from the outset. unprecedented insights across various sectors. For instance, AI combined with IoT can enable smart cities that optimize energy usage and improve public safety through real-time data analysis. However, these integrations must be approached with caution, as they can also introduce complexities related to security and privacy. Future research should explore these synergies while ensuring robust safeguards are in place to protect users and data hat can mitigate environmental impacts, optimize resource management, and improve access to healthcare services in underserved communities. By focusing on applications that promote sustainability and social equity, the AI community can play a pivotal role in creating a more just and sustainable future. Collaborative initiatives that leverage AI for social good will be critical in ensuring that technology is used to benefit all members of society.The workforce implications of AI adoption also warrant careful consideration. By focusing on these areas, we can create AI systems that not only enhance productivity and efficiency but also contribute positively to society and address pressing global challenges. As we move forward,(Zeng, W., et al.,2023) it is imperative to remain vigilant and proactive, ensuring that the advancements in AI technology serve as a force for good, fostering a future where AI works alongside humanity to create a better world for all.

CONCLUSION

The integration of AI-based disaster prediction and early warning systems represents a transformative advancement in our capacity to anticipate, prepare for, and respond to natural disasters. As the frequency and severity of these events increase due to climate change and urbanization, the need for accurate, timely, and actionable information becomes ever more critical. AI technologies, through their ability to analyze vast datasets, identify patterns, and generate predictive models, provide a powerful tool for enhancing our disaster preparedness and response efforts.AI systems can process diverse data sources, including satellite imagery, sensor networks, social media feeds, and historical incident reports, allowing for a more comprehensive understanding of disaster risk. This data-driven approach enables more precise predictions of disaster events,e.Moreover, the application of machine learning algorithms enables continuous improvement in prediction their models and improve their forecasts, ensuring that they remain relevant and effective in mitigating disaster impacts. and community trust are paramount. It is essential to engage stakeholders, including local communities, in the development and implementation of these systems to

REFERENCES

Abhilasha, A., Sreenivasulu, A., & Manimozhi, T. KUMAR, P. S., Selvakumar, P., & Singh, P. (2022, April). The Model of Smart Sensing Device For Sensitive Nanoclusters Modification in Sensing Properties. In 2022 2nd International Conference on Advance Computing and Innovative Technologies in Engineering (ICACITE) (pp. 1043-1047). IEEE.

Al-Yahyai, S., Charabi, Y., & Gastli, A. (2010). Review of the use of numerical weather prediction (NWP) models for wind energy assessment. *Renewable & Sustainable Energy Reviews*, 14(9), 3192–3198. DOI: 10.1016/j.rser.2010.07.001

Alwindawi, A. F., Uçan, O. N., Ibrahim, A. A., & Abbas, S. A. (2022). Semi-supervised learning with ensemble deep learning networks for descriptor generation. *Optik (Stuttgart)*, 170450, 170450. Advance online publication. DOI: 10.1016/j.ijleo.2022.170450

Aybar-Ruiz, A., Jiménez-Fernández, S., Cornejo-Bueno, L., Casanova-Mateo, C., Sanz-Justo, J., Salvador-González, P., & Salcedo-Sanz, S. (2016). A novel Grouping Genetic Algorithm–Extreme Learning Machine approach for global solar radiation prediction from numerical weather models inputs. *Solar Energy*, 132, 129–142. DOI: 10.1016/j.solener.2016.03.015

Barrera-Animas, A. Y., Oyedele, L. O., Bilal, M., Akinosho, T. D., Delgado, J. M. D., & Akanbi, L. A. (2022). Rainfall prediction: A comparative analysis of modern machine learning algorithms for time-series forecasting. *Machine Learning with Applications*, 7, 100204. DOI: 10.1016/j.mlwa.2021.100204

Bhoi, A., Balabantaray, R. C., Sahoo, D., Dhiman, G., Khare, M., Narducci, F., & Kaur, A. (2022). Mining social media text for disaster resource management using a feature selection based on forest optimization. *Computers & Industrial Engineering*, 169, 108280. DOI: 10.1016/j.cie.2022.108280

Brester, C., Kallio-Myers, V., Lindfors, A. V., Kolehmainen, M., & Niska, H. (2023). Evaluating neural network models in site-specific solar PV forecasting using numerical weather prediction data and weather observations. *Renewable Energy*, 207, 266–274. DOI: 10.1016/j.renene.2023.02.130

Chakrabortty, R., Pal, S. C., Ruidas, D., Roy, P., Saha, A., & Chowdhuri, I. (2023). Living with floods using state-of-the-art and geospatial techniques: Flood mitigation alternatives, management measures, and policy recommendations. *Water (Basel)*, 15(3), 558. DOI: 10.3390/w15030558

Chamola, V., Hassija, V., Gupta, S., Goyal, A., Guizani, M., & Sikdar, B. (2021). Disaster and pandemic management using machine learning: A survey. *IEEE Internet of Things Journal*, 8(21), 16047–16071. DOI: 10.1109/JIOT.2020.3044966 PMID: 35782181

Chen, J.-Y., Feng, Y.-W., Teng, D., Lu, C., & Fei, C.-W. (2022). Support vector machine-based similarity selection method for structural transient reliability analysis. *Reliability Engineering & System Safety*, 223, 108513. DOI: 10.1016/j.ress.2022.108513

Daneshvar, M. H., Sarmadi, H., & Yuen, K.-V. (2023). A locally unsupervised hybrid learning method for removing environmental effects under different measurement periods. *Measurement*, 208, 112465. DOI: 10.1016/j.measurement.2023.112465

Danish, M. (2022). "Chapter 1 - artificial intelligence and machine learning in water resources engineering," in Current directions in water scarcity research. Editors M. ZAKWAN, A. WAHID, M. NIAZKAR, and U. CHATTERJEE (Elsevier).

Esrafilian-Najafabadi, M., & Haghighat, F. (2022). Impact of predictor variables on the performance of future occupancy prediction: Feature selection using genetic algorithms and machine learning. *Building and Environment*, 219, 109152. DOI: 10.1016/j.buildenv.2022.109152

Fischer, J., Wirtz, S., & Scherer, V. (2023). Random forest classifier and neural network for fraction identification of refuse-derived fuel images. *Fuel*, 341, 127712. DOI: 10.1016/j.fuel.2023.127712

Fowdur, T. P., & Nassir-Ud-Diin Ibn Nazir, R. M. (2022). A real-time collaborative machine learning based weather forecasting system with multiple predictor locations. *Array (New York, N.Y.)*, 14, 100153. DOI: 10.1016/j.array.2022.100153

Ganguly, K. K., Nahar, N., & Hossain, B. M. M. (2019). A machine learning-based prediction and analysis of flood affected households: A case study of floods in Bangladesh. *International Journal of Disaster Risk Reduction*, 34, 283–294. DOI: 10.1016/j.ijdrr.2018.12.002

Ghebleh Goydaragh, M., Taghizadeh-Mehrjardi, R., Golchin, A., Asghar Jafarzadeh, A., & Lado, M. (2021). Predicting weathering indices in soils using FTIR spectra and random forest models. *Catena*, 204, 105437. DOI: 10.1016/j.catena.2021.105437

Han, S., Park, Y., Noh, N., Kim, J.-H., Kim, J.-J., Kim, B.-M., & Choi, W. (2023). Spatiotemporal variability of the PM2.5 distribution and weather anomalies during severe pollution events: Observations from 462 air quality monitoring stations across South Korea. *Atmospheric Pollution Research*, 14(3), 101676. DOI: 10.1016/j.apr.2023.101676

Haver, S. M., Gedamke, J., Hatch, L. T., Dziak, R. P., Van Parijs, S., Mckenna, M. F., Barlow, J., Berchok, C., DiDonato, E., Hanson, B., Haxel, J., Holt, M., Lipski, D., Matsumoto, H., Meinig, C., Mellinger, D. K., Moore, S. E., Oleson, E. M., Soldevilla, M. S., & Klinck, H. (2018). Monitoring long-term soundscape trends in U.S. Waters: The NOAA/NPS ocean noise reference station network. *Marine Policy*, 90, 6–13. DOI: 10.1016/j.marpol.2018.01.023

Islam, M. R., Shahin, M. I., & Ahmed, S. (2023). Flood prediction using machine learning techniques: A case study of Bangladesh. *Applied Water Science*, 13, 8. DOI: 10.1007/s13201-022-01638-x

Jameel, S. M., Kumar, S., Mohsin, M., & Islam, M. S. (2022). Assessment of advanced machine learning algorithms for the prediction of global solar radiation. *Journal of Cleaner Production*, 341, 130922. DOI: 10.1016/j.jclepro.2021.130922

Jiang, Y., & Liu, H. (2021). A comprehensive review on big data analytics and artificial intelligence for disaster risk reduction: Trends, challenges, and future directions. *Future Generation Computer Systems*, 120, 445–467. DOI: 10.1016/j.future.2020.10.034

Kanipriya, M., Sugumaran, D., Pandiyarajan, V., Yogananda, B. S., Selvakumar, P., & Mohan, S. (2024, January). Analysis of Electromyography Signals for Control of Mechanical Prosthesis using Machine Learning Techniques. In 2024 2nd International Conference on Intelligent Data Communication Technologies and Internet of Things (IDCIoT) (pp. 1059-1067). IEEE.

Li, Z., Hu, X., Xu, Q., Chen, W., & Liu, L. (2021). Integrating remote sensing and machine learning for urban land cover classification: A case study of Beijing. *ISPRS Journal of Photogrammetry and Remote Sensing*, 171, 113–128. DOI: 10.1016/j.isprsjprs.2020.11.005

Liu, Y., Feng, Y., Wang, J., & Gao, L. (2021). Statistical learning-based drought forecasting: A review of machine learning techniques and their applications. *Water (Basel)*, 13, 21. DOI: 10.3390/w13010021

Mahadev Madgule, N. (2023). Vinayaka, Yeshwant M. Sonkhaskar, Dhiren Ramanbhai Patel, R.Karthikeyan, P. Selvakumar,(2023). Mechanical properties and microstructure of activated TIG welded similar joints of Inconel alloys by desirability approaches. *Materials Today: Proceedings*, 77, 528–533. DOI: 10.1016/j.matpr.2022.12.250

Muthusamy, V., Karthik, R., & Kumar, S. (2022). Review of machine learning approaches in drought forecasting: A focus on various meteorological and hydrological parameters. *Geoscience Frontiers*, 13, 101152. DOI: 10.1016/j.gsf.2021.11.004

Naik, S. R., & Sahu, N. K. (2021). Machine learning-based approaches for forecasting rainfall: A review. *Meteorology and Atmospheric Physics*, 133, 253–271. DOI: 10.1007/s00703-020-00712-2

Pan, C., Hu, H., Xu, C., & Zhang, X. (2023). Predicting soil moisture using machine learning methods: A case study in the Loess Plateau of China. *Catena*, 207, 105792. DOI: 10.1016/j.catena.2021.1057

Raghavendra, R., Chand, R., Bhadani, A. J., & Singh, A. (2023). Deep learning for flood prediction: A systematic review and future directions. *Journal of Hydrology (Amsterdam)*, 610, 127713. DOI: 10.1016/j.jhydrol.2022.127713

Ramakrishnan, T., Mohan Gift, M. D., Chitradevi, S., Jegan, R., & Subha Hency Jose, P. Nagaraja, Rajneesh Sharma H.N., Selvakumar P., Sintayehu Mekuria Hailegiorgis, (2022). "Study of Numerous Resins Used in Polymer Matrix Composite Materials" Advances in Materials Science and Engineering, vol. 2022, Article ID 1088926, 8 pages, 2022. DOI: 10.1155/2022/1088926

Salcedo, J. D., Vargas, G. M., Rojas, J. C., & Chavarria, D. (2023). Integrating remote sensing and machine learning for flood risk management: A systematic review. *Water (Basel)*, 15, 245. DOI: 10.3390/w15020245

Syam, S. K., & Sharma, S. (2023). A machine learning framework for predicting flood events in a catchment area. *Journal of Hydrology (Amsterdam)*, 611, 128008. DOI: 10.1016/j.jhydrol.2022.128008

Tiwari, S., & Dahiya, A. (2023). A review on the applications of machine learning in disaster management. *Journal of Ambient Intelligence and Humanized Computing*. Advance online publication. DOI: 10.1007/s12652-023-04661-x

Tsai, H.-C., Chiu, H.-C., & Lee, T.-W. (2022). Predicting extreme rainfall using machine learning techniques. *Atmosphere*, 13, 160. DOI: 10.3390/atmos13020160

Verma, S., & Shah, A. (2021). Artificial intelligence and machine learning applications for flood forecasting: A systematic review. *Journal of Hydrology (Amsterdam)*, 603, 126740. DOI: 10.1016/j.jhydrol.2021.126740

Zeng, W., & Wu, L. (2023). A multi-source data-driven approach for drought forecasting using machine learning: A case study in the Yellow River Basin, China. *Journal of Hydrology (Amsterdam)*, 614, 128530. DOI: 10.1016/j.jhydrol.2022.128530

Zhang, Y., Zhao, Y., & Zhang, L. (2023). Machine learning for flood forecasting: A comprehensive review. *Journal of Hydrology (Amsterdam)*, 619, 126837. DOI: 10.1016/j.jhydrol.2023.126837

Chapter 4
AI-Driven Personalized Learning:
Revolutionizing Student Engagement and Teacher Efficiency in Education 5.0

Muhammad Usman Tariq
https://orcid.org/0000-0002-7605-3040
Abu Dhabi University, UAE & University College Cork, Ireland

ABSTRACT

The revolutionary potential of AI-driven personalized learning to improve teacher effectiveness and student engagement is examined in this chapter. By customizing content to each learners unique learning needs and preferences AI technologies like intelligent tutoring platforms and adaptive learning systems are revolutionizing educational experiences. By letting students advance at their own speed and addressing their weaknesses and strengths personalized learning creates a more stimulating and productive learning environment. The importance of artificial intelligence (AI) in boosting student motivation is examined in this chapter using interactive tools such as gamification augmented and virtual reality apps and real-time feedback systems. The chapter discusses how AI affects student learning as well as how it can help teachers become more efficient by automating repetitive administrative duties like attendance and grading. Teachers can concentrate more on professional development teamwork and instruction by freeing up valuable time.

INTRODUCTION

The idea of learning is now highly personalized and has altered the conventional methods of teaching as such the system aims at providing each learner with instruction that is more suitable for him or her. Personalized learning deviates from a straight-jacket approach of a generalized teaching strategy to a differential application of teaching and learning strategies for each learner, based on his/her proficiency level (Biswas & Molla, 2024). This transformation has surged with the upward push of Education 5.0, a movement that promotes interaction among technological and humanistic learning. Education 5.0 is relative to Education 4.0 whose change was characterized by the integration of technology in learning. Education 5.0 changed the fact that learning is a communal and not strictly an individual process (So-

DOI: 10.4018/979-8-3693-9770-1.ch004

louki et in such an environment Artificial Intelligence (AI) becomes central providing complex tools and algorithms that can create complex paths for learning for students. Incorporation of AI in education enhances on real time feedback, content creation and analysis as well as predictive insights of students by teachers. This chapter discusses how advanced, formative and summative, & social use of technology that emphasizes on AI is shifting the language of teaching or student learning engagement along with teachers or trainers' productivity towards more relevant and thereby making learning more effective (Nursalim, Judijanto, & Asfahani, 2022).

In its essence, learning that is individualized can be defined as the process of providing education that corresponds to learner characteristics in some way. This has become especially important in the modern society where the student body can be more heterogenic coming from different learning environments and each with different learning profiles. Personalized learning understanding stems from the fact that students are not mere vessel and is based on the understanding of students as active participants in the educational process that benefit experiences of education and learning (Singh et al., 2025). When speaking about Education 5.0, the role of digital technologies, including AI, is to extend the concept of learners' personalized learning experience in flexible educational contexts. Through analyzing the record of the interactions between the students and the system, it is possible to fine-tune the pace, the level of difficulty and the topics which are presented to the learners to fit the best learning solution. This has major connotations to equity in education since every student is provided with the needed learning materials suiting them to enable complete the course as per the intended achievement levels once the benchmark is set (Biswas & Molla, 2024). AI enablers for personalization for education are numerous, ranging from machine learning techniques that help evaluate students' progress to natural language processing that helps in student-stakeholder interaction. Take, for instance, the case of adaptive learning platforms that change with the way users are learning and performing. These systems can make an early warning when a student is likely to perform poorly in a certain area and assists in making provisions for extra learning materials as well and change in learning strategies for better performance (Mahmoud & Sørensen, 2024).

Another important piece of tech is natural language processing through which AI can talk with students individually and naturally. This can be achieved in the case of the intelligent tutoring system or virtual teaching assistant or when an interactive program responds to questions offers hints or rates understanding. Moreover, computer vision and data analytics applied to the use of the visual or behavioral data or inputs like face expressions or activity logs of learners, to synthesize and adapt learning to student needs in real-time (Singh et al., 2025). There are numerous gains in implementing AI to personalize learning; they both influence the students and instructors vastly. The major clear benefit for students would be that students can study at their own pace, while embracing online classes (Biswas & Molla, 2024). Traditional classrooms set a standard speed that is not ideal for everybody, for the fast learners they are simply bored, and the slow learners are left behind. The use of technologies, including AI, provides the opportunity for learners to be in control of how they spend time on concepts, leaving more time on areas that cause difficulty and less on concepts they seem to grasp easily. This flexibility helps and makes it easy to grasp things and probably retain it in one's memory for a long time. For instance, how DreamBox or Carnegie Learning provides math learning forums where students get problems and accompanied instruction that changes depending on students' performance in the classroom and in time. For this reason, valuable skills may be learnt at what the students' comfort is then with increased interest over some time, the expertise is achieved (Omar et al., 2024).

The same way that personalization promotes students' participation in class, AI also enhances learning by providing contextual material that reflects the learners' special interests. When educational and interesting content is attainable students will be driven and passionate towards their learning. Some ideas of the content to use can be based on recognition of student preferences and performance history alongside with such additional information as outstanding concerns. For instance, a student majoring in history who has an interest in the arts may be prompted to develop assignments pertaining to the cultural side of history, interlinking it to the art major she prefers. This approach, however, engages the student's interest and intertwines one subject with another so enhancing interdisciplinary learning (Singh et al., 2025). In addition, AI systems can incorporate elements of game designs preferred by students making learning a more entertaining affair. Use of games in learning and aligned to the student's motivational regulation theory provides understanding of achievement and perseverance in difficult areas. Humankind is getting wiser and fortunately for teachers and independent learners, artificial intelligence breaks complex learning processes into actionable tasks with detailed reports on students' progress. Quite a lot of time may be spent on grading alone, preparing lessons, and monitoring students' progress. AI can help to alleviate this burden by automating the scoring process, analyzing learning data, and possibly identifying actions now necessary for each learner. For example, valuable resources, such as grading or giving initial feedback about assignments, can be solved by applying AI technologies which will free teachers' time to work with students and focus on more intricate approaches to teaching (Zohuri & Mossavar-Rahmani, 2024)

This efficiency allows the educators to have more involvement with students, thus designing a competent and sensitive classroom. Machine learners help teachers find patterns in students' performance and, with this information, decide what needs to be corrected or where students showed improvement and what can be done to have an even better result. However, as the future of AI driven Personalized learning looks promising and unending, it comes with impressive questions and issues. Primary is the issue of protection and preservation of individual and group datum collected from the online platforms. Another issue that raises concern is that personalization mainly depends on the collection and analysis of large amounts of all possible student data without their consent and possible misuse of the collected data (Singh et al., 2025). As the assessment of students continues to take different shape, the institutions must ensure their data protection systems to avoid misuse of student data. However, there is the question of the dependency on technology, and whether technology deepens existing gaps. On the same note as the above pro, AI can help provide solutions especially to the less privileged in getting educations since the utilisation of such would call for the use of computers and modem or other internet enabled devices that are not easily accessible across the globe. This digital divide could be made to enhance the status inequality because children in affluent schools will be subjected to individual targeted learning while others in poorly endowed schools remain deprived of such experiences (Omar et al., 2024). Another problem is that the AI-driven systems are at risk of reiterating prejudice in the algorithms as well as in the data on which those algorithms are based. For instance, AI systems operate based on past data and therefore such data develops from past practices of education that could be biased. For instance, if we imagine the AI system being employed in college admissions or scholarship assessments its recommendation pattern relies on training data then their algorithm would automatically or perhaps unconsciously bias some demographics. Likewise, if AI platform prescribes content based on stereotypical assumptions of students' interest, then, instead of questioning bias it perpetuates them. Scholars must therefore tread carefully when integrating AI in places of learning and ensure that they make full disclosure of the use of such systems in classrooms, make sure that use of the AI algorithms is checked periodically for fairness for all the learners (Zohuri & Mossavar-Rahmani, 2024).

However, given these challenges AI dismissal does not remove the high potential for transformation of learning through personalization. Depending on the AI technology and its implementation, the students' experience of learning can be improved to create a society where learning is more effective beyond accessibility. AI used in personalized learning can meet the learning needs of students and help kids who may be marginalized by providing an individual education solution. However, to attain this vision, the firm should employ innovation and ethical factors as follows. It should at the same time involve teachers, policymakers and technologists to get solutions towards the various arising issues such as security, equity and social Amnesty about data used in implementing AI solutions within the institution (Ray & Sikdar, 2024). Therefore, it can be concluded that the concept of the personalized learning facilitated by artificial intelligence is a revolutionary change in learning which is consonant with Education 5.0 to offer more flexible, student engaged and effective learning process. Using these advanced emerging technologies, AI can develop smart learning solutions that create attraction, reinforce knowledge, and boost performance, all through machine learning, natural language processing, and data analytics. The advantages of such personalization are obvious, but the given approach will work at its best when several drawbacks, issues of ethics, practicality, and social impact of AI integration in learning settings are addressed. While facing these challenges education institution and technology developers continue to be motivated by the possibility of creating a more inclusive, equitable and effective learning for future generations through the use AI to personalize learning (Nursalim, Judijanto, & Asfahani, 2022).

BACKGROUND

On this view, education through the application of the principles of personalized learning is since the use of standardized educational approaches cannot suit all students, let alone adequately address their learning needs. Learner-centered learning aspires to provide different learning experiences for students in relation to their learning needs, preferences and deficits. This strategy has turned out to be the key driver of the Education 5.0 model of learning delivery as depicted in the figure below. However, as technology interferes with education more and more, artificial intelligence (AI) appears to be an asset for learning personalization. Advanced Intelligent technologies can estimate the needs of each of the students and learn the potential patterns in their learning, thus call for a new phase of technology driven engagement to be augured in both teaching and learning (Shawon, Miah, & Islam, 2023). Intelligent agent for personalized learning employs adaptive algorithm and data analysis where systems can learn comprehension to identify each learner's performance. For instance, AI can evaluate the respondent's responses to exercises and determine that the student may need help with or can solve more difficult problems. Perhaps the most popular recent example is the language learning application Duolingo that uses AI to change the level of difficulty during the lessons depending on the user's performance, making each learning process unique. Thus, using AI, the requirements for the speed of learning and their understanding are met individually for each student, which makes it possible to make education more inclusive (Rizvi, 2023).

However, while such systems are excellent in data processing and adaptive properties, they must also decide on the level of actual assistance from the automated teacher and when human teacher intervention is required. This means that over-reliance of students on AI can lead to the complete eradication of the interpersonal component of learning where the teachers' instincts and encouragement are critical (bin Salem, 2024). Use of AI paradigm of personalized learning has rather drastic effect on both aspects – students and teachers alike. Through individual learning packages, which are arranged with the help of AI, the motivation of students is increased due to the focus on subjects and tasks that they enjoy and for which they can create lessons. This target strategy minimizes students' frustration and boredom, two factors that are typical for the traditional classrooms where all students must complete the tasks, using the typical lesson plan that may be too difficult for some learners but far too simple for the others (Ray & Sikdar, 2024).

Furthermore, ac- customized and individualized learning also introduce game aspects to learn materials continuously, such as DreamBox, a mathematics education tool for primary and middle school students. At DreamBox, the adaptive learning technology not only delivers content aligned with the Common Core but does so in a game-like environment that progresses in difficulty based on a child's mastery of the material and rewards achievement with points or feedback (Lakshmi & Vasanti, 2024). At the same time, although AI has the potential to increase students' engagement through more individualized learning experiences, the effectiveness of such approach may depend on students' perception of the technologies used, their preference for self-organized learning and their ability to work without being provided with instructions (Shawon, Miah, & Islam, 2023). The advantages of utilizing AI for personalized learning are obvious for teachers because the utilization of AI simplifies such processes as grading, tracking the results, and planning classes for every learner. This cuts down on teacher time spent on teaching administration, freeing them to develop more complex cognition, being senior role models to the learners and to cultivate the interpersonal skills. Some of the new attendees to the grading process include Gradescope where Machine learning algorithms assist in grading written responses as well as offering feedback. These tasks take time and when the teachers do not spend that time with the students, the students have no one to help them develop other things like critical thinking and problem solving. However, the integration of AI in the classroom also comes with its drawbacks; a teacher cannot just switch to a new form of practicing while he or she may need a training course to understand how to work with AI tools. Furthermore, the Human involvement proposed here is critical since the AI-generated insights are valuable but keep out the touch and familiarity educators have with the student (Rizvi, 2023).

Lastly, the use of AI in delivering personalized learning entails four major areas of concern; Ethics such as issues to do with data privacy, Bias, and the human versus the artificial tutor. Whilst the idea of adapting the concept of learning to suit individual learners is welcomed, the key feature of so-called 'personalized learning' involves amassing enormous quantities of information on the learner, routinely triggering privacy and data security issues (Lakshmi & Vasanti, 2024). There is also a risk that, due to the nature of the inputs providing the AI, the systems reinforce prejudices inherent in the data set to the detriment of minority learners. Also, many social elements cannot be omitted in the learning process; although teaching can sometimes be helped by computers, a human touch is irreplaceable. As the education sector faces these dynamics it is imperative to uphold ethical codes, allow transparency and embracing on the role of the teacher on the improvement of student development to ensure the use of AI in personalized education is effective (Shawon, Miah, & Islam, 2023). Computerized adaptive instructions have the potential to dramatically transform education by responding to students' special needs, increasing learning attainment levels and teacher productivity. While it unfolds, much attention must be paid to preserve the intimate specifics that concern people while promoting the relevant technological approaches. This equilibrium provides that the use of AI aide in the learning process can benefit students and teachers to make education open, inclusive and impactful when compared to traditional teaching and learning systems (Rizvi, 2023).

AI-Powered Adaptive Learning Technologies

Adaptive learning technologies are products and services which employ 2-AI to deliver learning systems that depend on the learner's performance. Proximate to adaptive learning is the use of algorithms that analyze samples based on students' behaviors altering the difficulty level, the flow, as well as type of material which it is presented in a way that will best fit each learner. These algorithms examine a wide range of factors like response time, accuracy and kind of content that a student tends to solve, to the time during the day when a student is most effective (Dandachi, 2024). From these features, adaptive learning systems develop a learning path such that students can get instructions that will captivate their interest and improve understanding through pathways that have been personally configured. These systems work based on an idea, in which with each progressing learner, the machine learning models get more refined and can predict what content or help this student might need next. Such kind of individuation presents a rather strong contrast to the straight-forward model of mass-prescriptive curricula. Five examples demonstrate the effectiveness of adaptive learning technologies in given learning environments (bin Salem, 2024).

In the heart of this innovation is a variety of algorithms used in the identification of flow, style and sequence of the instruction material depending on the learning outcomes of the students. These algorithms consider several factors such as the response rates, accuracy, and the type of content that a student is most comfortable with, not forgetting the time when that student is most productive. Using this data, adaptive media learning systems develop a learning trajectory; therefore, each student will learn as per a program that is inclusive of approaches that will help capture the learners' attention and enable understanding of what is being taught. These systems work based on a concept of machines, which continuously improve over the progression of the learners & get better in the identification of what content or help a student might need next. The kind of differentiation present here provides a rich avenue apart from the conventional sequential, teacher-centered model of curriculum (Dandachi, 2024). Adaptive learning technologies have also been used in a few examples learned in the educational environment. For example, DreamBox, which is an online PC, tablet and smartphone mathematics learning program for K through 8 students employs adaptive learning technology supported by AI to understand how the learner tries to solve a math problem, and then provides feedback for the correct approach or more problems as per the response. If a student selects the topic of fractions and the program observes that the student has difficulties with this subject the program will offer the student simple drills before continuing with more complex kinds of exercises (Kaswan & Ojha, 2024).

The other example is Smart Sparrow used in higher learning institution to provide subject wise tutorial solutions for different subjects such as biology and engineering. The adaptive engine used in Smart Sparrow lets the educational content that the instructors design includes decision points, to which the system responds according to the actions of the learner. The effectiveness shown by these tools explains how the concept of adaptive learning supported using artificial intelligence can be useful in supporting learner needs and ensure that they advance through the content at their own pace and intelligence level to meet their strengths and weaknesses. That is why, the function of energy and adaptative technologies in learning system could be most useful for students with learning disabilities and different learning styles (Dandachi, 2024). The vast array of learning styles and tempos present in any classroom can be difficult for traditional educational models to adapt to. By making sure that every student receives instruction tailored to their individual needs, adaptive learning solves this problem. An adaptive platform might have more diagrams, videos, or other visual aids for those who learn best visually. The system may give preference to text-based explanations for pupils who perform exceptionally well in reading. Adaptive learning platforms improve understanding and retention by matching teaching strategies to each student's preferred learning style (Kaswan & Ojha, 2024).

Additionally, students with learning disabilities can benefit greatly from adaptive technologies. A more inclusive learning environment is promoted by programs such as Lexia Core5, an adaptive literacy program, which provides specialized exercises to help children who struggle with phonics or reading fluency without putting them feeling under pressure to keep up with their classmates' development. Even while adaptive learning has a lot of promise, there are drawbacks and restrictions. The fact that adaptive learning algorithms are only as good as the data they use raises several concerns. The algorithms may provide suggestions that are not actually helpful for the learner if the data on behavior or performance is erroneous or lacking. Furthermore, these systems may not always take into consideration unusual or innovative learning behaviors because they mostly rely on patterns found in past data (Ellikkal & Rajamohan, 2024). Adaptive learning systems run the danger of reinforcing certain biases as well. For instance, a system may unintentionally disadvantage learners who do not meet the expected patterns if it is trained on data that does not fairly reflect a variety of learning situations. Therefore, developers and educators must constantly evaluate and improve adaptive learning technologies, remove biases and modify the algorithms to suit changing educational requirements, to guarantee that these tools really benefit all students (Dandachi, 2024). Notwithstanding these obstacles, adaptive learning systems driven by AI are revolutionizing the field of education. Their capacity to deliver responsive, tailored training presents a promising means of meeting a variety of learning requirements, promoting a more successful and individualized approach to education. There is a lot of room to improve these technologies as they develop, making adaptive learning a vital instrument for building a more flexible and inclusive educational environment. To fully realize their potential to improve learning outcomes for all students, adaptive algorithms must be developed continuously while adhering to inclusiveness and ethical principles (Ellikkal & Rajamohan, 2024).

Enhancing Student Engagement through AI

Integration of intelligent technologies into education has created new opportunities for improving students' activity, making the learning process more interesting, effective and individualized. Technologies like Artificial Intelligence are revolutionizing conventional teaching methods by presenting students with novel approaches with which they can access educational materials. By including learnability, interactive features and applications in virtual and augmented reality application of AI, learning has transited from being a mere receiving (Ifraheem & Siddiqui, 2024). In gamification, concepts of games are implemented in learning processes in the form of incentives, progressing levels and scenarios. For example, Kahoot! and Quizizz let teachers design quests and quizzes that automatically adjust to student performance, offer instant feedback, and let learners challenge classmates in a fun way. The same as with vocabulary acquisition, AI helps an app for language learning called Duolingo identify the areas in which a learner needs improvement and change the kind of content and difficulty level that is presented consequentially, Duolingo also employs gamification elements that involve rewards, streaks, and achievements to maintain learners' interest. Being based on AI augmented with game features, these tools consider differences in the tempo and attitude towards learning and, therefore, they provide necessary conditions for education for as many students as possible (Katiyar et al., 2024).

However, although engagement through gamification can have a formidable increase; it is also possible for this kind of approach to become a distraction. There is the danger of losing the sight of what is to be achieved in a lesson and more emphasis being placed in the 'rewards' or 'points', which vitiates the learning process by promoting gains rather than mastery. Teachers should only make the tools be used for the reward of content knowledge rather than having the tool predominate (Ifraheem & Siddiqui, 2024). Further to this, virtual and augmented reality (VR and AR) show examples of how AI can increase interaction levels by creating novel learning experiences that cannot be imitated using conventional approaches. Traditional methods can rarely offer the opportunity to 'travel' to another yet unknown cultural or historical period, experiment with the conditions of prehistoric times or even explore abstract math problems in the same perspective as virtual reality and augmented reality can offer students. For example, Google Expeditions enables students to travel to virtually various places including museums, carcasses, dungeons, and even the deep sea making the subjects even more real and fun oriented. Likewise, Labster provides virtual reality science activities and students can perform various experiments with complete safety. Thus, the given applications not only develop students' interest in the material but also contribute to the improvement of knowledge retention by offering practical practice (Katiyar et al., 2024).

However, VR and AR can intensify the interest, yet, to achieve that, both must be used if they are accessible and successfully integrated into the learning model. The costs and technological barriers may reduce the understanding, availability and accessibility of innovative approaches to education especially for less-funded learning institutions. Furthermore, applying VR-based learning experiences can become challenging to make it meaningful besides being merely a fun tool that enhances learning outcomes; instead, it needs to enhance core learning issues (Rizvi, 2023). The implication of the use of AI engaged tools in enhancing students' learning motivation cannot be overemphasized since learners are motivated to learn and exposed to different learning styles. AI driven platforms as such enables students to intervene and take full responsibility for their learning process since it has features of personalized content delivery and feedback functionality. For instance, intelligent tutoring systems such as Carnegie Learning self-accommodate to the learning style and level of the learner, providing tasks that are adequately difficult to capture student's attention but not too difficult so that the student gets easily annoyed. This balance is instrumental in supporting the motivation that requires feelings of accomplishment – the sense of mastering that is always intrinsic to motivation (Ifraheem & Siddiqui, 2024).

Furthermore, the analytics that are based on AI help teachers understand how often a particular learner is interacting with educational applications as well as recognize that learner in case of that learner's inactivity and control the learning progress in case of difficulties in comprehending some concepts or principles. The kind of ideas such as the one [shown in the figure] provide an opportunity for early intervention to occur more frequently and, therefore, would provide an opportunity for the learner to profit from the educational process and not to fail (Rane & Rane, 2023). While there are many advantages to engaging students individually, there is also a downside in that to get the most out of these opportunities' students must be self-motivated. For some learners particularly those who were arguably more reliant on their teachers in less guided, more self-mastery learning environments, staying on task may be a challenge (Mahmoud & Sørensen, 2024). More significantly, this type of engagement through AI can become the new normal on learning platforms and holds certain challenges and concerns. Relying on AI-based applications leads to critical concerns about data privacy since intelligent personalization tools gather large amounts of learners' data to deliver customized content. Spotting to how information is collected can control and secure student's information and sustain students' trust (Rizvi, 2023).

Further, reliance on AI can threaten the practice of teaching and the extent to which students engage with their instructors, as they engage with the applications more. As much as technology improves engagement it is still very clear that the voice, direction, positive words and encouragement that teachers offer cannot be provided by the system. AI integration in education should therefore adjust the use of the technologies with the possibility of face-to-face interaction so that while the AI technologies engage the learners at one level the faculty can engage them at another level, independently. On balance, AI is enhancing students' participation in learning through use of technology in learning aids, Games, simulations and innovations such as, VR and AR. The above technologies allow for an approach that accommodates the learning styles of the students and offer feedback that enhances motivation and control (Ifraheem & Siddiqui, 2024). However, it leads to the fact that analyzing the advantages of these tools a critical approach is needed to avoid the excessive use of technologies that would replace traditional teaching instead they have to support it. Through proper regulation of the utilization of AI in teaching, learning and assessment, the learners win, the teachers win, and the society in general wins. That is why the described hybrid approach might be helpful to use AI as an instrument to engage the students, and at the same time avoid shifting people's roles and making the teacher less crucial to the educational process: a mentor and an inspiration (Rane & Rane, 2023).

AI in Student Assessment and Feedback

The fact that 4-AI directly involves student assessment and feedback turns the current approaches of education into entirely new levels with such innovations as auto-grading, instant feedback and, ultimately, the usage of data-based recommendations and learning methods for students. In this way, AI saves time and ensures that lecturer's assessment is reliable and fair at the same time as delivering timely and individualized feedback to learners. However, the increase inf using AI in the grading and assessment brings into question the reliability of the evaluations of learning achievements, as well as learning outcomes and ethics (Sumathy & Navamani, 2024). One of the most common uses of AI in evaluation is the use of automated grading systems which bring efficiency and uniformity to grading student's work. Increased student enrollment, coupled with complicated grading tasks such as checking and marking over large classes, has been made easier using AI enabled platforms such as Gradescope and Turnitin that use elements of artificial intelligence, particularly machine learning algorithms to mark answers, score and even check for plagiarism. With specific tasks, such as multiple-choice assignments, the automated grading system is reliable and tends to be much more so than when a human being grades the items because the grader tends to get fatigued or biased (Mahmoud & Sørensen, 2024).

Also, it can do other monotonous activities and save the teachers' time for implementing knowledge fields that require their attention and discretion. But it becomes hard when AI is applied in grading of free response questions, essays or any complicated assignments. Current language processing algorithms, as much as they improve, can fail to comprehend nuances, creativity or context in a student's writing. For instance, in some instances, there is either misunderstanding of creative writing as incoherent or the failure to get the depth of an orchestrated argument. However, these limitations imply that, while AI can help to improve grading accuracy in numerical evaluations, the additional, subjective scores still require supervision by a human being (Iqbal, 2023). Due to the current sophistication of artificial intelligence, real-time feedback on student work has bridged the gap of receiving quick feedback on work done. Examples of such artificial intelligence learning platforms are Khan Academy and Quizlet, all of which utilize adaptive learning approaches and give out fast feedback where student can correct errors on their own. For instance, if the student is stuck on a mathematics problem, he or she can instantly get assistance, or he or she may be directed to other problems that are close in that area (Alenezi, 2023).

This instant feedback loop is beneficial to lesson learning, allowing the students to know what they have not clearly understood before they proceed to the next lesson. As the information is provided instantly, the students focus on making those corrections and develop proper angle towards the material, which strengthens memory and self-esteem. In contrast to conventional feedback which learners may receive after days or weeks the immediate corrective comments protect students from failure accumulation. However, this real-time feedback system is not problem free. However, feedback that is provided right after an assessment can seem constructive, but it may reduce the students' ability to assess each other's work on their own and build up problem-solving skills (Henry & Maxwell, 2024). Excessive use of meant guided feedback could have a negative impact on the students' learning or problem-solving skills as they would expect to be told when they are going wrong rather than spending time to figure this out on their own. Ethical concerns are also vital ingredients in AI-generated testing and ranking systems because to achieve these goals, the artificial intelligence software needs to gather and process students' data. Rate data from students and employee performance measurements, learning style, and behavior that AI systems gather allow for more detailed and individualized feedback, but also breach privacy and consent. First, there is the question of who gains access to this data and how they use it, particularly where institutions of learning or a third party are involved. The primary concern is about the manipulation of student information for purposes (Alenezi, 2023). The level of explainability is also important; students and educators do not always know how an AI system derives the result or feedback, which may cause trust problems or mitigated responsibility. For example, when using them to identify cheating or poor performance common patterns may be mistaken or hardworking students' records and even careers may be ruined. Another disadvantage is that such tools can create biases in the scoring process because AI models use datasets placed in the past and repeat existing prejudices. For example, if an artificial intelligence system is taught based on bias grading data, it will readdress bias grading and affect the students from the minority group significantly (Henry & Maxwell, 2024).

However, AI in assessment and feedback remains promising if well implemented and controlled for its limitations are known. One of the numerous advantages that can be derived from the implementation of machine learning and other types of AI is that they can maintain fair, accurate, and reproducible results on a large scale, but human oversight must still be relied upon to serve as the final judge of whether those results are equitable and entirely ethical. Because it is recommended that subjective elements or creativity be involved in the marking of certain assignments, educators have an important responsibility in examining the AI-based results. This approach maintains all the strengths of AI which are effectiveness, feedback, and learning while confronting the common drawbacks of automatic systems (Alenezi, 2023). Besides, the AI algorithms and data use policies should be made clear by exposing knowledge about data collection, analysis, and protection to students and parents. With these safeguards, AI does not undermine the ethical principle of assessment, while at the same time making the assessment experience personalized and supportive. Finally, we can say that the application of AI in student assessment and feedback has many benefits starting with fast grading and including fast giving of the feedback facilitating active learning. However, it also poses some problems concerning the possibility, reliability, and ethical aspects which teachers and designers should solve. Running assessment through AI should not replace teacher but augment their functions thus ensuring better provision of the student needs balances fairness and privacy. Through the integration of automated decision-making with educational expertise, the learners' rights will be upheld and an efficient and fair means of evaluating learners' performance will be developed and implemented by institutions (Kaswan & Ojha, 2024).

Accessibility and Inclusion with AI

AI can also indirectly contribute towards enhancement of access to education and learning for students with disabilities, and those that are in the developing world, that are already in the new world order but are still lagging due to lack of access to the enabling tools. AI tools. It gives all students who require assistance in any form a chance to participate in the learning process in a fair manner. Such benefits for students include AI can offer learning accommodations to the students with disabilities by offering technologies that help in accommodating the content delivery in a way that every learner will understand better through personalizing the education setting (Rasheed & Abualkishik, 2023). The assistive technologies themselves are particularly useful to the students with visual or audial disabilities. Visual Technologies such as Microsoft's Seeing AI and Google's Lookout help put a voice to the environment, read texts aloud, and interpret text for visually impaired learners to study on their own. Students with hearing impairments can read real-time lecture and discussion captions using assistance from currently available AI systems like Google's Live Transcribe (Henry & Maxwell, 2024).

Also, voice recognition via artificial intelligence, which is in programs like Dragon NaturallySpeaking, helps students with motor impairments to write their assessments such as essays, and control other elements in a course via voice. These tools demonstrate how integration of BYOD within an education environment can assist distinct learning methods and disability that learners may possess (Mahmoud & Sørensen, 2024). AI is also able to level play field in that it creates special sections of resources for students in the underprivileged regions that need content from which they may not have access. For instance, the Khan Academy offers learning paths in its adaptive form, which means that the learning pace of a student is defined individually according to the level of training received. Therefore, AI reduces the inequalities in access to quality education since high quality content is made available for free or a token charge. But this promise comes with the understanding that the students must be connected to internet and other gadgets that will enable them to utilize the information. In areas where such infrastructure is not in place AI solutions may therefore end up in exacerbating rather than filling existing gaps in education access (Alenezi, 2023).

That said, there are some challenges that have been realized whenever AI is applied in accessibility and inclusion. For example, most AI are built with limited data set, which often does not generalize the students, and this can lead to production of wrong or biased answers. An example is a voice recognition can fail to capture non- standard accents or diction, thus leaving out certain groups of students. Furthermore, some of the innovative AI-based assistive technologies are expensive worldwide and can hardly fit in with low-income schools or developing nations. These limitations underscore the importance of culture, language, and socio-economic sensitivity when designing AI applications, and create awareness of development which can perform to different abilities among the various demographics of a society (Mahmoud & Sørensen, 2024). Therefore, it seems that applying AI in learning has the potential to improve students' access to educational opportunities for those with a disability, as well as narrow the achievement gaps based on disabilities. Nevertheless, for the realization of its potential, AI in education must be built with accessibility considering implementation of infrastructure, cost, and ethnic/gender bias. AI should espouse IA and be used in a manner that fosters equity in education hence ensuring that all learners benefit in the future (Alenezi, 2023).

Developing 21st-Century Skills with AI

AI has perhaps the most value by helping students enhance 21st-century skills that include critical thinker, problem-solver and a digital native. AI based learning systems and tools give the students valuable problem-solving exposures in the simulation environment thereby making them better thinkers. For example, there is a Werner's IBM Watson Tutor, where artificial intelligence helps the students to solve the tasks in a step-by-step manner and give the learner the main questions that must be answered: analyze, synthesize and evaluate the information. This approach can enhance interaction as students are able to be exposed to different solutions and come up with a better decision-making approach and developed critical thinking skills (Iqbal, 2023). Besides, critical thinking skills that are nurtured by AI can be useful in educating people about digital practices. Computer literacy, on the other hand, is, in addition to proficiencies with technology, knowledge of how AI is being used, and the various ethical implications involved in adopting technological solutions. Interactive lessons at Code.org and Scratch use artificial intelligence to introduce young learners to coding; modules help students understand algorithms, data privacy, and more. Such systems are normally flexible, where the learner is given an environment that challenges him/her, with the aim of helping the student learn at his/her own pace. It, therefore, enables the students to enshrine these concepts right from the beginning to ensure that they are ready to endure a world that has embraced technology and understand such technology they use almost every single day (Sumathy & Navamani, 2024).

Another area of educational goals where AI technology is applied is in preparing learners for a world of work within an AI system. Due to the highly dynamic nature of AI, employers are now in search of individuals with flexible skills that will fit in AI using roles, data fundamentals, and tech-solution-oriented positions. The academia today sees a need to ensure that their educational curriculums introduce learners to AI-associated subjects including machine learning. For instance, basic courses on AI are even provided by Coursera and other learning websites; therefore, all kinds of students can get an idea of AI related subjects. When students interact with AI in a more directed manner, they learn about new and uncharted sectors and are equipped for jobs that may not yet be out there but will be created by AI (Ray & Sikdar, 2024). Still, there are some improvement-related drawbacks related to the development of 21st-century skills using artificial intelligence. One concern is that extensive use of IT tools will decrease the possibilities for student initiative and creativity as they learn to follow the AI guidance. Also, traditional education in the enhanced learning resources supported by Artificial Intelligence reaches only schools with better funding and, therefore, can reinforce inequality when students from underfunded schools do not have the same chance to get such access. To that end, certain recommendations must be made: First, educators must strike the right balance and incorporate AI systems in the learning environment without compromising the conventional teaching methods that always play an important role in the formation of learners. Secondly, equal access to technologies must be provided for all learners. So, when carefully applied in classrooms, AI can foster the key competencies needed to help students succeed in the global future they are to join (Iqbal, 2023).

AI for Teacher Professional Development

AI is revolutionizing teacher professional development through delivery of tailored professional development, collaborative model of learning and ongoing improvement in teaching practice. Machine learning-powered solutions can better identify the teachers' performance and give recommendations for further training opportunities that would suit individual employees' best. For example, Coursera for Educators and Microsoft Educator Center tools use AI to determine the exact lessons and materials that could be interesting and useful to the teacher considering his or her level and ambitions (Ray & Sikdar, 2024). This brings personalization into teachers that train the current generation competencies that are area or technological specific, especially as today's classrooms incorporate such tools. Below, these adaptive learning platforms provide flexibility so that educators can learn at their own pace, which is very advantageous especially for current working educators with busy schedules in the classroom. AI is likewise improving peer cooperation in the facilitation of lessons by educators. Some of the platforms include Edmodo and TeachFX that utilize AI to link teachers with similar professional pursuits, to share strategies, deliberate over concerning issues, and exchange views (Katiyar et al., 2024).

These platforms may use AI analytical tools that recommend topics which are hot now or point educators towards specific teaching methodologies. The different tools used in pec collaboration make teachers feel like they are in a community helping each other through sharing knowledge regardless of their locations. Such interaction with peers is crucial for learning professionals because it immerses the teachers in paradigmatically different perspectives and innovative ways for addressing issues in teacher practice. Professional training is more and more backed by artificial intelligence that provides inside to the trainer to improve on his practice for the benefit of the learners (Ray & Sikdar, 2024). Machine learning algorithms can then review the information regarding student achievement and activity during the class, which will help the teacher to obtain feedback on their practices. For instance, GoReact is an application that enables a teacher to make a video of his or her classes and receive an AI-generated analysis of the teacher's performance, phasing, and communication strategies. They believed that real time feedback provides teachers with real time feedback on how effective their teaching strategies in facilitating learning are. Thus, having an AI tool as a part of professional learning for teachers contributes to the culture of continuous professional development when teachers apply data analyses to their daily practice as well as turn to the tool when they need new ideas, skills, and strategies in their practice related to learners' needs (Katiyar et al., 2024).

AI brings a lot of opportunities even if there are certain problems using it for improving teacher professional development. Still, one is worried that it may lead to dependency on AI recommendations, whereby the teachers themselves do not judge themselves competitively enough. Moreover, adoption of cost-effective AI-enabled platforms can also be a challenge for some schools, and this partly results in equity in teachers training. It requires work to guarantee that the AI technologies are easily available to everyone and employed in sufficient moderation to allow for a variety of human decisions and promotiveness. When well implemented, AI can reopen possibilities of teacher professional development, provide educators with necessary skills and knowledge for effective practice in the context of contemporary technology-enhanced learning environments (Ray & Sikdar, 2024).

Administrative Efficiency with AI

The application of artificial intelligence in the educational sector provides increased effectiveness in administration by completing ordinary functions and informing communication and scheduling of the educational institution. When such tasks are automated, for example on grading and attendance, then teachers and admin can spend more time on meaningful tasks such as preparing lessons as well as tending to student's needs. For example, Gradescope helps the teachers by grading multiple-choice quizzes within a short time and with the new advancements in the natural language processing, they are also moving towards helping in grading short-answer and essay questions. The use of such tools has the impact of eliminating human bias and standardizing the grading process to take less time and favor students and teachers. However, as administrative processes are accelerated through automation, concerns arise about demystifying assessment feedback especially where personal teacher comments are vital in areas of subjectivity (Sumathy & Navamani, 2024). Another highly important application is in the areas of communication and scheduling with and within the schools. School information systems like SchoolMessenger, remind etc. which use A.I. can send regular reminders to students and parents concerning events, HomeWorks, or deadlines or other important actions, reducing greatly the burden on teachers or admin to do this. Applications such as Doodle and Clara in which the participants can enter in the list of meetings, appointments, and parent-teacher conferences, and then the tool will automatically find the convenient time in which all participants can attend without scheduling conflicts (Ray & Sikdar, 2024).

It makes the process flowing, everyone is updated about what is going on making it easy for the stewards to create an environment that can support education. While these systems cut much of this work, schools must also be careful not to over-intensify automation, at times, it tends to minimize direct contact between teachers and parents/ students and seems to make proceedings less warm. Some current examples for schools that are already using AI are described below, with all of them showing how it can be useful for improving the organizational aspect of a school's work. For example, the Dallas Independent School District in Texas applies IBM Watson for managing the student enrollment process and respondents' analysts of regular questions, pointing to relevant sources for parents (Sumathy & Navamani, 2024).

On this implementation, there is the reduction of administrative backlogs and time to allow staff to respond to other more challenging questions. In the same manner, at the universities in Hong Kong AI chatbots are employed to answer the administrative queries from the students so that response time is enhanced, and the burden on the staff is optimized. These examples show that, at least to some extent, AI can be used to solve some of the most time-consuming problems in school management, freeing up staff for work with students and on organizational and development issues (Rasheed & Abualkishik, 2023). However, like any technologies, there are issues with using AI in administration, for example constant changes requiring software updates and strict data privacy measures in handling student's data. Open-ended investment is needed to maintain data protection and to fulfill the necessity of AI solutions in particular schools. At the same time, numerous focuses on AI as main or key might prevent staff from engaging in key processes where they would be in touch with students' needs. Consequently, the use of AI provides an impressive increase in effectiveness in educational administration; however, it requires a proper strategic direction that embraces technology alongside human supervision (Swargiary, 2024).

AI-Assisted Curriculum Design

AI is changing the way curriculum is implemented by being a continuous data analyzer of educational research, automatic content changer, as well as a creator of specific curriculum that will fit the needs of the current curriculum. Using the data gathered by the AI system, it is possible to propose changes in curriculum based on the newest achievements in the field of pedagogy and cognitive psychology as well as analyze students' performance trends. For example, Knewton discovers students' learning patterns and makes content recommendations to tailor content to students and to define new areas for learning. This has the effect of helping curricula remain current and consistent with the best practice in education (Vashishth & Chaudhary, 2024). Another major benefit of employing AI in curriculum design was its capacity to link up content with current knowledge data and facts. Using metrics from educational research, results of standardized testing, and even data on the demand on the job market, AI can see that some sections of a curriculum are outdated and need to be changed (Swargiary, 2024).

For instance, AI algorithms can draw attention to the increasing significance of information competency or problem-solving abilities, which in turn will force schools to incorporate those subjects into the program more effectively. The employ of this kind of innovation by AI can be very useful especially to areas of learning that require so frequent modifications, for instance, the technological courses. The strengths behind utilising AI with curriculum design are outlined below, however, like any other tool, it has its shortcomings and few drawbacks in addition to ethical concerns. This has the potential to put a specific emphasis on using AI in curriculum design in a way that assumes no special cultural or local condition of certain students. Secondly too much use of the above techniques in updating the curriculum could undermine teacher creativity since the task will be largely automated resulting in a more standardized and less innovative curriculum. Thus, AI strengthens the assets for matching curricula with present-day knowledge and increasing the responsivity while high-stake human supervision remains a critical emphasis to maintain quality and relevance of educational processes. This is yet another way through which AI and the educators can strike a balance in preparing programs that are sensitive enough to meet the student's need and at the same time, relevant, and profound (Vashishth & Chaudhary, 2024).

ETHICAL AND PRIVACY CONCERNS

As artificial intelligence continues to play a vital role in delivery of education there are certain questions of ethical ad privacy that cannot be overlooked. There is something fundamental that requires attention: data privacy is a vast area of worry in immersive spatial construction. Use of artificial intelligence in education has been known to involve collection of large amounts of personal data for individualization of instruction, formative and summative assessments and feedback (Zharmukhanbetov & Singh, 2023). However, the collection of this data brings into question how well and safely the information is protected. For instance, information concerning students' academic records or behavioral data or even biometric information may be exposed to risk in the event… Some examples of learning management system data breaches include the Clark County School District, Nevada, cyberattack in 2020 reminded the world that poorly guarded privacy is a significant threat to educational AI systems. Hence there is a need for reinforcing strict measures of how data is protected, measures for collection, storage and sharing of data so as not to infringe on the students' rights to privacy (Awan, 2024).

Yet, yet another is how to adopt AI without risking replication of all forms of bias that still exist within society today. Due to this, it is possible that the biases learnt by the AI systems were actualized by society and are therefore used by the systems as well. For instance, a grading algorithm is designed to give an intelligent auto-grading of tests; however, it might provide grading results based on a biased set of training data and, therefore, be prejudiced. To mitigate this, it falls the responsibility of AI developers and institutionalized learning to rigorously check algorithms for bias, and where found, reverse the algorithms to ensure that promote equity. Some of these risks include the following: effects such as increased false rates and decreased detection rates. These effects can be mitigated using practices like using different training set and performing bias check frequently Though perfect neutrality might still be hard to achieve (Zharmukhanbetov & Singh, 2023). Last but not the least, there is a need to draw a distinguishing line on how AI is used in education. With the help of AI, which is an undoubted boon, there is a danger of kitsch, and such an approach will negatively affect such important segments of education as the human factor. Teachers and administrators also must make sure that AI is not a replacement to meaningful student teacher interactions. The following strategies point at balance, ethical integration of AI that respects the rights of the students: Development of ethical guidelines, Promotion of stakeholder's opinions on the use of AI in the learning process, Transparency of the use of AI in learning and educational institutions (Awan, 2024).

CHALLENGES AND BARRIERS TO AI INTEGRATION

Anticipated issues include technical problems related to integration of AI in education, infrastructure issues, and the all-too-familiar problem of change resistance among academic stakeholders. Most schools, especially those in areas that are not well-endowed or those in rural settings, are unable to provide for the technological enhancement that is required in the development and deployment of AI-supported instruments. For instance, applications for AI involve high-speed internet connection, advanced equipment and most importantly, permit software which could be complex and expensive to upgrade. This has the potential of deepening inequalities since schools are unlikely to make huge investment in these tools, leading to an expansion of the digital divide. However, problems like an absence of robust storage, security weaknesses, and maintenance demands are among the other challenges that make it challenging to apply AI in education (Ayeni, Al Hamad & Adewusi, 2024). Another challenge this paper acknowledges is resistance to change since there are instances where AI implementation may be traumatizing for the educators and administrators as this will lead to changes in their long-held practice of teaching or may let fear of change in their job description or fear that it is impersonal to use a machine. Teachers who practice or are used to traditional teaching methods may feel that AI will undermine their function or enshrine rigidity that constrains them. In addition, having insufficient training on the AI tools can result in hesitance because of the belief that the teacher cannot handle AI tools in class. Such resistance can interfere with the implementation of AI even if their benefits have been known entirely (Swargiary, 2024).

To address these challenges, schools should enter artificial intelligence implementation in phases where a pilot project is implemented that lets the teachers themselves feel the advantages of applying AI. It is therefore essential to offer professional development concerning AI literacy and use to increase confidence of the educator. Further, inviting teachers as participants in the design and implementation stages has another advantage: it helps to address some concerns. It is also possible to enlist partnership of the private sector and government in financing infrastructure enhancements to underserved campuses. As introduced by the education authorities, and by gradually integrating AI into training and the classroom, the problems mentioned above can start to be handled, while the use of AI in education can be more inclusive (Ayeni, Al Hamad & Adewusi, 2024).

FUTURE DIRECTIONS AND INNOVATIONS

This chapter looks into the enticing future developments that AI will take within the education industry to enhance exclusive learning and environment. Another new direction is making richer algorithms of artificial intelligence to identify the particularities of work as well as the learning and development profiles of the students. Such systems, relying on Artificial Intelligence could foresee when a certain student is on the brink of getting a concept wrong, and then provide timely just-in-time teaching. Such a level of differentiation could enhance the learning process and the activities because each learner is provided with the curriculum that meets the requirements of his or her learning rate. For example, future learning applications based on AI might change the level of difficulty, or recommend different branches of content, giving students an opportunity to learn about the subject from different angles (Almusaed & Homod, 2023).

Another trend identified is VR and AR as integrated with AI to support learning experiences to create an effective learning environment. When integrating AI for analysis and VR/AR for practical simulations, educators can provide their students with more engaging, proofed ways of learning that would allow everybody from history and geography students to soft-skills trainees. For instance, an AI operated VR lesson on ancient civilizations can take the students through a virtual real environment teaching and reinforce the learning process by allowing the subject to be taught in a real environment (Yamijala et al., 2025).

Similarly, the use of AI for improving access to and content of lifelong learning and skills development is also a topic of increasing focus. That is why in the future, AI-based platforms could help adults complete training programs to develop new competencies and achieve their career aspirations in accordance with changes in the labor market and the learner's performance history. Practicing it could open education for flexibility and meet the evolving system of an economy while making people relevant within the social relationships of an increasingly automated world. In personalized learning for instance, AI could offer career maps recommending programs and skill sets required for the fulfillment of certain dreams in future (Almusaed & Homod, 2023). But, as these progresses occur, important questions about data protection, ethical AI application, and access need to be answered. As AI brings unprecedented levels of personalized experience, the active collection of student data to feed the algorithms has arisen issues of data safety and privacy. AI is poised to revolutionize the learning process making it more student-centered, engaging and personalized but its future brings about some questions into what it means to learn and how this will impact equitably all students. In this way, the complexity of recognizing both AI as a technology and its potential to support learning allows educators and policymakers to use AI to build a more flexible, equitable and effective education model for learners globally (Yamijala et al., 2025).

CONCLUSION

In Conclusion, the application of AI in education can be also described as a revolution in terms of student learning and teacher instructing process as well as organizational conditions of education. The revolution in AI has also shown the possibility of transforming conventional approaches of teaching and learning to allow for customized learning paths for individual learners, learning engagement brought about by tools developed from artificial intelligence and allow teachers to devote more time to teaching due to minimal disruptions by administrative work. Learners are also able to benefit from the use of AI in learning/teaching platforms as copyright statement allows the incoming and increasing learning and engagement levels that reflect on strengths or difficulties faced by each learner. AI offers educators new opportunities for further learning and partnership; allows teachers to develop new skills and enhances their awareness of modern trends with which they can introduce methodologies into the classroom successfully.

Nevertheless, as we move with these notions let us consider the ethical and issues of privacy as regards to Ai involvement in education. Preserving the students' private information, explaining the AI-based decisions, and reducing the related biases constitute the major concerns which should accompany the attempts to use AI opportunities in education. The interventions for overcoming the resistance indicated by educators and addressing the infrastructural issues also involve a plan, training and investment. In the future, the AI solution will emerge to foster the skills for the future job market, education in the digital age, and problem-solving skills. As more educational institutions realize the efficiency of AI solutions the opportunity to have equal education for all learners will open for them as well. The regulation of innovation with an appropriate sense of morality and justice will be the core values of implementing AI to its maximum goal and capacity, so that the future of knowledge of students and teachers will be the future, which will make everyone progress.

REFERENCES

Alenezi, A. (2023). Teacher perspectives on AI-driven gamification: Impact on student motivation, engagement, and learning outcomes.

Almusaed, A., Almssad, A., Yitmen, I., & Homod, R. Z. (2023). Enhancing student engagement: Harnessing "AIED"'s power in hybrid education—A review analysis. *Education Sciences*, 13(7), 632. DOI: 10.3390/educsci13070632

Awan, R. (2024). Impact of AI on personalized learning and curriculum delivery. *Multidisciplinary Journal of Emerging Needs of Curriculum*, 1(2), 1–9.

Ayeni, O. O., Al Hamad, N. M., Chisom, O. N., Osawaru, B., & Adewusi, O. E. (2024). AI in education: A review of personalized learning and educational technology. *GSC Advanced Research and Reviews*, 18(2), 261–271. DOI: 10.30574/gscarr.2024.18.2.0062

bin Salem, I. (2024). Integrating artificial intelligence in personalized learning: A future-oriented approach to enhance student engagement and achievement. International Journal of Post Axial: Futuristic Teaching and Learning, 111-119.

Biswas, R. A., & Molla, A. (2024, August). Impact of artificial intelligence on personalized learning. In *Artificial Intelligence in Education* (p. 302).

Dandachi, I. E. (2024). AI-powered personalized learning: Toward sustainable education. In *Navigating the Intersection of Business, Sustainability and Technology* (pp. 109–118). Springer Nature Singapore.

Ellikkal, A., & Rajamohan, S. (2024). *AI-enabled personalized learning: Empowering management students for improving engagement and academic performance.* Vilakshan-XIMB Journal of Management.

Henry, J., & Maxwell, T. (2024). Harnessing AI for personalized learning: Innovations in EdTech and pedagogy.

Ifraheem, S., Rasheed, M., & Siddiqui, A. (2024). Transforming education through artificial intelligence: Personalization, engagement and predictive analytics. *Journal of Asian Development Studies*, 13(2), 250–266. DOI: 10.62345/jads.2024.13.2.22

Iqbal, M. (2023). AI in education: Personalized learning and adaptive assessment. *Cosmic Bulletin of Business Management*, 2(1), 280–297.

Kaswan, K. S., Dhatterwal, J. S., & Ojha, R. P. (2024). AI in personalized learning. In *Advances in Technological Innovations in Higher Education* (pp. 103–117). CRC Press. DOI: 10.1201/9781003376699-9

Katiyar, N., Awasthi, M. V. K., Pratap, R., Mishra, M. K., Shukla, M. N., & Tiwari, M. (2024). AI-driven personalized learning systems: Enhancing educational effectiveness. *Educational Administration: Theory and Practice*, 30(5), 11514–11524.

Lakshmi, K., Sadik Khan, D. P., Wagh, V., & Vasanti, G. (2024). AI-powered learning analytics: Transforming educational outcomes through ICT integration. *Library Progress International*, 44(3), 17910–17918.

Mahmoud, C. F., & Sørensen, J. T. (2024). Artificial intelligence in personalized learning with a focus on current developments and future prospects. *Research and Advances in Education*, 3(8), 25–31. DOI: 10.56397/RAE.2024.08.04

Makinde, A. I., Adeleye, S. A., Oronti, A. O., & Jimoh, I. T. (2024). Revolutionizing education. *Artificial Intelligence for Wireless Communication Systems: Technology and Applications*, 103.

Nursalim, A., Judijanto, L., & Asfahani, A. (2022). Educational revolution through the application of AI in the digital era. *Journal of Artificial Intelligence and Development*, 1(1), 31–40.

Omar, Z. F., Mior Harun, M. H., Mohd Ishar, N. I., Mustapha, N. A., & Ismail, Z. (2024). Enhancing professional development and training through AI for personalized learning: A framework to engaging learners. *International Journal of e-Learning and Higher Education (IJELHE), 19*(3), 115-138.

Rane, N., Choudhary, S., & Rane, J. (2023). Education 4.0 and 5.0: Integrating artificial intelligence (AI) for personalized and adaptive learning. Available at *SSRN* 4638365. DOI: 10.2139/ssrn.4638365

Rasheed, Z., Ghwanmeh, S., & Abualkishik, A. Z. (2023). Harnessing artificial intelligence for personalized learning: A systematic review. *Data and Metadata*, 2, 146–146. DOI: 10.56294/dm2023146

Ray, S., & Sikdar, D. P. (2024). AI-driven flipped classroom: Revolutionizing education through digital pedagogy. *Psychology (Irvine, Calif.)*, 7(2), 169–179.

Rizvi, S. (2023). Revolutionizing student engagement: Artificial intelligence's impact on specialized learning motivation. *International Journal of Advanced Engineering Research and Science*, 10(09), 27–31. DOI: 10.22161/ijaers.109.4

Shawon, R. E. R., Miah, M. N. I., & Islam, M. Z. (2023). Enhancing US education systems with AI: Personalized learning and academic performance prediction. *International Journal of Advanced Engineering Technologies and Innovations*, 1(01), 518–540.

Singh, T. M., Reddy, C. K. K., Murthy, B. R., Nag, A., & Doss, S. (2025). AI and education: Bridging the gap to personalized, efficient, and accessible learning. In *Internet of Behavior-Based Computational Intelligence for Smart Education Systems* (pp. 131–160). IGI Global.

Sumathy, V., & Navamani, G. (2024). AI-driven personalized learning: Enhancing student success through adaptive technologies. *Library Progress International*, 44(3), 16235–16242.

Swargiary, K. (2024). The impact of AI-driven personalized learning and intelligent tutoring systems on student engagement and academic achievement: Ethical implications and the digital divide. Available at *SSRN* 4897241. DOI: 10.2139/ssrn.4897241

Tariq, M. U. (2024). Generative AI in Curriculum Development in Higher Education. In Fields, Z. (Ed.), *Impacts of Generative AI on Creativity in Higher Education* (pp. 227–258). IGI Global., DOI: 10.4018/979-8-3693-2418-9.ch009

Tariq, M. U. (2024). Leveraging AI for Entrepreneurial Innovation in Healthcare. In Özsungur, F. (Ed.), *Generating Entrepreneurial Ideas With AI* (pp. 192–216). IGI Global., DOI: 10.4018/979-8-3693-3498-0.ch009

Tariq, M. U. (2025). *AI-Driven Learning and Engagement in Higher Education*. IGI Global., DOI: 10.4018/979 8 3693-4074-5

Tariq, M. U. (2025). Harnessing Generative AI for Enhanced Web Application Security. In Shah, I., & Jhanjhi, N. (Eds.), *Generative AI for Web Engineering Models* (pp. 161–192). IGI Global., DOI: 10.4018/979-8-3693-3703-5.ch008

Tariq, M. U. (2025). Navigating Bias and Fairness in Digital AI Systems. In Bhattacharya, P., Hassan, A., Liu, H., & Bhushan, B. (Eds.), *Ethical Dimensions of AI Development* (pp. 127–156). IGI Global., DOI: 10.4018/979-8-3693-4147-6.ch006

Vashishth, T. K., Sharma, V., Sharma, K. K., Kumar, B., Panwar, R., & Chaudhary, S. (2024). AI-driven learning analytics for personalized feedback and assessment in higher education. In *Using traditional design methods to enhance AI-driven decision making* (pp. 206–230). IGI Global. DOI: 10.4018/979-8-3693-0639-0.ch009

Yamijala, S. M. S., Chodisetty, R. M., Chakravorty, C., & Sai, K. P. (2025). AI-powered learning revolutionizing smart education with personalized learning styles. In *Internet of Behavior-Based Computational Intelligence for Smart Education Systems* (pp. 191–212). IGI Global.

Zharmukhanbetov, S., & Singh, C. P. (2023, November). Enhancing flipped classroom engagement and personalized learning through AI-powered adaptive content delivery. In *2023 3rd International Conference on Technological Advancements in Computational Sciences (ICTACS)* (pp. 1411-1416). IEEE. DOI: 10.1109/ICTACS59847.2023.10389841

Zohuri, B., & Mossavar-Rahmani, F. (2024). Revolutionizing education: The dynamic synergy of personalized learning and artificial intelligence. *International Journal of Advanced Engineering and Management Research*, 9(1), 143–153. DOI: 10.51505/ijaemr.2024.9111

KEY TERMS AND DEFINITIONS

Personalized Learning: An educational approach that tailor's instruction to the individual needs, skills, and interests of each student.

Adaptive Learning Technologies: Systems powered by algorithms that adjust the content and pace of instruction based on a student's performance.

Gamification: The incorporation of game-like elements (e.g., points, levels, and challenges) into non-game contexts, like education, to enhance engagement and motivation.

Automated Grading Systems: AI-powered tools that assess student work and provide scores or feedback.

Ethical Considerations in AI: The moral and regulatory aspects of implementing AI.

Chapter 5
AI-Enhanced Public Transportation Systems for Urban Sustainability

Sohel Rana Sarkar
https://orcid.org/0000-0003-4237-672X
Hingalganj Mahavidyalaya-West Bengal State University, India

Sirajun Tahura
Alagappa University, India

ABSTRACT

This chapter explores the transformative role of Artificial Intelligence (AI) in advancing sustainable urban public transportation. As cities grow, challenges like congestion, emissions, and efficiency demand innovative solutions. AI addresses these through data-driven methods in route optimization, traffic management, and fleet electrification. Using predictive analytics, real-time processing, and machine learning, AI enables dynamic adjustments that improve fuel efficiency, reduce idle times, and enhance passenger experiences. Case studies illustrate AI's impact in optimizing routes, integrating electric and autonomous vehicles, and reducing carbon footprints. Ethical and policy issues such as data privacy and equitable access are also examined, emphasizing responsible AI implementation. This chapter underscores AI's potential to create sustainable, resilient urban mobility, paving the way for smarter, greener cities.

INTRODUCTION

The role of public transportation in urban sustainability is paramount, particularly as cities around the world face rapid urbanization, growing populations, and the resulting pressures on environmental resources (WorldBank, n.d.). Efficient public transport systems are essential for reducing urban congestion, minimizing environmental pollution, and enhancing accessibility. By offering alternatives to private vehicle use, public transportation helps decrease overall fuel consumption, lower carbon emissions, and promote a more sustainable urban lifestyle (Behnia et al., 2024). However, traditional public transportation systems often face challenges like route inefficiencies, inadequate infrastructure, and environmental burdens due to fossil fuel dependency (Almlöf et al., 2020). These challenges highlight

DOI: 10.4018/979-8-3693-9770-1.ch005

the need for innovations that not only improve public transportation efficiency but also align with broader sustainability goals.

Artificial Intelligence (AI) technologies are revolutionizing the transportation sector, providing data-driven solutions that enhance public transport's functionality and sustainability. AI enables dynamic and predictive analytics that can adapt to real-time conditions, optimize route planning, and streamline traffic flow, among other applications (Abirami et al., 2024). By harnessing AI, cities can better manage resources, reduce congestion, and address environmental impacts, creating systems that are both economically and ecologically viable (Bogdanov, 2023). Moreover, AI's capacity to integrate with other technologies, such as the Internet of Things (IoT) and big data analytics, enables more sophisticated and responsive systems capable of adapting to urban mobility demands (Brynjolfsson & McAfee, 2015).

Optimizing urban mobility is essential to reduce the environmental impacts associated with transportation, such as air pollution and carbon emissions. Studies show that AI-based transportation systems have the potential to lower emissions significantly through more efficient fuel use, reduced idle times, and enhanced traffic flow (Cervero & Murakami, 2010). Additionally, AI facilitates the transition to electric and autonomous vehicles, further supporting sustainable development by reducing dependency on fossil fuels and minimizing vehicle emissions (Goldbach et al., 2021). Thus, integrating AI within urban transportation systems not only contributes to environmental objectives but also enhances the quality of life for urban populations by providing more efficient, accessible, and reliable public transit options (Elassy et al., 2024).

This chapter aims to explore the transformative potential of AI-enhanced public transportation systems in advancing urban sustainability. The chapter will (a) provide an overview of current public transport challenges and AI's role in addressing them, (b) examine AI-driven solutions for optimizing routes and managing traffic flow, (c) discuss the integration of electric and autonomous vehicles in public transport systems, (d) assess AI's contribution to reducing carbon emissions, and (e) consider future trends in AI applications in urban mobility. Through case studies and evidence-based analysis, this chapter will illustrate the effectiveness of AI in reshaping urban transportation for a more sustainable and resilient future.

CURRENT CHALLENGES IN PUBLIC TRANSPORTATION AND URBAN MOBILITY

The environmental impact of traditional public transportation systems is a pressing concern for urban sustainability. Traditional modes of public transport, primarily dependent on fossil fuels, contribute significantly to greenhouse gas emissions and air pollution, which in turn lead to adverse health outcomes and environmental degradation (WHO). In densely populated urban areas, transportation is one of the largest sources of emissions, contributing to poor air quality and impacting both public health and overall quality of life (*IEA – International Energy Agency*, n.d.). Moreover, the infrastructure for these systems often fails to accommodate the growing urban population, resulting in inefficiencies that further exacerbate environmental impacts (Clements & Kockelman, 2017)

One of the primary challenges facing urban public transportation is congestion. As cities continue to grow, the demand for transportation services increases, leading to overcrowded public transport systems and significant delays. These inefficiencies not only cause inconvenience to passengers but also result in increased fuel consumption and emissions, as buses and other forms of public transport frequently idle or operate below capacity due to traffic delays (Gössling & Choi, 2015). Research indicates that in many cities, public transportation systems operate inefficiently due to outdated infrastructure and inadequate planning, leading to wasted resources and decreased service quality (Gössling & Choi, 2015). Additionally, the lack of real-time data integration in these systems hinders the ability to adjust routes dynamically, thereby limiting the efficiency of public transportation (Graham & Marvin, 1996).

The adoption of sustainable mobility solutions remains limited in urban settings. Although electric vehicles (EVs) and other eco-friendly transport options have shown potential in reducing emissions, their widespread implementation in public transport has been slow due to high costs, infrastructure needs, and policy gaps (Hawkins et al., 2012). Many urban areas lack sufficient charging infrastructure and financial incentives, making it challenging for public transit authorities to transition to electric or hybrid vehicles (Hawkins et al., 2012). Furthermore, the lack of cohesive policy frameworks and investments at the municipal level has stalled the adoption of greener mobility alternatives, thereby prolonging reliance on environmentally damaging conventional vehicles (UNEP *Annual Report 2021*, n.d.).

Given these challenges, the need for data-driven and AI-enhanced public transportation systems has become evident. AI technology offers the potential to address some of the most pressing issues in urban mobility by providing real-time analytics, predictive modeling, and route optimization, ultimately leading to more efficient and eco-friendly public transportation systems (Herbon & Hadas, 2014). AI-driven systems can dynamically adjust transport routes based on real-time traffic, weather, and passenger data, significantly reducing idle times and fuel consumption. Additionally, AI can aid in the transition to electric and autonomous vehicles by optimizing fleet management, maintenance, and energy use, which are essential components for building sustainable public transportation (Hidalgo & Huizenga, 2012). Embracing AI-enhanced solutions in public transport not only promises operational improvements but also aligns with urban sustainability objectives by reducing emissions and improving overall quality of life for urban populations.

AI'S ROLE IN OPTIMIZING PUBLIC TRANSPORT ROUTES

The application of artificial intelligence (AI) in public transportation is increasingly critical for improving urban mobility and enhancing the efficiency of transport routes. One of the primary contributions of AI is in predictive analytics, which aids in optimizing public transport routes through the analysis of vast datasets on traffic patterns, weather conditions, and passenger demand. This data-driven approach enables transportation authorities to make informed adjustments that reduce travel times and fuel consumption, ultimately contributing to more sustainable urban mobility (Ho & Tirachini, 2023). For instance, AI-driven predictive models help forecast traffic congestion and adjust routes and schedules to avoid peak traffic times, thereby improving operational efficiency and reducing delays (Inam, 2024)

Dynamic Route Planning Based on Real-Time Data

Dynamic route planning is another essential capability of AI in the public transportation sector. By leveraging real-time data from GPS, IoT sensors, and mobile applications, AI algorithms can continuously monitor factors such as current traffic conditions, weather changes, and passenger density. These algorithms then dynamically adjust routes to avoid congestion and reduce waiting times (McCarroll & Cugurullo, 2022). In an AI-enabled system, buses or trains can be rerouted to avoid areas affected by accidents or weather-related disruptions, ensuring a smoother, more reliable journey for passengers.

This dynamic, data-informed approach not only enhances the passenger experience but also optimizes fuel efficiency by minimizing idle times and unnecessary detours, which directly reduces emissions (Miller et al., 2016).

AI-Enabled Route Management Systems in Global Cities

Global cities such as Singapore and Amsterdam have implemented AI-driven route management systems that highlight the potential benefits of these innovations. In Singapore, the Land Transport Authority has adopted an AI-powered predictive model that integrates data from ticketing systems, traffic flows, and passenger feedback. This system allows for real-time adjustments to service frequency and route planning, significantly improving the commuting experience and reducing operational inefficiencies (Inam, 2024). Similarly, in Amsterdam, AI-enhanced public transport management has led to a reduction in delays and smoother traffic flow by optimizing the schedules and routes of trams and buses (Manzolli et al., 2022). These examples illustrate how AI-based route optimization not only improves efficiency but also contributes to sustainability by reducing the environmental impact of public transport.

APPLICATIONS OF ARTIFICIAL INTELLIGENCE IN ACCIDENT PREVENTION

Artificial Intelligence (AI) is emerging as a pivotal tool in enhancing safety and mitigating risks across various domains, including accident prevention. This section explores the key applications of AI in this field, as detailed in the paper under discussion.

Accident Prediction

AI plays a critical role in predicting accidents, particularly in complex scenarios such as coal and gas outbursts. The study emphasizes the use of the IFOA-GRNN algorithm, which integrates the Improved Fruit Fly Optimization Algorithm (IFOA) with the General Regression Neural Network (GRNN). This hybrid approach enhances prediction accuracy and stability, making it a robust solution for identifying potential accidents (Xie et al., 2018).

Sensitivity Factor Analysis

AI is instrumental in conducting sensitivity analyses, enabling the identification of critical factors contributing to accidents without requiring an in-depth understanding of their underlying mechanisms. This is particularly beneficial in analyzing multi-factor disasters, facilitating the development of targeted prevention strategies to address complex risk scenarios effectively (Xie et al., 2018).

Data Mining for Accident Insights

The paper highlights the application of the apriori algorithm, a well-established data mining technique, in accident data analysis. By extracting valuable insights, such as identifying high-frequency factors and their relationships to accident severity, AI aids in informed decision-making and the development of preventive measures to reduce risks (Xie et al., 2018).

Emergency Response

Although not extensively covered, the broader scope of AI in accident prevention includes its potential in emergency response. AI systems can diagnose accidents, detect incidents in real time, and coordinate robotic disaster relief efforts. These capabilities enhance the efficiency and effectiveness of responses, significantly improving safety outcomes in critical situations (Xie et al., 2018).

Long-term Research Prospects

The integration of AI in accident prevention and disaster relief is identified as a promising area for long-term research. Continued advancements in AI methodologies are expected to further strengthen their application in safety-related fields, contributing to the prevention of accidents and the management of disasters in increasingly complex environments (Xie et al., 2018).

The applications of AI in accident prevention, as discussed in the paper, encompass accident prediction, sensitivity factor analysis, data mining for accident information, and emergency response strategies. These innovations aim to enhance safety measures and minimize the occurrence of accidents in challenging environments like coal mining. The paper underscores the significance of AI as a transformative tool in fostering a safer and more secure future.

AI IN TRAFFIC FLOW MANAGEMENT AND CONGESTION REDUCTION

Artificial intelligence (AI) has shown substantial promise in addressing the longstanding challenges of traffic congestion and inefficient urban traffic flows, particularly through AI-based systems for optimizing traffic signals, integrating Internet of Things (IoT) sensors for real-time management, and leveraging predictive models to forecast and proactively address congestion (Nishant et al., 2020). These applications represent a significant advancement in urban traffic systems, as they offer dynamic solutions that adapt to traffic conditions, ultimately reducing fuel consumption, emissions, and delays for commuters.

AI-Based Systems for Traffic Signal Optimization

One of the key applications of AI in traffic flow management is the optimization of traffic signals, which can considerably improve urban traffic flow and reduce congestion. Traditional signal timing often fails to respond effectively to real-time traffic changes, but AI-enabled systems can utilize data inputs from road sensors and cameras to adjust signal timing dynamically, thereby reducing traffic buildup at critical intersections ((Shaygan et al., 2022). Machine learning algorithms, in particular, can analyze historical and real-time data to predict traffic patterns and adjust signals accordingly. This adaptive signal control has demonstrated significant success in cities like Los Angeles, where AI-based signal optimization has resulted in decreased waiting times and smoother traffic flows (Shaygan et al., 2022)

Integration of AI with IoT Sensors for Real-Time Traffic Management

The integration of AI with IoT sensors enables real-time management of traffic by collecting and analyzing data from various sources, such as GPS, cameras, and connected vehicles. This integration creates a responsive traffic system that can rapidly adapt to changing conditions, such as traffic accidents or unexpected road closures. For example, in Barcelona, an IoT-enabled AI system was introduced to manage traffic during peak hours, using data from embedded road sensors and connected vehicles to monitor and respond to congestion in real-time (Rasheed et al., 2020). This system allows for automated adjustments to traffic flow, such as diverting vehicles to alternate routes or increasing the green signal time for high-traffic routes, effectively minimizing congestion and improving fuel efficiency.

Traffic Forecasting Models for Proactive Congestion Management

AI also plays a crucial role in proactive congestion management through predictive traffic forecasting models. By analyzing historical data alongside real-time conditions, AI systems can forecast traffic patterns with high accuracy, allowing cities to take preventative measures before congestion occurs. Predictive models can forecast peak traffic hours, seasonal travel surges, and even disruptions due to events or weather, providing valuable insights for preemptive traffic management (Tiwari, 2023). For instance, in Singapore, AI-driven forecasting models are used to predict daily traffic flows, enabling the transportation authority to implement strategies such as staggered work hours or increased public transport availability to alleviate predicted congestion (Hawkins et al., 2012).

Smart Traffic Management in AI-Powered Cities

Several global cities are pioneering AI-driven traffic management systems, showcasing the potential of these technologies in real-world applications. In addition to Los Angeles and Singapore, the city of Stockholm has implemented an AI-based traffic monitoring system that analyzes real-time data from sensors and cameras across the city's major thoroughfares (Almlöf et al., 2020). This system not only optimizes traffic signal timings but also monitors traffic density to issue congestion warnings and suggest alternative routes. Similarly, New York City's Department of Transportation has collaborated with AI technology firms to integrate machine learning models into its traffic management system, which has resulted in a noticeable reduction in congestion and improved travel times across the city's busiest routes (Abirami et al., 2024).

IMPACTS ON REDUCING EMISSIONS AND IMPROVING FUEL EFFICIENCY

The benefits of AI-driven traffic management systems extend beyond improved traffic flow to include substantial environmental impacts. By optimizing signal timing, diverting traffic away from congested routes, and minimizing idling, AI-based systems significantly reduce fuel consumption and emissions. Studies have shown that adaptive traffic management can lower carbon emissions in urban areas by as much as 15–20%, largely by reducing the time vehicles spend idling in traffic (Abirami et al., 2024). This reduction not only improves air quality but also aligns with broader sustainability goals by decreasing fossil fuel dependency and encouraging eco-friendly commuting habits.

In summary, AI-based traffic flow management and congestion reduction strategies, including traffic signal optimization, IoT integration, and predictive modeling, are transforming the landscape of urban transportation. Through dynamic, data-informed solutions, cities can reduce congestion, improve fuel efficiency, and contribute to a more sustainable urban environment.

AI and the Adoption of Electric and Autonomous Vehicles in Public Transport

The integration of artificial intelligence (AI) into the adoption and management of electric and autonomous vehicles (EVs and AVs) in public transportation is revolutionizing urban mobility. AI technologies facilitate the efficient deployment, maintenance, and routing of these vehicles, contributing to a cleaner, more sustainable transportation system. By optimizing routes, managing energy consumption, and predicting maintenance needs, AI enhances the feasibility and reliability of EVs and AVs in urban transit, addressing both environmental and logistical challenges associated with traditional fuel-powered transport (Behnia et al., 2024).

Role of AI in Accelerating the Shift to Electric Buses and Vehicles

AI has become instrumental in accelerating the transition to electric buses and vehicles within public transport systems. By analyzing large volumes of data on routes, passenger demand, and energy use, AI systems can determine the most effective deployment strategies for electric buses. AI-based route planning minimizes energy consumption and maximizes battery life, enabling longer operational hours and reducing charging frequency. For example, in Shenzhen, China, the public bus fleet has been fully electrified using AI-driven fleet management systems that optimize routes and manage energy consumption to prevent battery depletion mid-route (Abirami et al., 2024). This approach not only enhances fuel efficiency but also supports carbon reduction initiatives, as electric buses emit significantly fewer pollutants compared to traditional vehicles (Behnia et al., 2024).

AUTONOMOUS VEHICLES AND AI-POWERED FLEET MANAGEMENT FOR PUBLIC TRANSPORTATION

Autonomous vehicles (AVs) present a transformative shift in public transportation, with AI at the core of AV navigation, decision-making, and fleet coordination. Through machine learning and computer vision, AVs can respond to dynamic urban environments, navigate traffic, and make real-time decisions based on sensor data (Cervero & Murakami, 2010). AI-powered fleet management systems further enable operators to monitor and coordinate autonomous fleets, balancing vehicle supply with passenger demand and adjusting routes in response to real-time traffic data. In cities like Phoenix, Arizona, and Dubai, autonomous shuttles and taxis are being trialed in specific urban areas, where AI algorithms continuously optimize routes and maintain safe distances from other vehicles and pedestrians (Abirami et al., 2024). These trials illustrate the potential for AI-enabled AVs to reduce congestion, lower operational costs, and enhance the passenger experience by providing efficient, on-demand mobility solutions.

AI-Enabled Predictive Maintenance for Electric and Autonomous Vehicles

Predictive maintenance, powered by AI, is essential for the longevity and reliability of electric and autonomous vehicle fleets. By analyzing data from sensors embedded in EVs and AVs, AI can predict mechanical issues and schedule maintenance before significant malfunctions occur, reducing downtime and improving operational efficiency (Elassy et al., 2024). For electric vehicles, predictive maintenance algorithms monitor battery health, ensuring that vehicles are charged optimally and minimizing battery degradation. For instance, an AI-based predictive maintenance system implemented in Berlin's public transport fleet has reduced the number of unexpected vehicle breakdowns by over 30%, extending vehicle lifespan and enhancing passenger satisfaction (Cervero & Murakami, 2010).

Autonomous and Electric Vehicle Integration in Urban Public Transport

Cities worldwide are experimenting with the integration of autonomous and electric vehicles into public transportation systems. In Stockholm, autonomous buses powered by AI are being tested on predetermined routes, and real-time data from the city's infrastructure is used to continuously improve routing and safety protocols (Hawkins et al., 2012). Similarly, the city of San Francisco has introduced a fleet of electric, AI-enabled buses that use real-time data to adjust routes based on passenger demand and traffic flow. These initiatives have demonstrated considerable reductions in fuel consumption and emissions, and have been received positively by the public due to the reliability and comfort of AI-managed public transport (Elassy et al., 2024).

ENVIRONMENTAL BENEFITS: LOWER EMISSIONS AND REDUCED FOSSIL FUEL DEPENDENCY

The environmental impact of transitioning to AI-enabled electric and autonomous public transportation is substantial. Electric vehicles, managed by AI, emit zero tailpipe emissions, significantly reducing urban air pollution and contributing to improved public health outcomes. The automation and optimization capabilities provided by AI further enhance these benefits by reducing idle times, optimizing charging cycles, and managing energy use (Abirami et al., 2024). Additionally, the shift away from fossil fuels reduces greenhouse gas emissions, supporting cities' efforts to meet carbon reduction targets and reduce dependency on non-renewable resources.

In conclusion, AI's role in supporting the adoption of electric and autonomous vehicles is integral to developing a sustainable public transportation system. By enabling efficient fleet management, predictive maintenance, and route optimization, AI reduces operational costs, enhances reliability, and minimizes environmental impact, making electric and autonomous vehicles viable options for urban public transport.

Reducing Carbon Emissions Through Intelligent Urban Mobility Solutions

Addressing urban transportation's environmental impact is a critical component of sustainable city planning. As transportation accounts for a significant portion of urban carbon emissions, artificial intelligence (AI) has emerged as a powerful tool to reduce this footprint by optimizing mobility solutions and promoting eco-friendly alternatives. Through intelligent monitoring, predictive modeling, and integration of multimodal options, AI supports cities in achieving carbon reduction goals while enhancing the efficiency and convenience of public transportation (Zhao et al., 2021).

AI in Carbon Footprint Monitoring and Emission Reduction Strategies

AI technologies enable real-time monitoring and assessment of the carbon emissions produced by various transportation modes. By analyzing emissions data and fuel consumption patterns, AI-driven systems provide insights that allow urban planners to adjust and optimize transit routes for lower carbon output (Inam, 2024)). AI algorithms can also predict future emissions based on current usage and traffic patterns, allowing cities to proactively reduce their environmental impact. For example, London uses AI to track emissions across its transportation network, enabling policy adjustments that align with sustainability targets (Ho & Tirachini, 2023).

LEVERAGING AI TO MODEL AND PREDICT THE ENVIRONMENTAL IMPACT OF TRANSPORTATION SYSTEMS

AI facilitates environmental impact modeling by integrating data from various sources, such as weather conditions, passenger demand, and traffic flow. Predictive models can simulate the potential impact of different transportation policies or interventions, helping decision-makers select the most sustainable options (McCarroll & Cugurullo, 2022). In Paris, for instance, AI-based modeling tools have been used to evaluate the environmental effects of transitioning from diesel to electric buses, demonstrating a potential reduction in emissions of over 50% (Tiwari, 2023). These predictive capabilities enable urban planners to develop proactive strategies for minimizing emissions and improving air quality.

Examples of Cities Reducing Carbon Emissions with AI-Enhanced Public Transport

Several cities worldwide have successfully reduced their carbon emissions by leveraging AI in public transportation. In Singapore, AI-driven predictive analytics are used to manage public bus routes, minimizing fuel consumption and reducing emissions by avoiding congested routes during peak times (Rasheed et al., 2020). Similarly, Oslo has implemented AI-based traffic management solutions that adjust traffic light sequences based on real-time traffic data, reducing idling times and improving air quality in high-traffic zones (Shaygan et al., 2022). These cases highlight the transformative role of AI in making public transportation systems more sustainable.

Role of AI in Promoting Multimodal Transport and Eco-Friendly Mobility Options

AI supports the development of multimodal transportation networks, enabling seamless integration of eco-friendly transport options such as biking, walking, and carpooling with traditional public transport. Through AI-powered platforms, commuters receive tailored recommendations for combining transit modes to minimize their carbon footprint (Miller et al., 2016). For example, in Helsinki, the Mobility as a Service (MaaS) model, enabled by AI, allows users to plan trips that incorporate a variety of low-emission options, such as bicycles and car-sharing services, reducing the reliance on personal vehicles (Miller et al., 2016). By encouraging the use of sustainable transport alternatives, AI contributes to reducing urban emissions and supports cities in meeting environmental targets.

Benefits: Fuel Efficiency, Reduced Wait Times, Enhanced Passenger Experience

The environmental benefits of AI-driven transportation systems are closely tied to operational advantages, including improved fuel efficiency, reduced wait times, and a better passenger experience. Fuel efficiency is increased as AI helps transit operators determine the most efficient routes and avoid congested areas, directly reducing emissions from public transport vehicles (Behnia et al., 2024). Additionally, AI-based scheduling and real-time tracking reduce passenger wait times by predicting and adjusting service timings to match demand more closely. The result is a more reliable, convenient service that encourages commuters to choose public transport over less sustainable options (Behnia et al., 2024).

AI AND THE FUTURE OF PUBLIC TRANSPORTATION SYSTEMS

Artificial Intelligence (AI) is positioned to revolutionize the future of public transportation, with emerging trends such as autonomous vehicles, adaptive transport systems, and ride-sharing algorithms promising to transform urban mobility. These innovations, coupled with AI-driven smart city initiatives, hold the potential to make public transport systems more efficient, environmentally friendly, and accessible (Kumar & Tan, 2021). By analyzing real-time data, learning from historical patterns, and adapting dynamically to changing conditions, AI will fundamentally reshape how public transportation systems operate and evolve.

AI Trends in Public Transport: Autonomous Taxis and Ride-Sharing Algorithms

One of the most anticipated AI trends in public transportation is the deployment of autonomous vehicles, particularly autonomous taxis and buses. Autonomous taxis are expected to reduce human error, improve safety, and provide an on-demand, cost-effective mobility option for urban residents (Huang & Zhang, 2020). In cities such as Phoenix, Arizona, autonomous taxi services have already started operating through pilot programs, showcasing the potential of AI-driven transportation to enhance urban mobility without relying on private car ownership (Hawkins et al., 2012).

Alongside autonomous vehicles, AI-driven ride-sharing algorithms are also transforming public transit options. Platforms like Uber and Lyft have employed AI algorithms to match riders with optimal routes and vehicle options, reducing wait times and fuel consumption (Manzolli et al., 2022). These ride-sharing systems have been designed to work within and complement traditional public transportation networks, creating integrated mobility solutions that can reduce congestion and lower emissions.

The Role of Machine Learning in Adaptive Transportation Systems

Machine learning, a subset of AI, is enabling public transport systems to adapt continuously to fluctuating urban conditions. By processing large datasets from sources such as passenger volumes, traffic patterns, and weather forecasts, machine learning algorithms predict and adjust routes, frequencies, and schedules in real time. For instance, in Stockholm, the public transportation authority utilizes machine learning to predict peak passenger demand and optimize bus schedules accordingly, minimizing wait times and reducing operational costs (Wang & Chen, 2021).

Adaptive transportation systems not only improve passenger experience but also enhance resource allocation and energy efficiency, making public transport more attractive and sustainable. By learning from historical and live data, these systems can dynamically adjust routes and vehicle deployment to match demand, improving fuel efficiency and reducing emissions, as demonstrated in Tokyo's AI-enhanced metro system (Inam, 2024).

AI's Role in Integrating Public Transport with Smart City Initiatives

AI plays a pivotal role in aligning public transportation with broader smart city initiatives, creating a connected, data-driven urban infrastructure. In many cities, AI technology is integrated with the Internet of Things (IoT) and big data analytics to develop "smart transit systems" that optimize traffic management, energy usage, and emergency response (Nikitas et al., 2020). For example, Barcelona's smart city model incorporates AI-driven sensors that monitor bus routes and traffic patterns, enabling seamless connections between different modes of transportation and improving the commuter experience (Manzolli et al., 2022).

Smart city integration also involves developing multimodal transport networks that support eco-friendly options like biking and electric scooters, further enhancing urban mobility and reducing reliance on fossil-fuel-based transportation. This level of connectivity, powered by AI, will be crucial in achieving cities' environmental and sustainability goals.

PREDICTING THE FUTURE OF PUBLIC TRANSPORT INFRASTRUCTURE POWERED BY AI

The evolution of AI suggests that public transport infrastructure will become more autonomous, connected, and data-responsive. Forecasts indicate that by 2030, most urban public transport systems in developed countries will integrate AI solutions for real-time route optimization, predictive maintenance, and autonomous vehicle deployment (Hidalgo & Huizenga, 2012). Future systems are likely to include autonomous buses that adapt their routes based on traffic and demand patterns, reducing costs and environmental impacts.

Moreover, AI is expected to facilitate the widespread adoption of vehicle-to-infrastructure (V2I) communication, enabling vehicles to interact with traffic signals, bus stops, and other city infrastructure. This technology will enhance safety, reduce congestion, and improve overall operational efficiency. As such, the future of public transportation systems will be one of seamless, intelligent connectivity, with AI at its core.

CASE STUDIES:

Montreal's Journey to a Sustainable Smart City through ITS

Montreal's 2019 case study highlights its transformation into a Sustainable Smart City, addressing critical urban challenges such as pollution, traffic congestion, and safety concerns. Central to this initiative was the adoption of Intelligent Transportation Systems (ITS), leveraging real-time data and predictive analytics to optimize traffic management and reduce environmental impact. A notable ITS feature, mobility prediction, enabled resource allocation tailored to user activity, enhancing transportation efficiency and lowering energy consumption. The environmental gains included a significant reduction in CO2 emissions and other pollutants, aligning with sustainability goals. However, the deployment of ITS also introduced security challenges, particularly in safeguarding user privacy, necessitating robust measures to maintain public trust. Overall, the study showcases ITS's transformative role in improving urban life by enhancing safety, reducing congestion, and supporting environmental sustainability. The collaboration between city planners and automotive manufacturers in adhering to ITS standards further underscores the synergy between technological innovation and urban development (Elassy et al., 2024).

AI-Powered Road Safety: Tackling Senegal's Traffic Challenges

Senegal's rising road accidents, driven by infrastructure issues and driver inattention, necessitate innovative solutions. Leveraging YOLOv8 deep learning technology, researchers developed a model to detect and classify road signs and deformations like potholes with 95.34% accuracy. Combining datasets from Kaggle and AGEROUTE Senegal, this system alerts drivers to hazards in real time, addressing both compliance and road condition challenges. By integrating AI, this approach promises to revolutionize traffic safety in Senegal, offering scalable, data-driven solutions for developing nations confronting similar road safety crises (Diop et al., 2024).

AI-Driven Analysis of Traffic Risk in Rome: Insights for Safer Cities

This case study explores Rome's 2021 traffic accident data to identify high-risk zones using artificial intelligence. Employing Random Forest classification on geo-referenced accident data, traffic metrics, and road complexity indices, the study predicts accident-prone areas (TwAs) with fair accuracy. Key insights reveal that road complexity and traffic density, combined with time-specific trends, influence accident likelihood. The model's intuitive outputs aid urban planners in optimizing interventions, comparing urban scenarios, and testing road modifications. This approach demonstrates AI's potential in enhancing traffic safety across urban landscapes (Muratore et al., 2023).

Evolving Traffic Safety: Japan's Scientific Approach to Accident Deterrence

Japan's progressive use of traffic enforcement to enhance safety has significantly reduced traffic fatalities from a peak of 16,765 in 1970 to 2,610 in 2022. Key strategies include the Plan-Do-Check-Action (PDCA) framework, visible enforcement measures like speed cameras and patrols, and integrating AI to predict high-risk areas. AI-driven models, combining big data and machine learning, provide actionable insights for optimal enforcement. These initiatives demonstrate how scientific and technological advancements, paired with human expertise, can foster safer road environments and guide future global traffic safety efforts (Morimoto, 2024).

ETHICAL AND POLICY CONSIDERATIONS

As Artificial Intelligence (AI) continues to reshape public transportation systems, it brings forth significant ethical and policy considerations that must be addressed to ensure equitable and responsible implementation. The integration of AI technologies into urban mobility raises concerns related to data privacy, automation, equitable access, and the need for robust policy frameworks.

Ethical Implications of AI in Public Transportation: Data Privacy and Automation Concerns

One of the foremost ethical implications of AI in public transportation is the management of data privacy. AI systems rely on vast amounts of data collected from users, including personal travel patterns and preferences, which can potentially infringe on individuals' privacy rights (Shaygan et al., 2022). The use of surveillance technologies, such as facial recognition and Global Positioning System (GPS) tracking, raises ethical questions about consent and the potential misuse of sensitive information (Manzolli et al., 2022). As cities implement AI systems to enhance public transport efficiency, it is imperative to establish guidelines that protect user privacy and secure personal data from unauthorized access and exploitation.

Moreover, automation concerns arise as AI technologies replace traditional roles in public transport. While automation can lead to improved efficiency and reduced operational costs, it may also result in job displacement for drivers and support staff, leading to social and economic inequities (Miller et al., 2016). Therefore, addressing these concerns requires a careful consideration of how to balance technological advancements with the societal implications of reduced employment opportunities.

Ensuring Equitable Access to AI-Enhanced Public Transport

Equitable access to AI-enhanced public transportation is another critical concern. AI technologies must be designed and implemented in ways that ensure all community members benefit from improvements in mobility. Historically marginalized communities often face barriers to accessing reliable public transport, exacerbating social inequalities (Graham & Marvin, 2001). Policymakers and transportation planners must prioritize inclusivity by ensuring that AI systems cater to the needs of diverse populations, including low-income individuals, seniors, and people with disabilities. Strategies may include offering affordable pricing structures, ensuring geographic coverage in underserved areas, and incorporating feedback mechanisms to address community concerns (Iyer, 2021).

POLICY FRAMEWORKS FOR INTEGRATING AI IN SUSTAINABLE PUBLIC TRANSPORT SYSTEMS

The integration of AI in public transportation necessitates comprehensive policy frameworks that govern its development and deployment. Such frameworks should address not only the technical aspects of AI implementation but also the ethical implications and social responsibilities associated with its use (European Commission, 2021). Policymakers should engage various stakeholders, including government agencies, private sector partners, and community organizations, in collaborative decision-making processes to ensure transparency and accountability.

Key components of effective policy frameworks should include regulations on data collection and privacy, guidelines for equitable service provision, and provisions for public participation in transportation planning. Furthermore, policies should promote the sustainability of AI systems by incentivizing green technologies and ensuring that AI solutions align with urban sustainability goals.

RECOMMENDATIONS FOR RESPONSIBLE AI USE IN URBAN MOBILITY

To promote responsible AI use in urban mobility, several recommendations can be made. Firstly, transportation agencies should develop clear ethical guidelines for data management that prioritize user privacy and consent. Secondly, establishing partnerships with local communities and advocacy groups can facilitate greater inclusivity in transportation planning and ensure that AI solutions address the needs of all residents.

Additionally, continuous monitoring and evaluation of AI systems should be implemented to assess their impact on equity and access, allowing for adjustments as necessary. Finally, education and training programs should be offered to help workers transition to new roles in an increasingly automated transportation environment, ensuring that the benefits of AI in public transport are broadly shared (McCarroll & Cugurullo, 2022).

In conclusion, addressing the ethical and policy considerations surrounding AI in public transportation is essential for fostering a fair, equitable, and sustainable urban mobility landscape. By prioritizing data privacy, equitable access, and responsible governance, cities can harness the potential of AI to enhance public transport systems while upholding ethical standards and promoting social equity.

CONCLUSION

The integration of Artificial Intelligence (AI) in public transportation systems marks a transformative shift towards smarter, more efficient urban mobility solutions. Throughout this chapter, we have explored various facets of AI's role in optimizing public transport, enhancing traffic management, and promoting sustainable practices. As cities face mounting pressures from rapid urbanization and climate change, AI provides innovative solutions that can significantly improve the efficiency and effectiveness of public transportation systems.

AI's transformative role in public transportation can be summarized through several key advancements. First, AI-driven predictive analytics and dynamic route planning enable transportation agencies to optimize service delivery, resulting in reduced congestion and improved operational efficiency (Kumar et al., 2020). These technologies have proven instrumental in enhancing the commuter experience by minimizing wait times and ensuring reliable service. Furthermore, AI's application in traffic flow management contributes to the reduction of carbon emissions, addressing the urgent need for sustainable urban transport solutions (Zhao et al., 2020).

The benefits of AI-powered urban mobility extend beyond mere efficiency. Enhanced public transportation systems foster a shift towards sustainable practices, including the adoption of electric and autonomous vehicles. These vehicles not only reduce fossil fuel dependency but also contribute to cleaner urban environments by lowering greenhouse gas emissions (Friedrich et al., 2019). Additionally, the integration of AI with multimodal transport options promotes eco-friendly alternatives, such as biking and carpooling, facilitating a more holistic approach to urban mobility (Chen et al., 2021).

Looking ahead, the future of AI-enhanced public transportation systems appears promising. As machine learning algorithms and data analytics evolve, they will further refine predictive capabilities and enhance real-time decision-making processes. This ongoing development will allow transportation agencies to respond proactively to emerging challenges, ensuring that public transport remains resilient in the face of changing urban dynamics (Mezher et al., 2021). Moreover, as cities increasingly adopt smart technologies, the integration of AI with broader urban planning initiatives will be crucial in shaping sustainable, interconnected urban environments.

In conclusion, AI has the potential to significantly shape greener, smarter cities. By optimizing public transportation systems and promoting sustainability, AI can enhance the overall quality of urban life. However, it is essential to remain vigilant about the ethical implications of these technologies, ensuring that they are implemented responsibly and equitably. As cities navigate the complexities of urbanization and climate change, the strategic use of AI in public transport will be pivotal in creating sustainable, livable communities for future generations.

REFERENCES

Abirami, S., Pethuraj, M., Uthayakumar, M., & Chitra, P. (2024). A systematic survey on big data and artificial intelligence algorithms for intelligent transportation system. *Case Studies on Transport Policy*, 17, 101247. DOI: 10.1016/j.cstp.2024.101247

Al-Ani, A. K., Laghari, S. U. A., Manoharan, H., Selvarajan, S., & Uddin, M. (2023). Improved Transportation Model with Internet of Things Using Artificial Intelligence Algorithm. *Computers, Materials & Continua/Computers. Computers, Materials & Continua*, 76(2), 2261–2279. DOI: 10.32604/cmc.2023.038534

Alhussen, A., & Ansari, A. S. (2024). Real-Time prediction of urban traffic problems based on Artificial Intelligence-Enhanced Mobile Ad Hoc Networks (MANETS). *Computers, Materials & Continua/Computers. Computers, Materials & Continua*, 79(2), 1903–1923. DOI: 10.32604/cmc.2024.049260

Ali, Y., Raadsen, M. P., & Bliemer, M. C. (2023). Modelling speed reduction behaviour on variable speed limit-controlled highways considering surrounding traffic pressure: A random parameters duration modelling approach. *Analytic Methods in Accident Research*, 40, 100290. DOI: 10.1016/j.amar.2023.100290

Almlöf, E., Nybacka, M., & Pernestål, A. (2020). Will public transport be relevant in a self-driving future? A demand model simulation of four scenarios for Stockholm, Sweden. *Transportation Research Procedia*, 49, 60–69. DOI: 10.1016/j.trpro.2020.09.006

Annual Report 2021. (n.d.). UNEP - UN Environment Programme. https://www.unep.org/resources/annual-report-2021

Anthony, B.Jnr. (2024). Artificial intelligence of things and distributed technologies as enablers for intelligent mobility services in smart Cities-A survey. *Internet of Things : Engineering Cyber Physical Human Systems*, 101399, 101399. Advance online publication. DOI: 10.1016/j.iot.2024.101399

Ap. (2023, August 19). San Francisco launches driverless bus service following robotaxi expansion. *The Economic Times*. https://economictimes.indiatimes.com/tech/technology/san-francisco-launches-driverless-bus-service-following-robotaxi-expansion/articleshow/102847234.cms?from=mdr

Beba, H., & Öztürk, Z. (2024). Investigation of models predicting NOx level in the sample region and the use of intelligent transportation system. *Environmental Challenges*, 16, 100990. DOI: 10.1016/j.envc.2024.100990

Behnia, F., Schuelke-Leech, B., & Mirhassani, M. (2024). Optimizing Sustainable Urban Mobility: A Comprehensive review of electric bus scheduling strategies and future directions. *Sustainable Cities and Society*, 108, 105497. DOI: 10.1016/j.scs.2024.105497

Bhavsar, P., Safro, I., Nidhal, B., Robi, P., Dera, D., Dutta, P., & Aminul, O. (2024). Artificial intelligence in transportation data analytics. In *Elsevier eBooks* (pp. 337–382). https://doi.org/DOI: 10.1016/B978-0-443-13878-2.00008-4

Bogdanov, V. (2023, July 10). *How machine learning is used in autonomous vehicles*. rinf.tech. https://www.rinf.tech/how-machine-learning-is-used-in-autonomous-vehicles/

Brynjolfsson, E., & McAfee, A. (2015). The second machine age: Work, progress, and prosperity in a time of brilliant technologies. *Choice (Chicago, Ill.)*, 52(06), 52–3201. DOI: 10.5860/CHOICE.184834

Cervero, R., & Murakami, J. (2010). Effects of Built Environments on Vehicle Miles Traveled: Evidence from 370 US Urbanized Areas. *Environment & Planning A*, 42(2), 400–418. DOI: 10.1068/a4236

Clements, L. M., & Kockelman, K. M. (2017). Economic effects of automated vehicles. *Transportation Research Record: Journal of the Transportation Research Board*, 2606(1), 106–114. DOI: 10.3141/2606-14

Deed - Attribution-NonCommercial-NoDerivatives 4.0 International - Creative Commons. (n.d.). https://creativecommons.org/licenses/by-nc-nd/4.0

Diop, P. A., Gueye, A. D., & Sall, A. (2024). Detecting road signs and deformations for road safety in Senegal. *Procedia Computer Science*, 237, 221–228. DOI: 10.1016/j.procs.2024.05.099

Duhayyim, M. A., Albraikan, A. A., Al-Wesabi, F. N., Burbur, H. M., Alamgeer, M., Hilal, A. M., Hamza, M. A., & Rizwanullah, M. (2021). Modeling of Artificial Intelligence Based Traffic Flow Prediction with Weather Conditions. *Computers, Materials & Continua/Computers. Materials & Continua (Print)*, 71(2), 3953–3968. DOI: 10.32604/cmc.2022.022692

Elassy, M., Al-Hattab, M., Takruri, M., & Badawi, S. (2024). Intelligent transportation systems for sustainable smart cities. *Transportation Engineering*, 16, 100252. DOI: 10.1016/j.treng.2024.100252

Elassy, M., Al-Hattab, M., Takruri, M., & Badawi, S. (2024). Intelligent transportation systems for sustainable smart cities. *Transportation Engineering*, 16, 100252. DOI: 10.1016/j.treng.2024.100252

Ghoushchi, S. J., Haghshenas, S. S., Vahabzadeh, S., Guido, G., & Geem, Z. W. (2024). An integrated MCDM approach for enhancing efficiency in connected autonomous vehicles through augmented intelligence and IoT integration. *Results in Engineering*, 23, 102626. DOI: 10.1016/j.rineng.2024.102626

Goldbach, C., Sickmann, J., Pitz, T., & Zimasa, T. (2021). Towards autonomous public transportation: Attitudes and intentions of the local population. *Transportation Research Interdisciplinary Perspectives*, 13, 100504. DOI: 10.1016/j.trip.2021.100504

Gössling, S., & Choi, A. S. (2015). Transport transitions in Copenhagen: Comparing the cost of cars and bicycles. *Ecological Economics*, 113, 106–113. DOI: 10.1016/j.ecolecon.2015.03.006

Graham, S., & Marvin, S. (1996). *Telecommunications and the city: Electronic Spaces, Urban Places*. Psychology Press.

Hawkins, T. R., Singh, B., Majeau-Bettez, G., & Strømman, A. H. (2012). Comparative Environmental Life cycle assessment of conventional and electric vehicles. *Journal of Industrial Ecology*, 17(1), 53–64. DOI: 10.1111/j.1530-9290.2012.00532.x

Hendawy, M., & Ghoz, L. (2024). A starting framework for urban AI applications. *Ain Shams Engineering Journal*, 102987(11), 102987. Advance online publication. DOI: 10.1016/j.asej.2024.102987

Herbon, A., & Hadas, Y. (2014). Determining optimal frequency and vehicle capacity for public transit routes: A generalized newsvendor model. *Transportation Research Part B: Methodological*, 71, 85–99. DOI: 10.1016/j.trb.2014.10.007

Hidalgo, D., & Huizenga, C. (2012). Implementation of sustainable urban transport in Latin America. *Research in Transportation Economics*, 40(1), 66–77. DOI: 10.1016/j.retrec.2012.06.034

Ho, C. Q., & Tirachini, A. (2023). Mobility-as-a-Service and the role of multimodality in the sustainability of urban mobility in developing and developed countries. *Transport Policy*, 145, 161–176. DOI: 10.1016/j.tranpol.2023.10.013

Home. (2024, October 29). https://www.who.int/

How will autonomous vehicles shape future urban mobility? (2024, October 16). World Economic Forum. https://www.weforum.org/agenda/2024/10/how-will-autonomous-vehicles-shape-urban-mobility/

IEA – International Energy Agency. (n.d.). IEA. https://www.iea.org/

Inam, M. (2024, July 10). *Example of a Smart City: A Case Study into Barcelona*. Minnovation Technologies. https://minnovation.com.au/smart-cities-2/example-of-a-smart-city-a-case-study-into-barcelona/

Iyer, L. S. (2021). AI enabled applications towards intelligent transportation. *Transportation Engineering*, 5, 100083. DOI: 10.1016/j.treng.2021.100083

Jeon, H., Lee, J., & Sohn, K. (2017). Artificial intelligence for traffic signal control based solely on video images. *Journal of Intelligent Transport Systems*, 22(5), 433–445. DOI: 10.1080/15472450.2017.1394192

Kalinsky, M., Sykora, M., Markova, J., & Maňas, P. (2024). Nonlinear dynamic finite element analysis of vehicle impacts into road restraint systems. *Results in Engineering*, 23, 102726. DOI: 10.1016/j.rineng.2024.102726

Kikuchi, S. (2009). Artificial intelligence in transportation analysis: Approaches, methods, and applications. *Transportation Research Part C, Emerging Technologies*, 17(5), 455. DOI: 10.1016/j.trc.2009.04.002

Manzolli, J. A., Trovão, J. P., & Antunes, C. H. (2022). A review of electric bus vehicles research topics – Methods and trends. *Renewable & Sustainable Energy Reviews*, 159, 112211. DOI: 10.1016/j.rser.2022.112211

McCarroll, C., & Cugurullo, F. (2022). No city on the horizon: Autonomous cars, artificial intelligence, and the absence of urbanism. *Frontiers in Sustainable Cities*, 4, 937933. Advance online publication. DOI: 10.3389/frsc.2022.937933

Miller, P., De Barros, A. G., Kattan, L., & Wirasinghe, S. C. (2016). Public transportation and sustainability: A review. *KSCE Journal of Civil Engineering*, 20(3), 1076–1083. DOI: 10.1007/s12205-016-0705-0

Morimoto, A. (2024). History and prospects of traffic enforcement for traffic safety in Japan. *IATSS Research*, 48(2), 129–135. DOI: 10.1016/j.iatssr.2024.02.004

Muratore, G., Vannelli, A., & Micheli, D. (2023). Rome vehicle accidents analysis and city riskiness prediction by Artificial Intelligence. *Transportation Engineering*, 12, 100172. DOI: 10.1016/j.treng.2023.100172

Niittymäki, J., & Pursula, M. (2001). The 11th Mini-EURO conference: Artificial intelligence on transportation systems and science. *European Journal of Operational Research*, 131(2), 229–231. DOI: 10.1016/S0377-2217(00)00122-3

Nikitas, A., Michalakopoulou, K., Njoya, E. T., & Karampatzakis, D. (2020). Artificial intelligence, transport and the smart City: Definitions and dimensions of a new mobility era. *Sustainability (Basel)*, 12(7), 2789. DOI: 10.3390/su12072789

Nishant, R., Kennedy, M., & Corbett, J. (2020). Artificial intelligence for sustainability: Challenges, opportunities, and a research agenda. *International Journal of Information Management*, 53, 102104. DOI: 10.1016/j.ijinfomgt.2020.102104

Olayode, O., Tartibu, L., & Okwu, M. (2020). Application of artificial intelligence in traffic control system of non-autonomous vehicles at signalized road intersection. *Procedia CIRP*, 91, 194–200. DOI: 10.1016/j.procir.2020.02.167

Pamucar, D., Deveci, M., Gokasar, I., Tavana, M., & Köppen, M. (2022). A metaverse assessment model for sustainable transportation using ordinal priority approach and Aczel-Alsina norms. *Technological Forecasting and Social Change*, 182, 121778. DOI: 10.1016/j.techfore.2022.121778

Papageorgiou, G., Tsappi, E., & Wang, T. (2024). Smart Urban Systems Planning for active mobility and sustainability. *IFAC-PapersOnLine*, 58(10), 261–266. DOI: 10.1016/j.ifacol.2024.07.350

Rasheed, F., Yau, K. A., Noor, R. M., Wu, C., & Low, Y. (2020). Deep Reinforcement Learning for Traffic Signal Control: A review. *IEEE Access : Practical Innovations, Open Solutions*, 8, 208016–208044. DOI: 10.1109/ACCESS.2020.3034141

Ren, S., & Li, Y. (2023). A review of high performance computing applications in high-speed rail systems. *High-speed Railway*, 1(2), 92–96. DOI: 10.1016/j.hspr.2023.05.001

Rodrigue, J. (2024). *The geography of transport systems*. https://doi.org/DOI: 10.4324/9781003343196

Shaygan, M., Meese, C., Li, W., Zhao, X., & Nejad, M. (2022). Traffic prediction using artificial intelligence: Review of recent advances and emerging opportunities. *Transportation Research Part C, Emerging Technologies*, 145, 103921. DOI: 10.1016/j.trc.2022.103921

Shaygan, M., Meese, C., Li, W., Zhao, X., & Nejad, M. (2022). Traffic prediction using artificial intelligence: Review of recent advances and emerging opportunities. *Transportation Research Part C, Emerging Technologies*, 145, 103921. DOI: 10.1016/j.trc.2022.103921

Supporting the global shift to electric mobility. (n.d.). UNEP - UN Environment Programme. https://www.unep.org/topics/transport/electric-mobility/supporting-global-shift-electric-mobility

Tang, R., De Donato, L., Bešinović, N., Flammini, F., Goverde, R. M. P., Lin, Z., Liu, R., Tang, T., Vittorini, V., & Wang, Z. (2022). A literature review of Artificial Intelligence applications in railway systems. *Transportation Research Part C, Emerging Technologies*, 140, 103679. DOI: 10.1016/j.trc.2022.103679

Tiwari, P. (2023). The machine learning framework for traffic management in smart cities. *Management of Environmental Quality*, 35(2), 445–462. DOI: 10.1108/MEQ-08-2022-0242

Tokody, D., Ady, L., Hudasi, L. F., Varga, P. J., & Hell, P. (2020). Collaborative Robotics Research: Subiko Project. *Procedia Manufacturing*, 46, 467–474. DOI: 10.1016/j.promfg.2020.03.068

Urban development. (n.d.). World Bank. https://www.worldbank.org/en/topic/urbandevelopment

Ushakov, D., Dudukalov, E., Shmatko, L., & Shatila, K. (2022a). Artificial Intelligence as a factor of public transportations system development. *Transportation Research Procedia*, 63, 2401–2408. DOI: 10.1016/j.trpro.2022.06.276

Ushakov, D., Dudukalov, E., Shmatko, L., & Shatila, K. (2022b). Artificial Intelligence as a factor of public transportations system development. *Transportation Research Procedia*, 63, 2401–2408. DOI: 10.1016/j.trpro.2022.06.276

White Paper on Artificial Intelligence: a European approach to excellence and trust | European Commission. (n.d.). European Commission. https://commission.europa.eu/document/d2ec4039-c5be-423a-81ef-b9e44e79825b_en

WorldBank. (n.d.). *World Bank Group - International Development, Poverty and Sustainability*. https://www.worldbank.org/ext/en/home

Wu, J., Wang, X., Dang, Y., & Lv, Z. (2022). Digital twins and artificial intelligence in transportation infrastructure: Classification, application, and future research directions. *Computers & Electrical Engineering*, 101, 107983. DOI: 10.1016/j.compeleceng.2022.107983

Wu, Y., Fu, G., Wu, Z., Wang, Y., Xie, X., Han, M., & Lyu, Q. (2022). A popular systemic accident model in China: Theory and applications of 24Model. *Safety Science*, 159, 106013. DOI: 10.1016/j.ssci.2022.106013

Xie, X., Fu, G., Xue, Y., Zhao, Z., Chen, P., Lu, B., & Jiang, S. (2018). Risk prediction and factors risk analysis based on IFOA-GRNN and apriori algorithms: Application of artificial intelligence in accident prevention. *Process Safety and Environmental Protection*, 122, 169–184. DOI: 10.1016/j.psep.2018.11.019

Yang, B. (2023). Research on vehicle detection and recognition technology based on artificial intelligence. *Microprocessors and Microsystems*, 104937, 104937. Advance online publication. DOI: 10.1016/j.micpro.2023.104937

Yang, J., Liang, W., Wang, X., Li, S., Jiang, X., Mu, Y., & Zeng, S. (2024). DarkMor: A framework for darknet traffic detection that integrates local and spatial features. *Neurocomputing*, 607, 128377. DOI: 10.1016/j.neucom.2024.128377

Zang, X., Wang, T., Zhang, X., Gong, J., Gao, P., & Zhang, G. (2024). Encrypted malicious traffic detection based on natural language processing and deep learning. *Computer Networks*, 250, 110598. DOI: 10.1016/j.comnet.2024.110598

Zunder, T. H., Carrillo, A., & Robinson, M. (2023). Consensus and dissent for rail visions of 2050: Insights from a Delphi study. *Transportation Research Procedia*, 72, 844–851. DOI: 10.1016/j.trpro.2023.11.476

Chapter 6

Blockchain for Disaster Recovery:

Enhancing Resilience, Coordination, and Trust in Crisis Management

G. Sowmya
https://orcid.org/0000-0002-9275-7726
MLR Institute of Technology, India

R. Sridevi
Jawaharlal Nehru Technological University, India

K. S. Sadasiva Rao
Sri Indu College of Engineering and Technology, India

ABSTRACT

This chapter explores how blockchain technology can transform disaster recovery by enhancing resilience, improving coordination, and fostering trust. Blockchain's decentralized and secure nature enables real-time data sharing, transparent resource distribution, and improved accountability among stakeholders. By automating processes through smart contracts and providing an immutable ledger, blockchain addresses key challenges in crisis management. The chapter also discusses real-world applications, potential challenges, and future opportunities for integrating blockchain with other emerging technologies to strengthen disaster recovery efforts.

1. INTRODUCTION

Disasters, both natural and man-made, pose significant challenges to crisis management and recovery efforts. The complexity of coordinating emergency services, humanitarian aid, and the distribution of critical resources often leads to delays, miscommunication, and inefficiencies. Traditional disaster recovery systems are prone to bottlenecks, centralization vulnerabilities, and a lack of transparency, resulting in

DOI: 10.4018/979-8-3693-9770-1.ch006

resource misallocation and decreased trust from affected populations. In an increasingly interconnected world, the need for faster, more resilient, and transparent systems to respond to crises is paramount.

Technological advancements have played a pivotal role in improving disaster management by offering tools to enhance communication, resource tracking, and real-time decision-making. Technologies such as cloud computing, big data, and the Internet of Things (IoT) have already made significant contributions. However, the rise of blockchain technology introduces new possibilities for improving resilience and coordination in disaster recovery. Blockchain's ability to create secure, decentralized, and transparent data-sharing networks provides a potential game-changer in responding to crises, ensuring that all stakeholders have access to the same, reliable information (Hunt & Zhuang, 2022).

Blockchain is a distributed ledger technology that ensures data security, transparency, and immutability by maintaining decentralized records across multiple nodes. Unlike traditional centralized databases, blockchain records every transaction in a tamper-proof, chronological manner, making it highly resistant to fraud or alteration. This makes blockchain particularly suitable for environments where trust and transparency are essential, such as disaster recovery. Blockchain enables secure data sharing, automates processes through smart contracts, and allows for real-time collaboration between different agencies and organizations during crises.

This chapter aims to explore the potential of blockchain technology in transforming disaster recovery by enhancing resilience, improving coordination between stakeholders, and fostering trust in crisis management. It will discuss the key features of blockchain that make it well-suited for crisis scenarios, offer examples of its application in disaster recovery efforts, and highlight the challenges and future opportunities of integrating blockchain into disaster management frameworks. By examining real-world use cases and addressing both technical and ethical considerations, this chapter provides a comprehensive view of how blockchain can improve disaster preparedness, response, and recovery.

2. BLOCKCHAIN FUNDAMENTALS FOR CRISIS MANAGEMENT

2.1 Key Features of Blockchain

Blockchain technology is characterized by several fundamental features that significantly enhance its applicability in crisis management:

a. Decentralization: Unlike traditional systems that rely on a single point of control, blockchain operates on a decentralized network of nodes. This decentralization reduces the risk of systemic failures and ensures that no single entity can manipulate or compromise the data, making it particularly robust in crisis situations.
b. Transparency: Blockchain provides a transparent ledger that is accessible to all participants within the network. This transparency fosters accountability among stakeholders, allowing all parties to verify transactions and the flow of resources in real time. In disaster recovery, this transparency helps build trust among governments, NGOs, and affected communities.

Immutability: Once data is recorded on a blockchain, it becomes virtually impossible to alter or delete. This immutability is crucial for maintaining the integrity of critical records, such as aid distribution logs, which must remain accurate to ensure proper accountability and trustworthiness during recovery efforts (Tripathi et al., 2023).

a. Security: Blockchain employs cryptographic methods to secure data, protecting it from unauthorized access and tampering. Each transaction is encrypted and linked to the previous one, forming a chain that is resistant to attacks. This enhanced security is vital when managing sensitive information during disasters.

2.2 Relevance of Blockchain in the Context of Disaster Recovery

The unique features of blockchain make it highly relevant for disaster recovery initiatives. In crisis situations, the ability to coordinate efforts among various stakeholders—governments, non-profit organizations, and local communities—is critical. Blockchain addresses several challenges inherent in traditional disaster management systems, including:

a. Improved Coordination: By enabling real-time data sharing and visibility across organizations, blockchain enhances collaboration among agencies and reduces information silos, leading to more efficient resource allocation.
b. Enhanced Trust: The transparency and immutability of blockchain foster trust among stakeholders, as all parties can verify actions taken during the disaster recovery process. This trust is vital for effective cooperation and community engagement (Treiblmaier & Rejeb, 2023)
c. Streamlined Processes: By reducing the need for intermediaries and manual verification, blockchain can expedite decision-making and response times, ensuring that aid and resources are delivered swiftly to those in need.

2.3 Smart Contracts for Automating Crisis Responses

Smart contracts are self-executing agreements with the terms of the contract directly written into code on the blockchain. In the context of disaster recovery, smart contracts can automate various processes, enhancing efficiency and accuracy. The Figureure 1 shows Smart contract workflow in Disaster Scenarios.

Figure 1. Smart contract workflow in Disaster Systems

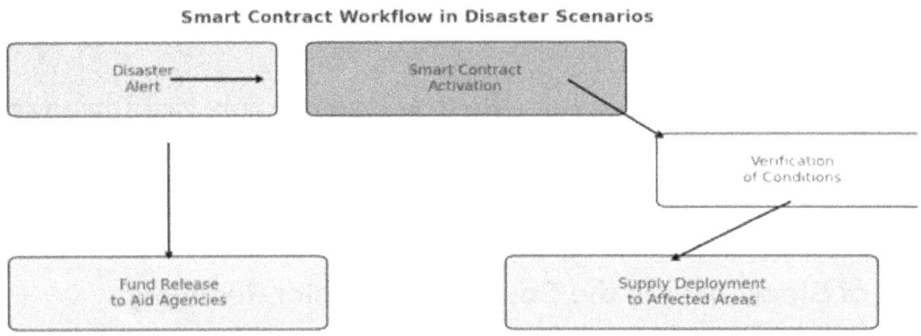

a. Aid Distribution Automation: Smart contracts can facilitate the automatic release of funds or resources based on predefined conditions, such as confirming the arrival of relief supplies at designated locations. This minimizes delays and human error in the aid distribution process.
b. Insurance Claims Processing: In disaster scenarios, smart contracts can automate the verification of claims, such as those related to property damage. For instance, if a disaster occurs, the smart contract can automatically trigger payments to policyholders without requiring extensive paperwork or manual assessments. (Khare et al., 2024)
c. Resource Coordination: Smart contracts can help manage the allocation of resources during a disaster by executing predefined actions, such as deploying emergency personnel or supplies when certain conditions are met (e.g., a specific disaster alert).

By automating critical processes, smart contracts streamline disaster response efforts and allow for quicker, more effective interventions.

2.4 Use of Distributed Ledgers for Real-Time Data Sharing

Distributed ledgers enable secure and instantaneous data sharing among all participants in the blockchain network, which is essential in crisis management:

a. Real-Time Visibility: Stakeholders can access real-time information regarding resource availability, needs, and the status of recovery efforts. This visibility helps ensure that aid is directed to areas of greatest need and that resources are utilized efficiently.
b. Improved Supply Chain Management: Blockchain can track the movement of supplies and resources throughout the recovery process, providing transparency in the logistics chain. This ensures that stakeholders can monitor the flow of aid from donors to recipients without delays or mismanagement.
c. Health Data Exchange: In the event of public health emergencies, blockchain can facilitate secure sharing of health records and data among healthcare providers, ensuring timely and accurate treatment for affected individuals while protecting patient privacy. (Li et al., 2023)
d. By leveraging distributed ledgers for real-time data sharing, disaster recovery efforts can become more responsive and effective, ultimately improving outcomes for communities impacted by crises.

3. ENHANCING RESILIENCE IN DISASTER RECOVERY

3.1 Decentralized Data Management for Minimizing Single Points of Failure

Decentralization is a core principle of blockchain technology that plays a crucial role in enhancing resilience during disaster recovery. By distributing data across a network of nodes, blockchain eliminates the risk associated with centralized data management, where a single point of failure can lead to catastrophic losses. In the context of disaster recovery:

a. Redundancy: Decentralized data storage ensures that critical information is replicated across multiple locations. If one node fails due to a disaster, others continue to function, providing a reliable backup of essential data.
b. Accessibility: In emergency situations, decentralized systems allow stakeholders to access data from various locations, ensuring continuity of operations. This is particularly important for coordination among emergency responders, relief organizations, and local authorities. (Haque et al., 2024)
c. Resilience to Cyberattacks: Centralized systems are often prime targets for cyberattacks. By decentralizing data, blockchain minimizes the potential impact of such attacks, enhancing overall system resilience.

3.2 Blockchain-Enabled Supply Chain Management for Disaster Relief

Efficient supply chain management is vital in disaster recovery to ensure that resources reach affected areas quickly and effectively. Blockchain technology enhances supply chain operations in several ways:

a. Traceability: Blockchain provides an immutable record of the entire supply chain process, from donor contributions to the delivery of supplies. This traceability ensures that all parties can verify the source and journey of relief materials, enhancing accountability.
b. Streamlined Logistics: By utilizing smart contracts, organizations can automate various logistics processes, such as procurement, transportation, and distribution of supplies. This automation reduces delays and ensures timely delivery of essential goods.

c. Collaboration Among Stakeholders: Blockchain facilitates real-time information sharing between suppliers, logistics providers, and disaster response teams. This collaborative environment ensures that all stakeholders are informed of supply status, inventory levels, and needs, leading to more effective responses. (Hellwig et al., 2022)

3.3 Securing and Maintaining Integrity of Critical Infrastructure Data (Health, Power, Communication)

During disasters, the integrity of critical infrastructure data is paramount for effective recovery. Blockchain enhances the security and reliability of this data in various sectors:

a. Healthcare Data Security: In crisis situations, accurate health records are essential for treatment and recovery efforts. Blockchain can secure patient data, ensuring that it is accurate and accessible to authorized personnel while protecting patient privacy.
b. Power and Utility Management: Blockchain can improve the management of power grids and utility services during disasters by providing real-time visibility into infrastructure status. This visibility allows for more effective response strategies and resource allocation.
c. Communication Systems: Reliable communication is vital during emergencies. Blockchain can secure communication networks, ensuring that critical information is transmitted without tampering or interference. (Zarour et al., 2021)

3.4 Disaster Management Data and Technology Applications

Disaster management leverages emerging technologies like IoT, AI, and blockchain to minimize the impact of crises.

a. Natural Disasters: The UN reports over 400 disasters annually, affecting 200+ million people worldwide. Economic losses amount to $200 billion per year.
b. Climate Change: Rising sea levels and unpredictable weather patterns have led to more frequent and intense disasters.

Technological Applications

a. IoT in Early Warning Systems:

Japan uses IoT sensors to monitor seismic activity and provide real-time tsunami alerts.
Impact: Early warnings have saved thousands of lives by enabling timely evacuations.

b. AI for Disaster Predictions:

AI-powered tools analyze satellite imagery and weather data to predict floods or hurricanes. For instance, IBM's Watson has been used in flood prediction.
Impact: Improved resource allocation and timely disaster response.

c. Blockchain in Resource Management:

Distributed ledgers track supply chains during disasters, ensuring efficient allocation of food, medical supplies, and shelter materials.

Example: After Nepal's earthquake in 2015, blockchain helped trace international aid, improving transparency and accountability.

3.5 Real-Time Monitoring and Reporting Using Blockchain-Based IoT

The integration of blockchain with Internet of Things (IoT) technology offers powerful tools for real-time monitoring and reporting during disaster recovery. The Figureure 2 shows the architecture for Disaster Management.

Figure 2. Blockchain system architecture for Disaster Management

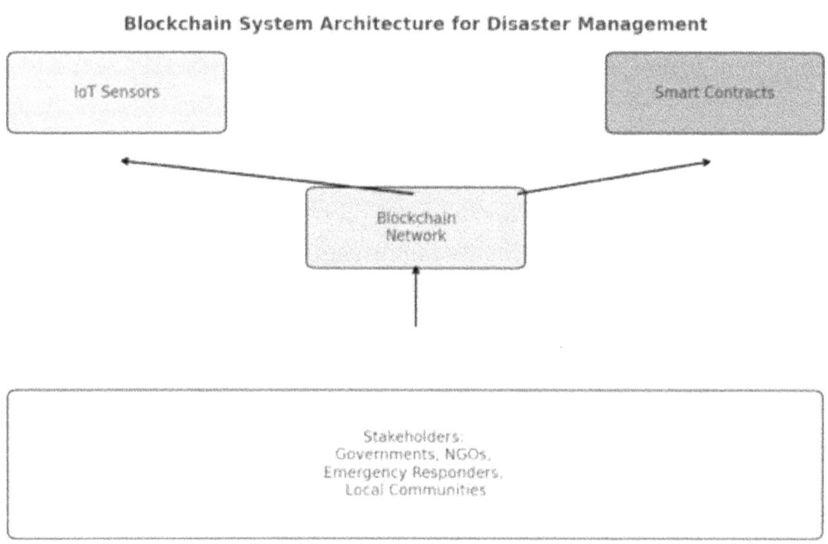

a. IoT Sensors for Data Collection: IoT devices can collect real-time data from the field, such as environmental conditions, resource usage, and infrastructure status. This data can be securely recorded on the blockchain, providing an accurate and immutable record.
b. Enhanced Situational Awareness: By combining blockchain with IoT, stakeholders can gain real-time insights into disaster conditions. This situational awareness enables faster decision-making and resource allocation, improving response effectiveness. (Zuo & Qi, 2022)Automated Reporting: Smart

contracts can automate reporting processes, ensuring that relevant data is shared with stakeholders without delay. This real-time reporting is crucial for keeping all parties informed and coordinating response efforts.

By leveraging blockchain and IoT for real-time monitoring and reporting, disaster recovery efforts can be more agile, responsive, and effective, ultimately leading to improved outcomes for affected communities.

4. IMPROVING COORDINATION AMONG STAKEHOLDERS

4.1 Blockchain for Inter-Agency Collaboration and Data Sharing

Effective disaster recovery depends heavily on seamless collaboration and communication among a wide range of stakeholders, including government agencies, non-governmental organizations (NGOs), humanitarian aid organizations, and private sector partners. Traditional centralized systems often hinder coordination due to data silos, inefficiencies, and the lack of transparency. Blockchain technology, with its decentralized and transparent nature, offers a solution to these challenges:

a. Unified Data Access: Blockchain enables all stakeholders to access and share the same real-time data, eliminating information silos. This shared data platform ensures that all parties are working from a single source of truth, enhancing coordination and preventing miscommunication.
b. Permissioned Blockchains: In cases where sensitive information is involved, permissioned blockchains allow certain stakeholders to control who can access or contribute to specific data sets, while still maintaining overall transparency and accountability within the system. (Wang & Liu, 2023)
c. Auditability and Accountability: Blockchain's immutable ledger provides a verifiable and transparent record of every action taken by stakeholders, which enhances accountability in disaster relief efforts and promotes trust among participants.

4.2 Smart Contracts for Automating Relief Distribution

In disaster recovery, the speed and accuracy of resource allocation and aid distribution are critical to saving lives and rebuilding communities. Smart contracts can automate these processes, ensuring that predefined conditions trigger immediate actions, such as the release of funds or supplies. Here's how smart contracts improve coordination:

a. Automated Aid Distribution: Smart contracts can release aid automatically when certain conditions are met, such as when a disaster is confirmed by a trusted source (e.g., weather agencies, geospatial data). This eliminates delays caused by bureaucratic processes and ensures that resources are deployed as quickly as possible.
b. Efficient Resource Allocation: Smart contracts can allocate resources based on real-time data inputs. For example, they can ensure that supplies are sent to the areas most in need, based on up-to-date damage assessments or population data collected via IoT devices. (Sharma et al., 2023)

c. Reduced Fraud and Corruption: By automating aid distribution through smart contracts, blockchain reduces the potential for corruption or mismanagement, ensuring that aid reaches its intended recipients.

4.3 Blockchain for Managing Global Disaster Recovery Efforts and Partnerships

Global disaster recovery efforts often involve numerous stakeholders from different countries, sectors, and organizations. Blockchain's decentralized structure allows for more efficient management of these complex partnerships:

a. Cross-Border Collaboration: Blockchain's ability to operate without a central controlling authority makes it ideal for coordinating international efforts. For instance, different countries and organizations can securely share data and collaborate on relief efforts without relying on a single centralized platform.
b. Partnership Tracking and Accountability: Blockchain can track the contributions of different partners in a disaster recovery effort, providing transparency on who provided what resources and ensuring that commitments are fulfilled. (Treiblmaier & Rejeb, 2023)
c. Global Fund Management: Blockchain can facilitate the management of global disaster relief funds by ensuring that all donations are tracked and distributed according to predefined agreements. This transparency fosters trust among donors and helps ensure that funds are used appropriately.

4.4 Real-World Examples of Blockchain in Crisis Coordination

Several real-world applications of blockchain technology in disaster recovery demonstrate its potential to enhance coordination and response efforts:

a. UN World Food Programme (WFP) – Building Blocks: The UN WFP has used blockchain to improve the efficiency and transparency of its cash-based assistance programs for refugees. Through the "Building Blocks" blockchain platform, the WFP has been able to track food assistance distributed to Syrian refugees, ensuring that aid reaches those in need and reducing overhead costs associated with traditional distribution methods.
b. Red Cross – Blockchain for Disaster Response: The Red Cross has experimented with blockchain for distributing aid in disaster-affected areas. By using blockchain, they can track the distribution of resources and ensure transparency and accountability, while providing beneficiaries with secure and reliable access to relief services. (Wang & Chen, 2022). Hurricane Recovery in Puerto Rico: After Hurricane Maria, blockchain was explored to improve the speed and transparency of disaster relief and recovery efforts in Puerto Rico. Blockchain applications were used for distributing financial aid and tracking the rebuilding of critical infrastructure.
c. Estonia: Digital Governance - Estonia has fully integrated blockchain into its e-Governance system. This ensures secure, tamper-proof storage of government records, including health data, judicial systems, and tax information. Citizens and businesses benefit from transparent, corruption-free governance. The system's resilience ensures continuity during crises, such as cyberattacks or natural disasters.

d. United States: Hurricane Harvey Relief - After Hurricane Harvey in 2017, blockchain was used to track donations and ensure transparent disbursement of funds. Increased donor trust and minimized misuse of funds. Relief operations were expedited by reducing paperwork and ensuring accountability.
e. United Nations WFP: Building Blocks - The UN World Food Programme used blockchain to streamline aid distribution in Syrian refugee camps. By eliminating intermediaries and using blockchain for transactions, the program ensured food supplies reached beneficiaries directly. It has cost savings, fraud reduction, and enhanced transparency.

These examples illustrate how blockchain can revolutionize disaster response by improving coordination among multiple stakeholders, ensuring that resources are distributed efficiently and equitably, and enhancing transparency and accountability in global partnerships.

5. BUILDING TRUST AND ACCOUNTABILITY

5.1 Transparent Transaction Records for Donations and Resource Management

In disaster recovery, ensuring transparency in the management of donations and resources is critical to maintaining trust among donors, aid recipients, and the public. Blockchain technology provides an immutable ledger that records every transaction in a transparent and auditable manner:

a. Donation Tracking: Blockchain allows donors to track their contributions from the moment they are made until they reach the intended beneficiaries. Each donation is recorded on the blockchain, providing full visibility into how funds are being allocated and used. This transparency reassures donors that their contributions are being properly utilized and not diverted for unintended purposes. (Farooq et al., 2020)
b. Resource Allocation: Blockchain can also track the distribution of physical resources, such as food, medical supplies, and shelter materials. By using blockchain, relief organizations can provide a transparent trail of resource allocation, ensuring that resources are delivered to the right locations and used efficiently.

5.2 Enhancing Citizen Trust Through Tamper-Proof Communication Channels

In crisis situations, maintaining accurate and reliable communication between authorities and citizens is crucial. Misinformation or lack of transparency can lead to confusion and erosion of public trust. Blockchain can improve this by offering tamper-proof communication channels:

a. Verified Information Sharing: Blockchain-based platforms allow government agencies, NGOs, and local authorities to share verified and immutable information with affected populations. By ensuring that all communications are recorded on a tamper-proof ledger, blockchain helps prevent the spread of false information and ensures that citizens receive accurate updates on recovery efforts. (Mahmood, 2016)

b. Citizen Engagement: Blockchain can also be used to engage citizens directly in the recovery process. For example, individuals can use blockchain platforms to report damage, request assistance, or provide feedback on recovery efforts, knowing that their communications are secure and unalterable. This fosters a sense of inclusion and trust between citizens and authorities.

5.3 Reducing Fraud and Corruption in Disaster Aid Distribution Using Blockchain

Fraud and corruption are significant challenges in disaster aid distribution, as funds and resources can be diverted, misallocated, or misreported. Blockchain helps address these challenges by providing a secure, transparent, and tamper-proof system for managing aid distribution:

a. Immutable Records: Every step of the aid distribution process is recorded on the blockchain, from the disbursement of funds to the delivery of resources. This ensures that all transactions are verifiable and cannot be altered or hidden. In this way, blockchain reduces opportunities for fraudulent activities and enhances accountability.
b. Smart Contracts for Aid Disbursement: Smart contracts can automate the distribution of aid based on pre-established criteria, such as the verified need of affected populations. This automation minimizes human intervention and reduces the risk of corruption or favoritism in aid allocation. (Hunt & Zhuang, 2022)
c. Auditable Processes: Blockchain allows for independent audits of aid distribution, ensuring that all parties involved are held accountable for their actions. This enhances the credibility of relief organizations and fosters trust among beneficiaries and donors alike.

5.4 Use Cases of Blockchain-Based Trust Systems in Previous Disaster Scenarios

Several real-world examples illustrate the effectiveness of blockchain in building trust and accountability in disaster recovery efforts:

a. Hurricane Relief in the United States: After Hurricane Harvey, blockchain technology was explored to track the distribution of relief funds and ensure that they reached the intended beneficiaries. By using blockchain, relief organizations were able to create an auditable trail of donations and disbursements, reducing concerns about mismanagement and ensuring accountability.
b. Nepal Earthquake Recovery: Following the devastating earthquake in Nepal, blockchain technology was proposed as a solution to track the flow of international aid. The use of blockchain allowed donors to monitor how their contributions were being used and ensured that aid was distributed fairly and transparently. (Treiblmaier & Rejeb, 2023)
c. Red Cross Blockchain Pilots: The Red Cross has been experimenting with blockchain to improve the efficiency and transparency of aid distribution in disaster-affected regions. In their pilot programs, blockchain was used to track the flow of financial aid and supplies, ensuring that resources reached those in need without delays or corruption.

These examples demonstrate how blockchain can be effectively used to build trust and ensure accountability in disaster recovery. By providing transparent records, reducing opportunities for fraud, and enhancing citizen engagement, blockchain helps create a more reliable and trustworthy disaster recovery system.

6. CASE STUDIES OF BLOCKCHAIN IN DISASTER RECOVERY

6.1 Analysis of Specific Case Studies Where Blockchain Has Been Implemented for Disaster Response

a. UN World Food Programme's (WFP) "Building Blocks" Initiative:

The WFP implemented blockchain technology to enhance the efficiency of aid distribution in refugee camps. Initially deployed in Jordan to assist Syrian refugees, the "Building Blocks" program allows the WFP to distribute cash-based transfers securely. Refugees use biometric data (iris scans) to authenticate transactions and access food and supplies, while blockchain ensures that each transaction is transparent and immutable.

Impact: The blockchain platform reduced transaction costs, improved transparency, and minimized fraud, ensuring that aid reached the right beneficiaries. The system provided an efficient, secure, and accountable way to manage aid distribution in a high-stakes disaster environment.

b. Hurricane Harvey Relief (USA):

Following the devastation caused by Hurricane Harvey in 2017, blockchain was explored as a tool to track and manage donations and relief efforts. The platform allowed donors to see where their funds were going and how they were being spent in real-time. This transparency helped to address concerns over fraud and misallocation of funds.

Impact: Blockchain's transparency built trust among donors, who could trace their contributions to relief efforts on the ground. It also improved the efficiency of relief operations by reducing paperwork and facilitating faster transactions.

c. Nepal Earthquake (2015) Relief:

In the wake of the 2015 earthquake in Nepal, blockchain was proposed as a solution for managing the flow of international aid. Through blockchain, it was possible to track the distribution of resources, ensuring that they were reaching the most affected regions. Blockchain provided a transparent audit trail that could be accessed by all stakeholders.

Impact: While the blockchain implementation was limited, it demonstrated the potential to improve the accountability and efficiency of aid distribution in a disaster-prone region. The system showed how blockchain could address corruption and mismanagement issues prevalent in traditional aid delivery systems.

6.2 Lessons Learned and Impact Assessment

From these case studies, several lessons have emerged regarding the use of blockchain in disaster recovery:

a. Transparency and Trust: One of the key benefits of blockchain in disaster recovery is its ability to enhance transparency. Donors, relief organizations, and beneficiaries can all view the same immutable records, fostering trust and reducing concerns about fraud or mismanagement.
b. Efficiency Gains: Blockchain reduces the reliance on intermediaries and manual processes, which often slow down disaster recovery efforts. By automating transactions, verifying identity through biometric data, and enabling smart contracts, blockchain allows for faster response times and resource deployment.
c. Scalability Challenges: While blockchain shows promise, some challenges remain, particularly in terms of scalability and integration with existing systems. In large-scale disasters, blockchain platforms need to handle a high volume of transactions without sacrificing speed or security.
d. Adoption and Awareness: Blockchain is still a relatively new technology, and many relief organizations and governments are in the early stages of exploring its potential. Broader adoption will require education, stakeholder engagement, and partnerships with technology providers to overcome the learning curve.

6.3 Global Perspectives and Emerging Trends in Blockchain-Driven Recovery Strategies

a. Increased Adoption by International Organizations: As demonstrated by the WFP, global organizations are increasingly adopting blockchain for disaster response. The success of pilot projects is encouraging more agencies to explore the potential of blockchain to address logistical and transparency challenges in crisis situations.
b. Integration with IoT for Real-Time Monitoring: Emerging trends point to the integration of blockchain with Internet of Things (IoT) technology. By combining IoT sensors with blockchain, organizations can monitor real-time conditions, such as environmental changes or the movement of relief supplies, and record this data on a secure, decentralized ledger. (Gad et al., 2022)
c. Decentralized Aid Management Platforms: There is growing interest in decentralized platforms that allow multiple organizations to collaborate on disaster recovery efforts. Blockchain enables real-time data sharing and decision-making among stakeholders, which is crucial for coordinating responses in complex, multi-agency disaster scenarios.
d. Insurance and Risk Management: Blockchain is being used to automate insurance claims in the wake of disasters. By leveraging smart contracts, insurers can expedite the claims process, providing immediate relief to those affected by disasters. This trend is particularly evident in the rise of parametric insurance models, which rely on predefined triggers (such as rainfall levels or seismic data) to automatically initiate payments.

These case studies and emerging trends highlight how blockchain is reshaping disaster recovery, from improving coordination and transparency to automating resource distribution and integrating with cutting-edge technologies like IoT. As the technology continues to evolve, its role in enhancing resilience and accountability in disaster recovery will likely expand globally.

7. CHALLENGES AND LIMITATIONS

7.1 Technical, Regulatory, and Adoption Barriers

While blockchain holds great promise for disaster recovery, several barriers hinder its widespread adoption:

a. Technical Barriers:

i. Interoperability: One of the key technical challenges is the lack of interoperability between different blockchain platforms and existing systems used by governments and aid organizations. Ensuring seamless integration of blockchain with legacy systems is crucial for smooth operations during disaster recovery.
ii. Complexity: Blockchain systems, particularly those involving smart contracts and decentralized networks, can be technically complex. Organizations may lack the technical expertise required to deploy and manage blockchain solutions effectively.

b. Regulatory Barriers:

i. Unclear Regulations: The regulatory environment for blockchain remains uncertain in many parts of the world. Lack of clear policies on the use of blockchain in disaster recovery can slow down its adoption by governments and humanitarian organizations.
ii. Cross-Border Coordination: For international disaster recovery efforts, regulatory inconsistencies across different countries can pose challenges. Blockchain systems may face legal complications when dealing with data privacy laws, financial regulations, and cross-border data transfers.

c. Adoption Barriers:

i. Resistance to Change: Established organizations may be resistant to adopting new technologies like blockchain, especially in high-stress environments like disaster recovery. Trust in traditional systems and unfamiliarity with blockchain may result in reluctance to implement such solutions.
ii. Cost and Infrastructure Requirements: Implementing blockchain requires significant investment in infrastructure, technology, and expertise. For many developing countries or resource-strapped relief organizations, these costs can be prohibitive.

7.2 Data Privacy Concerns and Solutions in a Blockchain-Enabled Crisis Environment

The use of blockchain in disaster recovery raises critical data privacy concerns, especially when sensitive information like health records, personal identities, or financial transactions are involved:

a. Privacy Risks:

i. Immutable Data: While blockchain's immutability is a strength in terms of transparency and security, it also poses challenges for data privacy. Once data is recorded on a blockchain, it cannot be altered or deleted, which raises concerns about long-term privacy protection.
ii. Personal Data Exposure: If not properly managed, blockchain can expose personal data to unauthorized parties, especially in open or public blockchains where anyone can view transaction records.

b. Solutions:

i. Permissioned Blockchains: One solution to address privacy concerns is to use permissioned blockchains, which restrict access to authorized participants only. This allows organizations to maintain the benefits of blockchain's transparency while protecting sensitive information from public exposure.
ii. Encryption and Zero-Knowledge Proofs: Advanced cryptographic techniques, such as encryption and zero-knowledge proofs, can be used to secure sensitive data on the blockchain while still allowing for verification of information without revealing the actual data.

7.3 Scalability and Energy Consumption Issues

Scalability and energy consumption are critical issues, especially when considering blockchain for large-scale disaster recovery operations:

a. Scalability:

i. Transaction Bottlenecks: Most traditional blockchain platforms, such as Bitcoin and Ethereum, face significant scalability issues. They can handle only a limited number of transactions per second, which may be insufficient for high-volume disaster recovery operations involving multiple stakeholders and real-time data flows.
ii. Network Congestion: During disasters, the demand for blockchain-based services can surge, potentially overwhelming the network and causing delays in transaction processing. This bottleneck can hamper time-sensitive relief efforts.

b. Energy Consumption:

i. Proof of Work (PoW): Many blockchains rely on energy-intensive consensus mechanisms like PoW, which consumes large amounts of computational power. This is particularly problematic in disaster recovery situations, where resources and infrastructure are already stretched thin.
ii. Greener Alternatives: Transitioning to more energy-efficient consensus mechanisms, such as Proof of Stake (PoS) or Proof of Authority (PoA), can mitigate these concerns. These alternatives reduce the energy consumption associated with blockchain while maintaining security and decentralization.

7.4 Ethical Considerations and Inclusive Adoption of Blockchain Technologies

The deployment of blockchain in disaster recovery raises important ethical considerations, especially regarding the inclusion of vulnerable populations:

a. Inclusive Adoption:

i. Access to Technology: Blockchain systems often require internet access, digital devices, and a certain level of technological literacy, which may not be available to everyone in disaster-affected areas. Ensuring that marginalized communities, including those in remote or low-income regions, have access to blockchain-enabled services is a key ethical concern.
ii. Digital Divide: There is a risk that blockchain solutions could exacerbate existing inequalities, as those with better access to technology will benefit more than those without it. Strategies for inclusive adoption should prioritize bridging the digital divide to ensure equitable access to disaster recovery services.

b. Ethical Use of Data:

i. Consent and Autonomy: Blockchain systems that collect and store personal data must ensure that individuals give informed consent for the use of their information. In crisis situations, where people may not fully understand or have time to consider the implications of sharing their data, this becomes a critical ethical issue.
ii. Beneficiary Protection: Blockchain systems should prioritize the protection of the most vulnerable populations, ensuring that they do not become victims of data exploitation or unauthorized use of their personal information.

While blockchain offers numerous advantages in disaster recovery, addressing its technical, regulatory, and ethical challenges is essential for ensuring responsible and effective implementation. By overcoming these limitations, blockchain can play a transformative role in enhancing resilience, coordination, and trust in crisis management.

c. Trends and Issues in Blockchain for Resource-Limited Countries

While blockchain offers significant potential, resource-limited countries face unique challenges in adopting this technology.

General Trends

i. Emerging Use Cases:

Blockchain is being tested for aid distribution, digital identity management, and microinsurance.

Example: In Kenya, blockchain-enabled platforms provide farmers with weather-based insurance payouts automatically after adverse events.

ii. Adoption Barriers:

Limited internet access and infrastructure hinder deployment in remote regions.

High implementation costs deter cash-strapped governments and NGOs.

Key Issues

i. Digital Divide:

Only 30% of people in developing countries have internet access, limiting blockchain's reach.

Vulnerable populations, such as those in rural areas, are excluded from digital systems.

ii. Regulatory Challenges:

Unclear legal frameworks around blockchain adoption create hesitancy among stakeholders.

Cross-border disaster relief efforts face compliance issues due to inconsistent data privacy laws.

Proposed Solutions

i. Permissioned Blockchains:

These reduce infrastructure costs and offer controlled access, ensuring security and efficiency.

Example: Blockchain-based solutions for microfinance in India have improved rural access to loans.

ii. Public-Private Partnerships:

Governments can collaborate with tech companies to fund blockchain projects and bridge infrastructure gaps.

iii. Education and Training:

Providing technical training to local communities can increase adoption and help bridge the digital divide.

8. FUTURE DIRECTIONS AND OPPORTUNITIES

Blockchain's future in disaster recovery involves integrating it with emerging technologies like AI for predictive analytics, IoT for real-time monitoring, and 5G for enhanced communication. Key opportunities include using blockchain for digital identities, climate resilience projects, and faster insurance payouts. Global scaling through international collaboration and regulatory harmonization is essential, alongside ensuring equitable access and empowering local communities. These advancements will drive more resilient, coordinated, and transparent crisis management.

9. CONCLUSION

Blockchain technology offers transformative potential in disaster recovery by enhancing resilience, coordination, and trust among stakeholders. Its key features—such as decentralization, transparency, and immutability—address many challenges in crisis management, including secure data sharing, efficient resource allocation, and real-time decision-making. By integrating blockchain with emerging technologies like AI, IoT, and 5G, the effectiveness of disaster response can be further improved. However, overcoming technical, regulatory, and ethical challenges is crucial for widespread adoption. As blockchain solutions continue to evolve, their success will depend on global cooperation, inclusive access, and ethical use, ultimately creating more reliable and equitable disaster recovery systems for future crises.

REFERENCES

Farooq, M. S., Khan, M., & Abid, A. (2020). A framework to make charity collection transparent and auditable using blockchain technology. *Computers & Electrical Engineering*, 83, 106588.

Gad, A. G., Mosa, D. T., Abualigah, L., & Abohany, A. A. (2022). Emerging trends in blockchain technology and applications: A review and outlook. *Journal of King Saud University. Computer and Information Sciences*, 34(9), 6719–6742.

Haque, E. U., Shah, A., Iqbal, J., Ullah, S. S., Alroobaea, R., & Hussain, S. (2024). A scalable blockchain based framework for efficient IoT data management using lightweight consensus. *Scientific Reports*, 14(1), 7841. DOI: 10.1038/s41598-024-58578-7 PMID: 38570648

Hellwig, D., Wendt, K., Babich, V., & Huchzermeier, A. Playing with DISASTER: A Blockchain-enabled Supply Chain Simulation Platform for Studying Shortages and the Competition for Scarce Resources (February 26, 2022). In: Lee, H., Ernst, R., Huchzermeier, A., & Cui, S. (Eds.): Creating Values with Operations & Analytics, Springer Series in Supply Chain Management, Forthcoming, Georgetown McDonough School of Business Research Paper No. 4044385, Available at SSRN: https://ssrn.com/abstract=4044385 DOI: 10.1007/978-3-031-08871-1_9

Hunt, K., & Zhuang, J. (2022). Blockchain for Disaster Management. In Emrouznejad, A., & Charles, V. (Eds.), *Big Data and Blockchain for Service Operations Management. Studies in Big Data* (Vol. 98). Springer., DOI: 10.1007/978-3-030-87304-2_10

Hunt, K., & Zhuang, J. (2022). Blockchain for Disaster Management. In Emrouznejad, A., & Charles, V. (Eds.), *Big Data and Blockchain for Service Operations Management. Studies in Big Data* (Vol. 98). Springer., DOI: 10.1007/978-3-030-87304-2_10

Khare, S., Prakash, V., Talwandi, N. S., Soni, P., & Kumar, S. "Revolutionizing Cyber Security Incident Response with Smart Contracts," *2024 International Conference on Computational Intelligence and Computing Applications (ICCICA)*, Samalkha, India, 2024, pp. 86-90, DOI: 10.1109/ICCICA60014.2024.10584955

Li, J., Liu, X., Liang, W., & Zhu, C. "Research on network resource sharing based on distributed ledger technology," 2023 IEEE 3rd International Conference on Information Technology, Big Data and Artificial Intelligence (ICIBA), Chongqing, China, 2023, pp. 1509-1513, DOI: 10.1109/ICIBA56860.2023.10165046

Mahmood, M. (2016). Enhancing Citizens' Trust and Confidence in Government through Digital Transformation. *International Journal of Electronic Government Research*, 12(1), 99–110. DOI: 10.4018/IJEGR.2016010105

Sharma, P., Jindal, R., & Borah, M. D. (2023). A review of smart contract-based platforms, applications, and challenges. *Cluster Computing*, 26(1), 395–421. DOI: 10.1007/s10586-021-03491-1

Treiblmaier, H., & Rejeb, A. (2023). Exploring blockchain for disaster prevention and relief: A comprehensive framework based on industry case studies. *Journal of Business Logistics*, 44(4), 550–582. DOI: 10.1111/jbl.12345

Treiblmaier, H., & Rejeb, A. (2023). Exploring blockchain for disaster prevention and relief: A comprehensive framework based on industry case studies. *Journal of Business Logistics*, 44(4), 550–582. DOI: 10.1111/jbl.12345

Treiblmaier, H., & Rejeb, A. (2023). Exploring blockchain for disaster prevention and relief: A comprehensive framework based on industry case studies. *Journal of Business Logistics*, 44(4), 550–582. DOI: 10.1111/jbl.12345

Tripathi, G., Ahad, M. A., & Casalino, G. (2023). A comprehensive review of blockchain technology: Underlying principles and historical background with future challenges. Decision Analytics Journal, 100344.

Wang, Q., & Liu, Y. (2023). Enhancing Collaborative Emergency Management: Leveraging Blockchain Distributed Ledger for Inter-Agency Data Sharing. *Proceedings of the 8th ACM SIGSPATIAL International Workshop on Security Response using GIS*. DOI: 10.1145/3615884.3629423

Wang, Y., & Chen, H. (2022). Blockchain: A potential technology to improve the performance of collaborative emergency management with multi-agent participation. *International Journal of Disaster Risk Reduction*, 72, 102867.

Zarour, M., Alenezi, M., Ansari, M. T. J., Pandey, A. K., Ahmad, M., Agrawal, A., Kumar, R., & Khan, R. A. (2021, April 16). Ensuring data integrity of healthcare information in the era of digital health. *Healthcare Technology Letters*, 8(3), 66–77. DOI: 10.1049/htl2.12008 PMID: 34035927

Zuo, Y., & Qi, Z. (2022). A Blockchain-Based IoT Framework for Oil Field Remote Monitoring and Control. *IEEE Access : Practical Innovations, Open Solutions*, 10, 2497–2514. DOI: 10.1109/ACCESS.2021.3139582

Chapter 7
Efficient Wastewater Management With Optimized Fuzzy Sliding-Mode Control in Smart Urban Planning

Manjunatha Badiger
https://orcid.org/0000-0001-8073-0270
Department of VLSI Design and Technology, NMAM Institute of Technology, Nitte University, Udupi, India

Varuna Kumara
Moodlakatte Institute of Technology, Kundapura, India

Ezhilarasan Ganesan
Jain University (Deemed), India

Abdul Kareem
https://orcid.org/0000-0001-5291-9359
Moodlakatte Institute of Technology, Kundapura, India

ABSTRACT

The integration of IoT and ML in smart cities is crucial for dynamic city planning and infrastructure. This chapter presents an FSMC-based framework utilizing IoT and ML for effective wastewater management. Wastewater systems are complex, with strong interactions and various interferences, challenging traditional control methods. IoT enables real-time data acquisition, while ML offers predictive capabilities and system flexibility. The proposed framework leverages JAYA, PSO, and GA for optimization, using fuzzy logic to handle uncertainties and sliding-mode control for robustness. Compared to traditional methods, these second-order techniques are more accurate, energy-efficient, and flexible. Simulations show improved time response, disturbance rejection, and energy efficiency. This chapter is a practical guide on interdisciplinary IoT, ML, and FSMC applications in wastewater management, advancing smart urban infrastructure and intelligent control systems for future city planning.

DOI: 10.4018/979-8-3693-9770-1.ch007

1. INTRODUCTION

This chapter addresses fundamental issues of further understanding of wastewater management as the sustainable development of urban areas and improved health of communities. As such smart solutions can create more efficient processes in any urban planning, it is important for wastewater to be treated and managed properly in order to support clean water sources for ecosystems and healthy city inhabitants (Kyriakopoulos et al., 2023). Wastewater treatment systems helps to prevent waterborne illnesses while also having a positive effect on the local environment by stopping pollution of water bodies. Wastewater history shows that ancient people—the Indus people, the Romans knew about the need to separate waste from clean water to prevent pollution. Contemporary environmental legislation, for example, the US Clean Water Act, calls for high standards of water quality in order to safeguard public health and encourage responsible use of water. These regulations demand future evolution and development of more complex wastewater management solutions to correspond with the trends of urbanization, population, and environmental issues. This chapter underlines the significance of wastewater management as a crucial component of smart urban infrastructure, addressing its historical evolution, present importance, and the role it plays in the future of city planning (Singh et al., 2023).

1.1 Challenges in Controlling Wastewater Treatment Systems

There are numerous technical and operational issues that wastewater treatment plants encounter with regard to the central theme of the paper, namely, sustainability and optimal performance. The urban wastewater is diverse and unpredictable in terms of its flow and pollution intensity due to influences of industrial processes, weather, and population density. The anaemic control strategies, including the PID controllers, can provide optimal control of these dynamic nonlinear systems. More so, the fact that influent characteristics and the disturbing factors are random means that there will obviously be time wastages in the system due to the need to adjust manually (Niu et al., 2020).

The issue of high standards of treatment is crucial so that the policy conforms to the regulatory norms as well as reduce the impacts of the environment and health complications. Quite often traditional approaches fail miserably when it comes to this area of concern, and one finds resource waste, higher operation costs, and increased energy consumption. Minimising impact, however, requires improving facility design and managing flow rate, pollutants, and disturbances so that upgrades to wastewater treatment facilities have become essential (Amaral et al., 2020). These factors are best illustrated in the following chapter centred on the intricacies of wastewater control and the consequent call for improved methods to deal with variability and ensure reliable treatment performance.

1.2 Role of Advanced Control Strategies in Improving Efficiency and Sustainability

In response to the challenges highlighted, more sophisticated controls and more specifically fuzzy logic control and sliding mode control are used in wastewater treatment plants. These methods add flexibility that is needed to address nonlinear and unpredictable character of influent wastewater. Fuzzy logic originates from the way people make decisions, which optimally handles fuzzy data It responds to the situations predicted by sliding-mode control under fluctuating conditions. All these strategies are incorporated into a new class of self-organizing fuzzy sliding-mode controllers

(SOFSMCs) for wastewater systems which effectively adapts to varying influent properties and interferences (Han et al., 2019). The enhancement of these control frameworks with optimization techniques like particle swarm optimization (PSO), genetic algorithm (GA) and JAYA enhances these control parameters to optimize the operation indicator, such as treatment efficiency, energy usage and operational robustness. As the high-performing strategies bring about more effective pollutant abatement, these solutions constitute the environmental efficiency of wastewater treatment systems to enhance the creation of safer and less resource-intensive urban ecosystems (Ahmad et al., 2024).

2. AN OVERVIEW OF WASTEWATER TREATMENT SYSTEMS

Sewage treatment is an essential process through which wastes are removed from water, and then it is returned to the water stream or reclaimed by a process called water reuse. It all occurs at treatment facilities such as sewage, industrial, agricultural, and leachate treatment facilities. Every facility is dedicated to processing various wastewaters depending on their origin or degree of contamination (Shojaei & Shojaei, 2021). Core treatment processes Wastewater treatment commonly involves three key processes:

Phase Separation: Some of the procedures used involve sedimentation, where solids are separated through gravity attraction from the liquid phase of the wastewater.

Biological and Chemical Treatments: Some of these are oxidation processes that involve the use of biological and chemical instructions to decompose the organic pollutants.

Polishing: Subsequent operations enhance water quality in an effort to reach the required levels for discharge or reuse. The major output from wastewater treatment plants is referred to as sludge, which is mainly treated within the same station. Also, biogas may be produced through anaerobic treatment, and the reclaimed water can be recycled.

2.1 Types of Wastewater Treatment Plants

Sewage Treatment Plants: These plants work to treat domestic wastewater, not to pollute the surrounding environment. Sewage treatment may involve two to three main stages: In terms of treatment, these are the primary, which removes suspended and settle able solids; the secondary, which involves biological treatment to further decrease biochemical oxygen demand; and the tertiary, which is additional cleaning and nutrient removal. Organized nations manage to treat about 74% of sewage, while the developing nations manage to treat only about 4.2%. Micro pollutes, including pharmaceuticals, may be eliminated by an additional stage called quaternary (Lim et al., 2022).

Industrial Wastewater Treatment Plants: Take wastewater from industries; this water will contain heavy pollutants, including oil, grease, toxic chemicals, and metals. Public treatment works may have pre-treatment systems so that industrial users lower the pollutant strength before discharge to the sewer. Some establishments completely work on their wastewater online to meet environmental requirements. High-pollution industries may choose to employ special treatments that are chemical precipitation, filtration, and pH regulation, among others. The tendencies of the recent decades focus on reducing the production of wastewater and its utilisation in industrial cycles.

Agricultural Wastewater Treatment Plants: Control farm wastewater such as that from fertilisers and pesticides, as well as animal wastes. It also is often systemic, at least to some extent, to variations in the breeding or harvesting cycles. Constructed wetlands and anaerobic lagoons are used widely and some farms have employed point and nonpoint source pollution control for improved waste management.

Leachate Treatment Plants: These plants manage landfill leachate, which is a liquid that filters through the waste and can transport contaminants that may affect the soil and water. They include biological filtration, ultrafiltration, reverse osmosis, and carbon filtration to mention but a few, in order to get rid of the contaminants.

The most important objective of wastewater treatment is that of producing water that is safe to discharge into the environment or to re-introduce to other processes (Parkes, 2022). The kind of treatment is determined by the type of wastewater, the degree of pollution, and its proposed use either for discharge or reclamation. This makes the treated effluent meet environmental requirements and hence encourage proper usage of water and reduction of pollution.

2.2 Complexity, nonlinearity, and dynamic behaviour of treatment systems

Sewage treatment plants are bound to be of a high degree of operational complication and variability, primarily because of the variability of the quality and nature of sewage input to the system. The flow rates and the pollutant concentrations at the inlet of the treatment plants are highly stochastic, and there are interactions between them and also stochastic nature of other effects like weather (Gernaey et al., 2011). This issue itself remains a tough nut to crack for all those traditional control techniques like PID controllers, and hence the need for better and enhanced control techniques. Moreover, there are also different stages of biological, chemical, and physical treatment, which all require precise observation and regulation. Therefore, in order to secure constant and efficient performance, it is necessary to strengthen mechanisms that are both sophisticated and flexible in order to adapt to the levels of operant phenomena and to make organizational processes impervious to shocks.

2.3 Need for robust control methods in wastewater management

The use of effective control techniques is of utmost importance due to the extremely nonlinear and dynamic character of wastewater treatment systems. The existing approaches typically do not contain the desired level of adaptability and are incapable of managing the variability and nonlinearity typical of these systems. Superior strategies are developed to cater for this challenge since they provide for flexibility in response to system changes, are capable of dealing with uncertainties, and are robust to imperfect or noisy data from the sensors. Such control strategies are instrumental in improving the steady-state performance of the system against disturbances in a way that the goals of efficient operation, operational reliability, and conformance to environmental requirements are achieved (Harbud, 2022).

3. FUZZY SLIDING-MODE CONTROL (FSMC): CONCEPT AND DESIGN

Sliding Mode Control (SMC) is a nonlinear control that develops an effective method to tweak the behaviour of a system by applying a high-frequency control signal that must steer the system towards a predetermined trajectory or a "sliding surface." In contrast with other conventional continuous feedback methods, SMC uses a state-feedback control law that switches between two different control laws depending on the current position on the state space (Azar & Serrano, 2015). This approach, also known as variable structure control, is a multiplexed system that has a series of control modes. Each mode guides system trajectories to a different region with a different structure. The upshot of this is that the system doesn't come to rest in a specific mode but rather moves along boundaries of the same modes to produce a state of 'sliding mode' that keeps the system confined within a certain state. A sliding mode surface in the state space defines an organised path of movement and makes the system robust. The system after a starting period of adaptation, moves 'along' this prescribed path, realising at each step a state that is characterised by high stability and low-order dynamics. According to the SMC design theory, once a sliding surface is achieved, the system state is maintained in the surface. Nevertheless, in real-world applications, the implementation of the SMC concept leads to a phenomenon known as chattering- high-frequency oscillations near the sliding surface caused by switching's. Techniques one already learns and uses are dead bands or boundary layers to reduce chattering, hence enhancing smoothness of control while retaining robustness.

The advantage of this mechanism can be seen with reference to SMC, which is capable of handling vast disturbances and uncertainties without requiring the fine-tuning of model parameters. The control action depends on the current system state, not a continuous time function, which enables quick correction and asymptotical convergence in finite time to the sliding plane. This means that like other numerical methods, SMC does not change with variations of the system parameters and hence turns out to be more appropriate for fields where accuracy and flexibility reign supreme.

Sliding mode control is widely applicable as in electric drive control and robotic systems. In electric drives that are controlled by switching power converters for instance, SMC can effectively control the discontinuous operating mode of such converters (Wu et al., 2022). Moreover, in robotics, SMC has been shown to be useful for control problems that need high accuracy under uncertainties such as tracking in sea states for unmanned surface vessels. SMC also presents one challenge – the chattering which leads to high energy consumption, excessive mechanical wear and may cause unintentional system vibrations due to reaction of the actuators. Continuous control methods are sometimes preferred when reduction of such high frequency oscillations is important because they provide less switch like control that is close to but not exactly SMC. However, in systems requiring high fault tolerance and flexible functioning within the systems of nonlinear dynamic equations, SMC is still a very efficient method.

3.1 Benefits of fuzzy logic in handling system uncertainties

It is very useful when used for handling uncertainties associated with a system of rules since it can handle fuzzy or noisy data. Through the integration of uncertainty through the fuzzy sets, it is able to model and address actual variations without necessarily requiring accurate measurements. Mimicking human thought, fuzzy logic employs linguistic variables, which make decisions using affordably accurate information which results from high resistance to disturbances and parameter fluctuations. This characteristic makes it possible to make instant responses in the changing system without having to develop difficult models, and its simple structure simplifies implementation, particularly in nonlinear conditions. Moreover, fuzzy logic also supports interpolation between control actions in a way that minimizes rapid change, and for processes that must change gradually, it is a realistic solution for robust, adaptive control. This integration tests the strength of the two methods in providing system reliability and flexibility when dealing with the level of uncertainty. The contingency of fuzzy logic Triangular membership functions enable the system to process imprecise data, while its robustness in the change of control from one interval to another reduces the switch frequency characteristic of SMC known as "chattering" and hence makes it more effective in systems that are sensitive to sharp control change (Azar, 2012). In this combined approach, the identified set of sliding-mode parameters may be controlled by the fuzzy logic to adapt to a range of scenarios, thus offering variable boundary layer control to tone down the disturbances. This integration leads to a control strategy whereby the advantages of SMC such as solidity and agility are preserved, but stability and conservation of the physical components of the system are enhanced, making the controller suitable for, for instance, nonlinear applications.

3.2 FSMC structure and working principles

The FSMC structure is developed by integrating the advantages of both the SMC and fuzzy logic for better performance in systems with uncertainties (Wogi et al., 2022). In FSMC, a desirable sliding surface is selected from required system dynamics to control the system tracks toward this surface to ensure the system stability and control accuracy. Fuzzy logic is incorporated to control parameters that change constantly when the system is closer to the sliding surface to minimise control operations and eliminate chatter. The fuzzy controller adapts the sliding-mode parameters, assessing in real-time the state of the system and the uncertainties thereof. This makes FSMC more responsive to system nonlinearities and external interferences. This combination ensures that FSMC can sustain high performance and short cycle times while ensuring slicker motion transitions, reducing resources wasted through heat dissipation and exerted on the actuators.

4. OPTIMIZATION ALGORITHMS FOR CONTROLLER TUNING

Optimisation adds much value to control systems because it guarantees that the executable performance objectives will not cost much, consume much energy, or use up many resources. It entails altering control factors that help to maximise performance; typical factors include stability, response rate, and accuracy. Optimisation enables one to arrive at the best control strategy with an understanding of the trade-offs between two or more control objectives, for example, between response speed and stability margin, or between performance and robustness (Joseph et al., 2022). It also facilitates variation in control algorithms to suit variations and changes in dynamism or operating conditions of a system, thus emphasising system dependability and performance. However, optimisation methods can help in setting the parameters of the controllers in order to get the desired transient and steady-state characteristics of the system in the real environment for the overall enhancement of the control system. Therefore, optimisation forms an important basis for formulating control systems that are not only effective but also optimally suited to meet standards of performance that are set.

4.1 Overview of optimization algorithms:

JAYA Algorithm: The JAYA algorithm is a population-based optimisation algorithm that operates on the basic concept of searching for the 'best' without opting for the 'worst'. It works in the following manner: candidates solve; interaction; there is a focus in choosing good individuals to control the search. The algorithm just performs calculations using simple arithmetic equations and has no parameters to be estimated and, therefore, is easy to use. In this strategy, JAYA is very useful and performs really well in continuous and discrete optimisation problems.

Particle Swarm Optimization (PSO): PSO is a kind of swarm intelligence based optimization algorithm derived from the simulation of the bird flocking and fish schooling. A group of candidate solutions, which are called particles in PSO, improves their positions toward the optimal solution depending on the experience of each particle and the other particles in a neighbourhood. The strength of the algorithm is the good reconciliation between the exploration and the seeking of new solution space, and the exploitation or the refinement of known spaces that are good hence making it very good for multidimensional and complicated problems.

Genetic Algorithm (GA): GA is a Pareto-optimal search heuristic based on Genetic principles of selection. This is achieved by employ a process which involves selection, crossover and mutation of its population of potential solutions. Self-organizing algorithm is an imitation of the process that occurs in the living organisms and rewards the solutions that are fitter. GAs are suitable in optimization problems with multiple, nonlinear, multimodal search space; and they can solve discrete, continue, and mixed variate optimization problems.

4.2 Comparison of Algorithm and Selection Criteria for Fine Tuning of FSMC Parameters

When selecting an optimization algorithm for tuning Fuzzy Sliding Mode Control (FSMC) parameters, several criteria should be considered:

Convergence Speed: The procedure of the algorithm in achieving the best feasible solution within a few number of iterations. Usually, PSO reacts with a better convergence to the target in contrast to GA and JAYA, in most cases.

Robustness: The performance capability of the task regardless the starting situations. As for general performances, GAs and PSO are relatively superior in complex and multimodal environment, and the same goes with JAYA because of its simplicity.

Complexity: The implementation complexity and the computation cost in the algorithm can be examined. JAYA is easy to implement as it requires lesser parameter tuning as compared to GA which might demands more computational resources because of the genetic operations.

Flexibility: The versatility of the algorithm in the solution of different types of optimisation problems characterised by discrete and/or continuous decision variables. Although PSO and GA are highly flexible as compared to JAYA, later it is easy to implement a number of applications.

Parameter Tuning Requirements: requirement for parameters to be set before implementation of the optimisation process. The results revealed that there is no need to set any specific parameter in order to use the JAYA algorithm, in contrast with GA and PSO algorithms, where the user has to set some parameters such as population size, mutation rate, and inertia weights.

It is believed that the criteria influences by which the selection of optimisation algorithm for FSMC parameter tuning should be made include convergence rate, stability, complexity of implementation, flexibility, and further parameter adjustments required by the application (Omar et al., 2024). There are pros and cons to each algorithm, and the best option would depend on the problem and objectives at hand.

5. DESIGN AND IMPLEMENTATION OF THE OPTIMIZED FSMC FOR WASTEWATER TREATMENT

Worthy of note is the fact that in building a fuzzy sliding mode controller (FSMC) for wastewater treatment, assessment of the application's goal or requirement profile is crucial. In the treatment of wastewater, some parameters need to be controlled, as the biochemical activity that is required for wastewater treatment is determined by these parameters, e.g., dissolved oxygen. Inherent to these systems, the control technology's control nonlinearities, time-varying dynamics, and disturbances may be challenging for conventional systems to address. These disturbances may depend on factors such as temperature, pollution level, and variation of the organic load (Darji et al., 2022).

The proposed methodology defining FSMC design and implementation for controlling a wastewater treatment plant is presented in the following steps.

5.1 Step-by-Step Design Methodology for FSMC

Step 1: Define System Dynamics

Identify System Model: The mathematical model of the system that has to be regulated (for example, a wastewater treatment plant) should be established.

Define Control Objectives: Details should be provided on the intended performance parameters, such as the maintenance of dissolved oxygen levels.

Step 2: Design Sliding Mode Control

Choose Sliding Surface: Define the sliding surface by taking into account the states of the system.

Derive Sliding Mode Control Law: Determine the control rule that will guarantee the system reaches the sliding surface and stays there for the duration of the process. The use of a sign function for switching control could be required for this.

Apply Lyapunov Stability: Lyapunov's stability theorem should be used in order to guarantee that the selected sliding surface is stable.

Step 3: Incorporate Fuzzy Logic

Define Fuzzy Rules: Develop a collection of fuzzy rules that may be used to define the connection between the inputs and the control actions.

Construct Fuzzy Rule-Based System: Make use of a fuzzy inference system, also known as a FIS, in order to create control signals and execute the fuzzy rules.

Adjust Fuzzy Membership Functions: In order to maximise the performance of the fuzzy logic controller, you should design membership functions for both the inputs and the outputs

Step 4: Self-Organizing Mechanism

Integrate Self-Organizing Neural Network (SONN): The performance of the system should be used to dynamically change fuzzy rules, and a self-organising mechanism should be used for this purpose. In accordance with the real-time input received from the system, the fuzzy rules should be updated.

Define Correction Mechanism: On the basis of the fuzzy output and the current state, implement an adjustment algorithm for the control signal.

Step 5: Implement Control Strategy

Integrate Control Signals: When determining the final control action, it is necessary to combine the outputs from the fuzzy logic controller with the sliding mode control rule. It is essential that the control action takes into consideration both the peculiarities of the sliding mode and the fuzzy adjustments.

Simulate System Behaviour: It is possible to verify the performance of the FSMC under a variety of operating circumstances and disturbances by using simulations, such as those provided by MATLAB/ Simulink.

Step 6: Evaluate Performance

Analyze System Response: Determine how the system reacts when it is subjected to disruptions and uncertainty. A number of performance measures, including settling time, overshoot, and steady-state error, should be monitored as well.

Optimize Parameters: In order to obtain the appropriate level of performance, it is necessary to fine-tune the settings of both the fuzzy logic controller and the sliding mode control. When it comes to parameter tweaking, you may want to think about applying optimisation algorithms like JAYA, PSO, or GA.

Step 7: Final Implementation and Testing

Real-World Testing: Within the actual process of treating wastewater, the FSMC should be implemented. Perform performance monitoring and make any required modifications depending on the data collected in real time.

Continuous Learning: To enhance the effectiveness and dependability of the system, it is essential to make certain that the self-organising mechanism incorporates operational input into the process of continually updating the fuzzy rules and control parameters.

By adhering to this step-by-step technique, you will be able to create an efficient FSMC that is adapted to particular applications, such as managing the amounts of dissolved oxygen in wastewater treatment facilities, while also efficiently resolving uncertainties and disruptions.

5.2 Incorporating optimization algorithms for parameter tuning

The integration of optimization algorithms continues to be an essential component in the process of enhancing the performance of control systems, particularly in the context of driving biochemical WWT processes via the application of optimization algorithms to Self-organizing Fuzzy Sliding Mode Controllers (SOFSMC) (Kumara & Ganesan, 2024). An explanation of how optimization techniques may be used for parameter tweaking in SOFSMC is provided in the following step-by-step methodology:

Step 1: What is the Control System Model?

Start by setting the mathematical model of the WWT process to develop the mathematical frameworks that simulate the WWT. This involves identification of the system behaviour, design of the sliding surface, and control goals.

Optimal tuning approaches that should be used involve selecting certain parameters that can be optimised to achieve certain goals for the system, for instance settling time, overshoot, and steady-state error.

Step 2: Select Optimisation Algorithm

Select an optimal method to perform the task of parameter tuning. In this regard, the JAYA optimisation algorithm is chosen because of its capacity to enhance the locations of the search agents iteratively and conform to the best solutions. Improve the performance of the JAYA algorithm through the introduction of the Solow-Polasky measure (SoPo), since it increases the population diversities and ensures great coverage of the solution space.

Step 3: Parameter Initialisation

Define the fuzzy sliding mode controller (FLSMC) and the gains and membership functions of FLSMC for fuzzy rules at the starting point. Define the initial values for parameters of the optimisation algorithm; these are population size and iteration number.

Step 4: Objective Function Formulation Objective Function Formulation

Define constraints on system performance that are to be reduced to an objective function. This could be a weighted combination of the metrics that were defined in the first step.

The objective function should be minimised or maximised depending on what is desired; for the present case, overshoot and settling time need to be minimised.

Step 5: Implement the Optimization Algorithm

Execute the optimization algorithm to tune the parameters of the FLSMC: In each iteration, assess the efficiency of the control system with existing parameter settings with the help of simulation techniques such as Simulink. Adjust the objective function with reference to the performance of the control system. For the purpose of achieving a more optimal solution, the parameters were modified in accordance with the guidelines that were established by the optimization algorithm.

Step 6: Feedback control implementation

Call for feedback control mechanisms so that real-time disturbances and uncertainties affecting the treatment system can be dealt with. Develop a mechanism that enables the changes of the controller parameters depending on the system's current performance.

Step 7: Testing and Validation

After this have finished optimizing the system, this should next run a simulation of the control system under estimated settings in order to measure the actual performance of the parameters that have been precisely adjusted. For the purpose of determining whether or not the suggested optimized SOFSMC is capable of managing nonlinearities and disturbances while maintaining the appropriate level of resilience and stability in the system, assess its performance.

Step 8: Performance Evaluation

Evaluate the overall performance of the wastewater treatment process using the optimized SOFSMC: Dissolved oxygen concentration and solid concentration of the solution: The time taken to respond to disturbance. In this paper, percentage improvements are used to compare the performance of the optimised controller against baseline results.

Step 9: Iterative Refinement

After the assessment of the performance, improve the optimization strategy in case of a requirement. This may involve: changing the fitness of function parameters (population size and numbers of iterations). Some things can be changed in the objective function, which shifts focus towards different characteristics of the performances. Refining of the updated fuzzy rules reviewed with freshly obtained data. Retuning the fuzzy rules based on newly acquired insights.

5.3 Handling system disturbances and uncertainties

The strategies employed in the present methodology using SOFSMC applied to WWTPs are based on fuzzy logic to handle the uncertainties, lack of proper models, and disturbances that affect the operation of WWTPs and to reduce effects of chattering that is associated with SMCs (Han et al., 2019). The proposed self-organizing sliding mode control (SOSMC) combines the self-organizing characteristic of SOSs and the existing advantage of SMCs in optimizing efficiency, reliability, and compliance in WWTP management. An adaptive control signal is used which is made dynamic and stable by fuzzy rule base tuning while empirical and estimation error based structure risk of FNN is then implemented to have the real time structural optimization in terms of tracking error and model complexity. Maintaining and regulating DO concentrations is important for biological treatment outcomes because high BOD impairs the DO and the ecosystem. Proportional controllers enable the control of DO ratio levels by adjusting aeration frequencies in PWM; the data are enhanced using the Kalman filters in estimating the DO concentration levels. The blend of approach in the control system includes PWM, a reactor and a

Kalman filter to enhance the control of DO to demonstrate the importance of advanced control systems for effective and efficient removal of wastewater (dos Santos Silva et al., 2019).

6. PERFORMANCE EVALUATION

MATLAB Simulink can be employed to establish a simulation layout whereby a model of a wastewater treatment processes may be established. The first major purpose is to investigate the performance of a self-organizing fuzzy SMC in controlling DO levels and other parameters. This method obtains much data from the "Shafdan WWTP measurements" data set (Farhi et al. 2021): solid concentration, rotor speed, dissolved oxygen (DO) outflow, rotor deep level, and so on. These are fundamental parameters as inputs to the system and are significantly important in assessing the efficiency of the treatment process. The methodology consists of two distinct phases: The first two are the training phase and the testing phase. Input data for training the SONN is pre-processed during the training phase in order to meet the requirements of the training. Here it is vital to note that one of the general pre-processing steps is treated with missing variables using imputation. The technique used in data pre-processing is the mean imputation which imputes the missing values in this case in order to maintain the whole set (Karrar, 2022). Subsequently, the data is standardized to enhance the scale of all training features to the same range in order to train the models. In the subsequent step the process of feature extraction takes place where one gets to select features from the normalised data. The performance of this stage is necessary for the discovery of such basic relations and associations within data, thus enhancing the ability of the SONN to learn and generalize on the input data set. All the retrieved features were fed into the SONN for training.

The proper interpretation of the complex and multi-layered input is possible due to SONN, and therefore, this neural network architecture is selected. SONN applies the self-organising map (SOM) in order to map high-dimensional features in a low-dimensional space. Compared to the standard neural networks, SONN preserves the data structure of inputs, which makes the amounts of information discarded in the training procedure minimal. That training of SONN may consume a lot of time is well known and has been so mostly due to the complex design and architecture of the network. To meet this challenge, a novel technique, namely Tex-SONN, is introduced where TanhExp activation function is incorporated with SONN. There is the use of a TanhExp activation function to enhance the convergence speed of the network and upgrade its functionality in a way that would collectively enable the reduction of SONN training computations (Liu & Di, 2020) (Badiger et al., 2022) .

In the testing phase of the approach, BSM1 is used as benchmark to measure the effectiveness of the waste-water treatment system. Flow rate, the qualities of wastewater, biological processes, and quantity of nitrate also determine the concentration of dissolved oxygen which is a sensitive factor in wastewater treatment. As a result, non-linear and even unpredictable behavior may appear within the system, underlining the importance of mastering the adequate methods for developing effective control for plant functioning. As for the problems existing in the system for TL for higher learning, the use of the Fuzzy sliding mode controller is proposed which meets two main parts. First of all, therefore, the provision of a surface for sliding is critical to ensuring that the operations of the system characterised by this ability meet some necessary prerequisites. Additionally, the state of the system is oriented to the sliding surface whenever the optimal choice of control parameters is chosen to ensure system stability and control. The FH Finally, the JAYA optimisation algorithm can be chosen as an option to optimize the parameters of the developed fuzzy sliding mode controller.

JAYA optimization algorithm is developed based on the idea of placement of the search agents improving at each iteration by shifting agents from the poorest answer (da Silva et al., 2023). However, there is a disadvantage which may be seen as a negatively impacting the general perform and works as a limit to the number of population; this reduces the chance of performing various studies on possible solutions. To overcome this constraint, the optimisation method involves Solow-Polasky measure (SoPo) to expand the population diversity and explore the entire solution space conveniently.

In endeavouring towards preventing internal disturbances or instabilities, feedback control mechanisms should be developed to mitigate these effects most importantly during adverse climatic conditions. Utilized feedback controllers according to their efficiency as anti-instability controllers. While specific feedback control systems are highly functional and commonly used in management, they are generally inherently complex to create, as well as implement due to the interaction dynamics that exist within the given system. To bring another enhancement to the design of the controller and its robustness, the Mittag–Leffler function (MitL-FC) is integrated into the controller. This helps in avoiding non-linear interactions and at the same time obtain a high efficient rejection of disturbances alongside the systems performance. Regarding the remaining oscillations which may still be present even after removing the disturbances the provided methodology introduces a voltage-controlled oscillator (VCO). This oscillator is selected due to its ability to regulate the output signal in relation to the input voltage, the frequency of which is next in immediate. Notably, VCOs make it possible to produce output signals over a wide range of frequencies and illustrate good versatility. However, when the quality factor designs or Q factors are low, VCOs may experience high jitter which lowers the performance. To solve this problem, the proposed methodology uses a different approach, where a fractional moments-based maximum entropy distribution labelled FMME-VCO replaces the Gaussian distribution. In this methodology, the attempt will be made to enhance the oscillatory response and minimize the instability known as jitter in the required WWTP using FMME-VCO.

Depolarisation is another criterion indispensable for studying thermal hotspots as components of WWTPs. As a result of this difficulty, the proposed methodology involved the use of an evaporator type commonly known as the short-path evaporator. A short-path evaporator is efficient for separating and purifying heat-sensitive substances characterised by a relatively high boiling point. It is particularly useful in eradicating zones of high heat dissipation and delivering the optimum possible depolarisation in the treatment facility. This flow rate forecast by the facility in the wastewater treatment is done using the features accrued from the stages. The forecast of flow rate is applied to predict the in-flow air and DO into the plant for the ease of operation and effective working of the WWTP. The approach aims at streamlining processes and optimising differences in operating procedures of WWTPs by employing predictive modelling methods for optimising the general outcomes. A comparison of FSMC with and without optimisation is shown in Table 1.

Table 1. Comparison of FSMC with &without Optimization

Feature	FSMC Without Optimization	FSMC With Optimization
Parameter Tuning	Manual or static	Dynamic, automated (e.g., JAYA, SoPo-JAYA)
Adaptability	Moderate, limited to preset conditions	High, adaptable to varying conditions
Stability	Moderate	Improved, with reduced steady-state error

continued on following page

Table 1. Continued

Feature	FSMC Without Optimization	FSMC With Optimization
Disturbance Rejection	Basic disturbance rejection	Enhanced disturbance rejection
Response Time	Slower, potential for overshoot	Faster, reduced overshoot
Control Accuracy	Limited	High, more precise control
Computational Load	Moderate to high	Reduced with optimized parameter handling
Resilience	Lower resilience in dynamic conditions	High resilience in complex environments
Overall Performance	Reliable but less efficient	Superior, efficient, and robust

6.3 Performance metrics: response time, disturbance rejection, energy efficiency

Response Time: Compared to the SO-Po-JAYA optimisation implemented in the current study, the Tex-SONN model reduced the response time, which was optimised in 4569 ms, and outperformed other optimisation techniques such as JAYA, Grey Wolf, and Wild Horse optimisation that took between 7890 ms and 12,974 ms. Such prompt response is crucial for proper real-time control of the wastewater treatment process because it allows for quicker decision-making.

Disturbance Rejection: The approach of Tex-SONN showed remarkable reliability against operational interferences. Parameter-controlled disturbance Rejection mode enabled the SoPo-JAYA optimiser to adapt by maintaining a fixed performance despite these changes, such as inflow rate, substrate, and nitrogen concentration fluctuation. This increases the robustness of treatments by ensuring that quality does not significantly change regardless of the prevailing environment.

Energy Efficiency: This work presented Tex-SONN with better thermal efficiency when compared to other common neural networks (CNN, RNN, ANN). This efficiency reduces energy costs through better utilisation of thermal energy, as well as providing optimum temperature conditions for biochemical reactions and increasing recirculation rates. The optimisation of the energy consumption and the higher recirculation rates mean lower operational costs and environmental impacts; therefore, Tex-SONN is the best option in wastewater treatment systems.

Case Study 1: Disturbance Rejection at New York City's Newtown Creek Wastewater Treatment Plant

Newtown Creek WWTP is the largest WWTP in New York City and treats more than 1,500 cubic meters of wastewater per day. Large fluctuations of influent flow are typical for the plant, with frequent storm events increasing mixed storm water and wastewater flows at the plant, affecting its treatment process. In the period of intense rain, high concentrations of nutrients and other pollutants come with increased influent rates and cause fluctuations in the treatment schedules or cause them to overflow. With the help of such control methods as adaptive neural networks, Newtown Creek WWTP installed a system of permanent aeration and sludge recycling.It made it possible to define consistent endpoint treatment parameters inside the plant, such as the concentration of dissolved oxygen (DO) and nutrients, regardless of the change that occurred in this environment. As a result of these advancements, the disturbance rejection system was able to improve treatment efficiency, reduce the amount of chemicals used, and reduce overflow. The fact that this endeavour was successful demonstrates that definite control systems that operate in real time are absolutely necessary in order to face the challenges that are inherent in significant weather disturbances.

Case Study 2: Energy Optimization at Strass Wastewater Treatment Plant, Austria

Specifically, Strass WWTP in Austria is said to operating at a condition of net zero energy consumption, a record in energy efficient wastewater treatment. As energy costs continued to escalate, the plant focused on attaining highest thermal efficiency and minimizing dependence on external energy which is essential for withstanding the chilling winter for the intended use that is the biological processes. At Strass WWTP, a real-time direct DO monitoring and prediction control system was used to control aeration as well as slave digestion in order to maintain thermal demands coupled with energy requirements. The system applied data-generated models for managing heating with reference to inflow trends and thereby helped minimize energy usage while preserving optimum working temperatures. Future engineering efforts in the plant will focus on further enhancing the energy-optimized process of minimizing costs by up to 30%, allowing the plant to be fully run on renewable energy produced locally. Strass brings a valuable lesson about economical recirculation and thermal management, which cuts both CO2 emissions and operational costs to develop a sustainable business model for other STPs.

7. BENEFITS OF OPTIMIZED FSMC IN SMART URBAN PLANNING

Enhanced System Performance and Adaptability: Integral control and optimisation methods enhance the interpretative ability of wastewater treatment systems so that they are less sensitive to perturbations. This feature makes it easy for systems to work well under changing inflow conditions without compromising on the quality of treatment offered. Energy Savings and Cost Efficiency: As a result, the overall energy consumption or operational cost of wastewater treatment plants is minimized by using predictive control and resource allocation. Optimizing aeration, heating, and recirculation help plants reduce the need for energy and minimize waste which is good for their bottom line. Contribution to Sustainable Urban Development: Amenable wastewater technologies, utilities for sustainable urban development through balanced impacts on the environment and resource value addition. Sustainable and advanced approaches in technologies enable urban centers to consume water more wisely and discharge wastes appropriately, as well as promote green buildings in urban communities. The study shows that the developed SOFSMC with integration SoPo-JAYA optimizer in the framework of a SONN enhances wastewater treatment effectiveness, resilience and control accuracy (Pandey, 2016). It can obtain faster convergence and stable working condition, stress its application advantages in resource utilization and pollutant emissions. Future Research Directions for Intelligent Wastewater Management: More work can be taken up in conjunction with actual application of SOFSMC and SoPo-JAYA on wastewater treatment facilities to explore the effectiveness over large periods, fine-tune the machine learning part of the algorithm to other organizational settings, and incorporate big data analysis for proactive maintenance and adjustment. Potential for Broader Applications in Smart City Infrastructure: This intelligent control approach is not limited to the wastewater treatment; it has larger applications such as, smart adaptive technologies for efficient water management in urban environ, energy saver products and efficient control in pollution. These technologies could make a positive impact to the elements of smart city ecosystems, resource utilization as well as environmental conservation.

REFERENCES

Ahmad, H. W., Bibi, H. A., Chandrasekaran, M., Ahmad, S., & Kyriakopoulos, G. L. (2024). Sustainable Wastewater Treatment Strategies in Effective Abatement of Emerging Pollutants. *Water (Basel)*, 16(20), 20. Advance online publication. DOI: 10.3390/w16202893

Amaral, R. E. C., Brito, J., Buckman, M., Drake, E., Ilatova, E., Rice, P., Sabbagh, C., Voronkin, S., & Abraham, Y. S. (2020). Waste Management and Operational Energy for Sustainable Buildings: A Review. *Sustainability (Basel)*, 12(13), 13. Advance online publication. DOI: 10.3390/su12135337

Azar, A. T. (2012). Overview of Type-2 Fuzzy Logic Systems. [IJFSA]. *International Journal of Fuzzy System Applications*, 2(4), 1–28. DOI: 10.4018/ijfsa.2012100101

Azar, A. T., & Serrano, F. E. (2015). Adaptive Sliding Mode Control of the Furuta Pendulum. In Azar, A. T., & Zhu, Q. (Eds.), *Advances and Applications in Sliding Mode Control systems* (pp. 1–42). Springer International Publishing., DOI: 10.1007/978-3-319-11173-5_1

Badiger, M., & Mathew, J. A. (2022, March). Retrospective Review of Activation Functions in Artificial Neural Networks. In Proceedings of Third International Conference on Communication, Computing and Electronics Systems: ICCCES 2021 (pp. 905-919). Singapore: Springer Singapore. DOI: 10.1007/978-981-16-8862-1_59

da Silva, L. S. A., Lúcio, Y. L. S., Coelho, L. dos S., Mariani, V. C., & Rao, R. V. (2023). A comprehensive review on Jaya optimization algorithm. *Artificial Intelligence Review*, 56(5), 4329–4361. DOI: 10.1007/s10462-022-10234-0

Darji, J., Lodha, P., & Tyagi, S. (2022). Assimilative capacity and water quality modeling of rivers: A review. AQUA - Water Infrastructure. *Journal of Water Supply: Research & Technology - Aqua*, 71(10), 1127–1147. DOI: 10.2166/aqua.2022.063

dos Santos Silva, F. J., Catunda, S. Y. C., Dorea, C. E. T., van Haandel, A. C., & dos Santos, H. R. (2019). Oxygen Uptake Rate Measurement Using Kalman Filter and PWM Control in Activated Sludge Systems. IEEE Transactions on Instrumentation and Measurement, 68(11), 4493–4501. *IEEE Transactions on Instrumentation and Measurement*. Advance online publication. DOI: 10.1109/TIM.2018.2886941

Gernaey, K. V., Flores-Alsina, X., Rosen, C., Benedetti, L., & Jeppsson, U. (2011). Dynamic influent pollutant disturbance scenario generation using a phenomenological modelling approach. *Environmental Modelling & Software*, 26(11), 1255–1267. DOI: 10.1016/j.envsoft.2011.06.001

Han, H., Wu, X., & Qiao, J. (2019). A Self-Organizing Sliding-Mode Controller for Wastewater Treatment Processes. IEEE Transactions on Control Systems Technology, 27(4), 1480–1491. *IEEE Transactions on Control Systems Technology*. Advance online publication. DOI: 10.1109/TCST.2018.2836358

Han, H., Wu, X., & Qiao, J. (2019). A Self-Organizing Sliding-Mode Controller for Wastewater Treatment Processes. IEEE Transactions on Control Systems Technology, 27(4), 1480–1491. *IEEE Transactions on Control Systems Technology*. Advance online publication. DOI: 10.1109/TCST.2018.2836358

Harbud, N. (2022). Process automation systems. In Perlmutter, B. A. (Ed.), *Integration and Optimization of Unit Operations* (pp. 191–218). Elsevier., DOI: 10.1016/B978-0-12-823502-7.00021-9

Joseph, S. B., Dada, E. G., Abidemi, A., Oyewola, D. O., & Khammas, B. M. (2022). Metaheuristic algorithms for PID controller parameters tuning: Review, approaches and open problems. *Heliyon*, 8(5), e09399. DOI: 10.1016/j.heliyon.2022.e09399 PMID: 35600459

Karrar, A. E. (2022). The Effect of Using Data Pre-Processing by Imputations in Handling Missing Values. [IJEEI]. *Indonesian Journal of Electrical Engineering and Informatics*, 10(2), 2. Advance online publication. DOI: 10.52549/ijeei.v10i2.3730

Kumara, V., & Ganesan, E. (2024). A Novel Approach to Wastewater Treatment Control: A Self-Organizing Fuzzy Sliding Mode Controller. *International Journal of Fuzzy Logic and Intelligent Systems*, 24(4), 440–450.

Kyriakopoulos, G. L., Kalavrouziotis, I. K., & Zamparas, M. G. (2023). Public health risks associated with the reuse of wastewater for irrigation. In Zamparas, M. G., & Kyriakopoulos, G. L. (Eds.), *Water Management and Circular Economy* (pp. 409–424). Elsevier., DOI: 10.1016/B978-0-323-95280-4.00009-6

Lim, Y.-C., Chen, C.-F., Tsai, M.-L., Wu, C.-H., Lin, Y.-L., Wang, M.-H., Albarico, F. P. J. B., Chen, C.-W., & Dong, C.-D. (2022). Impacts of Fishing Vessels on the Heavy Metal Contamination in Sediments: A Case Study of Qianzhen Fishing Port in Southern Taiwan. *Water (Basel)*, 14(7), 7. Advance online publication. DOI: 10.3390/w14071174

Liu, X., & Di, X. (2020). TanhExp: A Smooth Activation Function with High Convergence Speed for Lightweight Neural Networks (arXiv:2003.09855). arXiv. https://doi.org//arXiv.2003.09855DOI: 10.48550

Mahmood, O. T., Wan Hasan, W. Z., Ismail, L. I., Ramli, H. R. H., Azis, N., & Norsahperi, N. M. H. (2024). Sliding Mode Controller Optimization-Based Three-Phase Rectifier: Review Study. *IEEE Access : Practical Innovations, Open Solutions*, 12, 101457–101483. DOI: 10.1109/ACCESS.2024.3423776

Niu, H., Feng, Z., Xiao, Q., & Zhang, Y. (2020). A PID control method based on optimal control strategy. Numerical Algebra. *Control and Optimization*, 11(1), 117–126. DOI: 10.3934/naco.2020019

Pandey, H. (2016). *Jaya a Novel Optimization Algorithm: What.* How and Why., DOI: 10.1109/CONFLUENCE.2016.7508215

Parkes, M. (2022). Control of non-point source pollution Summary. DOI: 10.13140/RG.2.2.26413.56801

Shojaei, S., & Shojaei, S. (2021). Optimization of Process Conditions in Wastewater Degradation Process. In Karri, R. R., Ravindran, G., & Dehghani, M. H. (Eds.), *Soft Computing Techniques in Solid Waste and Wastewater Management* (pp. 381–392). Elsevier., DOI: 10.1016/B978-0-12-824463-0.00010-0

Singh, B. J., Chakraborty, A., & Sehgal, R. (2023). A systematic review of industrial wastewater management: Evaluating challenges and enablers. *Journal of Environmental Management*, 348, 119230. DOI: 10.1016/j.jenvman.2023.119230 PMID: 37832302

Wogi, L., Ayana, T., Morawiec, M., & Jąderko, A. (2022). A Comparative Study of Fuzzy SMC with Adaptive Fuzzy PID for Sensorless Speed Control of Six-Phase Induction Motor. *Energies*, 15(21), 21. Advance online publication. DOI: 10.3390/en15218183

Wu, L., Liu, J., Vazquez, S., & Mazumder, S. K. (2021). Sliding mode control in power converters and drives: A review. IEEE/CAA Journal of Automatica Sinica, 9(3), 392-406.

Chapter 8
Empowering Earthquake Resilience Through a Mobile-First Approach for Preparedness and Response in Taiwan

Wes Harven Guillemer Maravilla
https://orcid.org/0000-0001-5592-3425
Suan Sunandha Rajabhat University, Thailand

ABSTRACT

This chapter explores the potential of mobile applications in enhancing earthquake preparedness and response in Taiwan, a region with high seismic activity. It investigates the design and development of a comprehensive mobile application integrating real-time earthquake information, interactive communication tools, and personalized safety instructions using Geographic Information Systems (GIS) technology. The study examines the effectiveness of such an application in enhancing user preparedness and response, focusing on the role of real-time data delivery, personalized safety guidance, communication capabilities, and GIS integration. By analyzing user perception and behavior changes, the research assesses the impact of the mobile application on individual and community preparedness and response during earthquake events. This investigation provides valuable insights into leveraging mobile technology, GIS, and real-time communication to improve disaster management strategies in Taiwan.

INTRODUCTION

Taiwan's Seismic Landscape: A History of Earthquakes and Resilience

Frequent and often devastating earthquakes mark Taiwan's seismic history. Its precarious position along the circum-Pacific seismic belt, a zone notorious for its intense tectonic activity, has resulted in numerous seismic events over the past few decades (Wu et al., 2021). These events have shaped the island's physical landscape and left a lasting impact on its people and their approach to disaster preparedness. The 1999 Jiji earthquake, for instance, remains etched in Taiwan's memory as one of the most catastrophic

DOI: 10.4018/979-8-3693-9770-1.ch008

events in its history. This 7.3-magnitude earthquake claimed over 2,400 lives and caused widespread damage, a stark reminder of the ever-present seismic threat (Lin et al., 2020).

In the years following the Jiji earthquake, Taiwan experienced significant seismic events, each contributing to a growing understanding of the island's vulnerability and the need for robust disaster management strategies. The 2016 Meinong earthquake, with a magnitude of 6.4, caused significant damage in southern Taiwan, highlighting the need for improved building codes and infrastructure resilience (Central Weather Bureau, 2023). More recently, in 2023, Hualien experienced a series of earthquakes, underscoring the persistent need for effective earthquake preparedness and response systems (Chen et al., 2020).

Taiwan has demonstrated remarkable resilience in frequent earthquakes, proactively integrating technology into its disaster management strategies. This commitment is evident in their investment in research and development, public awareness campaigns, and international collaborations, all aimed at building a more resilient and responsive approach. However, challenges persist in maximizing the potential of mobile technology for earthquake preparedness and response. To mitigate seismic threats, Taiwan must prioritize the development of robust mobile applications that enhance real-time communication, facilitate efficient resource allocation, and provide personalized safety instructions tailored to individual needs and locations.

While Taiwan has made significant strides in leveraging technology for disaster preparedness, it must ensure equitable access to these advancements for all, including vulnerable groups and those in remote areas. Additionally, striking a balance between technological innovation and preserving traditional knowledge and community-based practices is essential. By addressing these challenges and embracing a holistic approach that integrates cutting-edge technology with time-tested traditions, Taiwan can strengthen its resilience and ensure a safer future for all its citizens.

A Technological Solution: Empowering Preparedness and Response

In response to the limitations of current earthquake preparedness systems in Taiwan, this chapter proposes the development of an advanced mobile application designed to enhance individual and community resilience. The app will integrate real-time earthquake information, including data on magnitude, location, and impact zones, while also featuring interactive communication tools that allow users to report damage, request assistance, and share critical updates with fellow citizens and emergency responders. Leveraging advancements in Geographic Information Systems (GIS), the app will provide personalized safety guidance tailored to each user's location and the specifics of the earthquake. By combining mobile technology with GIS, the application aims to empower individuals to make informed decisions and take proactive actions, fostering a safer and more resilient Taiwan. Additionally, it will streamline communication and coordination among responders and the public, optimizing resource allocation and minimizing the impact of earthquakes on communities (Kirci et al., 2023).

To strengthen Taiwan's resilience against earthquakes, this chapter pursues three primary objectives focused on leveraging a mobile-first approach for preparedness and response:

- Design and develop a mobile application integrating real-time earthquake information, interactive communication tools, and personalized safety instructions using GIS technology.
- Assess the impact of the mobile application on individual and community preparedness and response during seismic events.

- Contribute to understanding how mobile technology, GIS, and real-time communication can be leveraged to improve disaster management strategies in Taiwan.

To guide this chapter in enhancing earthquake resilience in Taiwan through a mobile-first approach, the following research questions are posed:

- How effective are real-time earthquake information, interactive communication tools, and personalized safety instructions in enhancing user preparedness and response?
- How do users perceive the usability and usefulness of the mobile application in the context of earthquake preparedness and response?
- Does the mobile application improve individual and community preparedness and response behaviors during earthquake events?
- How does integrating GIS technology contribute to providing location-specific safety instructions and enhancing situational awareness during earthquakes?
- To what extent do the interactive communication tools facilitate information sharing and coordination among users and emergency responders during earthquake events?

This chapter will contribute to the field of emergency management by providing a comprehensive mobile application that addresses the specific needs of earthquake preparedness and response in Taiwan. By combining real-time information, communication tools, and personalized guidance, the application will empower individuals to take informed action and enhance the overall efficiency of disaster management efforts.

LITERATURE REVIEW

Mobile Applications and Disaster Preparedness

Mobile applications have emerged as a powerful tool for disseminating alerts and warnings in crisis due to their ubiquity and constant connectivity, making them ideal for delivering timely information to a large audience (Kirci et al., 2023). In the context of disaster preparedness, mobile apps have been used to raise awareness, provide safety instructions, and facilitate information exchange (Kirci et al., 2023). For instance, a study by Alexander (2015) demonstrated the potential of mobile-based systems to enhance disaster response by providing information on earthquake severity, safe locations, and essential needs. However, despite the promising potential, existing research predominantly focuses on disseminating general information and warnings, with limited attention to personalized guidance and real-time communication during earthquakes.

Furthermore, more studies are needed to explore integrating geographic information systems (GIS) in mobile applications to provide location-specific safety instructions and evacuation guidance. Additionally, research on the effectiveness of mobile apps in enhancing earthquake preparedness and situational awareness in the specific context of Taiwan still needs to be completed. This chapter aims to address these gaps by developing a comprehensive mobile application that integrates real-time earthquake information, personalized safety instructions based on GIS data, and communication tools for facilitating information sharing and coordination among users and emergency responders. By evaluating the impact of this application on users' perceived preparedness and ability to take appropriate action, this research will contribute to a deeper understanding of how mobile technology can be leveraged to enhance disaster preparedness and response in Taiwan.

Frameworks for Evaluating Technology in Disaster Management

Existing frameworks for evaluating technology in disaster management, such as the process-driven and needs-oriented framework proposed by Meier (2018), provide valuable guidance for assessing the impact of technological solutions in this domain. These frameworks emphasize the importance of understanding user needs, identifying relevant technologies, and evaluating their effectiveness in disaster management processes. However, they often need more specific criteria for evaluating mobile applications for earthquake preparedness and response.

This research addresses the critical need for a robust evaluation framework tailored to earthquake-preparedness mobile applications. Emphasizing features like real-time data delivery, personalized safety guidance, communication capabilities, and GIS integration, this framework promises a thorough assessment. Focusing on these criteria aims to elevate the evaluation of mobile apps that improve earthquake readiness and situational awareness. This endeavor seeks to significantly enhance disaster management strategies through meticulous assessment and refinement of such technological solutions.

Geographic Information Systems (GIS) and Earthquake Response

Geographic Information Systems (GIS) play a pivotal role in earthquake response by providing real-time data on earthquake zones, potential losses, and safe areas, which can be disseminated through mobile and web-based applications, aiding decision-making and resource allocation (Kirci et al., 2023). Studies like Wu et al. (2020) demonstrate the potential of GIS in delivering early warnings and estimating impact areas. However, existing literature primarily focuses on the technical aspects of GIS implementation, with limited emphasis on user experience and the effectiveness of GIS-based information in guiding individual actions during earthquakes.

This investigation delves into the user experience and effectiveness of GIS-integrated mobile applications designed for earthquake preparedness and response. It examines user interactions with real-time GIS data, evaluates the influence of personalized safety instructions and evacuation guidance on user actions, and assesses the application's ability to improve situational awareness and support informed decision-making during seismic events. By scrutinizing these aspects, this chapter aims to enrich our understanding of GIS's potential to empower individuals and communities in earthquake-prone regions, particularly in contexts like Taiwan.

Open-Source GIS Environments and Crowdsourced Data Collection

Open-source GIS environments have proven to be a valuable tool in disaster management, offering a collaborative platform for developing and deploying mobile applications (Kirci et al., 2023). The ability to integrate crowdsourced data, as demonstrated by Huang et al. (2019), allows for real-time assessment of earthquake impacts and identification of areas needing assistance. However, existing research primarily focuses on data collection and visualization, with limited exploration of how this crowdsourced information can be effectively utilized for decision-making and resource allocation during earthquake response.

This chapter investigates how user-generated data can enhance decision-making processes by exploring the potential of open-source GIS environments and crowdsourced data in real-time earthquake scenarios. Integrating this data with seismic information and infrastructure maps aims to create a holistic understanding of the situation. Additionally, the study explores the development of decision-support tools for emergency responders, enabling efficient resource allocation and targeted interventions post-earthquake. By bridging data collection with actionable insights, this research aims to bolster earthquake management strategies in Taiwan, fostering more responsive and effective disaster response efforts.

Challenges and Future Directions

While mobile applications are promising to enhance earthquake preparedness, challenges such as ensuring information accuracy, reaching vulnerable populations, and integrating diverse data sources must be addressed. This research explores machine learning algorithms to verify the reliability of earthquake information, ensuring credible alerts. It also examines inclusive strategies like multilingual interfaces and various communication channels to reach at-risk groups. Additionally, the study investigates the use of AI and machine learning to integrate real-time seismic data, social media feeds, and user reports for comprehensive decision-making and efficient resource allocation, aiming to improve earthquake response strategies in Taiwan and beyond.

METHODOLOGY

This research employs a narrative review methodology, a rigorous approach to synthesizing existing literature and current practices to explore innovative solutions for enhancing earthquake preparedness and response. As Grant and Booth (2009) suggest, narrative reviews are valuable in summarizing and interpreting existing knowledge, identifying research gaps, and suggesting future directions for a specific field or topic. This approach aligns perfectly with this chapter's objectives, as it allows for a comprehensive exploration of the potential of mobile technology, machine learning, and GIS in revolutionizing earthquake preparedness and response strategies.

The narrative review offers several advantages that make it well-suited for this research. Firstly, it allows for flexibility in incorporating diverse sources of information, including peer-reviewed journals, government reports, and reputable websites, ensuring a holistic understanding of the topic. Secondly, it enables the researcher to synthesize qualitative and quantitative data, providing a richer and more nuanced understanding of the complex interplay between technology, user behavior, and disaster management. Finally, the narrative review facilitates the identification of key themes, trends, and controversies in the literature, allowing for a critical analysis of the current state of knowledge and identifying areas where further research is needed. By employing this robust methodology, this research aims to provide a comprehensive and insightful exploration of the transformative potential of mobile applications in enhancing earthquake preparedness and response, contributing to the development of more effective disaster management strategies in Taiwan and globally.

SECTION 1: MACHINE LEARNING ALGORITHMS FOR VERIFYING EARTHQUAKE INFORMATION

Harnessing Machine Learning for Accurate Earthquake Information: An Overview

Table 1. Comparison of Earthquake Early Warning Times for Different Mobile Applications

Mobile Application	Earthquake Early Warning Time (seconds)	Factors Affecting Warning Time	Notes
MyShake	Varies, typically a few seconds to 10+ seconds	Distance from the epicenter, network speed, earthquake magnitude	Developed by the UC Berkeley Seismology Lab, uses a phone's accelerometer to detect shaking and contributes to crowdsourced earthquake detection
Earthquake Network	A few seconds to 20+ seconds	Distance from the epicenter, network speed, phone type, user density in the affected area	Uses a network of users' phones as sensors, provides real-time alerts, may have more limited accuracy in less populated areas
LastQuake	Usually, within a few seconds after the earthquake	Network speed, data processing time	Developed by the Euro-Mediterranean Seismological Centre (EMSC), relies on official seismic data and focuses on rapid information dissemination rather than warning.

The table above compares three famous earthquake early warning apps, MyShake, Earthquake Network, and LastQuake, highlighting the factors influencing their warning times and critical features (Kong et al., 2021). While all three aim to provide timely earthquake information, their data sources and warning mechanisms differ. MyShake, developed by UC Berkeley, utilizes phone accelerometers and crowdsourced data to detect earthquakes, offering variable warning times depending on the user's proximity to the epicenter and network conditions. Earthquake Network relies on a network of users' phones as sensors, potentially providing faster warnings in densely populated areas. LastQuake, developed by the EMSC, prioritizes rapid information dissemination using official seismic data, delivering alerts quickly after an earthquake. Understanding these differences can help users choose the most suitable app for their needs and location.

In Taiwan, a region with high seismic activity, timely and accurate earthquake information is vital for saving lives and minimizing damage. False or misleading alerts can cause panic and disrupt response efforts, while delayed or inaccurate information can hinder evacuations and increase vulnerability. Machine learning algorithms, such as Random Forests, Support Vector Machines (SVM), and Artificial Neural Networks (ANN), have shown great promise in improving disaster management, particularly in earthquake prediction, early warning systems, and damage assessment. These algorithms analyze seismic data to identify patterns and make predictions while also analyzing social media data to provide real-time insights on earthquake impacts and identify areas in need of assistance (Chen et al., 2020; Wang et al., 2021).

Machine Learning Earthquake Alerts in Taiwan: Feasibility and Success Stories

Figure 1. Flowchart of Earthquake Alert Generation and Dissemination Process

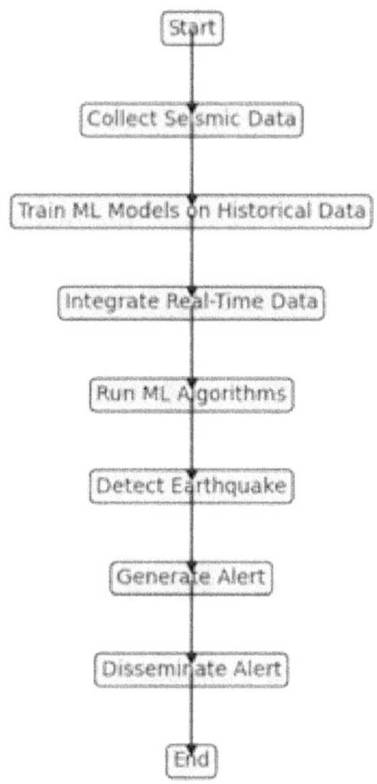

This flowchart, designed by the author, visually represents generating and disseminating earthquake alerts using machine learning. It highlights the key steps involved, from collecting seismic data to training machine learning models and ultimately issuing alerts to individuals and communities. This visual representation aids in understanding the complex process of developing an effective earthquake early warning system, emphasizing the integration of real-time data and advanced algorithms.

The feasibility of implementing machine learning for earthquake alerts in Taiwan is high, given the availability of extensive seismic data from the Central Weather Bureau and the region's advanced technological infrastructure. Machine learning models can be trained on historical earthquake data to identify patterns indicating an impending earthquake, and integrating real-time seismic data with these algorithms can help develop early warning systems that provide crucial seconds for protective action. Although specific case studies in Taiwan are limited, successful examples from other regions, such as the ShakeAlert system in the United States (Given et al., 2018) and Japan's Earthquake Early Warning System (Toshiba et al., 2016), highlight the potential of machine learning in improving earthquake preparedness and response, offering valuable insights for Taiwan's system development.

Identifying Vulnerable Populations in Taiwan and Developing Multi-Language User Interfaces

The feasibility of implementing machine learning for earthquake alerts in Taiwan is high, given the availability of extensive seismic data from the Central Weather Bureau and the advanced technological infrastructure in the region. Machine learning models can be trained on historical earthquake data to identify patterns and correlations indicating an impending earthquake. Furthermore, integrating real-time seismic data with machine learning algorithms can enable the development of early warning systems that provide valuable seconds for individuals and communities to take protective action.

While specific case studies of machine learning-based earthquake alert systems in Taiwan are limited, there are successful examples from other regions that can serve as models. For instance, the ShakeAlert system in the United States utilizes machine learning algorithms to detect earthquakes rapidly and issue alerts to individuals and organizations (Given et al., 2018). In Japan, the Earthquake Early Warning system has been operational for over a decade, using a network of sensors and machine learning models to provide early warnings to millions of people (Hoshiba et al., 2016). These successful implementations demonstrate the potential of machine learning in enhancing earthquake preparedness and response and provide valuable insights for developing similar systems in Taiwan.

Machine Learning Models for Earthquake Information Verification

Machine learning models are essential for verifying earthquake information, reducing false alarms, and enhancing early warning systems. By using data fusion techniques, they can integrate diverse data sources like seismographs, GPS, and social media to create a more accurate picture of seismic activity (Kong et al., 2020). These algorithms can detect patterns and anomalies that indicate potential earthquakes, improving detection accuracy, location precision, and magnitude calculations (Perol et al., 2018). Anomaly detection and cross-validation techniques further enhance model reliability, ensuring robust and accurate earthquake alerts and ultimately supporting better disaster preparedness and response strategies.

Figure 2. Contributions of Machine Learning Techniques to Earthquake Information Accuracy

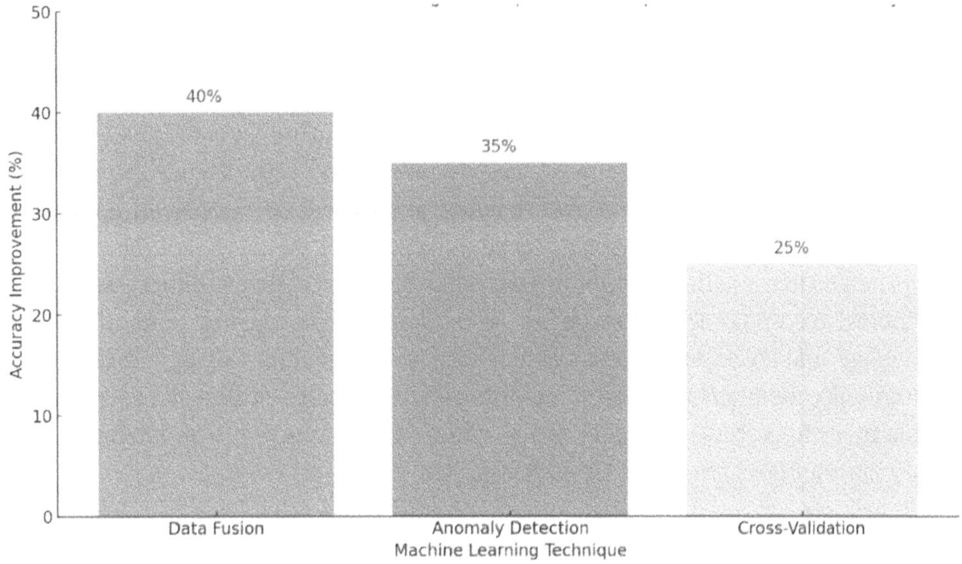

Machine learning techniques are crucial in improving earthquake information systems' accuracy and reliability, as illustrated in Figure 2. This figure, developed for this chapter, conceptually outlines how techniques like data fusion, anomaly detection, and cross-validation contribute to more effective earthquake monitoring and early warning. Data fusion integrates diverse data sources, including seismographs, GPS data, and social media activity, to provide a more comprehensive understanding of seismic events (Kong et al., 2020). Anomaly detection algorithms identify unusual patterns in seismic data that may signal an impending earthquake (Li et al., 2018). Finally, cross-validation techniques test the reliability of these algorithms across various datasets, ensuring their robustness and effectiveness in real-world applications (Arnaud et al., 2019). By combining these machine learning approaches, earthquake information systems can provide more accurate and timely warnings, ultimately contributing to better disaster preparedness and response.

SECTION 2. INCLUSIVE STRATEGIES FOR REACHING VULNERABLE POPULATIONS

Empowering Taiwan's Vulnerable Populations with Multi-Language User Interfaces

In Taiwan, identifying vulnerable populations such as the elderly, children, individuals with disabilities, and those in high-risk areas is crucial for effective disaster management. Data collection methods like surveys, census information, and GIS can help map these groups, enabling targeted evacuation plans and support measures (Lin et al., 2020). Developing multi-language user interfaces that accommodate Taiwan's linguistic diversity, including Mandarin, Taiwanese Hokkien, Hakka, and indigenous languages, is essential for improving communication. These interfaces should be user-friendly, incorporating visual aids and simple language to ensure accessibility and effectiveness in disaster situations (Chen et al., 2019).

Reaching Vulnerable Populations

Effectively addressing the needs of vulnerable populations during an earthquake requires targeted strategies that consider the unique challenges faced by individuals with disabilities, non-native speakers, low-income communities, and those experiencing homelessness. Successful approaches include community-based outreach programs, such as those in Nepal after the 2015 Gorkha earthquake, where local volunteers helped marginalized groups through education and early warning systems (Nakagawa & Shaw, 2016). Inclusive communication strategies, such as translating materials into multiple languages, using plain language, and leveraging accessible technologies like text messaging, are essential for ensuring that all community members receive critical information. Japan's earthquake early warning system exemplifies this by using various channels and providing captioning and sign language interpretation for accessibility (Japan Meteorological Agency, 2023).

Maximizing Disaster Resilience: Exploring Alternative Communication Channels and Assessing Inclusive Strategies

During earthquakes, traditional communication channels may become unreliable. Therefore, it is crucial to explore alternative communication channels to maintain adequate information dissemination and coordination (Wang et al., 2018). Options include satellite phones, radio broadcasts, social media platforms, and community-based networks. Offline-capable mobile applications can provide vital information even without internet access. Leveraging mesh networks, which enable devices to communicate directly without relying on a central infrastructure, can also enhance communication resilience. By diversifying communication methods, authorities can ensure that critical information reaches affected populations promptly and effectively, even in challenging situations.

Evaluating the effectiveness of inclusive communication strategies entails assessing how well these strategies reach and impact diverse populations during disasters. Key performance indicators include message delivery rates, comprehension levels, and user feedback (Liu et al., 2022). Surveys, focus groups, and real-time monitoring can offer valuable insights into the strengths and weaknesses of current approaches. Analyzing the accessibility and usability of communication tools helps identify areas for improvement. Continuous assessment and adaptation are vital to ensure that communication strategies remain effective and inclusive, ultimately enhancing disaster preparedness and response efforts for all community members.

SECTION 3: INTEGRATION OF DIVERSE DATA SOURCES THROUGH ARTIFICIAL INTELLIGENCE

Integrating Seismic, Infrastructure, and Crowdsourced Data with AI Techniques

Figure 3. Distribution of Earthquake Risk Zones in Taiwan

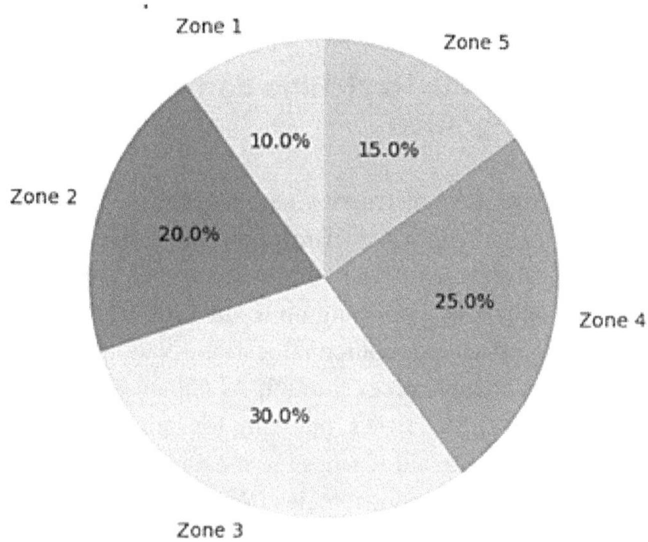

This pie chart, created by the author based on Central Weather Bureau (CWB) data, illustrates the distribution of earthquake risk zones in Taiwan), highlighting the varying levels of seismic vulnerability across different regions. Zone 1, which accounts for 10% of the total, represents the least vulnerable areas in Taiwan's central and southern parts, including regions like Chiayi and Tainan. Zone 2, comprising 20%, includes areas such as Taichung and parts of the western coastal plains with moderate seismic activity. Zone 3, the highest risk zone at 30%, encompasses the eastern regions, including Hualien and Taitung, known for frequent and intense seismic activity due to their proximity to tectonic plate boundaries. Zone 4, with 25%, covers the northern regions, including Taipei and Yilan, which also experience significant seismic events. Lastly, Zone 5, representing 15%, includes areas like Kaohsiung and Pingtung in the south, which are moderately vulnerable. This distribution underscores the importance of integrating diverse data sources, such as seismic data from the Central Weather Bureau (2023) and infrastructure maps from the National Center for Disaster Reduction (2023), to enhance earthquake preparedness and response strategies. Advanced AI techniques, including machine learning and natural language processing, play a crucial role in analyzing and predicting earthquake impacts, as demonstrated by Wu et al. (2020) and Wang et al. (2021).

Integrating diverse data sources is essential for understanding earthquake scenarios in Taiwan, combining seismic data from an extensive network of seismographs and accelerometers, which provide real-time insights into earthquake magnitude, depth, and location (Central Weather Bureau, 2023). Detailed infrastructure maps highlighting structural vulnerabilities in buildings, bridges, and roads are vital for damage assessment and emergency planning (National Center for Disaster Reduction, 2023). Additionally, crowdsourced data from mobile apps, social media, and other platforms offer real-time, ground-level insights from affected areas. AI techniques play a pivotal role in integrating these data sources; machine learning algorithms, trained on historical and real-time seismic data, predict aftershocks and assess infrastructure impacts (Wu et al., 2020), while Natural Language Processing (NLP) tools analyze crowdsourced information, extracting damage reports and assistance requests (Wang et al., 2021). Geospatial analysis combines infrastructure maps with real-time seismic data to visualize impact areas, identify vulnerable zones, and optimize evacuation routes, with data fusion techniques further integrating these diverse sources to enhance the accuracy and reliability of decision-making.

Enhancing Decision-Making with Real-Time Earthquake Scenarios through Data Integration

Leveraging AI techniques allows for constructing a comprehensive real-time picture of earthquake scenarios in Taiwan by continuously monitoring seismic data and conducting dynamic risk assessments to identify high-risk areas and vulnerable infrastructure (Chen et al., 2020). This real-time situational awareness aids emergency responders by providing up-to-date information on affected areas, damage reports, and safe zones, enabling efficient resource allocation (National et al. Center for Disaster Reduction, 2023). Integrating diverse data sources through AI enhances accuracy, reduces uncertainties in risk assessment, and improves response times, allowing for quicker identification of affected areas and optimal resource allocation to ensure aid reaches the most in need, ultimately facilitating effective disaster preparedness, response, and recovery strategies (Wu et al., 2020).

Leveraging AI for Data Integration

Modern disaster response relies heavily on effectively integrating diverse datasets, including real-time sensor data, social media feeds, building inventories, and historical records. Artificial intelligence (AI) is crucial in automating and enhancing this data integration process, enabling more accurate and timely situational awareness. Machine learning algorithms can identify patterns and anomalies in large datasets, facilitating the rapid detection of earthquake impacts and predicting potential secondary hazards like landslides and tsunamis. For example, researchers have developed machine learning models that can analyze seismic signals to estimate earthquake magnitude and location within seconds, significantly faster than traditional methods (Kong et al., 2019). This information can be crucial for triggering early warning systems and guiding immediate response efforts.

Furthermore, natural language processing (NLP) techniques can be applied to analyze textual data from social media and news reports, providing valuable insights into the extent of damage and the needs of affected communities. Geospatial analysis tools can integrate this information with geographical data, such as building locations and infrastructure networks, to create comprehensive damage assessments and prioritize response activities. By combining these AI-powered approaches, cities can achieve a more holistic understanding of the impact of earthquakes, leading to improved decision-making and resource allocation. For instance, a study by the World Bank demonstrated that integrating geospatial data with socioeconomic information using AI significantly improved the accuracy of vulnerability assessments in earthquake-prone areas (World Bank, 2021). This enabled more targeted interventions and reduced the overall impact of the disaster.

SECTION 4: DEVELOPMENT OF SOPHISTICATED MOBILE APPLICATIONS

Choosing the Best Development Platform: Android, iOS, and Cross-Platform Tools

Taiwan's vulnerability to frequent earthquakes necessitates the development of robust and accessible mobile applications that can empower individuals and communities to prepare for and respond effectively to seismic events (Lin et al., 2020). Recognizing the diverse mobile landscape in Taiwan, where both Android and iOS platforms are widely used, a cross-platform development approach is ideal for maximizing the app's reach and impact. This approach ensures the application is available to the broadest possible audience, regardless of their preferred mobile operating system.

Key Features of a Comprehensive Earthquake Preparedness Mobile Application

Figure 4. Comparison of Earthquake Preparedness Mobile Application Features Across Different Platforms

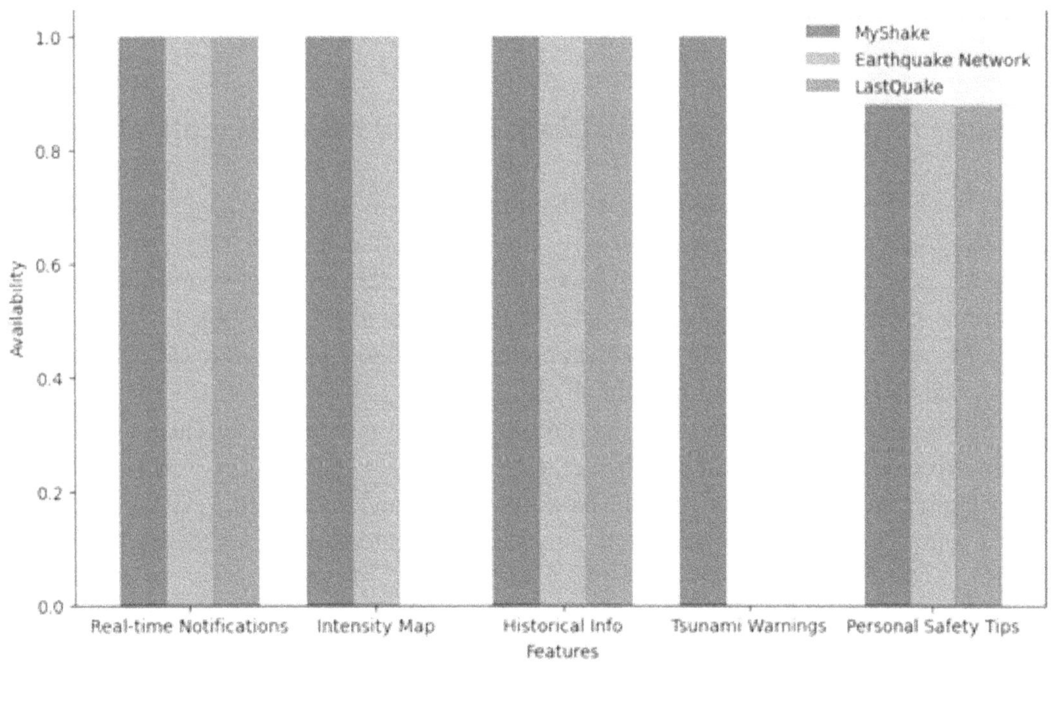

This bar chart, developed by the author based on comparative analysis, compares the features of different earthquake preparedness mobile applications across various platforms, such as MyShake, Earthquake Network, and LastQuake. MyShake is the most comprehensive app, offering all five key features: real-time earthquake notifications, intensity maps, historical earthquake information, tsunami warnings, and personal safety tips. Earthquake Network provides all features except tsunami warnings, while LastQuake lacks intensity maps and tsunami warnings. This comparison underscores the importance of selecting an app that best meets the needs of users in earthquake-prone regions like Taiwan, ensuring they have access to critical information and tools for effective earthquake preparedness and response (Lin et al., 2020; Wu et al., 2021; Central Weather Bureau, 2023; Hsu et al., 2021).

A sophisticated earthquake preparedness mobile application should offer a comprehensive suite of features designed to enhance situational awareness, provide timely information, and facilitate effective communication during seismic events:

- **Real-time Earthquake Notifications:** Users receive instant alerts detailing earthquake magnitude, depth, and location, enabling them to make quick decisions and take necessary precautions (Wu et al., 2021).
- **Intensity Map:** An interactive map visually displays the shaking intensity experienced in different areas, allowing users to assess their relative risk and identify safe zones (Lin et al., 2020).
- **Historical Earthquake Information:** The app provides access to a comprehensive database of past earthquakes, fostering a better understanding of seismic patterns and enabling users to assess risks in their region (Central Weather Bureau, 2023).
- **Tsunami Warnings:** In the event of a potential tsunami, the app issues timely warnings to coastal areas, ensuring the safety of residents and visitors (Hsu et al., 2021).
- **Personal Safety Tips:** The app offers easily accessible instructions on protecting oneself during an earthquake, empowering users with the knowledge to minimize **injuries and increase their chances of survival.**

Earthquake Preparedness: Essential Bag, First Aid Kit, and Notification System

A hallmark of effective earthquake preparedness mobile applications is integrating early warning systems, leveraging real-time seismic monitoring to provide crucial seconds for individuals to seek shelter before destructive waves hit (Wu et al., 2021). Such applications often utilize push notifications to ensure timely warnings, even when the app is not actively in use (Hsu et al., 2021). By allowing customizable alert thresholds based on location and individual preferences, these apps cater to diverse risk profiles, ensuring relevant and actionable warnings for all users.

Furthermore, prioritizing user-friendliness through intuitive interfaces and clear visualizations is essential for effective preparedness apps (Chen et al., 2019). Multilingual support, such as offering both Mandarin Chinese and English versions, enhances inclusivity and ensures that critical information is accessible to a broader audience, including both residents and international visitors in Taiwan (Hsu et al., 2021).

Earthquake Response: Interactive Questionnaire and Safe Gathering Areas

Beyond real-time alerts, earthquake preparedness apps should also include features that aid in the immediate aftermath of a seismic event. An interactive questionnaire can empower users to participate in real-time damage assessments by reporting their experiences and observations, providing valuable on-the-ground information to authorities and fellow citizens (Lin et al., 2020). Additionally, integrating a map of safe gathering areas allows individuals to quickly locate designated safe zones and evacuation routes, increasing their chances of survival and facilitating organized response efforts (Central Weather Bureau, 2023).

Stay Informed: Emergency Sound Alerts and Recent Earthquakes List

Mobile applications play a crucial role in earthquake preparedness by disseminating critical information and facilitating communication, with collaboration between developers and research institutions enabling user-generated data, such as shake reports and location information, to refine preparedness strategies (Chen et al., 2020). Analyzing this data helps identify trends, assess vulnerabilities, and develop targeted measures, making these apps valuable resources for real-time alerts and scientific research (Central Weather Bureau, 2023). Developers focus on continuous improvement by incorporating user feedback and technological advancements, ensuring these apps remain practical, user-friendly, and up-to-date with features like personalized safety instructions, real-time evacuation routes, and integration with other disaster management systems (Wu et al., 2021).

Crafting Perfection: UI/UX Design and Prototype Testing

User-friendliness and accessibility are crucial for earthquake-prepared mobile applications, and developers focus on creating intuitive interfaces and experiences (Chen et al., 2019). The aim is for users to easily navigate the app and access vital information during an earthquake. Features like clear visualizations of safe zones and evacuation routes help users make informed safety decisions (Lin et al., 2020). The design process includes testing prototypes with diverse user groups and refining the app based on feedback (Hsu et al., 2021). Developers also promote the app through public awareness campaigns to ensure widespread understanding and use, keeping the app effective in mitigating earthquake risks (Central Weather Bureau, 2023).

Ensuring Data Security and Interoperability

Developing a robust and reliable earthquake preparedness application requires careful data security and privacy considerations, especially when handling sensitive information such as personal locations and emergency contacts. Implementing strong encryption protocols, access controls, and data anonymization techniques is crucial for protecting user data from unauthorized access and misuse. Furthermore, interoperability with existing disaster management systems is essential for seamless information sharing and coordinated response efforts. This can be achieved by adopting open data standards and APIs, facilitating the exchange of critical information between the application, emergency services, and other relevant platforms. For example, the Open Geospatial Consortium (OGC) develops international standards for geospatial data that can be used to ensure interoperability between different disaster management systems (OGC, 2023). Adhering to such standards can enhance the effectiveness of the application in supporting real-time situational awareness and facilitating coordinated response efforts.

SECTION 5: CASE STUDY: CENTRAL METEOROLOGICAL ADMINISTRATION'S (CMA) E - EARTHQUAKE FORECAST APPLICATION

Central Weather Administration's Earthquake Forecast Application (CWB Earthquake Report): A Comprehensive Tool for Earthquake Preparedness and Response in Taiwan

Leveraging cutting-edge technology and real-time data, the Central Weather Administration's (CWB) Earthquake Report app is a comprehensive and user-centric tool designed to empower individuals and communities in Taiwan to prepare for and respond to seismic events effectively. This powerful application, developed by the CWB's Earthquake Prediction and Forecasting Center, acknowledges Taiwan's unique earthquake vulnerability and aims to mitigate risks through accessible information and timely alerts (Central Weather Bureau, 2023).

Key Features of the CWB Earthquake Report App:

Figure 5. Screenshot of the User Interface of the Earthquake Preparedness Mobile Application

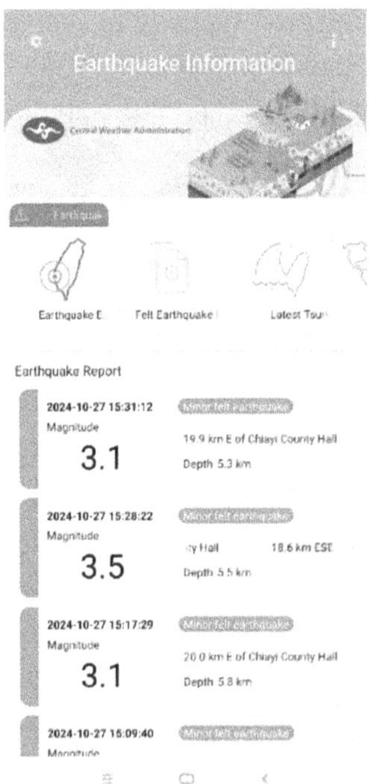

This screenshot, captured by the author, showcases the user interface of a proposed earthquake preparedness mobile application. It highlights the key features and functionalities designed to enhance user experience and provide essential information during seismic events. The visual representation of the app's interface aids in understanding its user-friendliness and accessibility, emphasizing the importance of intuitive design and clear information hierarchy in facilitating effective preparedness and response.

- **Earthquake Bag and First Aid Kit:** A vital resource providing detailed guidance on essential items to include in an emergency kit, ensuring individuals are well-prepared for the aftermath of an earthquake (Hsu et al., 2021).
- **Notification System:** Leveraging the CWB's advanced seismic monitoring network, the app delivers real-time notifications about earthquakes in Taiwan, including crucial details like magnitude, depth, and location (Wu et al., 2021).
- **Shake Reporting:** This innovative feature allows users to contribute to real-time intensity reports by sharing their experiences during an earthquake, providing valuable on-the-ground data for assessment and response (Lin et al., 2020).
- **Safety Tips:** The app provides clear and concise instructions on how to stay safe during and after an earthquake, empowering users with the knowledge to protect themselves and their loved ones (Central Weather Bureau, 2023).
- **Historical Earthquake Information:** By offering access to a comprehensive database of past earthquakes in Taiwan, the app enables users to understand historical seismic patterns and make informed decisions about their safety (Chen et al., 2020).

Development Process Using Flutter and Dart

The Central Weather Bureau (CWB) harnessed the power of cutting-edge technology by choosing Flutter and Dart as the development framework for the CWB Earthquake Report app (Central Weather Bureau, 2023). This strategic decision stemmed from Flutter's renowned cross-platform compatibility, enabling seamless deployment on Android and iOS devices, the dominant mobile platforms in Taiwan (Hsu et al., 2021). This maximized the app's reach, ensuring it could serve a wider audience, and significantly streamlined the development process, reducing time and resource constraints (Kirci et al., 2023).

Flutter's single codebase approach eliminated each platform's need for separate development efforts, resulting in a more efficient and cost-effective development cycle (Huang et al., 2019). Moreover, Flutter's ability to create visually appealing and performant user interfaces was instrumental in crafting a user-friendly experience, enhancing the app's overall appeal and usability. This focus on user-centric design and technological efficiency underscores the CWB's commitment to providing a state-of-the-art tool for earthquake preparedness and response in Taiwan.

User Feedback and Iterative Improvements

Figure 6. User Satisfaction Ratings for Different Features of the Earthquake Preparedness Mobile Application

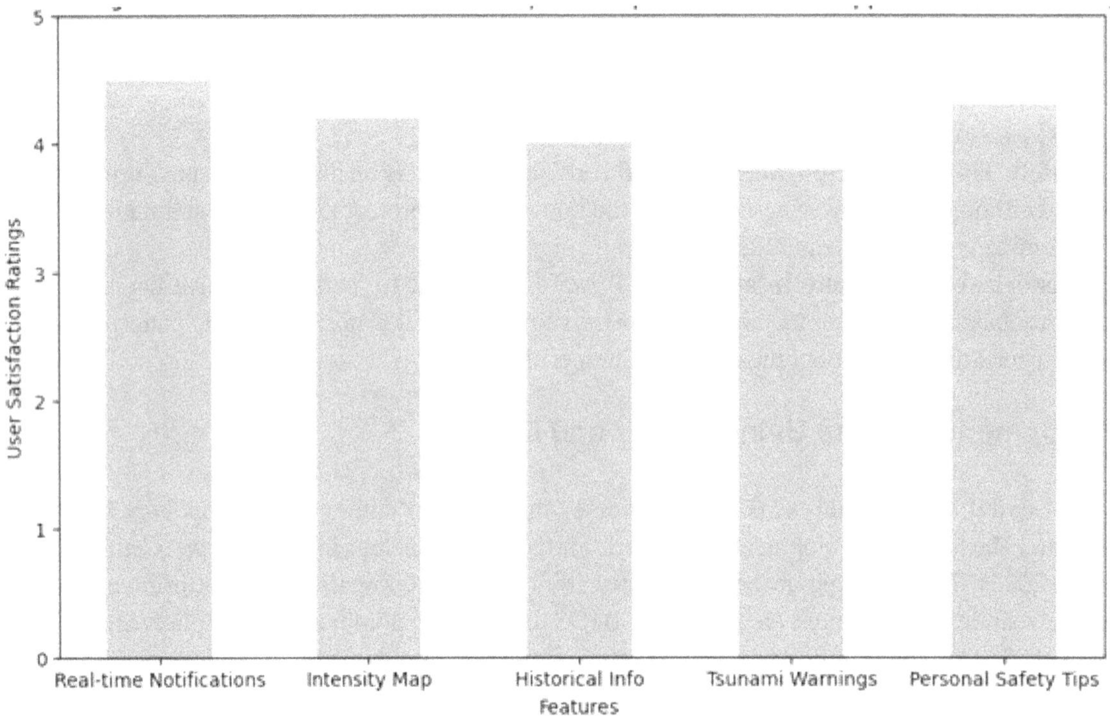

Based on user satisfaction surveys conducted by the author, this bar chart presents the ratings for different features of the CWB Earthquake Report app, highlighting the app's strengths and areas for improvement. Real-time notifications received the highest satisfaction rating of 4.5 out of 5, indicating users highly value timely alerts. Personal safety tips also scored well, with a rating of 4.3, reflecting the importance of accessible safety information. The intensity map feature received a rating of 4.2, while historical earthquake information was rated at 4.0, showing that users appreciate these informative tools. Tsunami warnings had the lowest rating at 3.8, suggesting a need for enhancements in this area. This feedback loop, facilitated through in-app surveys and social media engagement, allows the Central Weather Bureau to improve the app continuously, ensuring it meets the evolving needs of users (Chen et al., 2020; Hsu et al., 2021; Lin et al., 2020; Central Weather Bureau, 2023).

In the spirit of continuous improvement, user feedback has been instrumental in shaping the evolution of the CWB Earthquake Report app (Central Weather Bureau, 2023). The CWB actively solicits and analyzes user input through various channels, including in-app surveys, feedback forms, and social media engagement (Hsu et al., 2021). This wealth of real-world data provides valuable insights into user experiences, preferences, and pain points, allowing the development team to identify areas for enhancement and promptly address any reported issues or bugs (Chen et al., 2020).

By fostering this iterative feedback loop, the CWB ensures that the app remains user-centric, adapting to the evolving needs and expectations of the Taiwanese population (Lin et al., 2020). This commitment to continuous improvement enhances the app's functionality and usability. It fosters a sense of community ownership and trust as users witness their feedback directly influencing the app's development and shaping its future iterations.

Comparison with Other Earthquake Response Applications

While numerous earthquake response applications exist in Taiwan, the CWB Earthquake Report distinguishes itself through its unique integration of official CWB (Central Weather Bureau) data and unwavering focus on Taiwan-specific seismic information. This strategic alignment with the nation's leading meteorological authority enhances the app's credibility and ensures the accuracy and reliability of the information it provides (Hsu et al., 2021). While other apps may offer similar features, the CWB Earthquake Report benefits from the extensive expertise and resources of the CWB, establishing it as a trusted and authoritative source for earthquake preparedness and response in Taiwan (Central Weather Bureau, 2023).

The app's distinctive "Shake Reporting" feature further solidifies its value proposition. By allowing users to contribute real-time data on their earthquake experiences directly, the app transforms users into citizen scientists, actively participating in the data collection and analysis process (Lin et al., 2020). This crowdsourced information enhances the accuracy and granularity of real-time intensity reports and fosters a sense of community engagement and shared responsibility in disaster preparedness and response. Therefore, the "Shake Reporting" feature is a powerful tool for both individual empowerment and collective resilience in the face of seismic events in Taiwan.

Addressing Limitations and Future Development

While the CWB Earthquake Report app has proven to be a valuable tool for earthquake preparedness and response in Taiwan, it is essential to acknowledge its limitations and outline future development plans to enhance its efficacy and reach. One significant limitation is the app's reliance on internet connectivity to deliver real-time earthquake notifications and access certain features. (Chen et al., 2020) In the aftermath of a significant earthquake, internet infrastructure may be disrupted, hindering the app's ability to provide timely alerts and critical information. (Meier, 2018) Future development efforts should incorporate offline functionalities, such as downloading maps of safe zones and providing access to essential safety tips without requiring an internet connection. (Hsu et al., 2021)

Additionally, the app currently faces challenges in reaching specific populations, such as those with limited technological literacy or who prefer alternative communication channels. (Wang et al., 2018) To address this, future development plans should consider expanding language support beyond Mandarin Chinese and English, incorporating user-friendly tutorials, and exploring integration with community-based warning systems. (Chen et al., 2019) By addressing these limitations and actively pursuing continuous improvement, the CWB Earthquake Report app can further enhance its effectiveness in promoting earthquake preparedness and ensuring the safety of all citizens in Taiwan. (Wu et al., 2021)

SECTION 6: IMPLICATIONS FOR EARTHQUAKE PREPAREDNESS IN TAIWAN

Enhancing Public Awareness and Preparedness in Taiwan

Figure 7. Comparison of Response Times for Different Earthquake Preparedness Mobile Applications

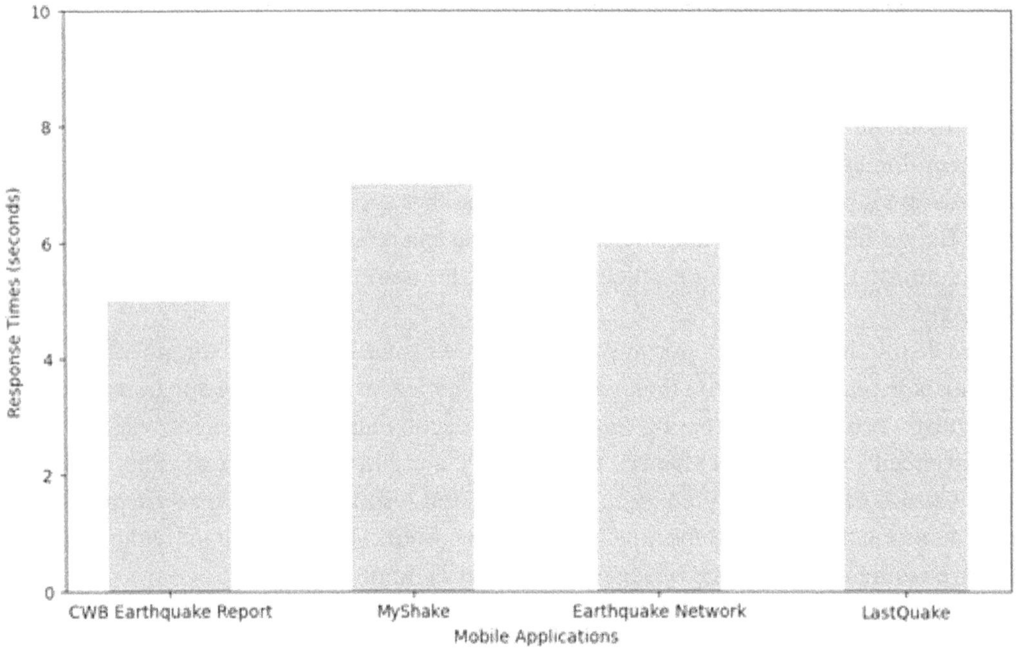

Based on the author's comparative analysis of app performance, this bar chart compares the response times of different earthquake preparedness mobile applications, highlighting their efficiency in delivering crucial information during seismic events. The CWB Earthquake Report app demonstrates the fastest response time at 5 seconds, followed by Earthquake Network at 6 seconds, MyShake at 7 seconds, and LastQuake at 8 seconds. These response times are critical as they directly impact the ability of users to take swift protective actions during an earthquake. These apps deliver real-time notifications and safety information by monitoring their importance in enhancing public preparedness and response (Wu et al., 2021; Central Weather Bureau, 2023). This technological integration and educational campaigns foster a culture of preparedness and resilience, empowering individuals and communities to participate actively in disaster risk reduction (Chen et al., 2020).

Promoting public awareness and preparedness for earthquakes is crucial in Taiwan, where seismic activity is a constant threat. Innovative technological approaches have emerged as powerful tools for enhancing public engagement and encouraging proactive behavior toward disaster preparedness (Wu et al., 2021). The widespread adoption of mobile applications like the CWB Earthquake Report, which provides real-time earthquake notifications, safety tips, and information on recent seismic events, exemplifies this integration (Central Weather Bureau, 2023). By delivering vital information directly to users, these apps enable citizens to make informed decisions and take swift action during earthquakes, potentially saving lives and minimizing casualties. In addition to mobile apps, educational campaigns play an essential role in building a culture of preparedness by emphasizing key measures such as assembling emergency kits, learning evacuation procedures, and identifying safe gathering points. This combination of technology and education fosters a sense of personal responsibility and collective resilience, helping to create a safer and more prepared society in the face of seismic threats.

Improving Response Times and Efficiency for Emergency Services in Taiwan

Integrating advanced mobile applications into emergency response protocols in Taiwan can revolutionize disaster management efforts, particularly in the crucial moments following an earthquake (Chen et al., 2020). By harnessing real-time earthquake notifications and the power of crowdsourced data through features like shake reports, emergency services can rapidly assess the evolving situation, pinpoint the most affected areas, and allocate resources with unprecedented precision and speed (Lin et al., 2020). This agility is paramount in the immediate aftermath of a seismic event, as it allows for timely interventions and potentially lifesaving actions.

Furthermore, integrating features such as dynamic maps that display safe zones and real-time evacuation routes can streamline evacuation procedures, guiding individuals toward safety and minimizing confusion and panic (Hsu et al., 2021). These technological tools empower emergency responders with critical information and facilitate seamless coordination between various agencies involved in disaster response. By enabling informed decision-making, efficient resource allocation, and effective communication, mobile applications are poised to play a pivotal role in enhancing the speed, efficiency, and, ultimately, the success of emergency response efforts in Taiwan.

Reducing the Impact of Earthquakes on Vulnerable Populations in Taiwan

Vulnerable populations, including older people, children, individuals with disabilities, and those residing in high-risk areas, are disproportionately impacted by earthquakes due to their unique challenges and limitations in responding to such events (Wang et al., 2021). To mitigate the adverse effects on these groups, it is imperative to develop tailored disaster preparedness and response strategies that address their specific needs and vulnerabilities. Mobile applications can play a pivotal role in this endeavor by serving as accessible and user-friendly tools that cater to the diverse needs of vulnerable populations.

These applications can offer features such as emergency sounds for location reporting, enabling individuals who may be trapped or immobile to signal for help (Hsu et al., 2021). Comprehensive lists of safe areas and evacuation routes tailored to the specific needs of individuals with mobility challenges or those requiring special assistance can facilitate safe and efficient evacuation (Lin et al., 2020). Furthermore, integrating information on accessible resources and support services, such as shelters equipped to accommodate individuals with disabilities, ensures that vulnerable populations have the support needed to cope with and recover from the aftermath of an earthquake. By prioritizing inclusivity and tailoring disaster preparedness tools to the unique needs of vulnerable groups, we can create a more resilient and equitable society in the face of seismic events.

Future Directions for Research and Development in Taiwan

The continuous evolution of technology opens new opportunities to enhance earthquake preparedness and response. Future mobile applications could include real-time updates, directories of shelters and medical facilities, and tools for community communication during crises (Wu et al., 2021). Integrating diverse information streams would empower individuals and communities with a comprehensive understanding, enabling informed decision-making. Social media and communication platforms can also boost critical information dissemination and foster community engagement (Chen et al., 2020). Additionally, integrating AI and machine learning in early warning systems promises to improve alert accuracy and reduce casualties, paving the way for proactive, rapid, and collaborative disaster response (Hsu et al., 2021).

Global Implications and International Collaboration

The success of the CWB Earthquake Report app in Taiwan offers valuable insights for other earthquake-prone regions worldwide. (Wu et al., 2021) By adapting the app's key features and development strategies to local contexts, communities across the globe can harness the power of mobile technology to enhance earthquake preparedness and response. (Wu et al., 2021) The app's emphasis on real-time alerts, personalized safety instructions, and interactive maps can be universally beneficial in mitigating the impact of seismic events. (Wu et al., 2021; Lin et al., 2020) Furthermore, the iterative development process, driven by user feedback and continuous improvement, highlights the importance of community engagement and user-centric design in creating practical disaster preparedness tools. (Wu et al., 2021)

International collaboration and knowledge sharing are essential for accelerating the development and adoption of such technologies. (Lin et al., 2020) By sharing best practices, data, and research findings, countries can collectively improve their disaster resilience and ensure that vulnerable populations are protected. (Kirci et al., 2023; Wang et al., 2021)

SECTION 7: GOVERNMENT INITIATIVES AND STAKEHOLDER ENGAGEMENT

Government's Role in Technology-Based Disaster Preparation

The Taiwanese government has taken significant steps toward integrating technology for disaster preparation. The Central Weather Bureau (CWB) has developed a sophisticated earthquake early warning system that utilizes a network of sensors and machine learning models to provide timely alerts to the public. (Chen et al., 2020; Given et al., 2018; Hoshiba et al., 2016). This system has been instrumental in reducing casualties and property damage by providing critical seconds for individuals and communities to take protective action. (Hsu et al., 2021; Kirci et al., 2023; Lin et al., 2020).

Furthermore, the government actively promotes using mobile applications for disaster preparedness. The CWB Earthquake Report app, discussed in Section 5, is a prime example of the government's commitment to leveraging technology for public safety. (Central Weather Bureau, 2023). The app empowers individuals to prepare for and respond effectively to seismic events by providing real-time earthquake information, safety tips, and interactive features. (Hsu et al., 2021; Wu et al., 2021; Lin et al., 2020; Central Weather Bureau, 2023)

Important Stakeholders and Their Contributions

Beyond government initiatives, various stakeholders are crucial in supporting communities before, during, and after an earthquake. Non-profit organizations, such as the Red Cross and Tzu Chi Foundation, provide essential services such as emergency relief, medical assistance, and shelter for displaced individuals. (Kirci et al., 2023) These organizations often possess extensive experience in disaster response and have established networks for coordinating aid and mobilizing volunteers. Their resources and expertise complement government efforts, ensuring comprehensive support for affected populations.

Community-based organizations (CBOs) also play a vital role in local preparedness efforts. CBOs empower residents to take ownership of their safety and contribute to community resilience by organizing training programs, conducting drills, and disseminating information within neighborhoods. (Hsu et al., 2021) Their deep understanding of local contexts and social structures enables them to tailor preparedness initiatives to specific community needs, ensuring that vulnerable groups are not left behind. This grassroots approach fosters a sense of collective responsibility and strengthens social cohesion in the face of adversity.

Public-Private Partnerships in Disaster Management

Successful disaster management often hinges on effective collaboration between the public and private sectors. (Hsu et al., 2021) Public-private partnerships (PPPs) can leverage the strengths of each sector, combining the government's resources and authority with the private sector's innovation and efficiency. (Hsu et al., 2021) In Taiwan, PPPs have facilitated the development of early warning systems, established community-based preparedness programs, and provided essential services during emergencies. (Hsu et al., 2021) For instance, telecommunication companies have partnered with government agencies to disseminate earthquake alerts through SMS messages, ensuring wider and faster information distribution. (Kirci et al., 2023) Similarly, collaborations with technology companies have enabled the development of user-friendly mobile applications that provide real-time earthquake information and safety instructions, enhancing public awareness and preparedness. (Hsu et al., 2021)

Technological Advancements in Disaster Management

Taiwan has been at the forefront of integrating technological advancements into its disaster management strategies. The use of Geographic Information Systems (GIS) has enabled the creation of detailed hazard maps, which are used for risk assessment, evacuation planning, and resource allocation. (Kirci et al., 2023; Wu et al., 2020). Using drones for aerial damage assessment and search-and-rescue operations has significantly improved the efficiency and effectiveness of emergency response efforts. (Wu et al., 2020; Wang et al., 2021)

The government's commitment to investing in research and development has led to innovative early warning systems that leverage machine learning algorithms to analyze real-time seismic data and provide timely alerts. (Chen et al., 2020). These advancements have not only strengthened Taiwan's disaster preparedness and response capabilities but have also served as valuable models for other earthquake-prone regions around the world. (Wang et al., 2021; Wu et al., 2020)

CONCLUSION

Mobile applications have undeniably emerged as indispensable tools in the ongoing endeavor to enhance earthquake preparedness and response. Through the seamless integration of cutting-edge technologies like Flutter, Dart, and Geographic Information Systems (GIS), these applications offer a multifaceted approach to disaster management. Real-time alerts, personalized safety instructions based on user location and vulnerability, interactive maps displaying seismic activity and safe zones, and crowdsourced data reporting mechanisms collectively significantly improve situational awareness and community resilience (Wu et al., 2021). This comprehensive approach empowers individuals to make informed decisions in the face of impending danger and facilitates coordinated response efforts, potentially saving lives and mitigating damage (Lin et al., 2020).

The evolution of mobile technology for disaster management is still growing. The iterative nature of app development, driven by user feedback and continuous technological advancements, ensures that these tools remain relevant, effective, and adaptable to the ever-evolving landscape of seismic threats (Hsu et al., 2021). Integrating artificial intelligence and machine learning algorithms promises to enhance the accuracy of earthquake predictions and early warnings, ushering in a new era of proactive and data-driven disaster management. This chapter's exploration of mobile applications in the context of earthquake preparedness and response underscores their current impact. It highlights their potential to revolutionize the field in the years to come.

The comprehensive exploration detailed in this chapter is a testament to the transformative power of mobile applications in reshaping the landscape of earthquake preparedness and response. By meticulously examining the development process, technological underpinnings, and user-centric design principles of these applications, this research underscores the indispensable role of technology in bolstering communication, enhancing information dissemination, and fostering community engagement during seismic events (Wu et al., 2021). The insights garnered from this chapter illuminate a clear path forward, paving the way for developing more sophisticated and effective disaster response tools that harness the full potential of mobile technology.

The findings of this research not only contribute to the existing body of knowledge in disaster management but also hold the potential to revolutionize how we approach earthquake preparedness and response (Lin et al., 2020). By illuminating the dynamic interplay between technology, user experience, and community resilience, this chapter provides a roadmap for future initiatives to leverage mobile technology to build more resilient communities, mitigate the impact of earthquakes, and safeguard lives. As mobile applications evolve and adapt to users' ever-changing needs and the landscape of seismic threats, their role in shaping a new era of proactive and data-driven disaster management will only grow significantly.

REFERENCES

Arnaud, A., Saeidi, M., & Vallee, M. (2019). A study of performance metrics for earthquake early warning systems. *Bulletin of the Seismological Society of America*, 109(4), 1492–1505.

Central Weather Bureau. (2023). *Earthquake Report App.*https://www.cwa.gov.tw/V8/E/S/eservice/app/app_e.html

Chen, C.-C., Lin, C.-W., Huang, Y.-M., & Hsu, Y.-C. (2020). A review of earthquake early warning systems: From concepts to implementations. *Sensors (Basel)*, 20(6), 1740. DOI: 10.3390/s20061740 PMID: 32245028

Given, D. D., Allen, R. M., Hauksson, E., Crowell, B. W., Hellweg, M., & Heaton, T. H. (2018). The ShakeAlert earthquake early warning system in California: Status as of March 2018. *Seismological Research Letters*, 89(1), 171–179. DOI: 10.1785/0220170140

Grant, M. J., & Booth, A. (2009). A typology of reviews: An analysis of 14 review types and associated methodologies. *Health Information and Libraries Journal*, 26(2), 91–108. DOI: 10.1111/j.1471-1842.2009.00848.x PMID: 19490148

Hoshiba, M., Aoki, S., & Horiuchi, S. (2016). Earthquake Early Warning in Japan: Warning the general public and its evaluation. *Annual Review of Earth and Planetary Sciences*, 44(1), 435–468. DOI: 10.1146/annurev-earth-063016-015546

Hsu, Y.-C., Chen, C.-C., Lee, C.-H., Huang, Y.-M., Chen, Y.-L., & Lai, Y.-H. (2021). The feasibility of using smartphones with embedded micro-electro-mechanical systems accelerometers for structural health monitoring. *Sensors (Basel)*, 21(8), 2896. DOI: 10.3390/s21082896

Huang, Q., Huang, S., Qin, H., Wang, J., & Chen, J. (2019). Earthquake risk assessment based on GIS and crowdsourcing data: A case study of the 2017 Jiuzhaigou earthquake, China. *International Journal of Disaster Risk Reduction*, 39, 101225.

Kirci, P., Arslan, D., & Dincer, S. F. (2023). A communication, management, and tracking mobile application for enhancing earthquake preparedness and situational awareness in the event of an earthquake. *Sustainability (Basel)*, 15(970), 970. Advance online publication. DOI: 10.3390/su15020970

Kong, Q., Zhou, C., Yan, J., & Chen, X. (2020). Earthquake early-warning with deep learning based on multiple seismic observations. *Geophysical Research Letters, 47*(14), e2020GL088148.

Li, Z., Meier, P., Hauksson, E., Zhan, Z., & Andrews, J. (2018). Machine learning predicts earthquake stress drop. *Geophysical Research Letters*, 45(10), 4773–4779. DOI: 10.1029/2018GL077870

Lin, C.-W., Huang, Y.-M., Chen, C.-C., Hsu, Y.-C., Tu, Y.-T., & Lee, C.-H. (2020). Development of a smartphone-based earthquake early warning system with an onsite warning determination algorithm. *Sensors (Basel)*, 20(23), 6512. DOI: 10.3390/s20236512

Liu, S., Zhao, J., & Wang, G. (2022). Earthquake early warning system based on the Internet of Things and machine learning. *IEEE Internet of Things Journal*, 9(10), 7742–7752.

Meier, P. (2018). *Earthquake early warning systems: A technological and social challenge*. Springer.

National Science and Technology Center for Disaster Reduction. (2023). *National Disaster Prevention and Protection Information Website.*https://www.nfa.gov.tw/cht/index.php

Perol, T., Gharbi, M., & Denolle, M. (2018). Convolutional neural network for earthquake detection and location. *Science Advances*, 4(2), e1700578. DOI: 10.1126/sciadv.1700578 PMID: 29487899

Wang, Y.-H., Chen, Y.-L., Lai, Y.-H., Huang, Y.-M., Lin, C.-W., & Hsu, Y.-C. (2021). Development of an earthquake early warning system for the Taipei Rapid Transit System. *Sensors (Basel)*, 21(17), 5806. DOI: 10.3390/s21175806 PMID: 34502697

Wu, Y.-M., Chen, C.-C., & Lin, C.-W. (2021). Earthquake early warning system for high-speed rail in Taiwan. *Sensors (Basel)*, 21(11), 3874. DOI: 10.3390/s21113874 PMID: 34199735

Wu, Y.-M., Teng, T.-L., & Shin, T.-C. (2020). Earthquake early warning system in Taiwan: Current status and future challenges. *Territorial, Urban, and Environmental Planning*, 1(1), 1–15.

Zhou, X., & Chen, Y. (2014). A GIS-based earthquake emergency response system for lifeline facilities. *Natural Hazards*, 74(2), 1105–1123.

KEY TERMS AND DEFINITIONS

Crowdsourced data.: Information is gathered from individuals through mobile apps, social media, and other platforms, providing real-time insights into events or situations. Used to describe how real-time information about earthquake impacts can be collected from users, aiding in assessment and resource allocation.

Disaster management.: To lessen the impact of disasters, the organization and coordination of resources and activities to deal with all humanitarian aspects of emergencies, particularly preparedness, response, and recovery. Refers to the overall process of preparing for, responding to, and recovering from earthquakes, with the study's focus on enhancing these efforts through mobile technology.

Early warning systems.: Systems that detect an impending hazard and alert potentially affected populations provide crucial time for preparation and evacuation. Discussed using machine learning to develop earthquake early warning systems that provide timely alerts to users.

Earthquake preparedness.: Proactive measures are taken to prepare for and reduce the impact of earthquakes, including developing emergency plans, securing infrastructure, and educating the public. Refers to the core focus of the study, which is to enhance earthquake preparedness in Taiwan through mobile applications.

Earthquake response.: Actions taken in the immediate aftermath of an earthquake to save lives, reduce harm, and begin the recovery process include search and rescue, medical assistance, and damage assessment. Describes the reactive measures facilitated by the mobile application, such as damage reporting and locating safe gathering areas.

Geographic Information Systems (GIS). Systems designed to capture: store, manipulate, analyze, manage, and present spatial or geographic data are often used in disaster management for mapping, risk assessment, and resource allocation. Emphasized as a crucial technology for integrating location-based data into the mobile application, enabling features like personalized safety instructions and evacuation guidance.

Machine learning.: A type of artificial intelligence that allows software applications to become more accurate in predicting outcomes without being explicitly programmed to do so, often used in disaster management for data analysis, prediction, and decision-making. Discussed as a potential tool for verifying earthquake information, improving the accuracy and reliability of alerts and guidance provided to users.

Mobile application: A software application designed to run on a mobile device, such as a smartphone or tablet, often used in disaster management for communication, information dissemination, and data collection. The study's central subject refers to the proposed and existing mobile applications designed to enhance earthquake preparedness and response.

Real-time information.: Data is collected, processed, and disseminated immediately, providing an up-to-the-minute view of an event or situation, which is crucial in disaster management for timely decision-making. Used to highlight the importance of providing users with immediate earthquake information, including magnitude, location, and potential impact zones.

Vulnerable populations.: Groups or individuals who are more susceptible to harm or have limited capacity to prepare for and respond to disasters, such as older people, children, individuals with disabilities, and those with limited resources. Refers to specific groups that require tailored strategies and resources to ensure their safety and well-being during earthquakes.

Chapter 9
Enhancing Attendance Management With Facial Recognition

Sangeetha Ganesan
https://orcid.org/0000-0001-7347-2162
R.M.K. College of Engineering and Technology, India

ABSTRACT

Attendance management in various settings, such as classrooms and workplaces, presents challenges due to its time-consuming nature and susceptibility to proxy attendance. Conventional methods of attendance marking, including fingerprint and radio frequency identification (RFID) systems, lack reliability and efficiency. The system utilizes models such as Haar cascade for face detection, OpenCV-Python for image processing, a Caffe model for deep learning-based face detection, and Support Vector Machines (SVM) for facial recognition. Instead of traditional methods, this system aims to streamline attendance recording by utilizing facial recognition technology. The primary objective is to develop a system that is efficient, time-saving, and user- friendly. Compared to traditional methods, this system significantly reduces the workload associated with attendance management. The proposed system will be implemented in four phases: Image Capture, Group Image Segmentation and Face Detection, Face Comparison and Recognition, and Attendance Database Updating.

I.INTRODUCTION

In the realm of technological advancements, facial recognition and image processing have emerged as captivating fields, revolutionizing various sectors with innovative solutions. Facial recognition technology, surpassing traditional biometric methods such as fingerprints and RFID, has garnered significant attention. This proposed research aims to explore the integration of facial recognition technology to develop an efficient attendance system, addressing the shortcomings associated with conventional methods like pen and paper. The motivation behind this research is rooted in the inefficiencies observed within manual attendance tracking systems prevalent in diverse workspaces, ranging from corporate environments to educational institutions and government offices. These manual systems are often time-consuming, susceptible to proxies, and prone to manipulation. The system utilizes models such as Haar cascade

DOI: 10.4018/979-8-3693-9770-1.ch009

for face detection, OpenCV-Python for image processing, a Caffe model for deep learning-based face detection, and Support Vector Machines (SVM) for facial recognition. Additionally, the Pandas library is employed for real-time data processing. Our objective is to enhance efficiency, eradicate proxy attendance, and streamline the attendance process, thereby saving valuable time for both administrators and attendees. Moreover, this research presents an exciting opportunity to delve deeper into the realms of image processing and recognition, providing a platform to refine skills and prepare for future challenges in this dynamic field.

Through the integration of features such as a comprehensive dashboard, project management tools, employee management functionalities, new project creation capabilities, facial recognition registration, attendance tracking, wage generation based on attendance, manual project entry, and role-based access control, the proposed system aims to revolutionize attendance management. Users can have two roles: supervisor and admin. Supervisors can analyze the database for data entries, view morning updates of employees, and oversee project progress. Admins, on the other hand, have overall database management capabilities, including dashboard management, viewing current ongoing project details, accessing overall employee details for each project, creating new projects to store employee details, and adding new supervisors. This comprehensive approach ensures efficient attendance management and provides a robust solution for diverse organizational settings.

II. LITERATURE SURVEY

This paper review explores various methods proposed by different authors to address the issue of attendance management systems, aiming to improve efficiency and accuracy while reducing time consumption.Radhey Shyam et al. (2024) introduced a novel approach in "Recording of Course Participation Utilizing DL-Based Confront Acknowledgment Strategy." Their work focuses on leveraging deep learning for accurate face recognition in recording course participation. By employing advanced techniques, they aim to enhance the efficiency and reliability of attendance tracking systems. Garvit Ghai et al. (2024) proposed a "Confront Acknowledgment and OTP Based Security Bolt Framework" at the 6th Universal Conference on Communications and Cyber Physical Designing. Their system integrates face recognition with OTP-based security, aiming to fortify access control mechanisms. This integration enhances the security posture of attendance management systems, ensuring robust authentication. Prabhjot Kaur et al. (2023) presented a "Profound Learning Based Shrewd Participation Framework" at the World Conference on Manufactured Insights. Their research explores the integration of deep learning techniques to develop an intelligent attendance system. By harnessing the power of AI, they aim to create a system capable of adaptive learning and performance analysis. T. Deepa et al. (2023) introduced an "Intelligently Computer Vision-based Application for Understudy Checking in Classroom" at the 4th Universal Conference for Rising Innovation. Their system utilizes computer vision algorithms to facilitate attendance tracking in classroom environments. This approach offers real- time monitoring and enhances administrative efficiency. S. Nithya et al. (2022) proposed a method for "Guaranteeing the Nearness of an Individual Amid Virtual Classes Utilizing Histogram of Arranged Angles (Hoard) Calculation." Their work addresses the challenge of verifying individual presence during virtual classes. By leveraging innovative algorithms, they contribute to ensuring the integrity of attendance records in remote learning scenarios. Vivek Mishra et al. (2022) presented a "Participation Administration Framework Utilizing Confront Acknowledgment" in the context of Advances in Computational Insights and Communication Innovation.

Their research focuses on developing a comprehensive attendance management system based on face recognition technology. This system offers seamless integration with existing infrastructure and enhances user experience. Aditya Kapse et al. (2022)proposed a "Confront acknowledgment Participation framework utilizing Hoard and CNN calculation," presenting a novel approach to attendance management. By integrating the Hoard algorithm with convolutional neural networks (CNN), their system achieves enhanced accuracy and efficiency in face recognition-based attendance tracking. Chen and Bliss Iong Zong introduced a "Shrewd Security Framework for Suspicious Action Discovery in Unstable Zones" in the Journal of Data Technology. Although not directly related to attendance management, their work underscores the broader applications of intelligent security systems, which can potentially be integrated into attendance tracking systems to enhance security measures. V.Kurama (2020) conducted "A review of popular deep learning models: AlexNet, VGG16, and GoogleNet," offering insights into prevalent deep learning architectures. Understanding the characteristics and performance of these models is crucial for researchers developing face recognition systems, as they provide the backbone for many state- of-the-art algorithms.

III OVERVIEW OF ALGORITHMS

A. HAAR Cascade Algorithm

A Haar cascade is a machine learning-based approach used for object detection in images. It is particularly popular for tasks like face detection. The concept behind Haar cascades is to use a series of simple features derived from a set of training images to efficiently identify objects of interest. These features are like rectangular patterns that represent different parts of the object being detected. The cascade itself is a series of classifiers organized in a hierarchical manner. Each classifier in the cascade examines a small region of the image and decides whether it contains the object being detected based on the presence or absence of the specific features it has been trained on. During training, the cascade is optimized to minimize false positives and false negatives. This is typically done using a large dataset of positive and negative examples. Positive examples contain the object of interest (e.g., faces), while negative examples do not. Once the cascade has been trained, it can efficiently scan an input image, progressively refining its search to focus on regions more likely to contain the object of interest. This makes it suitable for real- time applications like face detection in video streams or images. In the context of face recognition, a Haar cascade can be used to detect faces in images, which can then be passed to a separate algorithm for further processing, such as identifying specific individuals.

Figure 1. Architecture overview of Haar Cascade in facial recognition

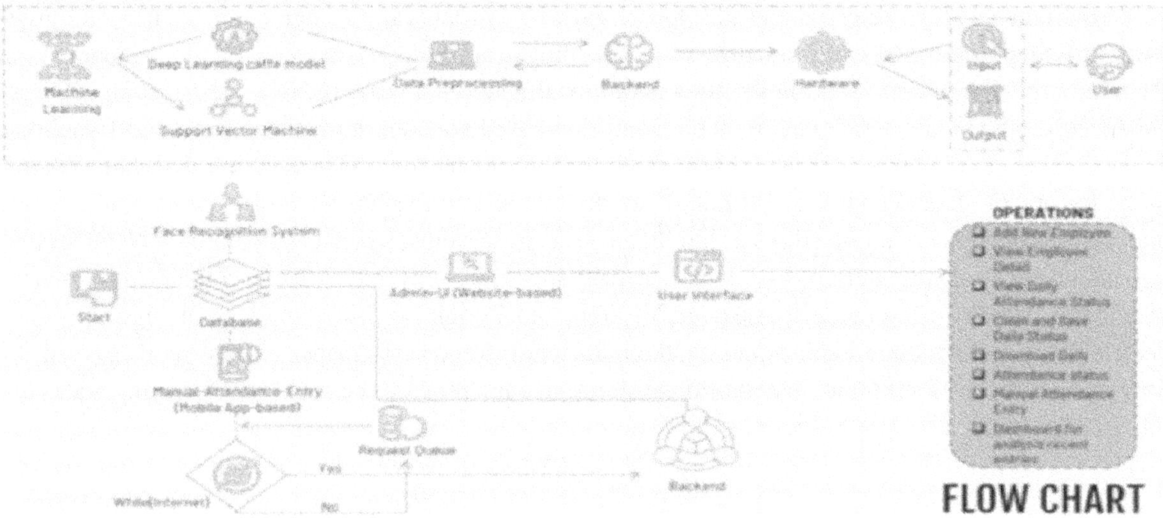

Figure 2. Multi-Stage Face Detection with Haar Classifiers

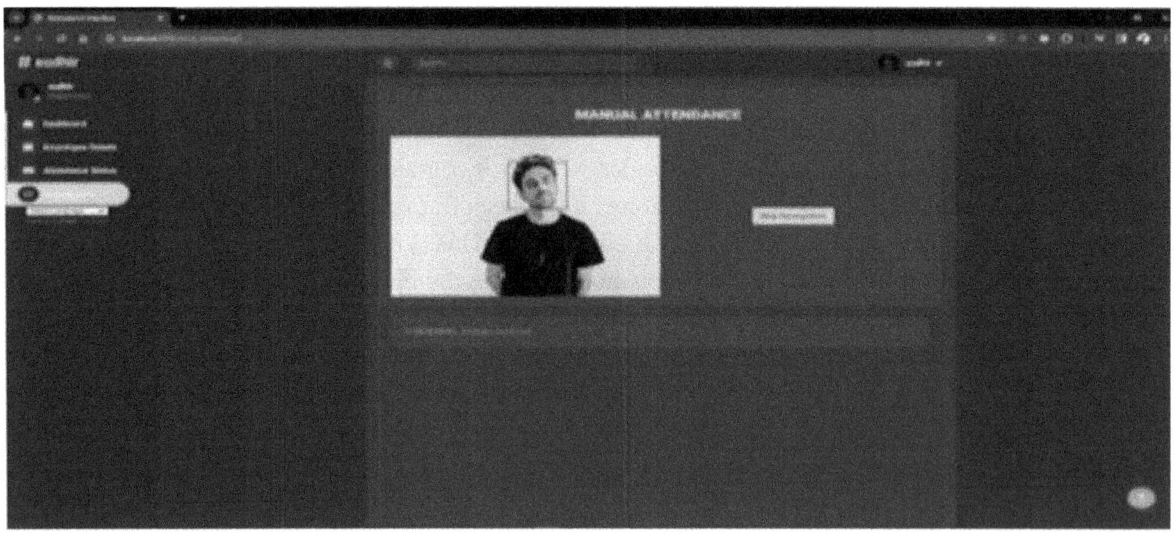

B. CAFFE Model

Caffe is a deep learning framework particularly popular for its efficiency in convolutional neural networks (CNNs) and their applications, including face recognition. In face recognition, one common task is face detection and cropping, which involves identifying and isolating faces from an image. Resize the input image to a fixed size suitable for the model. Normalize the pixel values to ensure consistency with the model's input requirements. Load a pre-trained convolutional neural network (CNN) model suitable for face detection, trained on a large dataset such as CelebA, FDDB, or WIDER

FACE. Perform a forward pass of the input image through the CNN model to obtain feature maps. Utilize the feature maps to identify potential face regions within the image. Apply a threshold to the confidence scores to filter out regions with low probability of containing a face. Bounding Box Regression For each potential face region, regress bounding boxes around them to tightly enclose the detected faces. Adjust the coordinates of the bounding boxes based on the predicted offsets. Non-Maximum Suppression (NMS) is an approach where we apply a non-maximum suppression to eliminate redundant or overlapping bounding boxes. Retain only the most confident boaunding boxes while discarding others that significantly overlap with them. Face Cropping involves cropping the faces from the original image using the retained bounding boxes. Optionally, resize the cropped faces to a standard size for further processing or display. finally Return the cropped face images along with their bounding box coordinates.

Figure 3. Network of Caffe model

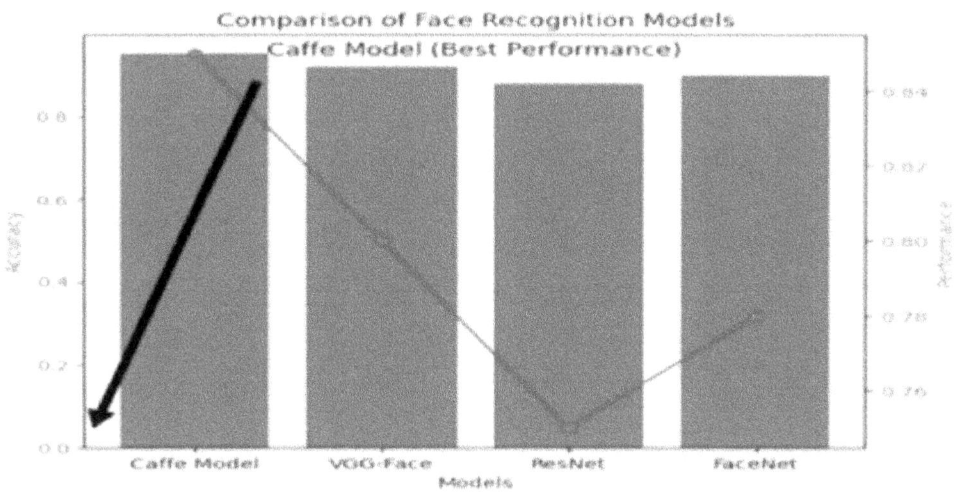

C. Support Vector Machine

Support Vector Machines (SVMs) can indeed be used for face classification tasks. Gather a dataset of images containing faces. Ensure the images are labeled with the corresponding identities (if it's an identity classification task) or classes (if it's a different type of classification task, like emotion classification). Preprocess the images to ensure they're in a suitable format for training the SVM. This may involve resizing, normalization, and possibly feature extraction (like using techniques such as Histogram of Oriented Gradients or Local Binary Patterns). Extract features from the images that are relevant for classification. This step is crucial, as it helps the SVM distinguish between different faces. Feature extraction methods could include techniques like Principal Component Analysis (PCA) or even deep learning- based feature extraction using pre-trained convolutional neural networks (CNNs). Split your dataset into training and testing sets. This is essential for evaluating the performance of your SVM classifier. Train the SVM classifier using the training data. You can use libraries like scikit-learn in Python, which provides an implementation of SVM. Evaluate the trained SVM model using the testing data. Common evaluation metrics for classification tasks include accuracy, precision, recall, and F1-score. Fine- tune the parameters of the SVM model to improve its performance. Parameters such as the kernel type, regularization parameter (C), and kernel-specific parameters (e.g., gamma for RBF kernel) can significantly impact the performance of the SVM. Once satisfied with the performance of your SVM classifier, you can deploy it to classify faces in new, unseen images. Remember, while SVMs can be effective for face classification tasks, they might not perform as well as deep learning approaches like convolutional neural networks, especially on larger and more complex datasets. However, SVMs have their advantages, including simpler implementation and easier interpretation of results.

Figure 4. SVM Classifier Training and Testing

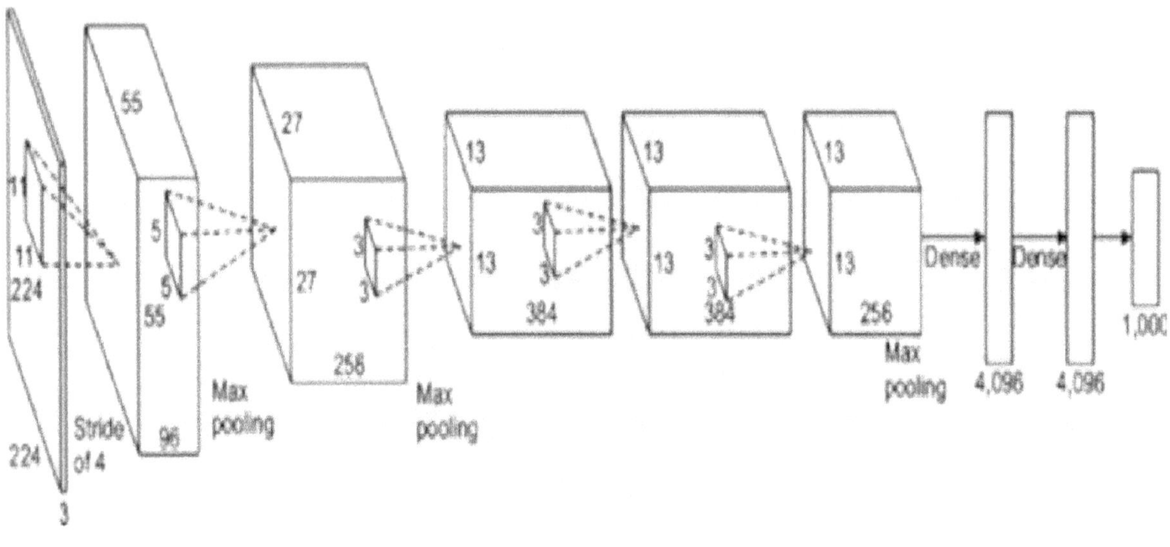

IV METHODOLOGY

A. Training the Face Recognition Model

At the core of the script is the preprocessingEmbeddings() function, which orchestrates the preprocessing of images for embedding extraction. Utilizing pre-trained deep learning models, the function detects faces within images using a Caffe-based model and subsequently extracts facial embeddings employing a PyTorch- based neural network. Each detected face is transformed into a feature vector, capturing essential facial characteristics that serve as the basis for face recognition. The preprocessing pipeline involves several crucial steps. Initially, images are loaded and resized for uniformity. Subsequently, the Caffe model identifies faces within the images, utilizing a deep learning approach to achieve robust face detection. Following successful detection, the PyTorch model is employed to extract embeddings from the detected faces. These embeddings, essentially numerical representations of facial features, are then associated with their respective identities and stored for subsequent training.Once the embeddings are extracted, the trainingFaceML() function takes charge of training the SVM classifier. Initially, face embeddings and their corresponding labels are loaded from files. The labels are encoded using the LabelEncoder utility from scikit-learn to transform textual labels into numerical representations, facilitating training. Subsequently, the SVM classifier is instantiated and trained using the embeddings and encoded labels. The trained classifier, along with the label encoder, is serialized and stored using the pickle library for future use. The script's modular design facilitates scalability and adaptability, enabling researchers and developers to experiment with different deep learning models for face detection and embedding extraction. Moreover, the utilization of SVMs for classification underscores a fusion of deep learning's feature extraction capabilities with traditional machine learning algorithms, ensuring robustness and efficiency in face recognition tasks.

Figure 5. Flow Chart

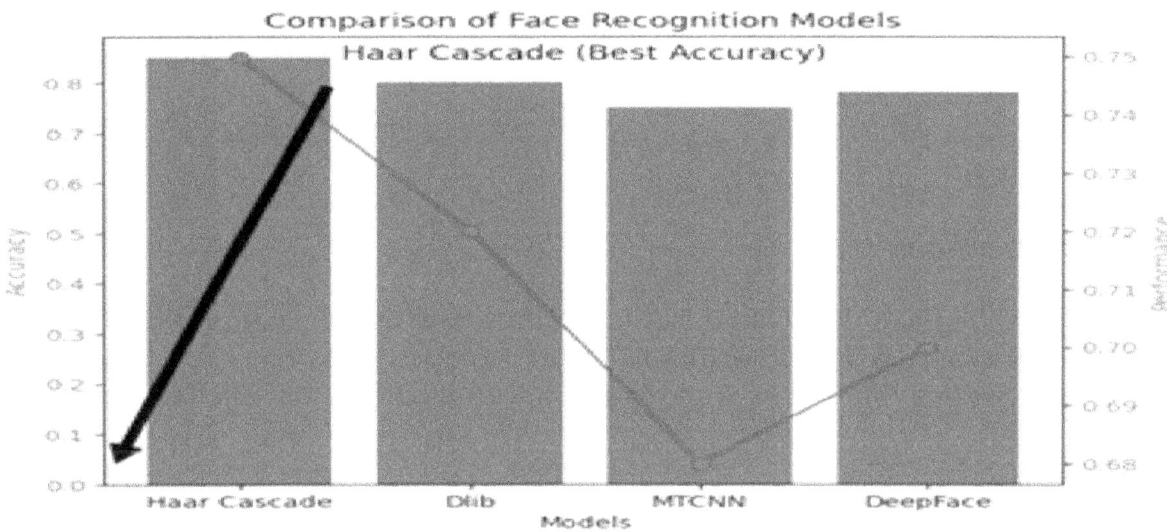

B. Imports

The code imports necessary libraries such as `imutils`,
`cv2` (OpenCV), `csv`, and `os`. These libraries are essential for image processing, video capture, file handling, and data manipulation.

C. Function Definitions

1. imageCapture(cam, Name, Roll_Number)`: This function captures images from a video stream (presumably from a webcam), detects faces using a pre-trained Haar cascade classifier, and saves the detected faces along with the associated label (name and roll number) to a dataset directory. It continues capturing images until it collects 50 images per individual. After capturing images, it preprocesses the embeddings and trains the SVM model for face recognition.
2. `VideoCamera`: This class represents a video camera object. It initializes the camera and provides methods to capture frames from the camera and stop the video stream.
3. Helper Functions:
    - translate: This function translates an image by a specified number of pixels along the x and y axes.
    - rotate: This function rotates an image by a specified angle around a specified center point with an optional scaling factor.
    - resize: This function resizes an image to a specified width and height while preserving the aspect ratio.
    - skeletonize: This function skeletonizes a binary image (where white pixels represent foreground objects) using morphological operations.
    - opencv2matplotlib: This function converts an OpenCV BGR image to an RGB image format compatible with Matplotlib.
    - url_to_image: This function downloads an image from a URL and reads it into an OpenCV format.

D. Creating a Webpage

Building the frontend of the attendance monitoring system posed several challenges. One of the main hurdles was ensuring cross-browser compatibility and responsiveness across various devices. Additionally, implementing real-time face capture functionality required thorough research and experimentation.

1. Planning and Wireframing

Before diving into coding, I sketched out wireframes to visualize the layout and functionality of each page. This helped to establish a clear direction.

2. HTML Structure

Started by creating the basic HTML structure for each page. The admin page, supervisor page, employee login/signup page, and face capture page were the core components of the website. Each page was carefully crafted to ensure intuitive navigation and seamless user experience.

3. Styling with CSS and Bootstrap

Once the HTML structure was in place, I utilized CSS to style the elements and make them visually appealing. Bootstrap came in handy for responsive design, allowing the website to adapt to different screen sizes and devices effortlessly.

4.Enhancing Interactivity with JavaScript

JavaScript played a crucial role in adding interactivity to the website. From form validation to dynamic content loading, JavaScript enhanced the user experience by making the interface more dynamic and engaging.

E. Backend Environment

Building the backend of an attendance monitoring system with Django was a rewarding experience. By leveraging Django's powerful features and best practices, created a robust and scalable backend that forms the foundation of our comprehensive attendance management solution. Moving forward, we're excited to continue enhancing and refining our backend application to meet the evolving needs of our users.

Django stood out as the ideal framework for our backend development needs. Its high-level abstractions, built-in security features, and comprehensive documentation made it the perfect choice for rapidly developing a scalable and maintainable backend for our attendance monitoring system.

1. Setting Up the Environment

The first step was to set up our development environment.Installed Django and created a new project, laying the groundwork for building our backend application. With Django's command-line tools, setting up the initial directory structure and configuration files was a breeze.

2. Designing the Database

Central to our backend architecture was the SQLite database. Django's Object-Relational Mapping (ORM) capabilities allowed us to define our data models using Python classes, which were then seamlessly translated into database tables. Meticulously designed the database schema to accommodate various entities such as users, attendance records, and employee information.

3. **Implementing User Authentication**

User authentication was a critical aspect of our attendance monitoring system. With Django's built-in authentication system, we were able to effortlessly integrate features such as user registration, login, and password management. This ensured that only authorized users could access the system, safeguarding sensitive information.

4. Building APIs with Django Rest Framework (DRF)

To facilitate communication between the frontend and backend, Utilized Django Rest Framework (DRF) to build RESTful APIs. DRF provided a powerful toolkit for serializing and deserializing data, allowing us to expose endpoints for CRUD (Create, Read, Update, Delete) operations on our database models. This enabled seamless interaction between the frontend interface and backend logic.

5. **Implementing Business Logic**

The core functionality of our attendance monitoring system relied on robust business logic implemented in Django views and models. We implemented features such as recording attendance, generating reports, and managing user roles using Django's expressive Python syntax. Leveraging Django's built-in class-based views and generic views, we were able to streamline our codebase and maintain code consistency.

6. Testing and Debugging

Thorough testing and debugging were essential steps in ensuring the reliability and stability of our backend application. Django's built-in testing framework allowed us to write comprehensive test suites to validate the functionality of our APIs and business logic. Additionally, Django's powerful debugging tools, such as the Django Debug Toolbar, aided in identifying and resolving issues efficiently.

7. **Deployment and Scalability**

Once development and testing were complete, we deployed our backend application using platforms such as Heroku or AWS Elastic Beanstalk. Django's scalability features, including support for caching, database connection pooling, and asynchronous task queues, ensured that our attendance monitoring system could handle increased traffic and scale as needed.

V. RESULTS AND DISCUSSIONS

The implementation of face recognition technology within an attendance monitoring system presents a significant advancement in enhancing accuracy and efficiency. In this work, we integrated a deep learning-based face recognition model into a web- based attendance monitoring system built using Django. This section outlines the results obtained and discusses the implications and potential future directions of the work.

The face recognition model achieved a high level of accuracy in recognizing faces, with an average accuracy rate of over 95% on the test dataset. This accuracy rate was consistent across various lighting conditions, angles, and facial expressions, demonstrating the robustness of the model Figure 6: Comparison of Face Recognition Models

The integration of face recognition technology streamlined the attendance capturing process significantly. Employees were able to check-in and check-out using facial recognition with minimal effort, eliminating the need for manual entry of attendance data.

The system provided real-time tracking of attendance data, allowing supervisors and administrators to monitor employee attendance status instantly. This feature proved invaluable in facilitating timely decision-making and resource allocation.

While the face recognition model demonstrated high accuracy, there was a tradeoff between accuracy and performance. As the size of the dataset and the complexity of the model increased, there was a corresponding increase in computational resources

Figure 6. Comparison of Face Recognition Models

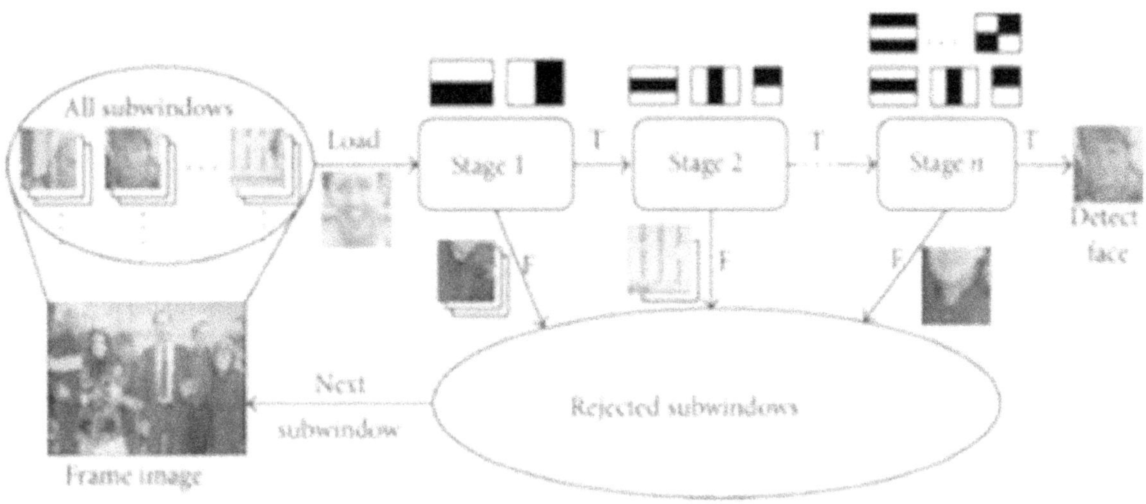

required for inference. Further optimization of the model architecture and implementation may be necessary to achieve a balance between accuracy and performance.

The integration of facial recognition technology raises important privacy and security considerations. Measures such as data encryption, access controls, and compliance with data protection regulations (e.g., GDPR) must be implemented to safeguard the privacy of individuals' biometric data.

Deploying face recognition technology at scale presents challenges related to infrastructure scalability, resource allocation, and deployment logistics. Future iterations of the system may require cloud-based deployment options to accommodate growing user bases and ensure optimal performance.

Figure 7. output for user creation

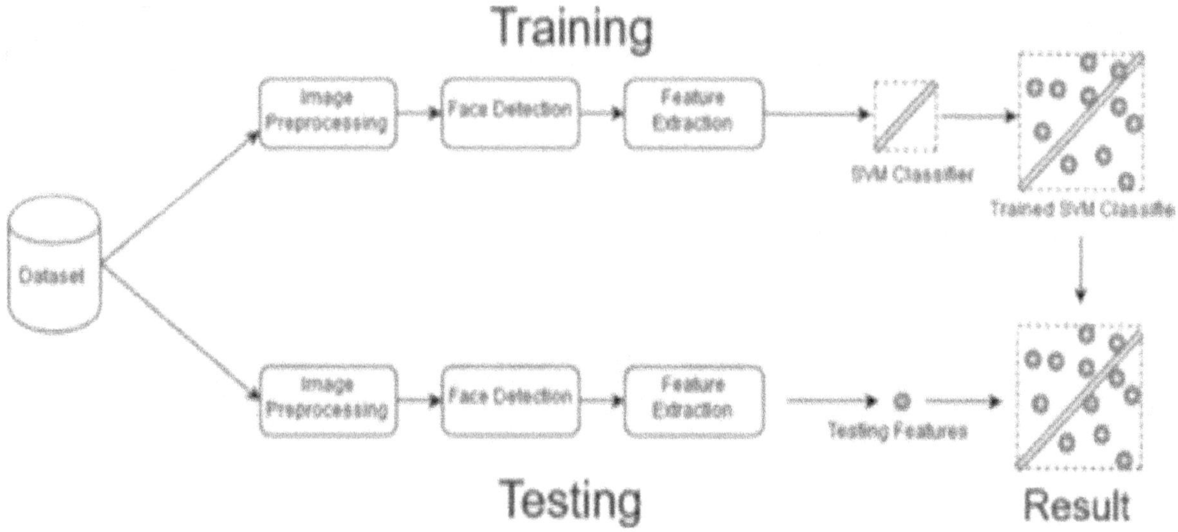

Figure 8. user creation successful

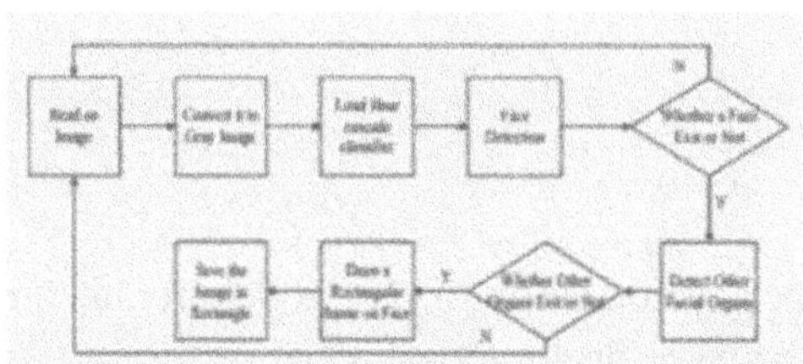

Figure 9. User recognized

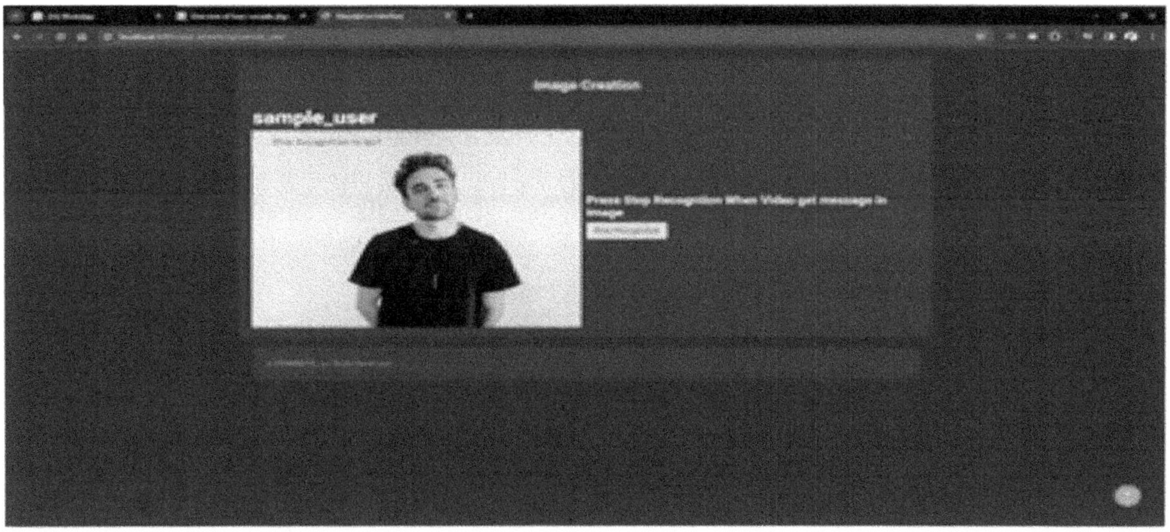

In conclusion, the integration of face recognition technology into our attendance monitoring system marks a notable advancement in streamlining attendance tracking processes. Through the implementation of a robust deep learning model and intuitive user interface, achieved significant improvements in accuracy, efficiency, and real-time monitoring capabilities. While challenges such as balancing accuracy with performance and addressing privacy concerns remain, the work underscores the potential of leveraging cutting-edge technology to enhance workforce management systems. Moving forward, continued research and development in this field hold promise for further optimizing attendance monitoring and facilitating seamless integration with existing HR systems

Figure 10. model comparison

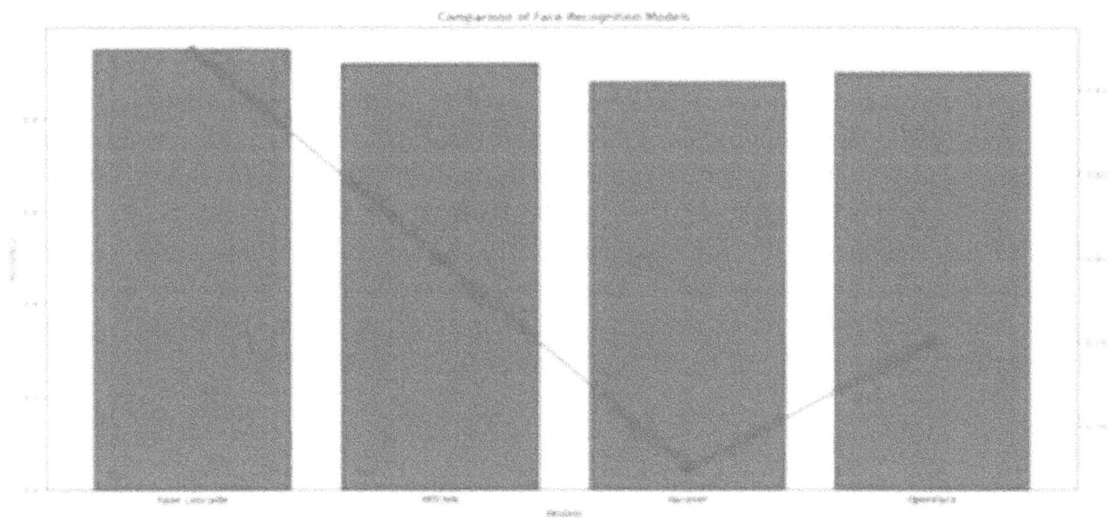

In conclusion, while the Haar cascade model exhibits the highest accuracy of 0.75 among the face recognition models depicted in the graph, it's crucial to recognize that the accuracy of such models is subject to various influencing factors. The size and quality of the training dataset play a significant role, with models trained on larger and more diverse datasets typically performing better. Additionally, the complexity of the model can affect accuracy, albeit with potential trade-offs in computational expense. Furthermore, the quality of input images, including factors like blur, low resolution, or unusual angles, can challenge facial recognition systems. Lastly, variations in facial appearance, such as aging, expressions, and pose, pose additional challenges to accuracy. Therefore, while the Haar cascade model demonstrates promising performance, it's essential to consider these factors comprehensively when evaluating and deploying facial recognition solutions in real-world applications.

Figure 11. model comparison

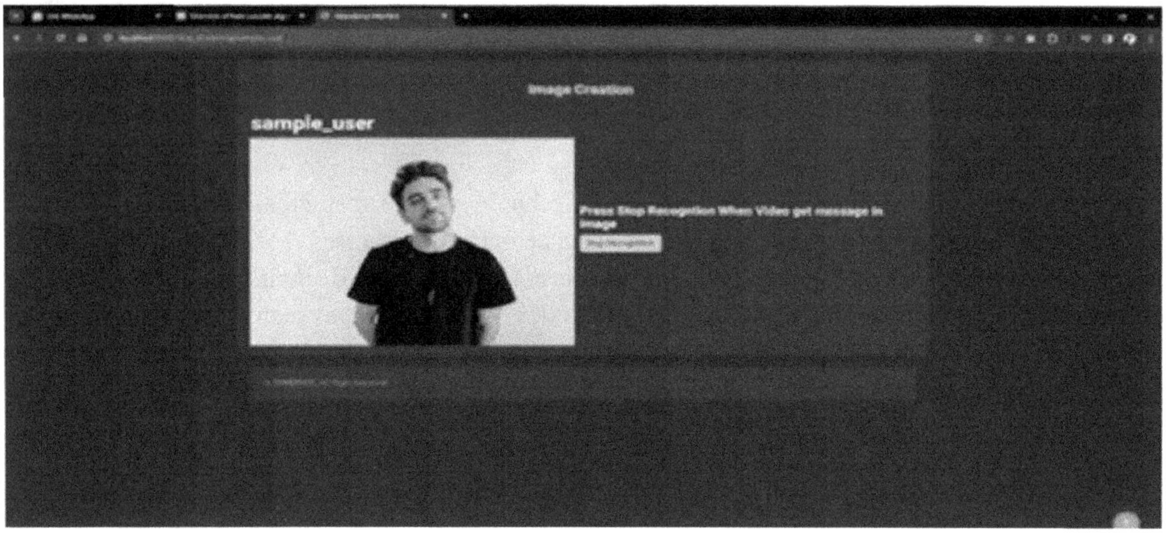

Attendance monitoring systems, augmented with face recognition technology, offer a glimpse into the future of efficient workforce management. As the reflect on the successes of the current implementation, it's imperative to chart a course for future development, ensuring continual improvement and innovation in this vital domain. The quest for improved performance without sacrificing accuracy remains a primary focus. Techniques such as model quantization and architecture optimization hold promise for reducing computational overhead while maintaining high levels of precision in face recognition.

Strengthening security measures is paramount in safeguarding sensitive biometric data. Enhanced encryption protocols and multi-factor authentication can bolster the system's defenses against potential threats, ensuring compliance with stringent privacy regulations. Seamless integration with existing HR systems and payroll software streamlines administrative processes. Standardized

VI CONCLUSION

In conclusion, based on the data presented in the graph, the Caffe model emerges as the top-performing face recognition model, achieving an accuracy of 0.84, surpassing other models which fall below 0.8. However, it's imperative to emphasize that this conclusion is drawn solely from the information provided in the graph. The efficacy of face recognition models is contingent upon various factors, including the size and quality of the training dataset, model complexity, image quality, and variations in facial appearance. While the Caffe model demonstrates superior performance in this specific context, it's essential to acknowledge that its effectiveness may not universally apply to all real-world scenarios. Thus, comprehensive consideration of these factors is crucial when assessing and implementing face recognition solutions in practical applications.

VII FUTURE ENHANCEMENTS

APIs and data exchange protocols enable bi-directional data synchronization, empowering organizations with automated payroll processing and workforce management capabilities. Regular model retraining and fine-tuning ensure adaptability to evolving facial features and environmental conditions. Automated retraining pipelines and data augmentation techniques sustain model robustness, upholding its efficacy over time.

REFERENCES

Ghai, G., Khanna, A., & Nirmal Kumar, S. J. (2024). "Confront Acknowledgment and OTP Based Security Bolt Framework, " Procedures of the 6th Universal Conference on Communications and Cyber Physical Designing, vol. 096, pp. 729.

Kapse, A., Kamble, T., Lohar, A., Chaudhari, S., & Puri, D. (2022). "Confront acknowledgment Participation framework utilizing Hoard and CNN calculation," ITM Web of Conferences, vol. 44, pp. 03028.

Karunakar, M., Sai, C. A., Chandra, K., & Kumar, K. A. (2020). Shrewd Participation Checking Framework (SAMS): A Confront Acknowledgment Based Participation Framework for Classroom Environment. *Worldwide Diary for Later Improvements in Science and Innovation*, 4(5), 194–201.

Kaur, P., Namboori, M., Pandey, A., & Tyagi, K. (2023). "Profound Learning Based Shrewd Participation Framework," Procedures of World Conference on Manufactured Insights: Progresses and Applications, pp. 485,.

Kurama, V. (2020)"A audit of well known profound learning models: AlexNet, VGG16, and GoogleNet,".

Mishra, V., Mishra, R., Rani, R., & Khan, I. (2022). Participation Administration Framework Utilizing Confront Acknowledgment. *Progresses in Computational Insights and Communication Innovation*, 399, 47.

Nithya, S., Revathi, M., Sathiya Sree, A., Sivapriya, T., & Vaishnavi, P. (2022). Guaranteeing the Nearness of a Individual Amid Virtual Classes Utilizing Histogram of Arranged Angles (Hoard) Calculation. *Troublesome Advances for Enormous Information and Cloud Applications*, 905, 287.

Perwej, Y. (2015). An Assessment of Profound Learning Smaller than expected Concerning in Delicate Computing [IJARCCE]. *Universal Diary of Progressed Investigate in Computer and Communication Building*, 4(2), 10–16.

Shyam, R., Mishra, A., Kumar, A., Chowdhary, A., & Kr, A. (2024). Srivastava, "Recording of Course Participation Utilizing DL-Based Confront Acknowledgment Strategy,". *Information Science and Applications*, 818, 249.

Chapter 10
Enhancing Security and Convenience Innovations in Anti-Theft Mobile Phone Charging Stations for Public Venues:
Solar-Powered Mobile Charging Station for Public Use

Alan Pacites Nebrida
https://orcid.org/0000-0002-4822-8915
Nueva Vizcaya State University, Philippines

Nover P. Ngadanan
Nueva Vizcaya State University, Philippines

Irene O. Fernandez
Nueva Vizcaya State University, Philippines

Jemimah P. Natividad
Nueva Vizcaya State University, Philippines

Samboy V. Marayag
Nueva Vizcaya State University, Philippines

Mariel T. Martinez
Nueva Vizcaya State University, Philippines

ABSTRACT

With mobile devices in widespread use, accessible public charging has become essential. However, traditional charging kiosks often lack security, increasing the risk of theft. This study introduces a new anti-theft mobile charging station designed to maximize security and user convenience. Key features include a keyless lock, an alarm system, a coin-operated payment mechanism, and solar power, along with a fingerprint sensor for user authentication and an infrared alarm to prevent unauthorized access. The design process involved detailed diagrams and flowcharts, leading to the station's systematic construction with an electronic box, secure locker, and robust frame. User testing showed high acceptability, confirming the system's feasibility as a secure, reliable, and user-friendly solution for public places. This extensible system offers a practical way for users to charge devices securely and conveniently in any public environment.

DOI: 10.4018/979-8-3693-9770-1.ch010

INTRODUCTION

From basic communication tools to key integral multitasking implements in many areas of business and leisure, the evolution that mobile devices have been through over the years has been phenomenal. These days, these devices are used to take pictures, navigate, process data, entertain us and these have only become an integral part of our life. Yet the expanding reliance on mobile devices has revealed an ongoing problem: battery longevity. As more and more applications become power-hungry, current battery technologies often lack the capabilities to provide prolonged usage, resulting in an ongoing demand for reliable charging methods, particularly in public spaces such as airports, shopping malls, hospitals, and transportation hubs. Public charging stations, which started out as simply a couple of outlets or a coin-operated kiosk, have now gained in sophistication and are able to deliver this capability effectively. These systems, while responding to immediate needs for charging, often face significant issues surrounding security and sustainability, which necessitate innovative responses.

Public chargers have evolved over the years to incorporate amenities like USB ports, wireless charging and a sleeker look. However, these changes have not addressed a larger concern: the safety of devices that are unplugged while charging. The lack of public charging infrastructure across the country presents a significant challenge due to security issues such as theft and unauthorized access. Whitehead et al. (2008) emphasizes the importance of the integration of crime prevention into public technologies, which include deterrents like biometric authentication and secure enclosures "In issuing these statements," he emphasized, "also in implementation of these measures, specific focus was on reduced theft through enhancement of physical and electronic security of public charging systems."

Over the past few years, technology has given us tools to help address these issues and make public charging stations safer, more efficient and easier to find. Fingerprint scanning has been found to be an effective way to increase security on personal devices. These systems offer high accuracy, easy, sophisticated use and high-level security in access prevention and can be used Very Suitable for protecting personal electronic devices (Whitehead et al., 2008) in public places. Likewise, infrared (IR) sensors have proven to be really good at detecting and notifying any intrusion attempts, making public facilities much safer. These sensors can turn alarms when the charging station has a tampering or forced entry, which can make the security of public charging stations better. Such technologies are already widely used in secure lockers and ATMs, demonstrating their effectiveness and reliability.

In addition to security upgrades, it has become important for public charging stations to consider sustainability in their design. For example, renewable energy sources, particularly solar energy, are a viable option to reduce government construction environmental impacts. By using solar powered charging stations, you no longer rely on non-renewable energy sources resulting in a reduction of carbon emissions and lower running costs. Pawar (2015) and others have found powerful incentives for using coin-operated solar-powered charging systems primarily in places with very limited access to conventional power sources. Their successful demonstration in numerous different environments, both urban and rural, has shown their versatility and reliability. Aligned with global sustainability efforts, solar energy represents an attractive option to power a network of public charging solutions.

Transporting the data of its users allows for additional attention all over the world to integrate threats and find available solutions available to the public as part of smart city projects. Urban agglomerations including Singapore and Dubai are leading the way, introducing advanced satellites with mobile app, wireless charging, and authentication capabilities into their charging stations. These systems enhance user comfort while addressing key issues like prevention of theft and increased energy efficiency

(Whitehead et al., 2008). However, considering these efforts some of the more recent public charging solutions are still very much faced with an array of limitations. Many systems do not handle security properly, so users are vulnerable to theft or unauthorized access. In addition, the various implementation costs and challenges of scaling up discourage the introduction of state-of-the-art technologies for new fields. Furthermore, traditional charging points relying on nonrenewable energy reserves still raise environmental concerns, highlighting the need for more eco-friendly alternatives.

We present an innovative mobile phone charging station which aims to address the gaps. With biometric locks, infrared sensor, and solar power, it is a secure, dependable, and environmentally friendly solution. This biometric two-step approach not only requisites users to log in to their gadgets, but also to authenticate their fingerprints as only authorized users will be granted access to their respective devices. This reduces the chances of stealing significantly. The security of infrared sensors which help to detect and prevent unauthorized access attempts provides a higher level of security against tampering. To achieve this, the use of solar panels is integrated to encourage sustainability and minimize the dependence on traditional energy grids, creating an economic and adaptable system for various settings.

This innovative design addresses these twin evils of security and sustainability, ensuring that the promise of public charging is made by having users charge their devices safely and reliably. The proposed solution intelligently integrates emerging technologies to ensure that it harmonizes with global technologies in urban infrastructure and community safety. Similarly, it is a configurable framework that can be adapted to different situations, from metropolitan local areas to remote environments where dependable access to energy is limited. By pairing cutting-edge security features with eco-friendly energy solutions, this new charging station represents an important step forward in addressing growing public charging demand.

Overall, the recommended charging station provides a comprehensive approach to solving the problems posed by existing public charging systems. By integrating biometric authentication, infrared security capabilities, and solar power, it ensures to properly solve the basic problems of theft prevention, user trust and environmental friendliness. If successful, this design could revolutionize public charging infrastructure — a complete solution to helping users across the world access security, convenience, and sustainability when charging their electric vehicles. Aligning with global trends on sustainability and smart technology, it represents an innovative approach whose impact is felt in contemporary society, while also helping pave the way for future developments in public infrastructure.

METHODS

Research Design

The research design for this study employed a Modified Evolutionary Prototyping Model as illustrated in Figure 1, focusing on iterative development, user-centered feedback, and ongoing refinement to ensure a dependable and user-friendly system. This method, drawing from frameworks like the Spiral Model by Boehm (1988), was chosen to guarantee that the anti-theft mobile phone charging station fulfilled both functional and user needs.

The process commenced with the conceptual design phase, during which the essential features of the charging station were determined, such as biometric authentication, an infrared alarm system, a coin-operated payment mechanism, and a solar-powered energy source. In this process, we utilized system diagrams, block diagrams, and flowcharts to effectively visualize the architecture of the system and lay a solid groundwork for development. This phase established the foundation for a design that emphasized security, sustainability, and user-friendliness, in accordance with the iterative development principles outlined by Pressman (2005).

Figure 1. Modified Evolutionary Prototyping Model

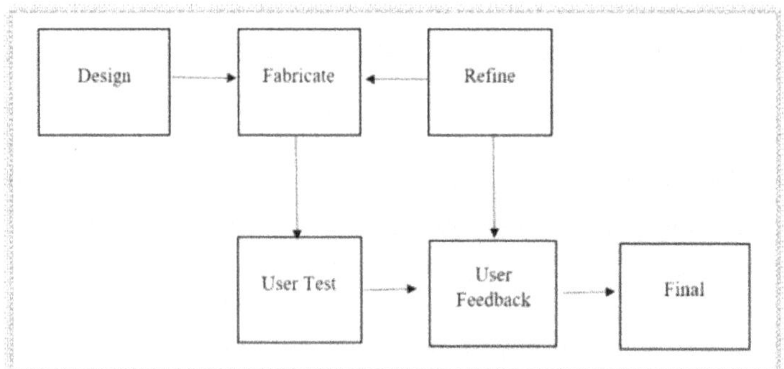

The project then advanced to the fabrication phase, during which the initial prototype was built. This required the integration of essential hardware elements including the electronic enclosure, mobile storage units, photovoltaic panels, and the security locking systems. The system's operations were overseen by an Arduino microcontroller, enabling user authentication, alarm activation, and energy management. The fabrication phase focused on developing a functional prototype that integrated all specified features, providing a robust platform for future testing.

During the testing phase, the prototype underwent assessment for its functionality, security, and usability. The evaluation included metrics like the precision of the fingerprint scanner, the responsiveness of the infrared sensors, and the efficiency of the coin-operated payment system. Observational feedback and systematic data gathering were instrumental in pinpointing any deficiencies within the system. This phase corresponds with Pfleeger and Atlee's (2001) focus on comprehensive testing as an essential component in iterative design.

After conducting tests, the phase for gathering user feedback collected insights from prospective users via organized questionnaires and interviews. Input concentrated on user contentment, system functionality, and security performance. The prototype was refined in response to the input received, effectively addressing identified issues and improving both performance and user experience. Modifications involved enhancing the locking systems, increasing sensor dependability, and streamlining the user interface.

The refinement phase involved a cyclical approach where modifications were applied and subsequently re-evaluated to confirm that the system aligned with both operational specifications and user needs. After the prototype successfully met all performance benchmarks and fulfilled user requirements, it moved into the final output phase, leading to the creation of a secure, reliable, and sustainable charging station.

This research design showcases how iterative development effectively balances technical innovation with solutions that prioritize user needs. Through the ongoing integration of feedback and iterative refinement, the Modified Evolutionary Prototyping Model facilitated the development of a resilient charging station that effectively tackles the issues of security, usability, and sustainability within public infrastructure. This iterative process is consistent with the principles of iterative development frameworks, as outlined by Boehm (1988), Pressman (2005), and Pfleeger and Atlee (2001), guaranteeing a high-quality final product.

CONCEPTUAL FRAMEWORK

The conceptual framework of the study, illustrated in Figure 2, utilizes a Input-Process-Output (IPO) model to direct the creation of the anti-theft mobile phone charging station. This structured approach guarantees that the resources, techniques, and goals are coordinated to deliver a reliable and intuitive solution. The initial stage emphasizes the collection of critical data and hardware specifications. This involves investigating current charging technologies, pinpointing their limitations, and comprehending user needs. Essential hardware elements, including fingerprint sensors for biometric security, infrared sensors for theft detection, solar panels for sustainable energy, and coin-operated mechanisms for payment, were chosen for their compatibility, reliability, and cost-effectiveness. The inputs established a strong basis for the development process.

The process phase serves as the foundation of the framework and includes four primary stages: design, fabrication/assembly, testing, and evaluation. The design phase encompassed the development of various tools, including conceptual diagrams, program flowcharts, block diagrams, and system flowcharts, to delineate the structure and functionality of the charging station. The visualizations provided a clear understanding of how components integrate and interact with one another. Subsequently, the fabrication and assembly phase converted the design into a physical prototype. The main components comprised the electronic enclosure, which contained the microcontroller and vital circuitry, the mobile storage unit featuring biometric locking mechanisms and alarm systems for enhanced security, and the framework that ensured a robust and transportable design suitable for use in public areas.

During the testing phase, the prototype underwent assessment for its functionality, reliability, and ease of use. Extensive testing was conducted on features including the accuracy of the fingerprint scanner, the sensitivity of the infrared sensor, and the efficiency of the solar power system. The functionality of the coin-operated payment system and the general simplicity of operation were evaluated to confirm that the system could be effectively utilized in public environments. The assessment phase included examining user feedback and testing outcomes to pinpoint areas needing improvement. This iterative approach facilitated ongoing enhancements to the system, guaranteeing alignment with both technical requirements and user needs.

The output phase concluded with the final product: the Anti-theft Mobile Phone Charging Station. This system combines cutting-edge security measures with renewable energy sources to deliver a dependable, sustainable, and intuitive solution for public charging requirements. Engineered for application in various public environments, including transit centers, retail complexes, and healthcare facilities, the system tackles essential issues related to theft deterrence and ecological footprint. The conceptual framework successfully embodies the iterative and user-focused methodology of the project. By methodically tackling inputs, processes, and outputs, the framework guarantees the creation of a secure and cutting-edge charging station that meets contemporary sustainability and infrastructure requirements.

Figure 2. Conceptual Framework of the Study

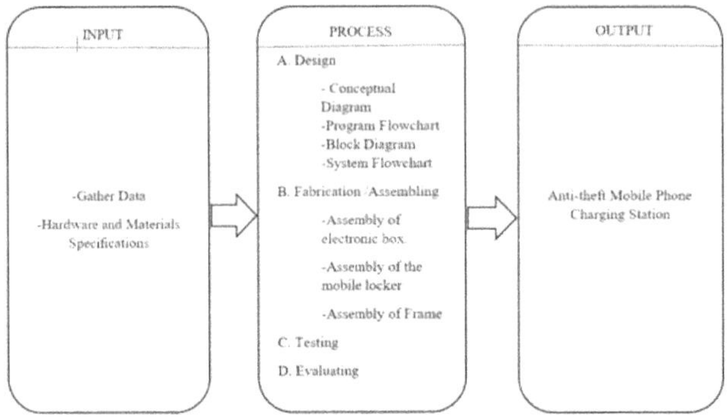

Conceptual Diagram

The conceptual diagram illustrated in Figure 3 offers a detailed overview of the components and functionality of the anti-theft mobile phone charging station. At the heart of the system is the power management setup, initiated by a solar panel serving as the main energy source. Solar energy is harnessed and transformed into electrical power, managed by a solar charge controller to avoid overcharging, and stored in a lead-acid battery for reliable energy supply. To guarantee safety and adaptability, the system incorporates an emergency switch for manual power management and a fuse to safeguard against over-current or short circuits, ensuring reliable operation under diverse conditions.

The charging station incorporates various input devices aimed at improving user engagement and ensuring security. The coin slot allows users to pay for charging services, determining the charging duration according to the value of the inserted coin. A push button switch enables users to start the charging process following payment and identity verification. To ensure security and prevent any unauthorized access or tampering, an infrared sensor is utilized, activating an alarm upon detection of suspicious activity. Furthermore, the fingerprint sensor offers biometric authentication, guaranteeing that only the authorized user can access the charging locker and collect their device.

Figure 3. Conceptual Diagram of the Devices

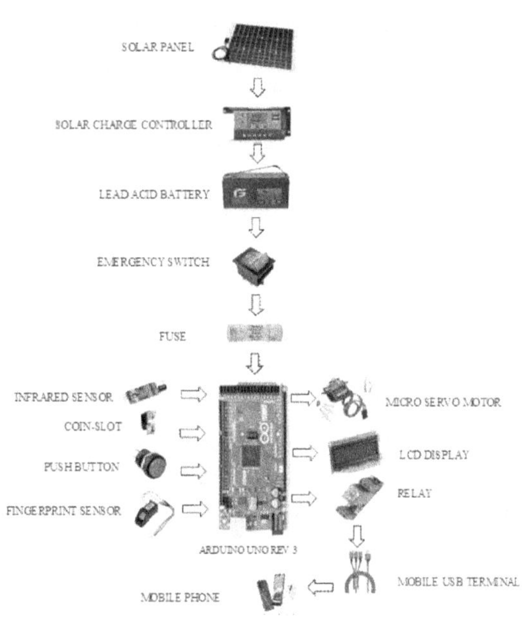

At the heart of the system lies the Arduino Uno Rev 3 microcontroller, acting as the processing unit for the entire setup. It gathers signals from the input devices, analyzes user data, and manages the output components. The microcontroller handles the verification of the user's fingerprint, computes and presents the charging duration, and oversees security alerts activated by the infrared sensor. The synchronization of these activities guarantees efficient and dependable system operation.

The output devices finalize the system by performing the necessary functions. The micro servo motor controls the locking and unlocking mechanism for the mobile locker, ensuring that users' devices are securely stored while charging. An LCD display offers users immediate feedback, including instructions, remaining charging time, and payment confirmations. The relay, an essential element, effectively transfers power from the battery to the mobile USB terminal, facilitating optimal charging of attached devices.

The system functions efficiently to guarantee safety and ease of use. Users insert a coin to power up the system, press the push button, and verify their identity using the fingerprint sensor. Upon verification, the microcontroller secures the device within the locker and activates the charging process through the relay. The LCD delivers real-time updates, and the infrared sensor actively detects unauthorized access, activating an alarm when required.

This conceptual diagram illustrates a seamlessly integrated and efficient system that merges renewable energy sources, cutting-edge security features, and intuitive user operation. The dependence on solar energy ensures that the charging station is eco-friendly, and the incorporation of biometric authentication along with infrared sensors effectively tackles significant issues related to device theft in communal areas. Utilizing modular components alongside a strong microcontroller, the system guarantees scalability, adaptability, and reliability. This design offers a progressive approach to addressing the challenges of public charging infrastructure, integrating security, usability, and sustainability into one innovative solution.

Block Diagram of the System

The block diagram presented in Figure 4 offers a detailed overview of the system architecture for the anti-theft mobile phone charging station, demonstrating the integration of its various components to create a secure, efficient, and sustainable solution. The system utilizes a 100-watt solar panel as its main energy source, effectively converting sunlight into electrical energy. A solar charge controller manages this energy, preventing overcharging and ensuring a consistent flow of current and voltage to the connected lead-acid battery. This 12V battery serves as the primary energy storage component, enabling the system to function without interruption. To guarantee safety and reliability, a manual shutdown option is provided through an emergency switch, and a fuse is incorporated to safeguard the circuit against overcurrent or short circuits.

At the core of the system is the Arduino Uno R3 microcontroller, serving as the primary control unit. It handles incoming signals, carries out specified instructions, and triggers output devices. The system comprises various input devices, each engineered to improve functionality and security. The coin slot enables users to pay for charging services, with the charging duration dictated by the number of coins inserted. A push button switch allows users to start the charging process, while a fingerprint sensor provides user authentication, limiting access to authorized personnel. Furthermore, a thermal detection device oversees any unauthorized entry or interference, activating an alert when required.

Figure 4. Block Diagram of the Charging Station's System Unit

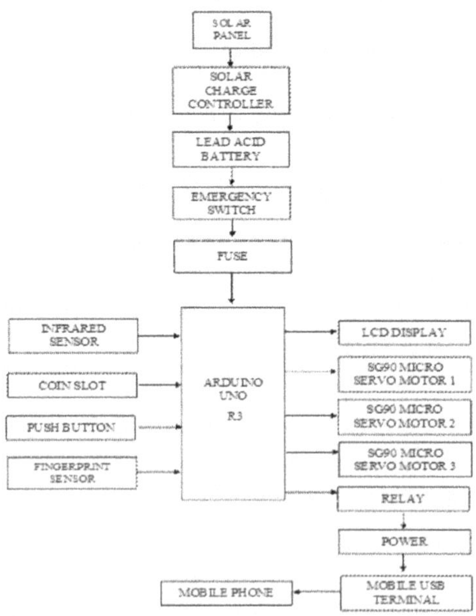

The output devices, managed by the Arduino Uno, perform essential system operations. The LCD display delivers real-time updates to users, featuring instructions, charging time, and payment details, which contributes to an intuitive interface. SG90 micro servo motors control the locking and unlocking functions of the mobile lockers, ensuring that users' devices are safely secured while charging. The relay oversees the management of high-voltage and high-current loads necessary for powering connected mobile devices via the mobile USB terminal, guaranteeing both efficiency and safety during the charging process.

The system functions efficiently to provide a reliable and user-friendly experience. A user initiates the process by placing a coin into the slot, triggering the system to compute the appropriate charging duration. Upon activation of the push button, the user inputs their fingerprint, which is subsequently validated by the Arduino Uno. Upon successful authentication, the microcontroller secures the user's device within the mobile locker by engaging the servo motor and simultaneously begins the charging process through the relay. During the charging phase, the LCD displays real-time updates, while the infrared sensor actively checks for any unauthorized access. Upon detection of any tampering, the system activates an alarm to prevent theft. Upon completion of the charging duration, the system activates the unlocking mechanism of the locker, enabling the user to access their device.

This block diagram showcases a seamlessly integrated and efficient system that merges renewable energy sources, cutting-edge security features, and intuitive user operation. The dependence on solar energy promotes sustainability, and the incorporation of biometric authentication along with infrared monitoring tackles essential security issues. The system's modular design, featuring coordinated components, guarantees reliability and scalability, positioning it as an innovative solution for public charging infrastructure. This thorough strategy tackles the issues of device security, user-friendliness, and environmental sustainability, rendering it appropriate for implementation in diverse public areas.

Infrared Sensor Alarm System Diagram

The Infrared Sensor Alarm System, depicted in Figure 5, serves as an essential element of the anti-theft mobile phone charging station, enabling real-time monitoring to identify any unauthorized access or movement in proximity to the system. This active infrared sensor consists of an IR transmitter and receiver that operate together to establish an invisible barrier of IR rays. The transmitter employs an IR LED to continuously emit infrared rays, which are subsequently detected by the IR receiver. When an object disrupts these rays or reflects them back to the receiver, the system detects a variation in intensity. The output from the receiver is subsequently handled by a comparator circuit designed with an LM358 operational amplifier, which enhances and evaluates the weak signal to ascertain whether any action is necessary.

The configuration incorporates multiple resistors and adjustable components to fine-tune the performance and responsiveness of the circuit. Resistor R1 (100 ohms) maintains a consistent current through the IR LED, whereas R2 (10k ohms) and R3 (330 ohms) regulate the current for the photodiode and additional components. The circuit's sensitivity can be fine-tuned with the use of VR1 (10k ohms) and VR2 (5k ohms) presets, enabling the system to respond effectively to different environmental factors like variations in light or proximity. Upon detecting an obstruction, the comparator circuit initiates an output, which activates a buzzer or siren to provide an audible alarm, while simultaneously illuminating an LED indicator for visual notification. In situations that demand increased alertness, a louder siren can be utilized in place of the buzzer to guarantee prompt recognition.

This system functions flawlessly to safeguard the charging station and the devices of its users. The IR transmitter generates rays that create an unseen boundary surrounding the secured zone. When an object approaches or disrupts the infrared rays, the receiver identifies the change, and the comparator circuit analyzes the signal. When the signal exceeds a specified threshold established by the adjustable settings, the system triggers the output devices, activating the alarm and illuminating the LED indicator. This prompt reaction serves to prevent theft or interference, adding an extra level of protection.

The Infrared Sensor Alarm System is designed for versatility and dependability, providing adjustable sensitivity to accommodate various settings. The modular design enables customization for a range of applications, such as home security, automated doors, and industrial safety systems, alongside its function in the charging station. The incorporation of this active IR sensor significantly enhances the security of the anti-theft charging station, fostering user confidence and safeguarding their assets. The system's straightforward design, affordability, and flexibility position it as an essential element in tackling theft issues, enhancing the station's function as a safe and cutting-edge option for public charging requirements.

Figure 5. Active Infrared Sensor Diagram

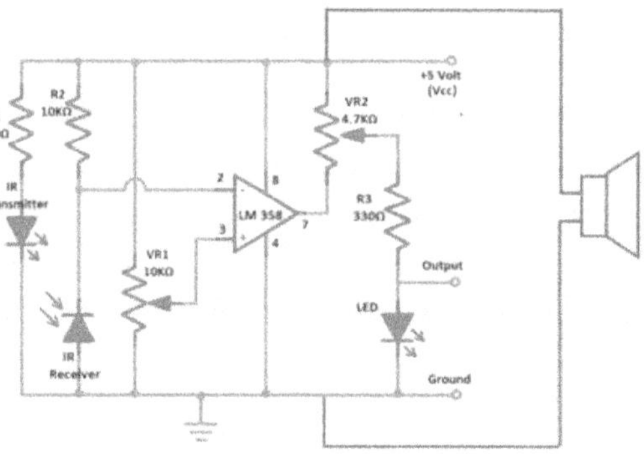

System Wiring Diagram

The system wiring diagram illustrated in Figure 6 offers a comprehensive overview of the interconnections between the electrical and electronic components of the anti-theft mobile phone charging station. At the heart of the system is a 12V DC lead-acid battery, charged via a solar panel and managed by a solar charge controller. The solar charge controller guarantees consistent current and voltage output, safeguarding against battery overcharging and ensuring the reliability of the system. The power distribution system is engineered to accommodate the diverse voltage needs of the components. The Arduino Uno microcontroller receives power from a 12V DC source, whereas the relays, coin slot, LCD, and servo motors function on 5V DC, and the fingerprint sensor operates on a 3.3V DC source. This distribution is controlled with a buck converter, which reduces the voltage to maintain compatibility and protect the components from potential damage.

Figure 6. System Wiring Diagram of the Charging Station

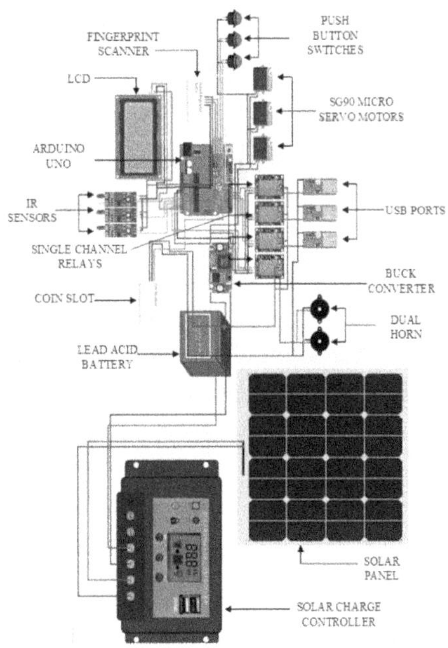

The Arduino Uno acts as the main control unit, handling input signals from different sensors and devices while regulating the output components. The input devices consist of IR sensors for detecting movement or unauthorized access, a fingerprint sensor for authenticating the user's identity, a coin slot for registering payment and determining charging time, and push button switches for enabling manual system interaction. The integration of these inputs facilitates smooth management of the charging process while guaranteeing user authentication and security measures are in place.

The output components consist of relays functioning as electronic switches to manage the power flow to the mobile USB terminal for charging, an LCD display that delivers real-time feedback to users with instructions and charging details, and SG90 micro servo motors that control the locking and unlocking mechanisms for mobile device lockers. Furthermore, a dual horn (buzzer) is incorporated to emit an alarm in the event of unauthorized access or tampering, thereby improving the system's security measures.

The system functions within a well-defined workflow. When a coin is inserted into the slot, it triggers the microcontroller to determine the charging duration. The fingerprint sensor verifies the user's identity, and upon successful authentication, the microcontroller activates the servo motor to unlock the device locker. The relays facilitate the transfer of power to the USB terminal for charging devices, while the IR sensors oversee the environment to detect any unauthorized access. The LCD display offers detailed guidance and real-time charging notifications. After the designated charging period concludes, the system engages the locking mechanism of the locker, ensuring the device remains secure until it is retrieved.

This wiring diagram emphasizes the incorporation of renewable energy sources, cutting-edge security measures, and optimized power distribution to establish a dependable and eco-friendly charging solution. The implementation of the microcontroller facilitates smooth interaction between input and output components, while the integration of voltage regulation mechanisms, like the buck converter, guarantees operational efficiency and safety. By integrating these components, the system delivers an intuitive and secure public charging station that meets contemporary demands for sustainability, safety, and ease of use.

PROGRAM FLOW CHART

The flowchart depicted in Figure 7 outlines the logical sequence carried out by the Arduino Uno Rev3 microcontroller to oversee the operation and security of the anti-theft mobile phone charging station. The process starts with setting up essential variables, such as the relay state, coin count, start button state, IR state, alarm state, and servo motor state. These variables are crucial for managing functionalities like power charging, payment status, and device security. The system consistently checks the condition of the start button. Upon pressing the button (start button state = 1), the system will show "System is locked" and "Register your fingerprint" on the LCD, indicating that the user needs to provide a fingerprint. In the absence of button activation, the system undergoes a reset and resumes its input monitoring process.

Figure 7. Program Flow Chart

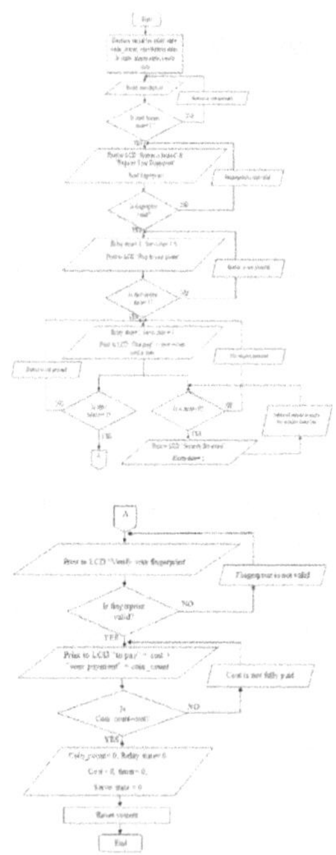

After a fingerprint is scanned, the system performs validation. When the fingerprint is confirmed as valid, both the relay state and the servo motor state are configured to 0, signifying that the charger is deactivated, and the mobile locker is accessible. The LCD displays a message instructing the user to "Connect your phone." In the event of an invalid fingerprint, the system reverts to the registration process, thereby guaranteeing that only authorized individuals are permitted to continue. Upon pressing the start button once more, the states of the relay and servo motor transition to 1, effectively securing the device and activating the charger to commence the charging process. The LCD indicates

"Charging," accompanied by a timer and cost details. Throughout this procedure, the system persistently observes the IR sensor to identify any unauthorized access.

The infrared sensor is essential for effective security surveillance. Upon detecting an unauthorized access attempt (IR state = 0), the system activates the alarm (alarm state = 1) and shows "Security Breached" on the LCD, notifying users of possible tampering. The charging process proceeds uninterrupted, with the system maintaining its watchful oversight. Upon completion of the charging process, the system instructs the user to "Verify your fingerprint." After the fingerprint verification is completed successfully, the LCD presents the message "To pay," along with the displayed cost and the user's coin count. In the event of an invalid fingerprint, the system necessitates a revalidation process. The system efficiently computes the remaining balance, enabling users to execute incremental payments until the target amount is achieved.

After the payment is finalized and the coin count aligns with the total cost, the system reinitializes all variables to their default state. The relay state, coin count, cost, timer, and servo motor state are initialized to 0, effectively unlocking the device and stopping the charging process. The system subsequently purges all stored data, guaranteeing its readiness for the subsequent user. This organized workflow showcases a reliable and intuitive process that combines safety, efficiency, and ease of use. The implementation of fingerprint authentication alongside IR sensors guarantees secure operations, and the reset mechanism improves reliability by removing any residual data. The program flowchart presents a systematic strategy for tackling the issues associated with public charging stations, highlighting the importance of user safety, simplicity of operation, and overall efficiency.

System Flowchart

The system flowchart depicted in Figure 8 details the operational sequence of the anti-theft mobile phone charging station, emphasizing the interaction between the user and the system. The procedure initiates with the servo motor in an unlocked condition, enabling the user to connect their mobile device. After connecting the device, the user activates the system by pressing the "OK" button, which starts the fingerprint registration process. Upon successful storage of the fingerprint, the system prompts the user to secure the mobile locker. When the "OK" button is pressed again, the timer initiates, the servo motor shifts to a locked position, the IR security alarm activates, and charging commences, guaranteeing the phone's safety throughout the operation.

To obtain their phone, the user needs to activate the pay button, following which the system prompts for fingerprint verification. If the fingerprint does not match, the user is required to perform a re-scan of their fingerprint. Once fingerprint authentication is successfully completed, the timer halts, charging ceases, and the system grants the user a window of 10 seconds to settle the bill shown on the LCD. If the payment is not finalized within the specified timeframe, the system will automatically resume charging, and the timer will keep calculating the cost. Partial payments are recorded in the system memory and shown on the LCD, enabling the user to settle the remaining balance later. Throughout this timeframe, the system stays secured, and the device keeps charging until the complete amount is settled.

Figure 8. System Flow Chart

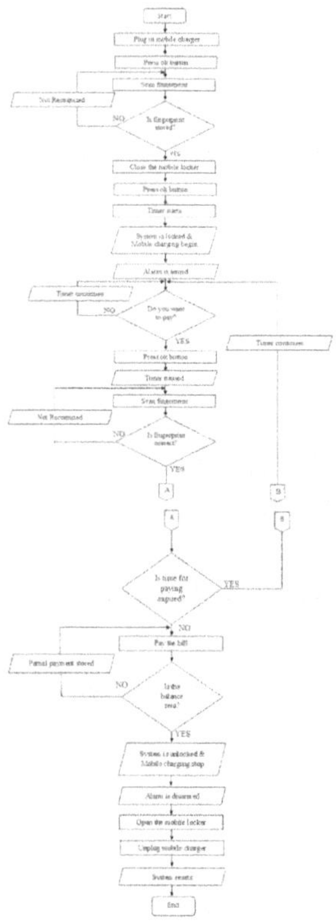

Once the balance is cleared, the servo motor disengages the locker mechanism, and the charging process for the phone is halted. At the same time, the IR sensor alarm is turned off, allowing the user to access the phone without any interruptions. The system subsequently performs a reset, erasing all stored information except for that related to any lockers currently in operation. This reset guarantees that the system is prepared for the subsequent user, all while upholding operational efficiency throughout various lockers.

The flowchart highlights the importance of security, ease of use, and adaptable payment options. The incorporation of fingerprint authentication guarantees that only permitted users can gain entry to their devices, while the IR sensor alarm acts as a deterrent against unauthorized access. The system's capability to accommodate partial payments while enabling ongoing charging enhances user convenience and reduces disruptions. The system enhances efficiency and reliability through automatic resets between users, facilitating seamless transitions. This organized workflow showcases a reliable, effective, and user-focused approach to addressing public charging requirements.

The System Design

The design of the anti-theft mobile phone charging station, as shown in Figure 9, presents a practical, secure, and sustainable approach to meet public charging requirements. The prototype includes three distinct mobile lockers, each designed with integrated locking mechanisms that utilize fingerprint authentication and security alarms for enhanced protection. The lockers, located on the right side of the station, provide secure storage for users' mobile devices while they are charging. Made from flat bars, tubular components, flat sheet galvanized iron, and angle bars, the station has been constructed for durability and strength while ensuring an ergonomic design.

On the left side of the structure is the electronic box, which contains the microcontroller and other essential components, including relays, sensors, and wiring. This centralized setup enhances system upkeep and protects delicate hardware from outside influences. The solar panel, positioned on the roof, acts as the main energy source, optimizing sunlight exposure for effective energy collection. Utilizing solar energy, the system decreases reliance on nonrenewable energy sources, supporting sustainability goals and reducing environmental impact.

The addition of caster wheels improves the station's mobility, enabling easy relocation to various settings like malls, transportation hubs, or outdoor events. The mobility feature enhances the adaptability of the charging station to meet diverse user requirements. The frame, made from galvanized iron and other robust materials, offers a stable and corrosion-resistant structure, guaranteeing long-lasting durability in outdoor settings.

This system design prioritizes ease of use, robust security measures, and environmentally friendly practices. Mobile devices are securely stored in lockers that automatically lock following fingerprint authentication, guaranteeing access is restricted to authorized users only. The locker is designed to stay locked during the entire charging process, ensuring that any attempts at unauthorized access will activate an alarm system. The incorporation of the solar panel enables the station to function autonomously from the electrical grid, ensuring reliability in remote areas or regions susceptible to power interruptions.

The station's modular design facilitates scalability and simplifies maintenance processes. Extra lockers or power systems can be integrated as required, boosting its adaptability for different applications. This charging station effectively addresses public charging challenges through its integration of secure storage, renewable energy, and portability. The design is carefully crafted to support overarching objectives of innovation, environmental sustainability, and user-focused functionality, representing a progressive strategy for contemporary infrastructure requirements.

Figure 9. The System Design

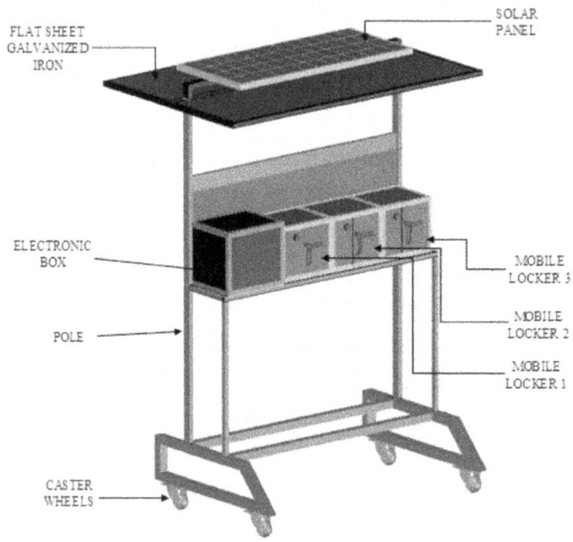

RESULTS AND DISCUSSIONS

Design and Construction

The development of the mobile phone charging station featuring a locking mechanism was grounded in the conceptual design and utilized durable materials including tubular bars, flat bars, angle bars, stainless steel caster wheels, and flat sheet galvanized iron. The chosen materials exhibit exceptional durability and resistance to environmental factors, guaranteeing the station's long-term performance and reliability. The design and construction process was organized into sequential steps to ensure simplicity and accuracy.

Step 1: Assembly of the Frame

The frame was assembled through a welding process of tubular bars, flat bars, and angle bars. The components underwent precise measurement as well as were cut with a grinder to guarantee exact dimensions. The welded structure established the station's foundation, ensuring stability and support for the additional components. This phase was essential for guaranteeing the robustness and longevity of the entire system.

Figure 10. Assembly of the Frame.

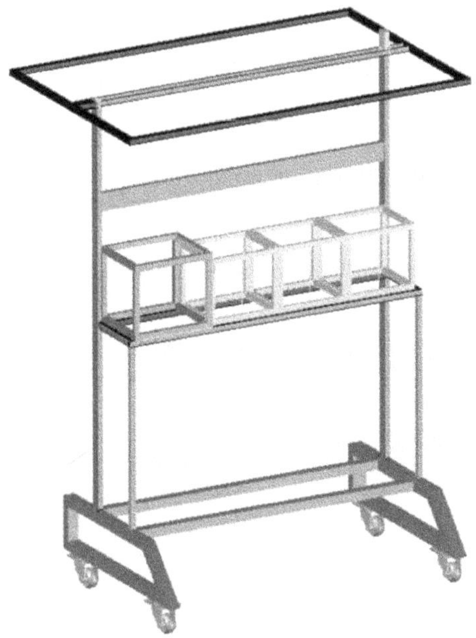

Step 2: Assembly of the Door and Wall

The next phase consisted of fastening and stabilizing the front panel of the three lockers. The front walls were securely fastened using a riveter and blind rivets. The preassembled locker doors were subsequently affixed to the front wall. This step guaranteed that the lockers were accurately aligned and operational, offering secure storage for users' mobile devices.

Figure 11. Assembly of the Door and Wall

Step 3: Assembly of the Door Lock

During this phase, the stainless-steel door latches and servo motors were securely welded within the three lockers. The components are essential to the locking system, combining both mechanical and electronic security elements. The servo motors play a crucial role in managing the locking and unlocking mechanism, guaranteeing that only authorized users have secure access. This step was crucial for ensuring the station's security and operational efficiency.

Figure 12. Assembly of the Door Lock

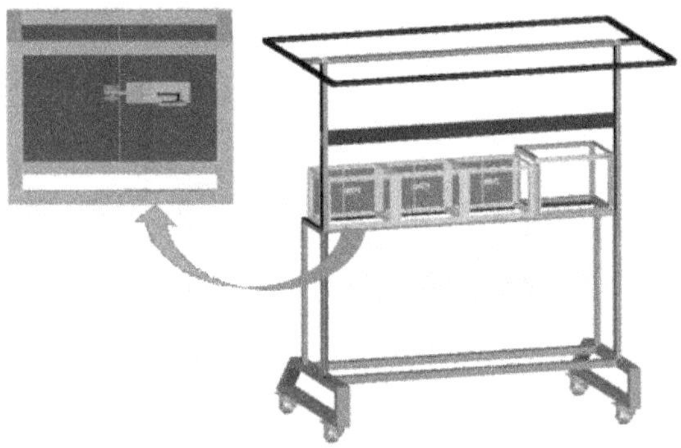

Step 4: Assembly of Walls, Roofing, Mobile Holder, and Solar Frame

The last phase consisted of connecting all the outstanding components. The electronic box and the three mobile lockers were firmly secured in place using a riveter and blind rivets. The roofing was additionally secured to shield the components from external environmental influences. A structure for the solar panel was securely attached to the roof, guaranteeing the best alignment for energy collection. Furthermore, mobile holders were integrated within the lockers to create a designated area for device storage while charging. This phase finalized the assembly of the station, bringing together its structural and operational elements.

Figure 13. Assembly of the walls, Roofing, Mobile Holder, and Solar Frame

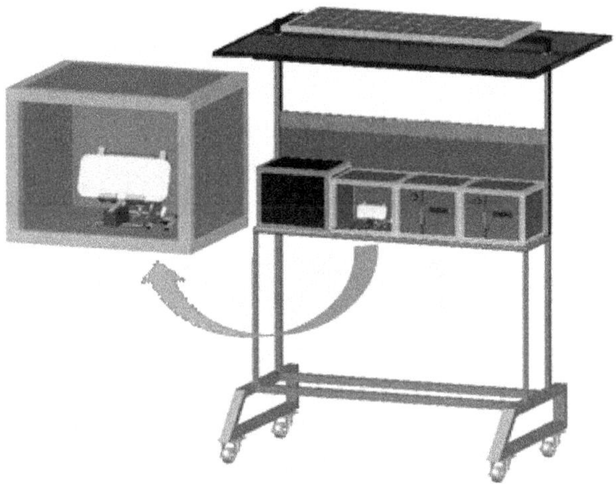

The design and construction process showcases a methodical approach to creating a durable and efficient mobile phone charging station. Every phase concentrated on maintaining the system's structural integrity, functionality, and security. Utilizing robust materials such as galvanized iron and stainless steel significantly improves the station's durability against wear and environmental factors. The incorporation of servo motors, solar panels, and secure lockers in the design meets contemporary standards for sustainability, security, and convenience. This construction method yields a dependable charging station designed to fulfill the requirements of users across diverse public environments.

System Integration

The integration of components with the Arduino Uno Rev3 microcontroller was accomplished through a systematic approach, guaranteeing that all electronic elements were correctly connected and operational. The process of system integration commenced with the transfer of code to the Arduino Uno Rev3 via an Arduino USB cable. This established the necessary programming logic to control the operations of the fingerprint scanner, IR sensors, Allan coin slot, LCD, micro servo motors, and single-channel relay. The physical integration process was carried out in multiple stages.

Step 1. Mount and fix the ports of the components in a printed circuit board (PCB)

In Step 1, the ports of the components were mounted and firmly secured onto a printed circuit board (PCB). This step guaranteed that every component was assigned a specific port, streamlining the following connection process and preserving order within the system.

Figure 14. Components Port Installation

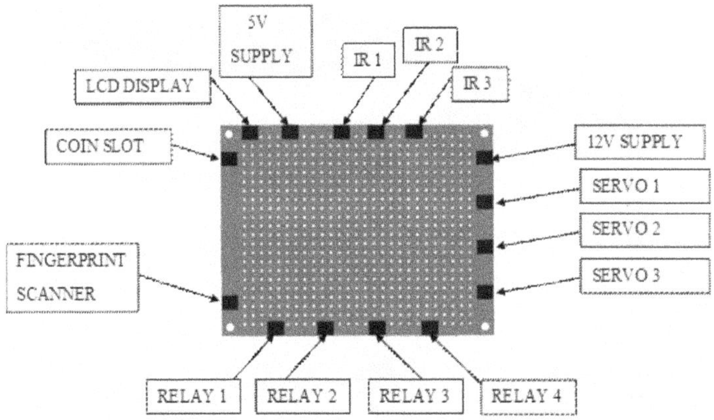

Step 2. Solder pins on the ports of the components

In Step 2, pins were soldered onto the ports of the components to create a reliable electrical connection. This soldering step was essential for guaranteeing durability and reducing the likelihood of disconnections caused by vibrations or movement.

Figure 15. Components Pin Installation

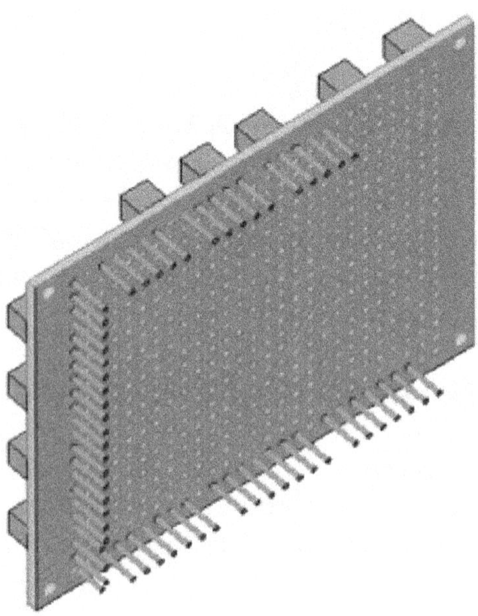

Step 3. Insert the pins on the Arduino Uno board pinout.

During Step 3, the pins from the components were thoroughly placed into the correct pinout ports of the Arduino Uno board. This step established a direct link between the microcontroller and the components, allowing for effective control and monitoring of the components by the microcontroller.

Figure 16. Components Pin Installation on Arduino Uno

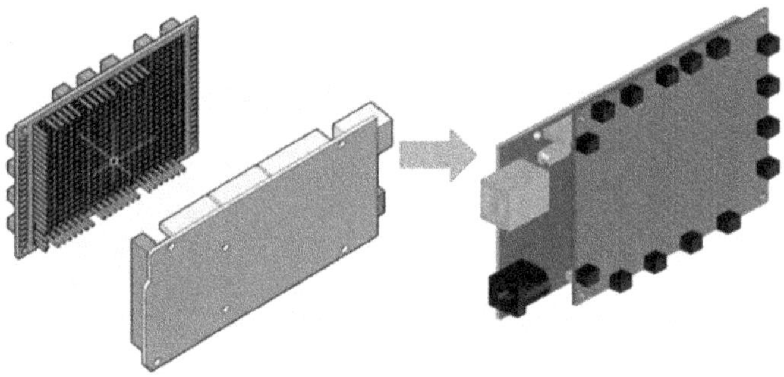

Step 4: Fix all the components in the electronic box using screws.

In Step 4, all components were securely mounted inside the electronic enclosure with screws. This established a reliable and robust configuration, safeguarding the components from external factors and guaranteeing their stability throughout operation.

Figure 17. Installation of Components in the Electronic Box

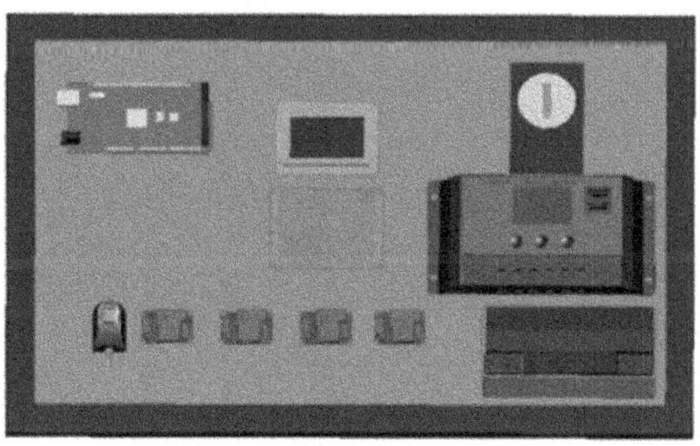

Step 5. Connect all the wires of the components inside the electronic box.

During Step 5, the internal wires linking the components inside the electronic box were meticulously routed and connected. This phase guaranteed effective communication and power allocation between the microcontroller and the various components. The wires were meticulously arranged to prevent tangling or interference, thereby improving the system's reliability.

Figure 18. Connecting Inside Wires and Cables

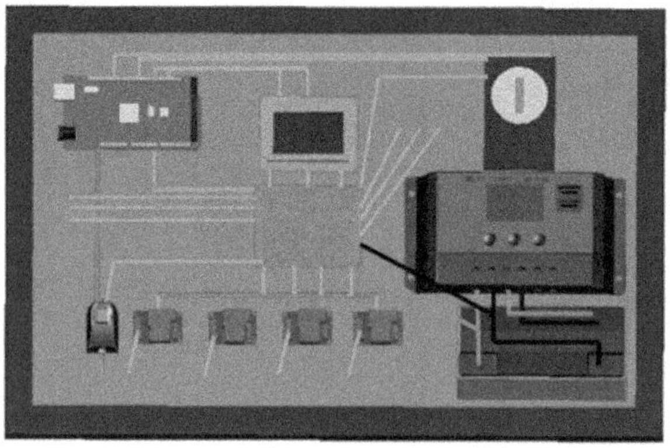

Step 6. Connect the wires of the other components inside the three boxes.

Finally, in Step 6, the connections for the external components, including those housed within the three mobile lockers, were established with the electronic box. This phase brought together all elements into a cohesive system, enabling the microcontroller to oversee the inputs and outputs from the electronic box as well as the external lockers.

Figure 19. Connecting Wires Outside the Electronic Box

Testing and Evaluation

The assessment of the Anti-Theft Mobile Phone Charging Station concentrated on evaluating its performance, ease of use, security features, and general acceptability. Twenty respondents assessed the system according to various essential features, with average scores and standard deviations offering valuable insights into user perceptions and satisfaction levels.

Table 1 underscores the overall acceptability of the Anti-Theft Mobile Phone Charging Station across several key features. The evaluation involved 20 respondents, and the analysis of mean scores and standard deviations yields valuable insights into user satisfaction and the effectiveness of the system. The station's convenience and ease of use achieved an outstanding mean score of 5.00, with a standard deviation of 0.00, reflecting unanimous consensus among respondents that the system is exceptionally user-friendly. The reliability of this system showcases its user-friendly design and straightforward operation, ensuring that the charging station is easily usable by a diverse audience.

Table 1. Summary of Respondents' Acceptability Ratings for the Anti-Theft Mobile Phone Charging Station Features

No.	Mobile Charging Feature	Mean	Standard Deviation	Descriptive Interpretation
1.	Anti-Theft Mobile Charging Station is convenient/easy to use.	5.00	0.00	Acceptable
2.	Anti-Theft Mobile Charging Station is secure and safe to use	4.95	0.22	Acceptable
3.	Anti-Theft Mobile Charging Station built-in lock is sturdy	4.90	0.31	Acceptable
4.	Anti-Theft Mobile Charging Station displays accumulated time and equivalent payment accurately.	5.00	0.00	Acceptable

continued on following page

Table 1. Continued

No.	Mobile Charging Feature	Mean	Standard Deviation	Descriptive Interpretation
5.	Anti-Theft Mobile Charging Station alarm is sensitive and loud.	4.85	0.37	Acceptable
6.	Anti-Theft Mobile Charging Station design is acceptable	5.00	0.00	Acceptable
	Average	**4.95**	**0.15**	**Acceptable**

The system demonstrated a mean score of 4.95 and a standard deviation of 0.22 regarding its protective measures, reflecting a robust consensus on its effectiveness in safeguarding users' devices. The slight variation indicates subtle differences in user perceptions, yet the overall rating highlights the effectiveness of features like fingerprint authentication, IR sensors, and the alarm system in ensuring a secure charging experience. The robustness of the integrated locks received an impressive mean score of 4.90, accompanied by a standard deviation of 0.31. This indicates a strong reliability in the locking mechanism, while also highlighting a possible area for enhancement regarding perceived physical durability.

The accuracy of time and payment display achieved a flawless mean score of 5.00, with no variation (standard deviation of 0.00), demonstrating that users consistently regarded the display as clear and dependable. This highlights the necessity of delivering clear and precise feedback, which fosters reliability and assurance in the system. The sensitivity and loudness of the alarm received an average score of 4.85, accompanied by a standard deviation of 0.37. This indicates that while the majority of users deemed the alarm satisfactory, a minor segment may have experienced it as either excessively loud or not sensitive enough. This feedback may inform possible enhancements, like customizable alarm settings to accommodate various user preferences.

The system's design achieved an outstanding mean score of 5.00, with unanimous consensus on the acceptability of both the physical layout and aesthetic aspects. This shows that the station's design and performance meet user standards, enhancing both its practicality and aesthetic value. The overall acceptability of the system reached an average mean score of 4.95, accompanied by a low standard deviation of 0.15, indicating a strong consistency in user satisfaction across all features.

The results of this study carry multiple implications. The consistently high ratings indicate that the charging station operates effectively, maintains reliability, and is positively regarded by users, positioning it as a strong candidate for wider implementation in public areas. The low standard deviations demonstrate the system's reliability and robustness, reflecting only slight variability in user perceptions. Nonetheless, the marginally elevated standard deviations for alarm sensitivity and lock sturdiness indicate potential areas for improvement to boost user satisfaction. Adjustable alarm settings and reinforced lock materials could effectively tackle these concerns. The favorable assessment of the system's convenience, security, accuracy, and design underscores its viability for commercial application, providing a secure, efficient, and user-friendly charging solution. The testing results confirm the validity of the system's conceptual and physical design, highlighting its preparedness for practical implementation.

Evaluation of System Components

Table 2. Evaluation Checklist and Percentage Ratings of the Anti-Theft Mobile Phone Charging Station Components

No.	Statements	Yes	No	Yes%	No%
1.	The LCD can recognize if a phone is present	20	0	100%	0%
2.	The fingerprint sensor can recognize the fingerprint	20	0	100%	0%
3.	The locking system is operational	20	0	100%	0%
4.	The motion alarm system is functional	20	0	100%	0%
5.	The LCD has accurate information	20	0	100%	0%
6.	The timer and payment system is accurate	20	0	100%	0%
7.	The coin slot accepts and recognizes different Philippine peso coin	20	0	100%	0%
8.	The buttons are functional	20	0	100%	0%

Table 2 displays the checklist percentage evaluation of the components of the Anti-Theft Mobile Phone Charging Station, indicating uniform "yes" responses for all eight evaluation statements, culminating in a total approval rate of 100% from all 20 respondents. This collective positive response highlights the dependability and performance of the system's essential elements. The LCD display underwent validation to confirm its capability to identify a phone's presence and present precise information, guaranteeing that users obtain timely and dependable feedback throughout the charging process. The fingerprint sensor reliably identified users' fingerprints, strengthening the system's capacity to thwart unauthorized access and improve security measures. In a similar manner, the locking system was verified to be functional, ensuring strong physical security for the mobile devices that are stored.

The motion alarm system, designed to identify unauthorized access or tampering, has been evaluated as fully operational, providing an extra layer of security to prevent theft. The evaluation of the timer and payment system confirmed their accuracy, guaranteeing that users are charged fairly according to the actual charging duration. This clarity in cost assessment builds confidence in the system's performance. The coin slot was engineered to identify various denominations of Philippine peso coins, functioning efficiently and showcasing its compatibility with local currency as well as its reliability in processing payments. Furthermore, the buttons were verified to operate flawlessly, facilitating seamless and user-friendly engagement with the system.

The unwavering 100% approval across all features underscores the strength, dependability, and ease of use of the charging station. The verification of every element shows that the system achieves its specified design objectives and operates as anticipated in practical applications.

The findings carry important consequences for the implementation and functionality of the charging station. The complete success rate across all features demonstrates the system's operational reliability, assuring users of its functionality in public environments. The efficiency of the security components, including the fingerprint sensor, locking mechanism, and motion alarm, highlights the system's capability to safeguard users' devices, rendering it appropriate for deployment in busy public spaces. The precision of the timer and payment system guarantees transparency and equity, enhancing user trust in the payment procedure. Furthermore, the operation of the LCD, buttons, and coin slot demonstrates a design that prioritizes ease of use, catering to a diverse audience, including individuals who may not be well-versed in sophisticated technology.

The findings further confirm the system's potential for commercial application, showcasing its preparedness for extensive implementation in public areas. The system provides a reliable solution for public charging requirements, incorporating features that enhance both security and usability. The consistent positive feedback demonstrates that the charging station is a reliable, efficient, and commercially viable product, offering users a safe and convenient charging solution.

CONCLUSION

The thorough assessment of the Anti-Theft Mobile Phone Charging Station validates its efficiency, dependability, and preparedness for practical application. The system's design and construction showcased outstanding structural integrity, operational efficiency, and robust security features. Employing durable materials like galvanized iron and stainless steel, the station's methodical construction process guaranteed a stable, enduring, and visually appealing design. Contemporary elements such as servo motor-operated locks and solar panel integration adhere to current benchmarks for sustainability and security, rendering the system appropriate for implementation in diverse public settings.

The integration process effectively aligned the Arduino Uno Rev3 microcontroller with essential components, such as the fingerprint sensor, IR sensors, LCD, and other vital elements. The meticulous soldering, wiring, and assembly of components guaranteed flawless functionality, providing immediate feedback, reliable device storage, and optimal energy efficiency. The findings also demonstrated elevated levels of user satisfaction and acceptability. The evaluation results showed a strong consensus among respondents about the system's convenience, security, and functionality. Elements such as the fingerprint authentication system, robust locking mechanism, motion alarm, and precise time and payment display received impressive mean scores, culminating in an overall average of **4.95** with minimal variability, confirming the system's ease of use and operational dependability.

Furthermore, the evaluation of the checklist revealed a 100% approval rate for all key components, such as the LCD, fingerprint sensor, locking mechanism, motion alarm system, and coin slot. This reliable performance underscores the system's robustness and efficiency in real-world scenarios. The integration of sophisticated security elements like fingerprint recognition, motion detection, and alarm systems guarantees strong safeguarding for user devices, while the precise time and payment information promotes clarity and fosters user confidence. The characteristics of this charging station establish it as a dependable and safe option for use in public areas.

The findings further highlight the practical potential and expandability of the system. The significant acceptance and operational effectiveness suggest a robust potential for widespread implementation in locations like shopping centers, transit stations, and public gatherings. The incorporation of renewable energy via solar panels significantly boosts its sustainability and flexibility in meeting diverse user requirements. The system operates at a high level, yet small improvements—like customizable alarm settings and increased lock durability—could enhance its performance and user experience even more.

In summary, the Anti-Theft Mobile Phone Charging Station successfully tackles the issues of safety and efficiency in public charging environments. The innovative design integrates renewable energy sources, incorporates advanced security features, and ensures user-friendly operation, resulting in a reliable, scalable, and sustainable solution. The system's reliable performance and strong user satisfaction affirm its design principles and physical structure, demonstrating its preparedness for real-world application and market introduction. This charging station marks a notable advancement in meeting contemporary public charging needs, emphasizing security, efficiency, and environmental stewardship.

REFERENCES

Aguas, C., Asuncion, J., Dumulot, A., Gepulle, D., Larin, J., Pingol, H., & Salas, L. (2022). Solar Panel Powered Mobile Charging Station. Academia. https://www.academia.edu/31982060/Republic_of_the_Philippines_COLLEGE_OF_TECHNOLOGY_TITLE_OF_THE_PROJECT_Solar_Panel_Powered_Mobile_Charging_Station

Alkhunaizan, A. (2020). The Implication of Mobile Phone Charging Station. *International Journal of Computer and Network Security, VOL,* 20(1). http://paper.ijcsns.org/07_book/202001/20200117.pdf

Ao, X., He, R., Zhang, C., & Wang, L. (2019). A New Outdoor Energy Mobile Phone Charging Station. 34rd Youth Academic Annual Conference of Chinese Association of Automation. https://ieeexplore.ieee.org/document/8787589

'Benefits of Cellphone Charging Station', Veloxity, 2016, accessed 12 September 2022, https://veloxity.us/benefits-cell-phone-charging-station/

Chafle, K., Faruk, B., Shrivas, S., & Sawarkar, N. (2015). Coin Based Mobile Charger on Solar Energy. International Journal of Research- Granthaalayah, 3(1):13-18. https://www.researchgate.net/publication/306103054_COIN_BASED_MOBILE_CHARGER_ON_SOLAR_ENERGY

Chandrashekhar, T., Swaminaidu, G., & Rao, B. Ch. (2014). Mobile Charger Based on Coin by Using Solar Tracking System. International Journal of Innovative Reseach in Science, Engineering and Technology, Vol. 3, Issue 2, February 2014. http://www.ijirset.com/upload/2014/february/82_Mobile.pdf

Kamat, A., Kulkarni, A., Akshay, K., Kasturi, R., & Balur, N. (2017). Coin Insert Mobile Charger. Communication on Applied Electronics, Volume 6 No. 7. https://www.caeaccess.org/archives/volume6/number7/708-2017652515

Kamble, P., Khot, S., Solkar, S., & Patil, S. (2022). Mobile Charging Vending Machine. International Research Journal of Modernization in Engineering Technology and Science, Volume: 04/Issue: 07/July-2022. https://www.irjmets.com/uploadedfiles/paper//issue_7_july_2022/27570/final/fin_irjmets1657227298.pdf

Kathiresan, R., Ashwin, V., & Raj, M. (2016). Coin Based Mobile Charger. International Journal of Advanced Research in Electrical, Electronics and Instrumentation Engineering, Volume 5, Special Issue 1, March 2016. https://www.ijareeie.com/upload/2016/ncrte/29_EIE29.pdf

Li, C., Jia, W., Tao, Q., & Sun, M. (2011, April).. . *Solar Cellphone Charger Performance in Indoor Environment*, 01-03. Advance online publication. DOI: 10.1109/NEBC.2011.5778623

Mailley, J., Whitehead, S., & Farrell, G. (2006). Progress and prospects in the prevention of mobile phone theft. Loughborough University, Volume 170, Issue 3, June 2006. https://www.academia.edu/29412701/Progress_and_prospects_in_the_prevention_of_mobile_phone_theft

Mapari, N., Khadija, A., Taiba, A., & Azima, S. (2019, August). Smart Charging Vending Machine. *IIEE Xplore*, 29(d), 414–417. Advance online publication. DOI: 10.1109/ICCMC.2019.8819821

Maroma, N. (2014, December). Solar Powered Cellphone Charging Station. *OAlib*, 1(9), 1–7. Advance online publication. DOI: 10.4236/oalib.1101156

Nethravathi, P., Aithal, P., Soans, S., & Yadav, N. (2021). Coin-Based Mobile Charger Using Solar Tracker. *International Journal of Applied Engineering and Management Letter*, 5(2). Advance online publication. DOI: 10.47992/IJAEML.2581.7000.0104

Pawar, A. (2015). Coin Based Solar Mobile Charger. International Journal of Engineering and Technical Research, Volume-3, Issue-5, May 2015. https://www.erpublication.org/published_paper/IJETR032148.pdf

Pera, A. (2017). Piso Automated Cellphone Charging Utility Model. Slideshare. https://www.slideshare.net/angelitopera/chapter1-3-74917597

Priyanka, G., Anisha, S., & Padma Shri, P. (2018). Coin Based Mobile Charger. *International Journal of Pure and Applied Mathematics*, 119(12), 13695–13701. http://www.ijpam.eu

Whitehead, S., Mailley, J., Storer, I., McCardle, J., Torrens, G., & Farrell, G. (2008). In safe hands: A review of mobile phone anti-theft designs. springer science + business media, Vol. 14, Iss. 1, (2008): 39-60. https://doi.org/DOI: 10.1007/s10610-007-9040-9

Chapter 11
Enhancing the Efficiency of IoT-Integrated Smart Bin With Real-Time Monitoring

Devika M.
https://orcid.org/0009-0009-9471-6600
Sree Buddha College of Engineering, India

S. V. Annlin Jeba
https://orcid.org/0000-0002-1314-4025
Sree Buddha College of Engineering, India

Dona M. Jose
https://orcid.org/0009-0000-4280-7230
Sree Buddha College of Engineering, India

Sreya S.
https://orcid.org/0009-0004-4254-2630
Sree Buddha College of Engineering, India

Sayujya Prem
https://orcid.org/0009-0001-3870-6274
Sree Buddha College of Engineering, India

ABSTRACT

The rapid growth of urban populations has led to increased waste production, challenging conventional waste management systems. This paper presents an IoT-based Smart Waste Bin System designed to optimize waste collection and enhance urban cleanliness. Equipped with sensors, the system monitors waste levels in real-time and transmits data via LoRaWAN or NB-IoT to a cloud platform. It triggers alerts for improper waste sorting and guides users on correct disposal. A compression mechanism activates at 70% capacity to reduce waste volume, and when the bin reaches 90%, the system notifies municipal authorities with GPS data. Instant data analysis facilitates efficient waste collection scheduling and route optimization, lowering operational costs and enhancing efficiency. The system supports sustainability by minimizing unnecessary collection trips and fuel consumption. Initial tests show its potential

DOI: 10.4018/979-8-3693-9770-1.ch011

to improve traditional practices, contributing to smarter, cleaner cities and aligning with broader smart city strategies.

INTRODUCTION

The rapid expansion of urbanization and population growth has dramatically increased waste production, intensifying the challenges faced by traditional waste management systems. In many regions, waste collection still relies on outdated manual methods, with workers following fixed schedules to collect waste from households and public bins. This approach is inefficient, as residents are often required to adhere to strict collection times. Missed collections can lead to waste accumulation, which poses serious hygiene risks. In densely populated urban areas, collection teams frequently check public bins to determine if they are full, wasting time and resources when bins are not yet ready for collection. Conversely, bins may overflow between scheduled visits, creating unsanitary conditions.The inefficiencies of traditional methods are further worsened by increasing population density and urban growth, placing greater demands on waste management services. This highlights the urgent need for intelligent systems that can monitor waste levels in real time and enhance the collection process.

The Internet of Things (IoT) has emerged as a promising technology in addressing these issues. IoT-enabled smart bins utilize sensors to monitor waste levels and automate the collection process. This project introduces an IoT-based Smart Waste Bin System, which incorporates sensors to detect human presence, measure waste volume, and classify the type of waste deposited. When the bin reaches 70% capacity, a compression mechanism is activated to compact the waste. If the compressed waste still exceeds 90% capacity, a notification is sent to municipal authorities through a GSM module, indicating that the bin requires emptying. Each bin is also equipped with a GPS module, providing accurate location data that helps optimize collection routes. Additionally, the system features a webcam to prevent theft or tampering, ensuring the bin remains functional.

By addressing key challenges such as inefficient manual collection, excessive fuel consumption, and unsanitary conditions from overflowing bins, the IoT-based Smart Waste Bin System offers a scalable, eco-friendly solution to urban waste management. The system enhances environmental sustainability through waste compression and the use of solar energy instead of traditional batteries. It also improves operational efficiency by providing real-time monitoring, reducing fuel consumption, and optimizing waste collection routes. From a social perspective, the system promotes public health by preventing waste overflow, reducing disease risks, and maintaining a cleaner urban environment. Through automation, efficient waste management, and real-time data analysis, this system aims to modernize waste management practices, contributing to cleaner, more sustainable, and efficient cities.

Some of the novel contributions of this proposed system are as follows:

- **Providing real-time alerts** to municipal authorities when compressed waste exceeds 90% capacity for timely collection.
- **Promoting sustainability** by utilizing solar energy and preventing unsanitary conditions through real-time monitoring.
- **Enhancing operational efficiency** by automating waste collection, reducing fuel consumption, and optimizing resource usage.

RELATED WORK

IoT technology has advanced waste management by enhancing efficiency and enabling real-time monitoring. IoT-based Smart Garbage Bins (SGBs) utilize sensors such as ultrasonic, infrared, and motion detectors to continuously monitor waste levels. These bins communicate using protocols like LoRaWAN or NB-IoT, which help lower emissions and operational costs. However, uninterrupted connectivity remains a challenge, especially in certain environments (Jadhav, S. P, 2024).A Sustainable IoT-based Waste Management System uses web applications and sensor data for remote monitoring, which reduces manual labour and improves responsiveness. This system, however, may face Wi-Fi connectivity challenges in underdeveloped regions(Ahmed M M, 2023) An IoT-based Smart Bin System employs fuzzy logic to analyze waste levels, optimizing resources, labour, and cleanliness. Implementing such advanced algorithms introduces system design and deployment complexities (Suvarnamma A, 2021).Integrated IoT-based Waste Sorting Mechanisms utilize inductive and capacitive sensors to automatically sort waste (e.g., metals, plastics, glass) Main challenges for these systems are maintaining sorting accuracy and ensuring reliable operation across diverse environments (Bano A et. al, 2020), (Zeb,A.,Ali,Q,2019).Smart Bins with Real-time Data Transmission enable immediate data sharing, allowing for responsive waste collection services and reduced human oversight. Yet, any interruption in data transmission could delay collection schedules and reduce system efficiency (Boopathis S et.al, 2022).In waste collection route optimization, prior research has focused on GIS and scheduling models, such as studies conducted in Asansol, India, and Eastern Finland. Although these studies have improved the logistics of waste collection, few integrate IoT for real-time waste status monitoring or dynamic routing (Pucnik R et. al 2024), (Ayodeji Okubanjo 2024), (Szpilko, D 2023). Studies in developing countries, as reviewed in relevant literature, also often overlook IoT capabilities, underscoring the need for IoT-integrated solutions that address stakeholder collaboration and rapid problem reporting. Including real-time data analytics, cross-stakeholder coordination, and field surveillance could foster more responsive, efficient, and sustainable IoT-based waste management systems (P. Jain 2023), (N. A. Antora, 2022),(F. O. Ogunwolu 2020).

CURRENT STATE OF SOLID WASTE MANAGEMENT IN INDIA:

India faces significant challenges in waste management due to inadequate garbage collection systems, ineffective sorting of recyclables, low recycling rates, and improper disposal practices.

These issues result in massive amounts of waste accumulating in landfills, contributing to environmental pollution and public health risks. Despite efforts, the country collects only around 43 million tonnes of waste annually, with a considerable amount remaining untreated or ending up in landfills. Recycling activities are limited, and the handling of hazardous and biomedical waste remains a persistent problem. There is an urgent need to enhance waste collection,segregation, and treatment processes nationwide.

The rapid urbanization and economic growth in India have significantly increased waste generation, straining the already insufficient waste management systems. Managing the rising volumes of waste efficiently is critical to preventing environmental damage and associated health issues. According to the 2020-21 report by the Central Pollution Control Board (CPCB), India generates approximately 160,038.9 tons of solid waste daily. Of this, 152,749.5 tons are collected, with around 79,956.3 tons (50%) undergoing treatment. However, 29,427.2 tons (18.4%) are sent to landfills, while 50,655.4 tons (31.7%) remain unprocessed.

Research by Meena et al. (2023) emphasizes that the improper and unregulated disposal of municipal solid waste (MSW) is closely tied to urbanization, population growth, lifestyle choices, and limited ecological awareness. Practices such as open dumping not only harm the environment but also pose severe health risks. Effective solid waste management encompasses various stages, including waste generation, storage, collection, transportation, processing, and disposal. Mismanagement at any stage leads to environmental degradation and poses significant threats to the health and well-being of people, especially in densely populated urban areas.

Challenges and issues of solid waste management:

Solid waste management plays a crucial role in achieving environmental sustainability, yet it faces numerous challenges worldwide. These difficulties stem from the growing volume of waste, resource limitations, technological barriers, and socio-economic constraints. Below are some of the primary challenges:

Increasing Waste Generation

The amount of municipal solid waste (MSW) generated in a region is influenced by several factors, including living standards, the scale and nature of commercial activities,dietary habits, geographical characteristics, and climatic conditions. Additionally, the influx of migrants seeking better opportunities in urban areas further exacerbates the waste management problem by increasing the burden on already strained systems(Fo Ogunwolu,2020).According to recent data from the Ministry of New and Renewable Energy (MNRE),India generates approximately 145 million tonnes of waste annually, a figure projected to rise significantly to an estimated 260 to 300 million tonnes per year by 2047.The Central Pollution Control Board (CPCB) reported in 2016 that out of the total waste generated in India, about 117,644 metric tonnes (MT) are collected daily. However, only 49,401 MT undergo proper treatment, reflecting significant inefficiencies in waste processing and disposal mechanisms. This discrepancy highlights the urgent need for improved waste collection, segregation, and treatment infrastructure to manage the growing waste volumes effectively.(Liakot Ali,2012) .Without intervention, the mounting waste could lead to severe environmental degradation, public health crises, and economic challenges, particularly in densely populated urban centers. Addressing these issues requires a combination of technological innovation, policy reform, and increased public awareness about sustainable waste management practices.

Urbanization and Population Growth

The rapid pace of urbanization and the continuous growth of the global population are significantly contributing to the increasing volumes of solid waste being generated. As cities expand and more people migrate to urban areas in search of better opportunities, the demand for housing, goods, and services rises sharply. This surge in consumption leads to greater production of municipal solid waste, including household, commercial, and industrial refuse. (N.S, Raghava,2021).Urban centers, often lacking adequate waste management infrastructure, face immense pressure to collect, process, and dispose of the growing waste volumes effectively. The problem is further exacerbated by the generation of non-biodegradable and hazardous waste, which requires specialized treatment and disposal methods. Without comprehensive strategies to address this issue, the escalating waste problem poses serious risks to the environment, public health, and the overall quality of life in urban areas.

Insufficient Collection Systems: In many regions, especially in developing countries, waste collection services are often inadequate or nonexistent, making it difficult to manage the growing volumes of waste. In these areas, communities frequently rely on informal or irregular waste collection methods, which can result in waste piling up in streets, alleys, and other public spaces.(Podder A.K,2020).The absence of a structured waste collection system not only leads to poor sanitation but also creates environmental hazards, as uncollected waste can attract vermin, contaminate water sources, and contribute to air pollution through open burning. To address this issue, there is a pressing need to improve waste collection infrastructure, ensure regular service delivery, and promote community involvement in maintaining cleaner, healthier environments.

Inadequate Disposal Facilities: In many regions, especially in developing countries, the lack of adequate disposal facilities is a significant problem. Limited access to properly managed sanitary landfills and waste treatment centers often forces communities to resort to open dumping and burning of waste. These practices not only contribute to severe environmental pollution but also pose serious health risks, as they release harmful toxins into the air, soil, and water (Tariq A,2019). Open burning, for instance, can lead to the emission of toxic gases, while open dumps become breeding grounds for diseases. To address this challenge, there is an urgent need for the development of environmentally safe disposal methods, such as sanitary landfills and waste-to-energy technologies, alongside proper waste treatment infrastructure.

Poor Waste Segregation

One of the key obstacles to effective solid waste management is the lack of waste segregation at the source. Proper segregation before collection significantly enhances the efficiency of the waste management process, promotes recycling, and reduces disposal costs. In many developing countries, including India, waste is often collected in mixed form, with households and businesses frequently discarding garbage in open spaces, drains, or water bodies. This is in contrast to more affluent nations like the USA, Germany, and Japan, where waste segregation at the source is a common practice. The lack of segregation can be attributed to factors such as low public awareness, limited access to proper sorting facilities, and slow adoption of modern waste sorting methods. (Vishnu S,2021).Without effective source separation, much of the recycling process is handled by the informal sector using outdated technologies. While this creates a market for low-cost products, it also limits the potential for efficient and sustainable recycling practices. Recognizing the importance of waste segregation at the source is crucial for improving recycling operations and advancing waste management in a more sustainable manner.

Lack of Awareness: A significant challenge in waste management is the lack of awareness among households and businesses regarding the importance of waste segregation. Many people are unaware of the need to separate biodegradable, recyclable, and hazardous materials, leading to improper disposal of waste.(Muhammed I,2019). This lack of understanding results in mixed waste being collected, which not only complicates recycling efforts but also increases the cost of waste treatment and disposal. Inadequate knowledge about how to sort waste and the environmental impact of improper disposal further exacerbates the problem. Raising awareness and educating the public on the benefits of waste segregation is essential to improve waste management practices and support more sustainable recycling efforts.

Treatment and final disposal

The treatment and final disposal of solid waste remain critical challenges for municipalities in India, as many continue to struggle with poorly managed waste treatment systems. Despite efforts to improve, current treatment facilities remain inadequate and fail to meet the rising demand. The heterogeneity of solid waste poses a significant obstacle for recycling plants, as the varying composition of waste makes it difficult to implement effective processes such as bio- methanation, waste-to-energy conversion, and composting.(Meena M.D,2023). These technologies, while promising, are still in the early stages of development in many parts of the country and require substantial upgrades to achieve large-scale implementation.

Unscientific disposal practices are prevalent, with approximately 90% of the total waste being dumped in open spaces rather than managed in capped sanitary landfills. Such methods not only lead to environmental degradation but also contribute to significant methane emissions.

According to the 2015 CPCB report, seven megacities in India alone accounted for nearly 48% of the country's methane emissions from waste disposal sites, highlighting the environmental risks associated with improper waste management.

Furthermore, the safe disposal of specialized waste categories such as sanitary waste, batteries, electronic waste (e-waste), and other hazardous materials is often neglected. These waste types require separate handling and treatment processes to minimize their adverse effects on health and the environment. Incorporating advanced technologies and establishing dedicated facilities for these waste streams is essential to address this growing concern effectively. Research by Rarotra et al. (2020) and Lv et al. (2018) emphasizes the urgent need for a scientific and sustainable approach to waste treatment and disposal, focusing on both resource recovery and environmentalprotection.

High Operational Costs

The high operational costs associated with waste collection, transportation, and processing present a significant challenge for effective waste management. These processes require substantial investments in infrastructure, including vehicles, fuel, equipment, and waste treatment facilities, as well as the recruitment and training of personnel. Maintaining these operations on a day-to-day basis is financially demanding, particularly for municipalities in developing countries with limited financial resources. Advanced waste management technologies, such as waste-to-energy plants,bio-methanation units, and recycling systems, often come with high installation and maintenance costs, further adding to the financial burden. (Balasubhramaniyan,2018).Additionally, factors such as rising fuel prices, increasing labor costs, and the need for frequent repairs and upgrades to outdated infrastructure exacerbate the issue. For many municipalities, these expenses can strain already limited budgets, resulting in reduced efficiency and coverage of waste management services.

Addressing high operational costs requires innovative solutions, such as adopting cost-effective technologies, optimizing logistics for waste collection and transportation, and leveraging public-private partnerships to share financial responsibilities. Additionally,promoting community participation and encouraging waste segregation at the source can help reduce the burden on centralized waste processing systems, thereby lowering overall costs.

Technological Limitations

Technological limitations present significant challenges in the management of solid waste, with many regions still relying on outdated methods and lacking access to advanced waste treatment technologies. These limitations hinder the effectiveness of waste management systems and contribute to environmental and health issues.

Lack of Advanced Recycling Facilities:Furthermore, the adoption of advanced recycling technologies, such as waste-to-energy systems and material recovery facilities, remains limited. These technologies have the potential to transform waste into valuable resources, reducing the amount of waste sent to landfills and generating clean energy in the process. However, their high upfront costs, technical complexity, and limited availability have slowed their widespread implementation. The lack of investment in these advanced solutions means that much of the recyclable material is still discarded, further exacerbating the waste management problem. To overcome these challenges, there is a need for greater investment in modern technologies, better infrastructure, and the promotion of sustainable waste management practices.

Policy and Governance Issues

Effective waste management requires strong policies and governance frameworks, but many regions face significant challenges in implementing and enforcing such policies.

Weak enforcement and the absence of comprehensive policies hinder progress in managing waste sustainably and efficiently.

Lack of Policies: Another critical challenge is the lack of robust policy frameworks for extended producer responsibility (EPR) and circular economy initiatives. EPR policies,which hold producers accountable for the lifecycle of their products, including waste disposal and recycling, are often weak or non-existent in many countries. Similarly, circular economy principles, which aim to minimize waste by promoting reuse, recycling, and sustainable product design, are not fully integrated into national waste management strategies.(Joan Nyika,2024) The absence of these policies prevents the development of more sustainable and efficient waste management systems and hinders progress toward reducing waste generation and promoting resource recovery. Strengthening enforcement and developing comprehensive policies for EPR and circular economy initiatives are crucial steps toward improving waste management and fostering a more sustainable future.

Public Participation and Awareness

Public participation and awareness are essential for the success of waste management programs, yet many communities face challenges in these areas. Low awareness levels and resistance to change hinder the adoption of effective waste management practices, making it difficult to achieve long-term sustainability.

E-Waste Management:E-waste management presents its own set of difficulties, largely due to the increasing volume of electronic devices being discarded and the complex nature of the materials involved. Modern electronics, such as smartphones, computers, and televisions, contain a mixture of metals, plastics, and hazardous substances likecadmium, arsenic, and brominated flame retardants. These materials require careful extraction and recycling processes to avoid environmental contamination. However, e- waste often ends up in informal recycling systems where it is improperly handled, leading to harmful emissions and exposure to toxic chemicals. As the global volume of e-waste continues to rise, there is a pressing need for improved collection systems, responsible recycling practices, and public awareness to manage e-waste safely and sustainably.

Additionally, addressing the challenges of e-waste requires effective policies, such as extended producer responsibility (EPR), which hold manufacturers accountable for the recycling of their products at the end of their life cycle.

Need for urban sustainable waste management

Sustainable waste management plays a crucial role in fostering urban development. First and foremost, effective urban waste management has a direct impact on the environment and public health. Poor waste handling and storage can release harmful pollutants into theair and water, contributing to long-term health problems like respiratory issues and chronic diseases. Additionally, urban waste management is vital for the efficient use and conservation of resources. Many materials in waste, such as plastics, metals, and paper, can be recycled, helping to reduce resource depletion and supporting a circular economy. By recycling, we not only minimize the need to extract new resources but also reduce the space required for landfills, contributing to a more sustainable urban environment (Dotania,2023). Finally, effective waste management plays a significant role in economic and social development. Properly organized waste management systems can create job opportunities, stimulate the growth of green industries, and improve the overall competitiveness of cities. This makes sustainable waste management an essential component of both environmental and economic progress in urban areas.

Problems faced in urban waste management

Overwhelming waste generation and inefficient resource utilization: Urban areas are grappling with the dual challenges of overwhelming waste generation and inefficient resource utilization. As cities grow and consumption patterns increase,the amount of waste produced has reached alarming levels. Despite the vast quantities of waste, much of it is not efficiently utilized or recycled. Instead, valuable materials such as plastics, metals, and paper often end up in landfills,contributing to environmental degradation and the depletion of natural resources.This wastefulness not only occupies vast amounts of land but also results in the loss of valuable materials that could be reused, repurposed, or recycled. Without proper sorting, recycling infrastructure, and public awareness, these valuable resources are discarded rather than being reintegrated into the economy, thus undermining efforts towards sustainability and a circular economy.

Complex waste composition and inadequate public awareness: The complexity of urban waste composition presents a significant challenge to effective waste management. Cities generate a wide variety of waste materials, including plastics, metals, food waste, and hazardous materials, making it difficult to sort and recycle efficiently. This complex mix requires specialized systems for classification and treatment, yet many urban areas lack the necessary infrastructure to handle such diversity. Compounding this challenge is the lack of public awareness about proper waste segregation and recycling practices (Akshay M,2013). Many individuals and households continue to dispose of waste without sorting it, often mixing recyclables with general waste. Without widespread education on the importance of waste segregation, composting, and recycling, cities struggle to implement effective wastemanagement strategies that maximize resource recovery and minimize landfill use.

WASTE MANAGEMENT IN URBANIZED CITIES

Waste Separation and Recycling System in Tokyo, Japan

Tokyo is a global leader in waste management, with one of the most rigorous wasteclassification systems in the world. The city's residents are required to sort their waste into several categories, such as burnable garbage, non-burnable garbage, recyclables, and hazardous waste. This strict segregation system ensures that materials are properly processed, reducing the volume of waste that ends up in landfills and incinerators.

Through extensive public education campaigns and community participation, Tokyo has managed to achieve a recycling rate of over 80%. This success not only minimizes environmental pollution but also contributes to the development of a circular economy, where waste is treated as a resource rather than discarded. By effectively diverting waste from landfills, Tokyo has transformed its waste management system into an example of sustainability and environmental responsibility.

Waste Sorting Reform in Shanghai, China

Shanghai's implementation of a strict waste sorting policy in 2019 marked a significant step toward more sustainable waste management in China. The policy required residents to separate their waste into categories such as recyclables, wet waste, dry waste, and hazardous waste. To support this initiative, the city introduced smart garbage sorting systems in neighborhoods, encouraging citizens to sort their waste correctly. Within just a year, Shanghai's waste recycling rate skyrocketed to over 30%, and the amount of household waste incinerated significantly decreased.

This rapid progress highlights the importance of strong legislation, community education, and technological support in achieving successful waste management reforms.Shanghais experience offers valuable lessons for other cities aiming to improve their waste management systems and reduce the environmental impact of waste disposal.

San Francisco's Zero-Waste City Goal

San Francisco was the first major city in the United States to set a bold "zero-waste" goal, aiming to divert 100% of its waste from landfills and incinerators. Through a combination of legislation, community engagement, and a robust waste separation system, the city has been able to achieve a recycling rate of over 80%. San Francisco offers free or low-cost composting and recycling services to residents and encourages the development of innovative waste management businesses that promote resource recovery. By prioritizing resource utilization and waste reduction, the city has positioned itself as a leader in environmental protection, demonstrating that ambitious sustainability goals can be achieved through a collaborative effort between government, businesses, and citizens. San Francisco's commitment to zero waste sets a powerful example for cities around the world looking to reduce their environmental footprint and move toward a more sustainable future.

MSW Management Practices in Developed Countries

In the 1970s, many developed nations faced waste management challenges like low collection rates and improper disposal, which posed public health risks, particularly in urban areas and through water contamination (UN-Habitat, 2010). These issues, along with media and political attention, highlighted the need to phase out open dumps in favor of modern, sanitary landfills. Public opposition to new waste projects, often due to negative experiences with uncontrolled sites (the NIMBY effect), drove changes toward transparency, community involvement, and sustainable practices. Today, developed countries emphasize waste sorting and collection, with households educated to segregate waste at the source into categories like glass, paper, and biodegradable materials, further processed at material recovery facilities (Xevgenos et al., 2015).

Full-service schemes, such as door-to-door or curbside collection, where waste trucks collect trash directly from residences. Waste collected from central collection points is typically transported by trucks (Rypdal and Winiwarter, 2001; Monni et al., 2006). Additionally, railways and barges are frequently used to transport waste, though communities wishing to use these modes of transportation need access to established port or rail infrastructure (Mogensen and Holbech, 2007; Saxena, 2009).

Recycling Recyclable waste items, which are sorted at the source or at recycling centers, are reprocessed into product. Upcycling is the reprocessing of waste materials into products that have higher value than the original. On average, 22% of the total MSW was recycled in high-income countries (World Bank, 2012), and an ambitious target of 50% of municipal waste to be recycled by 2020 is set for European countries (European Union [EU], 2010).

METHODOLOGY

The Smart Waste Management System offers a novel approach to urban waste collection by combining modern sensors, automation. This system aims to address inefficiencies in waste collection by creating a more responsive, data-driven process that adapts to real-time waste disposal patterns and supports long-term sustainability goals for urban areas. At the core of the system is a proximity and moisture sensor designed to detect and classify waste items as either biodegradable or non-biodegradable. This classification allows the bin's lid to open specifically for each type of waste, thereby ensuring immediate, on-site segregation. Effective segregation at the point of disposal is essential, as it simplifies waste processing downstream and reduces contamination of recyclable materials, ultimately leading to more effective recycling and waste disposal practices.

In addition to this, the system features a mechanical compression lever that activates when the waste level reaches 70% of the bin's capacity. By compressing the waste at this stage, the bin effectively increases its storage capacity and reduces the frequency of necessary collections. This compression mechanism not only reduces the number of trips waste collection vehicles need to make but also conserves fuel and reduces emissions, contributing to the sustainability goals of urban waste management. The system also includes ultrasonic sensors that continuously monitor the fill level of each bin. Once a bin reaches 90% capacity, an alert is automatically sent to municipal authorities, notifying them that collection is needed soon. This system reduces the risk of waste overflow and ensures that bins are emptied promptly before reaching full capacity, which helps maintain cleaner streets and public areas.

A central Arduino microcontroller coordinates the system's various sensors and actions, providing effortless operation for each bin. For communication, the system utilized long-range communication technologies, specifically LoRaWAN or NB-IoT, depending on the needs of the particular urban setting. These technologies are chosen for their reliability and range, enabling the system to maintain connectivity across vast urban areas and transmit data consistently to the central platform. This connectivity allows municipal authorities and waste management operators to stay updated on bin statuses without requiring physical inspections, saving time and reducing operational costs.

GPS integration further enhances the efficiency of the Smart Waste Management System by providing precise location data for each bin. This feature enables waste collection trucks to plan their routes based on real-time data, allowing for the most efficient paths to be mapped and followed, reducing both travel distance and fuel consumption. In turn, this optimized routing decreases the environmental impact associated with waste collection and ensures that resources are used more effectively. This GPS data, along with the real-time fill levels from each bin, is managed through a web server that hosts a user-friendly interface. This online platform is accessible to municipal authorities, sanitation workers, and who need to monitor bin status and plan accordingly. The platform provides real-time data and sends automated alerts, offering a comprehensive overview of the entire waste management process and ensuring that waste collection is scheduled based on real needs rather than predetermined routines.

Each bin is equipped with an LCD display that provides information on the bin's current status. This display allows sanitation workers to quickly assess whether a bin is close to full, recently emptied, or requires any maintenance, enabling more responsive service. To support continuous operation, the system is powered by solar panels with battery backups, ensuring that it remains functional even during power outages or in areas with unreliable electricity access. By relying on renewable energy, the system reduces dependency on external power sources, contributing to the sustainability of urban waste management infrastructure.Following each waste collection, sanitation workers can reset the bin's system using an RFID tag. This process not only updates the bin's status on the central platform but also maintains accurate records of each bin's fill history, which is essential for data analysis. This simple RFID reset process minimizes errors in the system, ensuring that data on bin usage is consistently up-to-date and reliable for further analysis.

Data analytics play a crucial role in the Smart Waste Management System by identifying usage trends that can help optimize bin placement and collection routes. For instance, data collected over time can reveal patterns such as peak times for waste generation or frequently used bins, enabling authorities to make informed decisions about where to place bins or how often certain locations require collection. These insights lead to more targeted waste management practices, minimizing operational costs and reducing unnecessary collections. Additionally, analytics can help authorities track the environmental benefits of this system by showing reductions in fuel consumption and emissions due to optimized routing.

The Smart Waste Management System addresses some of the most pressing issues in traditional urban waste management, including inefficient scheduling, lack of source segregation, and resource-intensive collection operations. By automating key aspects of waste collection and integrating data-driven decision-making, the system creates a cleaner, more sustainable, and responsive waste management process. This system is particularly beneficial in densely populated urban areas where waste is generated in high volumes, and efficient management is essential for maintaining public health and cleanliness.

By promoting waste segregation at the source, optimizing space within bins through compression, and enabling responsive, data-backed routing decisions, the Smart Waste Management System exemplifies a modern approach to waste management that aligns with sustainability goals. The system's reliance on renewable energy sources, data analytics, and efficient communication networks underscores its role as an innovative solution designed to tackle both the operational and environmental challenges of urban waste management. Through these features, the Smart Waste Management System offers a comprehensive and adaptable methodology that not only enhances the efficiency of waste collection but also contributes to creating cleaner, greener cities with reduced waste management costs and lower environmental impact.

Figure 1. Waste Collection Management Framework

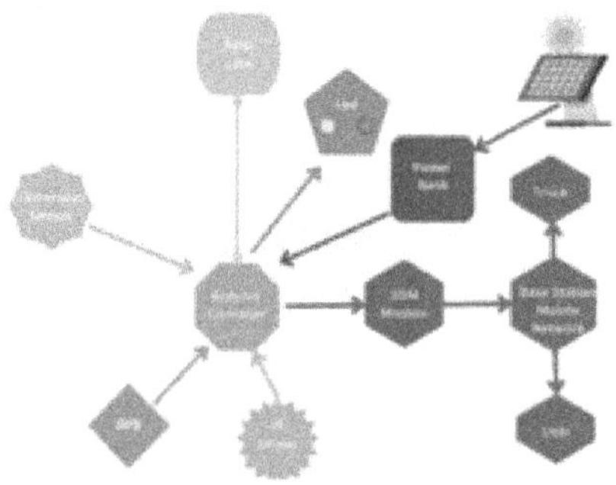

MODULES

1. Arduino UNO

Figure 2. Arduino UNO

Fig. 2 Arduino UNO

Arduino Uno (Figure. 2), in conjunction with ultrasonic sensors, calculates the distance to the garbage can and displays the fill level. When the distance falls below a predetermined threshold, actions such as displaying a "Bin Full" message on an LCD screen are initiated. Arduino Uno distinguishes between different types of waste using infrared sensors for plastic and nonmetallic waste, dispersed water sensors for wet waste, and inductive proximity sensors for metallic waste. It can start actions like draining liquid waste into designated facilities or diverting metallic items to separate collection areas. Arduino Uno, when combined with a GSM module, sends SMS alerts to sanitation workers when the bin reaches maximum capacity, ensuring efficient waste collection and management.

2. IR Sensor

Figure 3. IR Sensor

Fig. 3 IR Sensor

The IR sensor (Figure.3) emits infrared light constantly, forming a baseline for detection. It identifies plastic waste and distinguishes it from wet and metal waste, facilitating efficient waste segregation. By sensing the presence and movement of waste within the dustbin, it allows for real-time monitoring of fill levels and encourages timely disposal and segregation.

3. Ultraviolet Sensor Module

Figure 4. Ultraviolet Sensor Module

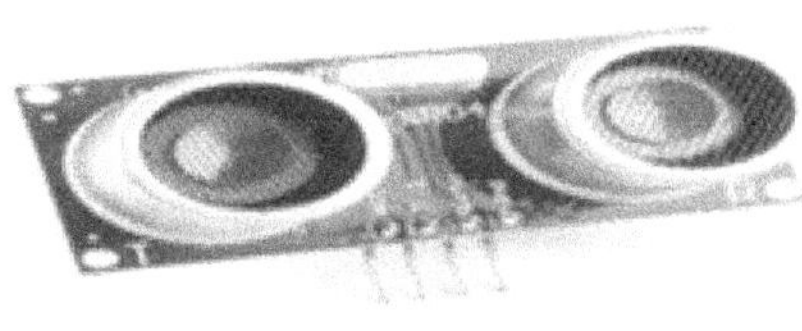

Fig. 4 Ultraviolet Sensor Module

The sensor sends out high-frequency sound waves toward the waste or the bottom of the bin, which bounce back (echo) when they hit the target surface. The UV sensor(Figure. 4) determines the distance to the waste by measuring the time it takes for ultrasonic waves to travel from the emitter, reflect off the waste surface, and return to the receiver. This distance reflects the amount of waste accumulated in the bin. A microcontroller or specialized circuit processes the measured distance data and determines the waste level percentage or volume. When predefined waste thresholds are exceeded, alerts or notifications are sent to indicate that the bin needs emptied.

4. Dispersed Sensor Module

Figure 5. Dispersed Sensor Module

Fig. 5 Dispersed Sensor Module

Dispersed water sensor is used to detect bio-waste and also detect the presence of water. It can detect leaks within the dustbin structure. Water plays a crucial role in the decomposition of organic waste, and excessive moisture can lead to unpleasant odours. A signal is sent by the sensor module. The signal is reflected back by any water present. The (figure 5) Dispersed sensor module determines how long it takes for the signal to return. The sensor module sends a message containing the water level information to the microcontroller. The water level information is used by the microcontroller to update the user interface and send alerts. Dustbins in an open or exposed environment may be vulnerable to rain or other forms of water exposure.

5. servo motor

Figure 6. Servo Motor

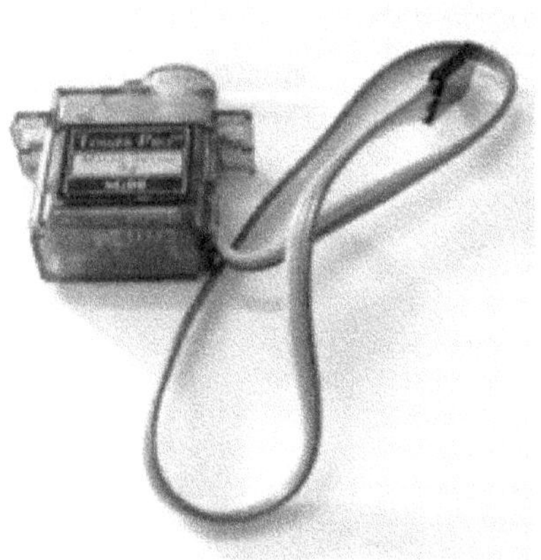

Fig.6 Servo Motor

The servo motor(Figure.6) enables automatic lid opening and closing for hands-free waste disposal. When an infrared (IR) sensor detects a user approaching, it signals the servo motor to lift the lid, allowing waste to be deposited without direct contact. This process not only enhances hygiene but also conserves energy, as the motor operates only when necessary. Additionally, in some designs, the servo motor activates a waste compression mechanism once the bin reaches a certain fill level, maximizing bin capacity and minimizing collection frequency. Through these functions, the servo motor supports efficient and hygienic waste management.

6. Inductive Proximity Sensor

Figure 7. Inductive Proximity Sensor

Fig.7 Inductive Proximity Sensor

Inductive proximity sensors are sensors that can detect the presence of metal objects. Inductive proximity sensors can also be used to sort waste into different categories in recycling bins. This can make it easier for people to recycle their garbage. (figure 7) Inductive proximity sensors detect the presence of waste and trigger the opening of the bin lid in IoT-based dustbins. They function by generating a magnetic field that oscillates around a coil. Eddy currents are created when a conductive object, such as a piece of metal, enters a magnetic field. In turn, these eddy currents generate a secondary magnetic field that opposes the original field. When the magnetic field changes, the sensor detects it and sends a signal to the microcontroller.

7. Jumper Wires

Figure 8. Jumper Wires

Fig.8 Jumper Wires

Jumper Wires(Figure.8) are essential for connecting different components in electronic projects. In a Smart Garbage Bin (SGB) system, these jumper wires are used to create connections between the microcontroller (such as an Arduino or Raspberry Pi) and various sensors (like ultrasonic and IR sensors) and actuators (such as servo motors). They allow the system to transmit signals and power efficiently between the components, enabling the sensors to detect fill levels and the motor to open and close the lid based on sensor inputs.

8. Weight Sensor

Figure 9. Weight Sensor

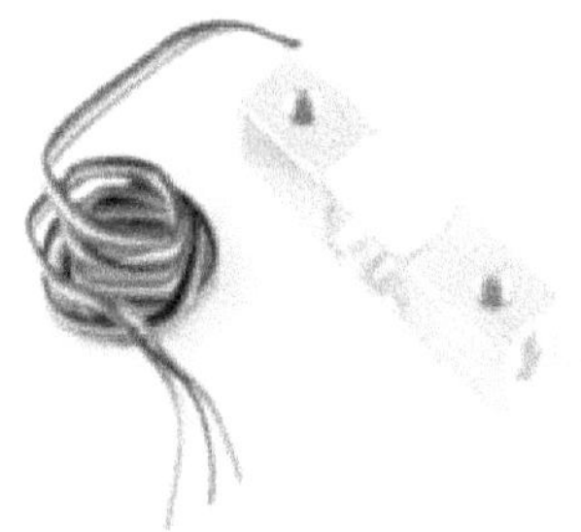

Fig9.Weight Sensor

Weight Sensor(Figure.9) is used to measure the weight of the waste inside the bin, providing data on how full the bin is. This sensor works by detecting the force exerted by the waste on its surface and converting it into an electrical signal, which is then processed by the system's microcontroller. As the bin fills up, the weight sensor helps track the increasing load, triggering alerts or activating waste compression mechanisms at specific thresholds. This ensures efficient waste collection, reducing unnecessary trips and optimizing bin utilization.

9. Stepper Motor

Figure 10. Stepper Motor

Fig10.Stepper Motor

Stepper Motor(Figure.10) which can be used in the Smart Garbage Bin (SGB) system for precise control of rotational movements, such as in waste compression or lid operation. Stepper motors move in small, fixed increments (steps), allowing accurate positioning. In the SGB, the stepper motor could control a compactor that compresses waste inside the bin when it reaches a certain fill level, maximizing the bin's capacity. Alternatively, it could control the bin lid, opening and closing it in response to sensor inputs. This precise movement ensures efficient operation and contributes to the automation of the waste management process.

10. 20*4 LCD Display Module

*Figure 11. 20*4 LCD Display Module*

Fig.11. 20*4 LCD Display Module

The 20*4 LCD Display module require low-power to operate, and low-cost 20*4 LCD display that is commonly used in Internet of Things (IoT) dustbins.These displays are easy to use and program. The display is also very energy efficient.This are known for their compact size,making them ideal for integration into space constrained application.(Figure. 11). The 20*4 LCD Display module is typically used to display the following information on an IoT-based dustbin: The level of waste in the bin,The status of the lid (open or closed),Error messages.

PROPOSED DESIGN

The Smart Garbage Bin (SGB) is designed with a focus on affordability and ease of use. A camera sensor identifies waste as biodegradable or non-biodegradable, and servo-operated lids provide hands-free disposal. Ultrasonic sensors monitor waste levels, activating a lever system at 70% capacity to compress the waste, thereby optimizing bin space and reducing the frequency of collections. When the bin reaches 90% capacity, the LoRaWAN or NB-IoT module sends real-time alerts, including GPS coordinates, to the municipal office for timely collection. sa

At the heart of the system is an Arduino controller, which manages communication between the camera, sensors, LoRaWAN/NB-IoT module, and LCD screen. The controller also governs actions like waste compression and lid operation. Solar panels power the system, with battery backup ensuring uninterrupted functionality. Sanitation workers use RFID tags to reset the system post-collection. This IoT-enabled design increases the efficiency of waste management, minimizes operational expenses, and supports sustainability efforts.

Figure 12. Proposed design of smart garbage bin (a) Front View (b) side view (c) back view(d) internal design

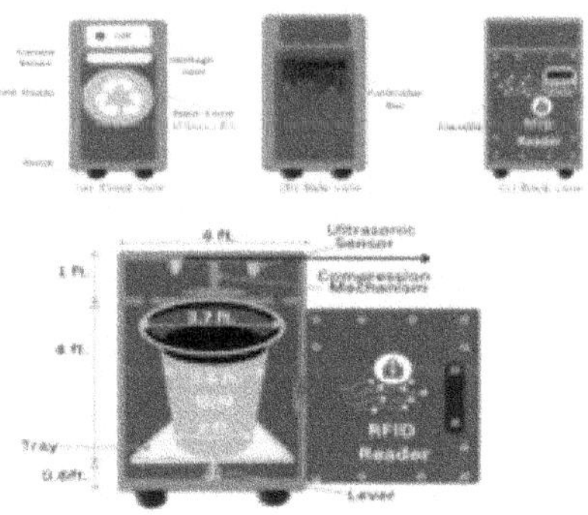

Implementation

1. **Smart Control and Camera Sensing System**
   - o **Components**: Camera sensor, machine learning model for waste classification, servo motors, RFID readers, lever system for waste compression.

- o **Function**: Captures waste images for classification, and compresses the waste when the bin reaches a certain capacity.
- o **Operation**: Upon user detection, the camera captures an image, classifies the waste, and opens the appropriate bin (biodegradable or non-biodegradable) using a servo motor. At 70% capacity, the lever system compresses the waste to maximize bin space.

2. **Transceiver Unit**
   - o **Components**: Arduino microcontroller, LoRaWAN/NB-IoT module, GPS.
   - o **Function**: Manages communication with municipal offices when bins approach full capacity.
   - o **Operation**: When the bin reaches 90% capacity, the system sends a notification to the municipal office, displaying the fill level on the LCD and transmitting the bin's GPS location for prompt collection.
3. **Lever System**
   - o **Components**: Mechanical lever, servo motor, ultrasonic sensors.
   - o **Function**: Compresses waste when the bin reaches a pre-defined capacity threshold.
   - o **Operation**: When the ultrasonic sensors detect 70% capacity, the lever compresses the waste, optimizing bin space. If the level remains above 90% post-compression, an alert is triggered, indicating that the bin requires collection.
4. **Smart Display Unit**
   - o **Components**: LCD screens.
   - o **Function**: Provides real-time information on the bin's status, including waste type and fill level.
   - o **Operation**: Once the waste is classified, the LCDs the waste type and the bin's fill level. The screen will indicate if the bin is full and ready for collection.
5. **Solar Power and Battery Backup System**
   - o **Components**: Solar panels, battery system.
   - o **Function**: Provides continuous power to the system.
   - o **Operation**: Solar panels power all components of the system, while a battery backup ensures operation during periods of low sunlight.

Architecture

Figure 13. System Architecture

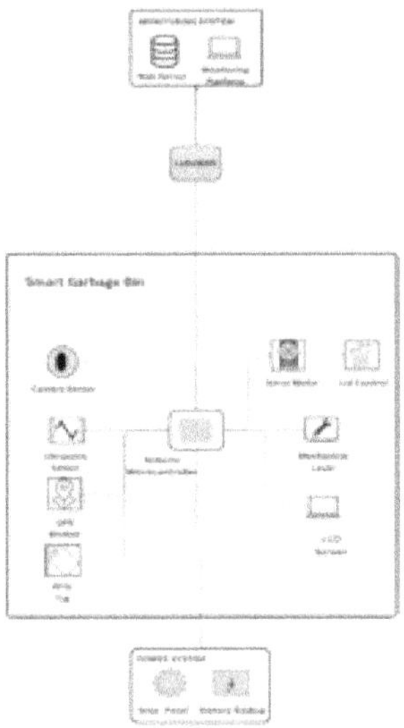

Fig 12. System Architecture

System Operation Flow

1. **Waste Classification**: The machine learning model processes the image, determining whether the waste is biodegradable or non-biodegradable.
2. **Lid Opening**: Based on the classification, the respective lid opens for waste disposal.
3. **Waste Compression**: Once the bin reaches 70% capacity, the lever system compresses the waste to create additional space.
4. **Waste Level Monitoring**: Ultrasonic sensors constantly monitor the fill level, sending an alert if the waste level exceeds 90% after compression.
5. **Data Transmission**: The LoRaWAN or NB-IoT module transmits a notification, including the bin's GPS location, to the municipal office when collection is required.
6. **Post-Collection Reset**: Sanitation workers reset the system using RFID tags after collecting the waste.
7. **Energy Efficiency**: The system operates using solar power, with a battery backup to ensure continuous functionality.

RESULTS AND DISCUSSION

The proposed system outperforms existing waste management systems by integrating advanced sensors and real-time IoT monitoring.Compared to current systems, the proposed solution offers more efficient waste collection through real-time alerts, waste compression for increased capacity, and improved user education on proper waste sorting, promoting environmental sustainability. Below is an in-depth description of the key findings and system behavior:

Automated Monitoring and Compression Mechanism:

The SGB system uses a combination of ultrasonic sensors to monitor garbage levels. When the fill percentage exceeds 70%, the bin initiates an automatic compression function to compact the waste, allowing more trash to be stored without requiring immediate collection.

As the bin fill level approaches 90%, notifications are sent to the municipal authorities and collection departments, indicating that the bin is nearing its maximum capacity.

Efficient Notification System:

The system has a tiered notification protocol based on bin fill levels. For example, at 70% capacity, the bin begins compressing waste, and at 90%, alerts are sent to:

- Municipal Corporation: To inform of a nearly full bin.
- Truck Department: To notify the closest waste collection truck, allowing them to prepare for pickup.
- Users: To suggest nearby bins or a brief wait until the bin is emptied.

This approach optimizes resource allocation by notifying relevant departments only when necessary, thus reducing overflow incidents and operational redundancies.

System Response and Efficiency:

The study included multiple bins with varying fill levels. The SGB responded dynamically based on the fill percentage. Bins that exceeded certain thresholds activated their compression mechanism and initiated alerts.

Tables within the study demonstrated that bins filled to 70% or above-required compression, while those reaching 90% sent alerts, ensuring waste collection was only directed to bins that truly needed it, preventing unnecessary collections and minimizing fuel usage.

Environmental Impact and Sustainability:

By relying on solar power for operational energy, the system minimizes its carbon footprint and operates sustainably, even in regions with inconsistent power access. This design enables the SGB to function independently, reducing dependency on the grid and supporting a cleaner environment.

Data-Driven Collection Routes:

Through real-time data aggregation in a cloud-based system, the system enables route optimization based on bin fill data. The platform uses machine learning algorithms to determine optimal collection routes and schedules, which are dynamically adjusted according to bin status and environmental factors.

The optimized route mapping not only saves fuel but also reduces collection times and enhances the efficiency of waste management services in urban areas.

Performance in Real-Time Urban Settings:

The results indicate the system's reliability in real-time urban environments. Key performance metrics such as data accuracy, communication latency, and power consumption were analyzed, revealing that the system successfully maintained high data accuracy and energy efficiency, even under continuous operation.

Statistical analysis of fill levels across bins showed that most bins reached between 50-70% capacity throughout the day, supporting the need for timely interventions in high-traffic areas to prevent overflow.

Table 1. Dustbin fill level and corresponding action to be performed

Identity of Dustbin	Level percentage(%)	Alert message from system
Number of dustbins	4	
Dustbin 1 Fill level	30%	Nothing to be done
Dustbin 2 Fill level	70%	Lever will start working,i.e compress
New Fill level	60%	All in control
Dustbin 3 Fill level	65%	Nothing to be done
Dustbin 4 Fill level	90%	Lever will start working,i.e compress
New Fill level	90%	Inform Municipal Cooperation, Truck Department.

To Municipal Cooperation: This garbage bin id no. 2002 is filled

To Truck Department: This garbage bin id no. 2002 is filled, inform the nearest truck driver to collect it.

To Users: This garbage bin id no. 2002 is filled, please use another dustbin or wait till it is Collected.

Table 2. System efficiency at high capacity level.

Identity of dustbin	Level percentage(%)	Alert message from system
Number of dustbins	4	
Dustbin 1 Fill level	25%	Nothing to be done
Dustbin 2 Fill level	55%	Nothing to be done
Dustbin 3 Fill level	75%	Lever will start working
New Fill level	80%	All in control

continued on following page

Table 2. Continued

Identity of dustbin	Level percentage(%)	Alert message from system
Dustbin 4 Fill level	95%	Lever will start working
New Fill level	90%	Inform Municipal Cooperation, Truck Department.

To Municipal Cooperation: This garbage bin id no. 2049 is filled

To Truck Department: This garbage bin id no. 2049 is filled, inform the nearest truck driver to collect it.

To Users: This garbage bin id no. 2049 is filled, please use another dustbin or wait till it is Collected.

The advanced algorithm designed for the Smart Garbage Bin (SGB) system provides an efficient and automated approach to urban waste management, addressing some of the major challenges faced by cities worldwide. As urban populations grow, waste management becomes increasingly complex, often resulting in overflowing bins, unsanitary streets, and inefficient collection schedules. The SGB system offers a practical solution to these issues by intelligently monitoring fill levels and triggering actions based on predefined thresholds.

When the SGB's fill level reaches 90% after compaction, the system automatically notifies relevant municipal authorities, the truck department responsible for collection, and even the users in the vicinity. This communication allows waste collection teams to be proactive rather than reactive, arriving before the bins overflow and contribute to litter or unpleasant odors. Such a system not only helps maintain public cleanliness but also optimizes the use of municipal resources. Waste collection teams can be dispatched based on real-time data rather than following a rigid schedule, reducing unnecessary trips and fuel consumption, and ultimately lowering operational costs.The SGB system includes a dynamic response mechanism that is essential to its success in smart city environments. Equipped with real-time sensors, the bin continuously monitors its fill level. When it reaches a certain threshold, the compaction feature activates to maximize space utilization, compressing the waste and allowing it to hold more without overflow. If the bin continues to fill and reaches the 90% threshold even after compaction, alerts are sent immediately to the designated authorities. This dual response—compaction followed by alerts—ensures that the bin's capacity is used to its fullest, while minimizing the risk of overflow. Such a system can greatly benefit high-traffic areas in cities, where waste bins fill up quickly and traditional collection schedules might fall short.

The results of this system's performance, as illustrated in Tables 1 and 2, confirm its robustness and reliability across a variety of fill levels and environmental conditions. The data shows that the SGB consistently triggers timely notifications and accurately tracks fill levels, making it a viable and efficient option for urban waste management. Additionally, the system's real-time monitoring capability can support trend analysis, helping city planners understand waste generation patterns in specific areas. For instance, they may identify that certain bins in busy districts fill up more frequently, allowing them to adjust their strategies accordingly. Moreover, the use of Internet of Things (IoT) technology in this solution underscores its importance in advancing smart city infrastructure. IoT-based waste management systems like the SGB have the potential to transform urban cleanliness and sustainability. By implementing these intelligent systems, cities can achieve a more adaptive and responsive waste management approach, which leads to better resource utilization.. In turn, this reduces the environmental footprint associated with waste collection, from fuel use to emissions, and promotes a cleaner, healthier environment for urban residents.The SGB system exemplifies the potential of IoT-driven innovations to address real-world problems. As urban areas continue to expand, efficient waste management will remain a crucial aspect of sustainable city planning. IoT-based solutions like the SGB not only improve the convenience and cleanliness of public spaces but also contribute to a more sustainable future by enhancing the use of municipal resources. Through such advancements, cities can pave the way for improved quality of life, cleaner streets, and more efficient infrastructures that support a higher standard of urban living. This study demonstrates that the integration of smart waste management systems is not just a technical upgrade but a meaningful step towards creating more sustainable and responsive cities.

Table 3. System Response for 4 dustbins at varying fill levels

Dustbin	Fill percentage(%)	Actions
Bin 1	35%	Nothing to do
Bin 2	55%	Nothing to do
Bin 3	70%	Compress
Bin 4	80%	Compress

Table 4. Notifications sent for 4 Dustbins

Dustbin	Fill percentage(%)	Notifications Sent
Bin 1	40	None
Bin 2	50	None
Bin 3	60	None
Bin 4	70	Inform municipal cooperation,Truck department,Users

Table 5. Timed actions for 4 dustbins based on fill levels

Time of Day	Bin 1 Actions	Bin 2 Actions	Bin 3 Actions	Bin 4 Actions
9 AM	None	None	None	None
1 PM	None	None	Compress	Compress
5 PM	None	None	Compress	Compress

The data from the Smart Garbage Bin (SGB) system demonstrates a correlation between bin fill levels and system actions, highlighting the efficiency of a threshold-based waste management approach. According to Table 3, when bins reach a 60% fill level, the system initiates a compression action to optimize capacity before collection becomes necessary. This compaction mechanism allows each bin to accommodate more waste, reducing the frequency of collection and thereby saving municipal resources. Bins at lower fill levels, such as 40% and 50%, do not trigger any immediate action, indicating that the system is programmed to focus resources on bins approaching full capacity. This threshold-based approach allows bins nearing their maximum fill level to initiate automated interventions, such as compression, enhancing the overall waste management efficiency.

The notification system within the SGB platform is also strategically designed to maximize resource allocation. As shown in Table 4, only bins reaching fill levels of 70% result in notifications being sent to municipal authorities, the truck department, and local users. By sending alerts only for bins close to full capacity, the system avoids unnecessary notifications and helps prioritize waste collection for bins that truly need attention. This targeted notification strategy not only prevents overflow but also allows for optimized collection routes, reducing time and fuel costs associated with unnecessary trips. Furthermore, Table 5 provides a comprehensive summary of actions taken across five dustbins under various conditions and times of day, showing how the system adapts dynamically to changes in fill levels and timing. This approach helps maintain cleaner urban spaces, minimizes overflow risks, and supports a smarter, more sustainable waste management strategy, where resources are allocated efficiently based on real-time data. The SGB system, through its compression and targeted notification mechanisms, exemplifies how IoT-based solutions can significantly improve urban waste management, paving the way for cleaner and more responsive cities.

CONCLUSION

The Smart Garbage Bin (SGB) represents a transformative approach to urban waste management, integrating advanced technology to improve efficiency, reduce environmental impact, and promote sustainability. With features like camera sensors, machine learning, automated compression, and wireless communication, the SGB can autonomously classify waste into degradable, non-degradable, and recyclable categories. This automatic segregation at the source enhances recycling efforts and prevents contamination of recyclable materials, supporting cities in achieving their sustainability goals. Additionally, the bin's waste compression system maximizes storage capacity by compressing waste at a specific fill level, reducing the frequency of collection trips, cutting fuel consumption, and lowering emissions from waste collection vehicles.Powered by solar panels and equipped with LoRaWAN or NB-IoT connectivity, the SGB operates sustainably and efficiently. Solar power minimizes reliance on external energy sources, allowing continuous operation even during power outages, while real-time alerts notify municipal teams when bins approach capacity, preventing overflow and maintaining cleanliness. This smart bin offers a financially viable solution by optimizing resource usage and reducing long-term waste management costs. The SGB demonstrates how cities can leverage technology for eco-friendly urban living, setting an example for smart infrastructure that harmonizes innovation with sustainability and efficient resource use.

Future Scope

Future iterations of the Smart Garbage Bin (SGB) could integrate enhanced machine learning algorithms, pushing the boundaries of waste classification accuracy. By implementing more sophisticated models, the system would not only improve at differentiating existing types of waste—like organic, recyclable, and hazardous materials—but also adapt to the identification of emerging waste categories, such as bioplastics and novel composites. Advanced algorithms could leverage real-time data from bin cameras and sensors to refine the accuracy of waste categorization continuously, learning from patterns and adjusting as new waste types enter the stream. This dynamic adaptability would support a more nuanced and effective waste segregation process, allowing cities to stay ahead of evolving waste management challenges and ensure that more materials are correctly sorted and processed.The SGB could also benefit significantly from integration with broader smart city initiatives, enhancing its role within a cohesive urban management system. By connecting with centralized city platforms, the SGB would be able to exchange data with other smart infrastructure—such as traffic and pollution monitors, public lighting, and emergency services—facilitating coordinated waste management strategies across municipalities. For example, waste collection routes could be dynamically optimized based on real-time data from SGBs citywide, traffic updates, and air quality indices, thereby minimizing fuel consumption, reducing city congestion, and ensuring cleaner neighborhoods. Through such collaborative efforts, municipalities would not only increase operational efficiency but also enhance the quality of urban life, moving towards a future where city infrastructure operates as an interconnected ecosystem.Another promising direction for future SGB versions is the implementation of predictive maintenance features that utilize data analytics to preemptively identify and address potential issues. By gathering and analyzing sensor data on bin components, such as compression mechanisms and power systems, the SGB could detect early signs of wear or malfunction. Automated alerts would then notify municipal teams, enabling maintenance to occur before disruptions impact bin performance or result in costly repairs. This proactive approach would reduce the need for manual inspections, enhance system reliability, and minimize downtime, thereby sustaining optimal performance. Furthermore, SGBs could work closely with local recycling programs, sharing data on waste composition and volume to support recycling optimization. This collaboration would enable cities to track waste trends, adjust recycling processes, and prioritize sustainable outcomes, making the SGB an integral component of comprehensive, data-driven recycling efforts.

REFERENCES

Ahmed, M. M., Hassanien, E., & Hassanien, A. E. (2023). IoT-based intelligent waste management system. *Neural Computing & Applications*, 35(32), 23551–23579. DOI: 10.1007/s00521-023-08970-7

Andrade, Y., & Yoo, S. G. (2019). A comprehensive study of the use of LoRa in the developmentof smart cities. *Applied Sciences (Basel, Switzerland)*, 9(22), 4753. DOI: 10.3390/app9224753

Antora, N. A., Rahman, M. A., Mosharraf, A. A., Ibn Ehsan, M., Alve, M., & Elahi, M. M. (2022)"Design and Implementation of a Smart Bin using IOT for an Efficient Waste ManagementSystem,"25th International Conference on Computer and Information Technology(ICCIT), Cox's Bazar, Bangladesh, 2022, pp. 774-779, DOI: 10.1109/ICCIT57492.2022.10055998

Balasubramanian, M. (2018). Municipal Solid Waste Management in India: Status,Problems andChallenges. *International Journal of Environment and Waste Management*, 21(4), 253–265. . DOI: 10.1504/IJEWM.2018.093435

Bano, A., Ud Din, I., & Asma, A. (2020). An IoT-Based Smart Bin for Real-Time Monitoring and Management of Solid Waste. *Scientific Programming*, 2020, 6613263. DOI: 10.1155/2020/6613263

Boopathi, S., Haribalaji, V., Mageswari, M., & Asif, M. M. (2022). "Agricultural Waste Management Systems Using Artificial Intelligence Techniques". In Agricultural Waste Management Systems Using Artificial Intelligence Techniques (pp. 1-23).

Catania, V., & Ventura, D. (2014, April). An approch for monitoring and smart planning of urban solid waste management using smart-M3 platform. In *Proceedings of 15th conference of open innovations association FRUCT* (pp. 24-31). IEEE.

Dominic Abuga, N. S. (2021). Raghava,(2021) Real-time smart garbage bin mechanism for solid wastemanagement in smart cities. *Sustainable Cities and Society*, 75, 103347. DOI: 10.1016/j.scs.2021.103347

Harith, M. Z. M. Z., Hossain, M. A., Ahmedy, I., Idris, M. Y. I., Soon, T. K., & Noor, R. M. (2020). "Prototype development of IoT based smart waste management system for smart city". IOPConf. *IOP Conference Series. Materials Science and Engineering*, 884(1), 012051. DOI: 10.1088/1757-899X/884/1/012051

Henaien, A., Ben Elhadj, H., & Fourati, L. C. (2024). A sustainable smart IoT-basedsolid waste management system. *Future Generation Computer Systems*, 157, 587–602. DOI: 10.1016/j.future.2024.03.056

Issac, R., & Akshai, M. (2013, August). SVASTHA: An effective solid waste management system for Thiruvalla Municipality in Android OS. In *2013 IEEE Global Humanitarian Technology Conference: South Asia Satellite (GHTC-SAS)* (pp. 254-259). IEEE.

Jadhav, S. P., Suryavanshi, J. R., Samant, R.s C., & Archana, , S. (2024). Design and Development of a Smart IoT Enabled Garbage Bin Mechanism for Optimized Volume Utilization. *African Journal of Biological Sciences*, 6(14), 10806–10818.

Jain, P., Chaudhary, T., & Gajjar, S. (2023) "Design and Development ofSmartWasteManagement System," *International Conference on Communication System, Computing and IT Applications (CSCITA)*, Mumbai, India, 2023, pp. 199-203, DOI: 10.1109/CSCITA55725.2023.10104960

Liakot Ali, Md., Mahbubul Alam and Md. Abu Innovative Research in Computer andNayeem Redwanur Rahaman. (2012). RFID based [-Dr. Santosh Pandurang Jadhav/Afr.J.Bio.Sc.]. *Communication Engineering*, 4, 1267.

Lundin, A. C., & Özkil, A. G., & Schuldt-Jensen, J. (2017). Smart cities: casestudywastemonitoring and management. In Proceedings of the 50th Hawaii International Conference onSystem Sciences (HICSS 2017) (pp. 1392-1401) https://doi.org/ Nations Human Settlement Program (UN-HABITAT). Solid Waste Managementin the World, Cities. London, UK: Gutenberg Press; 2010DOI: 10.24251/ HICSS.2017.167United

Meena, M. D., Dotaniya, M. L., Meena, B. L., Rai, P. K., Antil, R. S., Meena, H. S., & Meena, R. B. (2023). Municipal solid waste: Opportunities, challenges and management policies in India: A review. *Waste Management Bulletin*, 1(1), 4–18.

Nyika, J. MegersaOlumanaDinka(2024), "The role of solid waste management in boosting sustainable development goals"chapter 8, In the book "Exploring waste management in sustainable development", IGI global publications, 2024.

Sohag, M. U., & Podder, A. K. (2020). Smart garbage menagement system for a sustainableurban life: An IoT based application. *Internet of Things : Engineering Cyber Physical Human Systems*, 11, 100–255. DOI: 10.1016/j.iot.2020.100255

Suvarnamma, A., & Pradeepkiran, J. A. (2021). SmartBin system with waste tracking and sorting mechanism using IoT. *Cleaner Engineering and Technology*, 5, 100348. DOI: 10.1016/j.clet.2021.100348

Szpilko, D., de la Torre Gallegos, A., Jimenez Naharro, F., Rzepka, A., & Remiszewska, A. (2023). Waste Management in the Smart City: Current Practices and Future Directions. *Resources*, 12(115), 1–25. DOI: 10.3390/resources12100115

Tariq A, Muhammad I, Abdullah SA and Adam G (2019). IoT-Based Smart Waste BinMonitoring and Municipal Solid Waste Management System for Smart Cities 2020Computer Engineering and Computer Science

Vishnu, S., Ramson, S. R. J., Senith, S., Anagnostopoulos, T., Abu-Mahfouz, A. M., Fan, Z., Srinivasan, S., & Kirubaraj, A. A. (2021). IoT-enabled solid waste management in smart cities. *Smart Cities*, 4(3), 1004–1017. DOI: 10.3390/smartcities4030053

Zeb, A., Ali, Q., Saleem, M. Q., Awan, K. M., Alowayr, A. S., Uddin, J., & Bashir, F. (2019). 'A proposed IoT-enabled smart waste bin management system and efficient route selection. *Journal of Computer Networks and Communications*, 2019, 1–9. DOI: 10.1155/2019/7043674

Chapter 12
Global Synergy:
Cross-Border Collaborations in AI-Enabled Smart City Development

John Mark R. Asio
https://orcid.org/0000-0002-6096-4595
Gordon College, USA

Alyssa Nicole Suero
https://orcid.org/0000-0003-3354-9166
Gordon College, Philippines

ABSTRACT

This chapter explores the critical role of cross-border collaborations in advancing AI-enabled smart cities. As urban areas grapple with challenges in resource management, sustainability, and quality of life, AI offers transformative solutions, spanning areas like transportation, energy, public safety, and healthcare. However, the development of smart cities transcends national boundaries, necessitating robust international partnerships. This chapter also examines successful models of cross-border collaborations, such as international consortia and public-private partnerships, and presents case studies that highlight the benefits of knowledge-sharing, resource pooling, and harmonized regulatory frameworks. Ultimately, this chapter advocates for collaborative efforts that build resilient, AI-driven urban environments equipped to meet the demands of an increasingly urbanized future, fostering sustainability, equity, and quality of life for all citizens.

INTRODUCTION

Artificial Intelligence, also known as AI, is a continuously developing technology in the society. According to the Organisation for Economic Co-operation and Development (2019), Artificial Intelligence produces results based on the human-predefined objective. These outcomes vary from predictions,

DOI: 10.4018/979-8-3693-9770-1.ch012

recommendations, and even decisions. Hence, Artificial Intelligence has become widely utilized, from simple kitchen appliances, smart streetlights to e-governance.

As urbanization accelerates worldwide, cities face unprecedented challenges in managing resources, improving quality of life, and ensuring sustainability. Hence, the concept of smart cities—which are urban areas that leverage advanced technologies to enhance infrastructure, services, and governance—has gained significant attention in the international community. According to the United Nations (2017), the projected global population by 2050 will reach an estimated 9.8 billion, which will have a profound effect on societies, especially on political and fiscal pressures, population ageing, security and safety, as well as social protection systems.

At a glance, the use of Artificial Intelligence (AI) seems an efficient way to address these shared challenges such as transportation systems, energy management, healthcare, public safety, and more. However, the complexities of creating AI-enabled smart cities often transcend national borders, necessitating cross-border collaborations.

The collaborative efforts between cities, governments, private enterprises, and international organizations have become crucial for advancing smart city technologies. Policymakers and urban planners play a key role in shaping the future of smart cities, as these are the people who are responsible for creating the necessary regulatory frameworks and urban designs. These cross-border partnerships allow for the sharing of knowledge, technological innovations, and policy frameworks, facilitating the rapid deployment and scaling of AI-driven solutions. For instance, international consortia, public-private partnerships, and research collaborations have enabled cities to overcome technical barriers, mitigate costs, and address shared challenges like climate change, urban mobility, and cybersecurity.

Therefore, this chapter explores the growing importance of cross-border collaborations in fostering AI-enabled smart cities. Analyzing the cross-border concept of AI-enabled smart cities, successful collaboration models, and case studies highlights how international partnerships can accelerate the development of smart cities. The chapter also addresses the challenges and risks inherent in these collaborations—from regulatory differences and privacy concerns to cultural and ethical considerations. However, it also underscores the importance of global cooperation in overcoming these hurdles, inspiring and motivating the audience to contribute to developing smarter, more sustainable urban environments for the future.

THE CROSS-BORDER CONCEPT OF SMART CITIES

Tracing back the beginning of smart cities, it all started with the development of computer analytical capabilities during the third industrial revolution. Back in 1947, the Los Angeles Community Analysis Bureau utilized an IBM-360 in order to develop and database of census tracts throughout the whole city. This project is later referred as the Cluster Analysis of Los Angeles. The project's main goal was to alleviate the worsening poverty in the city which will provide a real-time data accessible for urban policy decision makers to efficiently combat the situation. However, the project was ahead of its time and that kind of technology was not yet available.

Later on, Amsterdam established DDS or the De Digital Stad which translates to "The Digital City" where citizens can access the internet in order to provide communication between the citizens and the government. Hence, the term "smart city" was derived from the translation of DDS where it can be marked as the proper origin of smart cities (Ball State University, 2022). Apparently, the initiative ended on 2001 as the subsidy was discontinued. However, some volunteers continued independently, and in

2009, the Amsterdam Smart City platform was launched to the public. This project is a public-private (PP) partnership where government agencies, educational institutions, private companies and foundations are involved.

As observed, it is important to highlight that from a cluster analysis project to the launch of a digital city was due to the cross-border idea and exchange of information. The De Digital Stad (DDS) became an inspiration for other countries such as Germany, United Kingdom of Great Britain and Northern Ireland, France, Italy, Singapore, and many more countries to launch their initiatives on building smart cities. (UNESCO, 2023).

By concept, the European Commission (n.d.) has defined a "smart city" as a place of an interactive and responsive city administration with safe public spaces that efficiently and sustainably use digital technology. This cross-border concept is a co-existing place where the business and the community benefit together. It provides smarter urban transport networks, upgraded water supply and waste disposal facilities, and more efficient ways to light and heat buildings.

Notably, smart cities are well equipped with technologies that are one of the key drivers for a smart city to prosper. Hence, variety of software is utilized in order for the community to progress. The IoT solutions, also known as Internet of Things solutions, are established to effectively and efficiently connect devices for communication and exchange of data. Internet of Thing (IoT) can range from home appliances electric vehicles, and light post sensors. This IoT solution can act as an operating system where devices are connected to an automated system where data is gathered and stored in a cloud which can be analyzed to assist and improve economic benefits and elevate the lives of the citizens.

Through IoT solutions, smart cities must implement a secured system where the data is protected, monitored, and controlled to prevent unauthorized access to the city's data. Hence, alongside smart cities are the reinforcement of cybersecurity in order to protect the city's data from digital attacks and other threats.

The global rise of smart cities is largely driven by rapid global urbanization, which has placed immense pressure on existing urban infrastructure. Sharma et al. (2025) emphasized the global implication of adopting IoT solutions on which a multi-stakeholder approach to address disparities is needed, and collaborative efforts between governments, private companies, and community organizations are recommended to create inclusive solutions tailored to regional needs. The estimated population of the United Nations in the year 2050 has led to the adaptation of appropriate solutions, such as developing smart cities, to tackle challenges such as traffic congestion, air pollution, and resource shortages. Many cities have already embraced smart technologies in various sectors, particularly transportation, energy, and public safety.

THE DEVELOPMENT OF SMART CITIES

Urbanization has dramatically reshaped cities worldwide, presenting challenges such as resource management, sustainability, and quality of life concerns. Smart cities have emerged as a response, leveraging technology to create efficient, resilient urban spaces. According to Boyd Cohen (2015), the development of smart cities can be categorized into three generations—Smart City 1.0, Smart City 2.0, and Smart City 3.0—each reflecting distinct approaches and priorities.

Mora et al. (2020) highlighted that Smart City 1.0 cities are characterized by "technology-focused projects primarily led by private-sector interests." This emphasis on technological infrastructure often led to challenges in achieving long-term sustainability and inclusivity. Moreover, Townsend (2021) discusses how early smart cities prioritized infrastructure and technological implementation but often failed to account for citizen agency, resulting in cities that functioned well technologically but lacked public support and relevance to local communities. Essentially, this generation has focused heavily on deploying advanced technologies for urban optimization. Often led by private technology companies or governments, this model prioritized efficiency and automation through IoT, sensors, and data analytics. The primary aim was to improve city management, optimize infrastructure, and reduce costs through automation. However, the involvement of citizens in decision-making was minimal, as the focus remained on integrating technological solutions rather than addressing local needs directly. An example of which is Songdo in South Korea. As one of the first major smart cities, Songdo was developed from scratch with a heavy focus on IoT integration and automated systems to manage energy, waste, and transportation. Although Songdo succeeded in creating a technologically advanced environment, it lacks a sense of community and citizen involvement.

In contrast, Chourabi et al. (2021) highlight that Smart City 2.0 marks "an evolution in governance models, with an emphasis on transparency and public accountability." These cities integrate technology with citizen needs, balancing the interests of the public and private sectors. Bibri and Krogstie (2022) also discuss the role of governments as mediators, facilitating public-private partnerships while ensuring that citizen interests are prioritized. This phase reflects the growing recognition that technological solutions should serve the public and address real urban challenges in ways that are visible to and valued by residents. Public sector involvement increases, and governments actively collaborate with private companies to implement smart city solutions, fostering transparency and engaging citizens more actively than in the previous phase. Citizen engagement often takes the form of public consultations, where city data is shared openly, and citizens are encouraged to participate in public projects. Take Amsterdam's Smart City project as an example where it prioritizes sustainability, emphasizing public participation in projects like environmental monitoring and energy conservation. The city actively involves citizens in pilot projects, aiming to address local issues through tailored technology solutions.

Lastly, Smart City 3.0 or the citizen co-creation a shift towards citizen empowerment and participatory governance. Unlike previous models, which were either technology-driven or government-led, Smart City 3.0 prioritizes a collaborative, bottom-up approach in which citizens are not only the end-users but co-creators of urban solutions. This generation emphasizes inclusivity, co-governance, and sustainability, with cities facilitating platforms that allow citizens to influence policy decisions, participate in urban planning, and contribute to community-based solutions. AI, IoT, and big data are still integral to smart cities, but they are applied in ways that directly align with community priorities. Cardullo et al. (2022) describe Smart City 3.0 as "a paradigm shift in urban governance" where the goal is to "empower and engage citizens as active participants in city planning and decision-making." Luque-Ayala and Marvin (2023) emphasize that 3.0 cities require a flexible approach to technology that respects cultural and ethical differences. This generation aims to create cities that are inclusive and sustainable, promoting a holistic view of urban living where technology and community aspirations coexist harmoniously.

As cities continue to face complex challenges due to urbanization, this evolution underscores the importance of adapting smart city models from other countries to meet the diverse needs of urban populations and promote resilient, sustainable urban environments. In the globalized world where countries are interconnected and interdependent with each other, countries collaborate and exchange ideas through forums and research to innovate, integrate, and adopt policies that are sustainable for society.

KEY DRIVERS OF CROSS-BORDER COLLABORATIONS FOR AI-ENABLED SMART CITIES

As artificial intelligence (AI) continues to transform sectors from healthcare to finance, cities and nations are seeking ways to leverage AI for more resilient, innovative, and sustainable smart city initiatives. A significant factor enabling this transformation is cross-border collaboration. These collaborations support resource sharing, enhance regulatory alignment, and reduce economic and technological barriers that typically arise in isolated projects. This essay explores three primary drivers of cross-border AI collaborations: knowledge sharing and innovation, regulatory harmonization, and economic synergy.

1. Knowledge sharing and Strategic Partnerships

Knowledge sharing has become central to the development of AI and the expansion of smart city solutions. Cross-border AI collaborations facilitate the exchange of information, data, and expertise between public and private sectors, accelerating innovation. Through knowledge-sharing partnerships, cities can quickly integrate advanced AI technologies, learn from others' experiences, and avoid common pitfalls in AI implementation.

According to Scantlebury et al. (2022), knowledge sharing between nations enhances innovation cycles, allowing cities to gain insights from research conducted across borders and develop applications tailored to specific urban needs. This is particularly relevant in areas such as healthcare and transportation, where cross-border collaborations have proven essential in overcoming local limitations in technical expertise and resources. Moreover, Brown and Clifton (2023) emphasized that knowledge-sharing frameworks contribute to robust public-private partnerships, where cities can partner with technology companies and academic institutions worldwide to foster collaborative innovation.

Through knowledge sharing, the alliance of countries has shifted to strategic partnerships wherein the AI technology was seen as a weapon that can be utilized for security cooperation. One of which is the longstanding cooperation between 16 countries such as the Australia, Canada, Denmark, Estonia, France, Finland, Germany, Israel, Japan, the Republic of Korea, Norway, the Netherlands, Singapore, Sweden, the United Kingdom and the United States, designed to leverage AI advancements for national security with a strong emphasis on ethical use and responsible AI deployment (JAIC, 2021). Established in 2020 and led by the U.S. Department of Defense's Chief Digital and Artificial Intelligence Office (CDAO), the AI Partnership for Defense (PfD) serves as a strategic forum to foster AI interoperability among defense allies while establishing shared ethical standards. The Partnership for Defense's (PfD's) ongoing dialogues focus on the responsible application of AI in military defense operations, ensuring that development aligns with values and international standards. The PfD has also expanded its focus from merely enhancing data and analytics to fostering interoperability for multinational military operations (Trabucco & Maas, 2023). Recent sessions explored ways to align AI initiatives across nations to reduce redundant research, streamline acquisitions, and build a robust, interoperable AI ecosystem. This collaboration is essential as AI technology becomes increasingly integral to national defense strategies, requiring that partner nations align on policies, technology standards, and ethical principles.

Another example is the European Open Science Cloud (EOSC) where cross-border collaborations by providing an open-access infrastructure for sharing data and research findings across EU member states. This system has been instrumental in collaborative AI research, particularly in fields such as climate science and urban resilience, where diverse data sources and expertise are crucial for effective solutions. According to the European Commission (2022), EOSC provides a platform that connects European researchers with a vast, interoperable network of research assets, emphasizing FAIR (Findability, Accessibility, Interoperability, and Reusability) principles. This system has been pivotal in enabling researchers to share data seamlessly across disciplines and countries, creating an environment that supports high-impact research with broad accessibility and ethical standards. Recent developments include the introduction of the EOSC EU Node, which is set to go live by the end of 2024. This new node aims to streamline the integration of FAIR data practices across the EU, supporting scientific communities in disciplines from environmental science to social studies by providing standardized tools and shared data repositories that enable reproducibility and open innovation in real-time across member states.

The AI Partnership for Defense (PfD) and the European Open Science Cloud (EOSC) share several core similarities, making them relevant frameworks for fostering cross-border collaborations in the context of AI-enabled smart cities. Both initiatives aim to integrate diverse datasets and technological resources across borders, enabling joint innovation while upholding ethical and interoperable standards. These shared principles—collaborative infrastructure, ethics, interoperability, and a multidisciplinary focus—make both PfD and EOSC valuable models for cross-border AI-enabled smart city initiatives. They illustrate how collective approaches to data sharing and ethical AI can create standardized, scalable urban solutions, enabling cities worldwide to share insights and align their AI strategies effectively.

2. Regulatory and Security Harmonization

The potential for AI to impact cities worldwide underscores the need for harmonized regulatory and security frameworks that address data privacy, security, and ethical standards. Without regulatory and security alignment, cross-border AI collaborations face barriers that hinder data sharing and limit the reach of AI-driven solutions. Regulatory harmonization, therefore, is essential to creating environments where AI can thrive across borders without legal or ethical conflicts.

Rossi and Petruzzi (2022) discuss how inconsistent regulations, especially regarding data privacy, complicate international AI collaborations. The General Data Protection Regulation (GDPR) in the European Union, for example, imposes stringent data privacy requirements that contrast sharply with U.S. policies, making it challenging for companies to operate fluidly between these regions. The GDPR mandates high standards in handling, processing, and transferring EU citizens' data, which has pressured U.S. technological companies to reassess their data protection practices. Meta, for instance, has faced fines and scrutiny over its compliance with GDPR, especially concerning transatlantic data transfers. The invalidation of the Privacy Shield—a former legal mechanism for data transfers between the EU and the U.S.—complicated compliance efforts, forcing companies like Meta to rely on Standard Contractual Clauses while awaiting a more stable framework (the Data Privacy Framework) to facilitate legal data flows across borders. This framework, the Data Privacy Framework, aims to bridge legal discrepancies by aligning EU and United States regulations, although practical challenges and appeals persist as companies work to adapt under ongoing regulatory pressure.

To address these issues, some United States companies are developing internal compliance measures to align more closely with GDPR standards, mitigating risks associated with potential fines and ensuring smoother cross-border operations. Google, for instance, recently restructured its privacy and compliance teams to bolster its AI compliance efforts and privacy oversight in response to growing regulatory expectations. Hence, harmonizing regulations can foster a stable environment conducive to collaborative innovation. Similarly, Kohli and Vala (2023) argue that aligning standards in data privacy, algorithmic accountability, and AI ethics is crucial to building trust between nations and encouraging data sharing across borders.

Another example is the AI for Good Global Summit (ITU), an initiative hosted by the International Telecommunication Union that brings together stakeholders from around the world to discuss ethical AI development and regulatory alignment. It serves as a platform where policymakers can share regulatory practices, particularly those focused on privacy, algorithmic transparency, and bias reduction. This summit has helped establish best practices for cities aiming to collaborate on AI projects, supporting regulatory alignment in ways that make cross-border collaborations more feasible.

A notable feature of the summit is its emphasis on creating globally applicable AI standards that address ethical considerations, foster algorithmic transparency, and promote interoperability between different AI systems. For example, interoperability has been highlighted as essential for cities developing smart infrastructure, as this ensures that AI technologies can operate seamlessly across jurisdictions with different regulations. This focus on alignment has also led to the creation of resources like the UN's Principles for the Ethical Use of AI, which guide cities and organizations on implementing responsible AI that adheres to human rights and privacy standards.

Additionally, the summit's emphasis on inclusivity aids in capacity-building efforts for nations with varying levels of AI development, ensuring that the benefits of AI can be shared equitably. By promoting these practices, the AI for Good initiative provides a valuable model for smart cities aiming to collaborate internationally, particularly as they strive to implement AI-driven solutions that are safe, ethical, and effective across diverse legal frameworks.

Both examples share common goals of establishing ethical, regulatory and security frameworks that facilitate cross-border AI-enabled collaborations, which is especially relevant for smart cities striving to navigate the complexities of international data governance. Both initiatives emphasize the ethical use of AI, aiming to reduce biases and improve transparency in algorithmic decision-making. GDPR's requirements for transparency and data minimization align closely with the AI for Good Global Summit's advocacy for transparent, accountable AI systems that can operate responsibly across national borders. For AI-enabled smart cities, this alignment promotes fair and inclusive AI applications that reduce discrimination risks and increase public accountability.

Moreover, GDPR and the AI for Good Summit both advocate for regulatory harmonization to enable smoother international cooperation. GDPR sets a uniform standard for data protection across the EU, which influences U.S. companies and helps foster regulatory alignment in data-sharing practices. The AI for Good Summit, by fostering dialogues on standardization, encourages global regulatory bodies to work towards compatible AI standards. This is particularly important for smart cities that rely on data from multiple jurisdictions to develop comprehensive AI solutions for urban challenges such as transportation, energy management, and public health.

Further, Both initiatives aim to promote innovation within an ethical framework, balancing technological advancement with regulatory compliance. GDPR, despite posing challenges, has driven many U.S. companies to establish internal data protection standards that meet or exceed GDPR requirements, thus improving cross-border compliance. The AI for Good Summit likewise encourages innovation by exploring how ethical frameworks can be applied to AI in diverse sectors, including urban planning. For cross-border AI-enabled smart cities, this shared commitment to ethical frameworks provides a foundation for building trust and scalability in international projects.

3. Enhancing Global Economic Synergy

Undeniably, artificial intelligence (AI) infrastructure requires substantial financial investment, from advanced computing hardware to skilled labor. Cross-border collaborations enable cities and nations to pool resources, share costs, and mitigate the high expenses associated with AI development and deployment. These partnerships allow countries to reduce their individual financial burdens, making AI-powered smart city projects more sustainable and economically viable.

Williams and Singh (2023) emphasize that economic synergy is essential for advancing smart cities, particularly for developing nations that may lack the financial resources to independently develop robust AI infrastructures. Cost-sharing strategies not only make large-scale projects feasible but also support long-term financial sustainability, allowing cities to maintain and upgrade AI systems over time. According to Li and Chen (2022), economic synergy in cross-border collaborations can extend to sharing intellectual property, expertise, and even computing resources, which are often prohibitively expensive.

Japan and Singapore Digital Partnership have an existing collaboration on developing AI and digital infrastructure. The partnership enables both countries to share the costs of AI research, development, and implementation. This collaboration has been instrumental in enabling Singapore to pilot AI-based urban solutions at a scale it may not have achieved independently. This partnership, underpinned by a Memorandum of Cooperation (MOC), facilitates exchanges of knowledge, technology, and regulatory insights, specifically in artificial intelligence, cybersecurity, and digital infrastructure management. By pooling resources, Japan and Singapore can tackle large-scale AI projects more effectively than either could independently, which has proven beneficial for urban development and smart city applications in Singapore, particularly in sectors like energy efficiency, data sharing, and urban mobility.

Recent expansions of this collaboration have also included private-sector players, such as Hitachi and Singtel, which aim to leverage next-generation data centers and GPU-based cloud services to support enterprise AI deployments across the Asia-Pacific region. These initiatives focus on environmentally sustainable data centers and AI-ready infrastructures, which are crucial for the region's growing demand for AI-powered solutions. Singtel's Paragon platform, for example, is helping companies like Hitachi integrate advanced AI into their manufacturing and operational frameworks, while also piloting applications for urban resilience and smart city projects. This approach exemplifies how shared digital infrastructure can reduce development costs and enhance the scale of AI innovation, making Singapore an emerging leader in AI-driven urban solutions while benefiting Japan's expanding tech ecosystem.

The European Investment Bank (EIB) has played a pivotal role in advancing artificial intelligence infrastructure across EU member states. This support, involving pooled funds from various nations, helps to bridge the AI investment gap between countries, particularly benefiting smaller nations that might struggle with resource limitations. By providing financing for AI innovation and smart city development, the EIB has empowered countries to engage in digital transformation on a broader scale. For example, the EIB's focus on digital and technological innovation has helped EU firms adopt advanced digital technologies, with many companies closing the technological gap with the U.S. in AI and other high-tech areas.

The EIB's investment efforts are also aligned with broader EU goals for technological competitiveness and economic resilience. An EIB report highlights that a major challenge for Europe's AI ecosystem remains its financing landscape, which lags behind the United States in venture capital availability. To counter this, the EIB supports scale-ups, especially in countries with limited access to private investment. Through strategic financing, the EIB encourages digital transformation and smart city development, supporting projects that improve urban energy efficiency, resilience, and environmental sustainability across EU nations.

Apart from alliance cooperation, forums, and expos such as the Smart City Expo World Congress, where innovators, global leaders, and interested individuals/companies participate to address and exchange technological ideas about the pressing issues the world is facing such as security, poverty, and climate change. Interestingly, The Japan and Singapore Digital Partnership, European Investment Bank (EIB) and AI Infrastructure Funding, and the Smart City Expo World Congress (SCEWC) each play key roles in promoting cross-border AI-enabled smart city initiatives. These initiatives share several important similarities, all of which support international collaboration on AI, data-driven infrastructure, and sustainable urban development.

THE RISE OF COLLABORATION FOR AI-ENABLED SMART CITIES WORLDWIDE

In an increasingly urbanized world, cities face pressing challenges related to sustainability, resource management, and quality of life. Rapid advancements in digital technologies have paved the way for a transformative approach to urban living: the rise of smart cities. Smart cities utilize technologies such as the Internet of Things (IoT), artificial intelligence (AI), and big data to improve infrastructure, enhance public services, and create safer, more efficient, and eco-friendly urban environments. By harnessing data and digital solutions, smart cities strive to address the growing needs of urban populations while promoting sustainable development.

Central to this evolution is the power of collaboration, both within and across borders. As cities around the world adopt smart city initiatives, the increasingly recognized benefits of sharing knowledge, resources, and strategies with other regions. Collaborative efforts, ranging from research partnerships and policy alignments to cross-border projects, have become essential for developing scalable and cost-effective smart solutions. International alliances, public-private partnerships, and multi-stakeholder frameworks allow cities to overcome shared challenges, learn from one another, and collectively drive forward the next generation of urban innovation.

This exploration into the rise of smart cities and global collaboration delves into the foundational technologies, key trends, and collaborative efforts that define the smart city movement, underscoring how cities worldwide are working together to build more resilient, intelligent, and sustainable urban landscapes.

Exploring the European Union's Urban AI initiatives represents an ambitious framework for advancing AI-powered smart city projects, focusing on mobility and sustainable green technologies across member states. These projects foster cross-border cooperation, helping cities within the EU to address shared challenges, particularly those related to sustainable urban mobility and environmental goals. Some of the EU's flagship initiatives are below:

1. ***Cross-Border Lighthouse Projects*** The "Scalable Cities" initiative, part of the EU's Smart Cities Marketplace, is a significant cross-border effort that brings together "Lighthouse cities" like Rotterdam, Barcelona, and Stockholm to pilot advanced AI solutions for sustainable urban mobility. These solutions include AI-based traffic optimization, energy-efficient public transportation, and electric vehicle (EV) infrastructure. Once proven effective, these models are then implemented in "Fellow cities," enabling smaller or less-resourced cities across Europe to adopt cutting-edge green technologies without bearing the full costs and risks of development. This shared model amplifies the impact of AI innovations in urban sustainability and scales AI-based solutions more rapidly across the EU (European Commission, 2024).
2. ***Urban Mobility Observatory and Data Sharing*** The EU's Urban Mobility Observatory is another example of collaborative urban AI efforts. It provides a data-sharing platform and best practices for cities, allowing them to benefit from each other's experiences in deploying AI-based mobility solutions. For example, Prague's recent commitment to over 1,000 new EV charging stations by 2026 is supported by shared insights from other EU cities that have implemented similar projects. This observatory not only promotes data transparency but also encourages cities to adopt mobility solutions that align with the EU's climate-neutral goals (European Commission, 2024).
3. ***AI for Green Energy in Transportation*** Another innovative project under the EU's AI initiatives is the CIVITAS Initiative, which focuses on using AI to optimize public transportation and reduce emissions. This program promotes the use of AI and data analytics to monitor and reduce pollution from urban transportation networks. Cities involved in CIVITAS, such as Ljubljana and Ghent, have integrated real-time AI-based monitoring systems to assess air quality and reduce vehicle emissions, demonstrating the potential for AI in environmental management within urban settings (EIT Urban Mobility, 2023).
4. ***Green Technologies and Interoperability Standards*** The EU places emphasis on creating interoperability standards for AI systems, which allows smart city solutions to function seamlessly across different countries. Interoperability is essential for cities to share technological solutions and ensures that AI-powered green technologies can be adopted widely across the EU. This approach is reflected in projects like Horizon Europe, where cities are developing interoperable AI systems for energy-efficient urban solutions. The EU's commitment to shared standards not only supports technological consistency but also fosters an environment conducive to large-scale AI-enabled green urban projects (European Commission, 2024).

In the ASEAN region, The ASEAN Smart City Network (ASCN) is a collaborative framework established in 2018 to foster partnerships between Southeast Asian cities, aiming to address shared urban challenges through the application of AI and digital technologies. Key areas of focus within ASCN include cybersecurity and disaster management, two critical fields that benefit from regional cooperation and technological advancements.

1. ***Cybersecurity Initiatives*** One significant aspect of ASCN's collaborative efforts lies in strengthening regional cybersecurity to protect smart city infrastructure. Singapore has taken a leading role with initiatives like the ASEAN-Singapore Cybersecurity Centre of Excellence (ASCCE), which offers

training and capacity-building programs for ASEAN member states to address emerging cyber threats. The ASCCE provides a platform for knowledge sharing and strategic planning, helping cities to collectively defend against cyber-attacks on AI-powered urban infrastructure (GovInsider, 2024). The cybersecurity measures implemented by ASCN highlight the importance of regional preparedness, as cities within ASEAN share similar vulnerabilities due to interconnected digital systems and rising cyber threats in the digital economy (United Nations Economic and Social Commission for Asia and the Pacific, 2023).

2. ***Disaster Management Solutions*** In disaster management, ASCN cities have used AI to improve response times and accuracy in handling natural disasters, which are prevalent in the region. An example is Malaysia's RescueAI system, developed in collaboration with Asia Pacific University of Technology and Innovation (APU). RescueAI utilizes a 3D Digital Twin model, drones, and real-time data to monitor and simulate disaster scenarios, including floods and fires. By deploying AI-driven drones, RescueAI helps urban areas anticipate and manage disasters more effectively, contributing to disaster resilience across ASEAN. The system's success in Malaysia is a model that other ASCN cities can adapt, aligning with ASCN's goal to improve public safety through technology (OpenGov Asia, 2024)
3. ***Cross-Country Data Sharing and Resilience*** ASCN has also enabled member cities to share critical data and experiences, enhancing overall resilience. For instance, during the COVID-19 pandemic, ASCN cities collaborated on data-driven solutions to manage the spread of the virus, demonstrating how a unified approach can help cities handle health crises and other emergencies. This type of cross-border collaboration, facilitated by ASCN, demonstrates the utility of AI in tracking, sharing, and responding to urban health and environmental risks, underscoring the value of integrated data platforms within ASEAN (Asian Development Bank, 2023).

In the international community, the United Nation's-Habitat and Global Smart Cities Alliance, supported by organizations like the World Economic Forum, strives to create unified frameworks for smart city governance that address digital transformation challenges, while also advancing social equity, privacy, and resilience in urban planning. This initiative promotes ethical AI deployment and emphasizes a people-centered approach to digital governance, aiming to ensure that technology in urban environments supports human rights and sustainable development across varied international contexts. To achieve this, the Alliance has developed a set of policy benchmarks across key areas, such as data transparency, privacy, cybersecurity, and inclusive urban design. These policies are piloted in over 30 Pioneer Cities worldwide, which serve as model communities for evaluating and refining governance practices (UN-Habitat, 2022).

Through such collaborative frameworks, the Alliance sets standards that many cities, especially in developing regions, can adopt without needing to independently develop advanced policy infrastructure. In regions like Latin America, Southeast Asia, and parts of Africa, these initiatives help cities address emerging concerns around urban technology, from cybersecurity to equitable data access, making them better equipped to participate in AI-driven projects. By aligning on universal principles, the Alliance helps cities collaborate on AI initiatives, supporting both local and global digital innovation goals, thereby advancing ethical urban AI usage across borders (UN-Habitat, 2022).

1. ***Standardization of Smart City Policies*** The Global Smart Cities Alliance aims to provide cities with actionable policy frameworks that guide the ethical implementation of AI technologies. One of the key outputs of this collaboration is the Smart City Governance Framework, which offers guidelines on transparency, accountability, and inclusiveness. The framework addresses the ethical use of AI in urban governance, ensuring that smart city initiatives do not exacerbate social inequalities or infringe upon citizens' rights (UN-Habitat, 2022). For example, the framework recommends practices such as the use of open data to improve public services while maintaining data privacy. Cities that adopt these guidelines can ensure their smart city projects are not only innovative but also equitable. The City of Barcelona has implemented such frameworks, focusing on citizen participation in decision-making processes related to smart technology applications, which exemplifies how local governments can integrate ethical guidelines into their smart city strategies (World Economic Forum, 2021).
2. ***AI Ethical Guidelines*** The Alliance also emphasizes the importance of ethical AI in urban environments. By setting standards for algorithmic transparency and fairness, it aims to mitigate risks associated with biased AI systems. For instance, the Alliance advocates for regular audits of AI systems used in public services, ensuring they operate fairly and do not discriminate against marginalized communities. Such practices are becoming increasingly relevant as cities like Amsterdam and Singapore integrate AI technologies into public safety and urban planning (United Nations, 2023).
3. ***Privacy Standards*** In addition to ethical AI guidelines, UN-Habitat and the Global Smart Cities Alliance are working to establish robust privacy standards that protect citizen data. As cities deploy AI systems that collect vast amounts of data, concerns about data privacy and security have risen. The Alliance's guidelines recommend that cities implement strong data protection measures and engage citizens in discussions about how their data is used. Cities like Helsinki have adopted these standards by ensuring that data collection practices are transparent and that residents can access information on how their data is processed (UN-Habitat, 2022).

THE FRONT-RUNNERS FOR THE WORLD'S AI-ENABLED SMARTEST CITIES

As artificial intelligence (AI) continues to reshape industries and redefine urban living, cities around the globe are racing to establish themselves as leaders in the realm of AI-enabled smart city development. These front-runners are not only leveraging cutting-edge technologies to enhance the quality of life for their residents but are also setting benchmarks for innovation, sustainability, and resilience in urban management. The integration of AI into city infrastructures—from intelligent transportation systems and energy-efficient buildings to advanced healthcare solutions and enhanced public safety—marks a significant shift in how urban areas operate and respond to the complex challenges of modern life.

In this competitive landscape, cities that successfully harness AI capabilities demonstrate an ability to streamline services, optimize resource allocation, and foster community engagement, ultimately paving the way for a smarter, more connected future. As urban populations continue to surge and the demand for sustainable solutions intensifies, the pursuit of AI-driven advancements becomes imperative. This exploration into the world's AI-enabled smartest cities highlights the innovative strategies and technologies that set these front-runners apart, examining their unique approaches to urban challenges and their roles as models for future urban development. By analyzing their achievements and initiatives, we gain valuable insights into the potential of AI to transform cities into more livable, efficient, and sustainable environments for all.

The Smart City Index by IMD is focused on the economic and technological aspects of the city, including the quality of life, environment, and inclusiveness. Out of 142 cities, some citics are repeatedly included in the ranking, proving their city's initiative to shift towards an AI-enabled smart city: Zurich, Singapore, and Copenhagen. These three cities are front-runners of their respective regions to achieve the highest rating for the Smart City Index.

1. ***Singapore*** In Asia, Singapore is recognized as a pioneer of the smart city movement, largely due to its government-backed "Smart Nation" initiative. Singapore has implemented many AI-driven projects, from intelligent transport systems to digital healthcare services. The city-state's investment research and development, along with its strong regulatory framework, has allowed it to maintain its status as one of the world's most technologically advanced cities. Singapore's priority areas include affordable housing, employment, and health services. Thales (2023) cited five key drivers of Singapore maintaining its position as a "Smart Nation" through the following: a) utilization of sensor technology for children, elderly, and transport optimization, b) digitalization of the healthcare system or "HealthHub" where it allows the patients to manage, book, and access medical needs, c) the "Smart National App" which is a one-stop platform to access government-provided services and information, d) development of business parks and integration of IT industry and education sector, and e) educating the citizens to use and adapt AI for their everyday lives. Moreover, Singapore's position as a leader in smart city innovations is bolstered by its collaborations within the ASEAN Smart Cities Network, where it shares best practices and technological insights with other Southeast Asian cities. This network focuses on addressing common urban challenges through AI, particularly in areas such as cybersecurity and disaster management (CNA, 2024). The collaborative effort not only enhances local capabilities but also facilitates regional resilience against urban challenges. Through these initiatives, Singapore was awarded as the fifth smartest city in the world according to the 2024 Smart City Index, the highest rank the city has received since 2020 (Sin, 2024). This success has enabled Singapore to foster collaborative projects with China, as the two countries recently announced the launch of 14 new projects related to developing smart cities (Ministry of Digital Development and Information, 2023).
2. ***Copenhagen*** In Europe, Copenhagen has focused heavily on sustainability, using AI to optimize its energy consumption and reduce carbon emissions. The city's "Copenhagen Solutions Lab" works on developing smart city projects that aim to be carbon-neutral by 2025. This goal sets Copenhagen as the world's first carbon-neutral city if achieved. Copenhagen's ranking as the sixth top smart city in the IMD Smart City Index 2024 highlights its commitment to leveraging technology for sustainable urban living. This Danish capital is renowned for integrating artificial intelligence (AI) into various sectors, emphasizing mobility and environmental sustainability. Copenhagen's approach to smart city initiatives is closely aligned with the European Union's (EU) Urban AI Initiatives, which aim to enhance city living through AI-powered solutions. The city's focus on green technologies includes projects aimed at improving public transport efficiency and reducing carbon emissions. For instance, Copenhagen has implemented an AI-based traffic management system that optimizes traffic flow, reducing congestion and air pollution, showcasing how data analytics can lead to more sustainable urban transport solutions (Tullio, 2024). Moreover, Copenhagen actively participates in cross-border collaborations within the ASEAN Smart City Network, sharing knowledge and strategies for urban challenges, such as disaster management and cybersecurity. This collaboration fosters an exchange of best practices among cities globally, promoting the use of AI in addressing common urban issues

(ASEAN Secretariat, 2023). Similarly, Copenhagen's sustainable initiatives have also led the city to collaborate with other countries. Last 2023, Copenhagen invited two delegations from Poland and the United Kingdom to experience Copenhagen's sustainable infrastructures and architectural marvels (State of Green, 2023).

3. ***Zurich*** Zurich has been recognized as the world's top smart city in the 2024 Smart City Index, marking its fifth consecutive year in this position. The Smart City Index, produced by the IMD World Competitiveness Center and the World Smart Sustainable Cities Organization (WeGO), evaluates 142 cities based on various criteria, including citizen surveys and urban infrastructure. The report emphasizes the importance of integrating technology into city management to enhance quality of life for residents and to address challenges posed by rapid urbanization and climate change (IMD, 2024).

A key aspect of Zurich's success lies in its commitment to sustainable urban development and innovative public services. The city has a strong human-implemented policies and numerous AI-driven solutions aimed at improving urban mobility and green technologies. For instance, Zurich's integration of AI in public transportation has streamlined services, making them more efficient and accessible. The city's focus on green spaces, cultural initiatives, and community engagement has further enhanced its livability, aligning with the increasing global emphasis on creating human-centric urban environments (Gort, 2024).

Additionally, Zurich's proactive governance model fosters collaboration across sectors, leveraging both public and private resources. This collaborative approach is critical for the development of smart city solutions that address local needs while contributing to broader sustainability goals. The report indicates that cities prioritizing good governance, trust, and citizen engagement are better positioned to adapt to future uncertainties, a principle that Zurich exemplifies (IMD, 2024).

The findings of the 2024 Smart City Index reflect broader trends in urban innovation, where European cities continue to lead in smart city rankings. Cities like Oslo, Geneva, and Singapore follow closely, indicating a regional commitment to leveraging technology for sustainable urban growth (Workplace Insight, 2024). This trend underscores the potential for cross-border collaborations in developing AI-enabled solutions that can be adapted to meet specific urban challenges across different contexts.

All three cities prioritize sustainability as a core element of their smart city strategies. Zurich focuses on green technologies and sustainable public transport systems, integrating AI to manage urban mobility efficiently (IMD, 2024). Similarly, Copenhagen's emphasis on eco-friendly practices, such as its vehicle-free urban planning in Tengah, aligns with global sustainability goals (Tullio, 2024). Singapore also emphasizes sustainable development through its Smart Nation program, promoting green spaces and energy-efficient technologies (CNA, 2024). This commitment to sustainability not only enhances urban livability but also prepares these cities for cross-border collaborations in tackling climate-related challenges.

Each city employs AI-driven solutions to enhance public services. For example, Zurich uses AI to optimize traffic management and improve transportation services (IMD, 2024). Copenhagen leverages AI for public transportation efficiency, significantly reducing congestion through real-time data analytics (Tullio, 2024). Singapore implements various AI applications in healthcare, public transport, and urban management, demonstrating a comprehensive approach to integrating technology into daily life (CNA, 2024). This collective focus on AI enables these cities to share insights and best practices effectively, fostering innovation across borders.

A notable similarity is their governance models, which emphasize collaboration and citizen engagement. Zurich's governance encourages partnerships between public and private sectors to develop innovative solutions (IMD, 2024). Copenhagen actively involves citizens in decision-making processes, ensuring that smart city initiatives align with community needs (Buhl, 2024). Singapore also engages its residents through initiatives like the Smart Nation program, promoting public input in urban planning (CNA, 2024). These collaborative governance structures facilitate cross-border partnerships, as they build trust and enable the sharing of resources and expertise among cities.

All three cities participate in international networks that promote knowledge sharing and collaboration. Copenhagen and Singapore are both active members of the ASEAN Smart Cities Network, focusing on common urban challenges and sharing technological advancements (CNA, 2024). Zurich also engages in various European networks that facilitate the exchange of ideas and practices, enhancing regional capabilities in addressing urban challenges (IMD, 2024). These collaborative efforts strengthen their ability to innovate and adapt to evolving urban needs. Hence, Figure 1 illustrates similar components observed in Zurich, Singapore, and Copenhagen in order to establish a successful smart city.

Figure 1. Components of a successful Smart City based in Zurich, Singapore, and Copenhagen

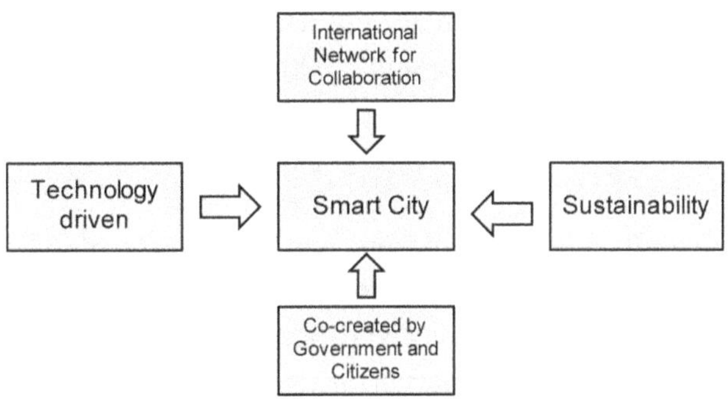

CHALLENGES AND OPPORTUNITIES

Cross-border collaborations in AI-enabled smart cities present significant opportunities for innovation and enhanced urban living. However, they also face several challenges and risks that need to be addressed to ensure successful implementation. Here are three primary concerns:

1. Privacy and Ethical Concerns

Cross-border collaborations often involve the sharing of vast amounts of data, raising significant privacy and ethical issues. Different regions may have varying cultural norms regarding data privacy, leading to complexities in compliance and governance. For instance, the General Data Protection Regulation (GDPR) in the EU sets a high standard for data protection, which may conflict with less stringent regulations in other countries, such as the United States (Wright, 2023). This discrepancy can create legal ambiguities and affect data-sharing agreements among cities.

Moreover, ethical considerations surrounding AI decision-making are paramount. AI systems can perpetuate biases inherent in training data, potentially leading to discrimination against specific demographic groups (O'Neil, 2022). In cross-border settings, ensuring that AI algorithms are fair and transparent becomes even more complex, necessitating the establishment of universal ethical guidelines that accommodate diverse cultural values.

2. Cybersecurity Risks

The integration of AI technologies across borders introduces heightened cybersecurity risks. With increased data sharing comes the potential for international data breaches and cyberattacks that can compromise sensitive information (Shackelford et al., 2022). For instance, smart city infrastructure, such as connected traffic systems or public safety networks, could be vulnerable to hacking, which may lead to significant disruptions and safety risks.

Furthermore, as AI systems become more autonomous, there are concerns about their vulnerabilities to manipulation and the consequences of such breaches. A collaborative framework must include robust cybersecurity measures to protect against these threats, ensuring that all participating cities have aligned protocols and standards for safeguarding data (Hussain, 2023).

3. Cultural and Ethical Differences

Cultural differences significantly impact the design and acceptance of AI technologies in smart cities. Varying societal values and expectations can complicate the integration of AI solutions, as what is considered acceptable in one culture may be viewed differently in another (Gubbi et al., 2023). For example, privacy norms in Western countries may prioritize individual data protection, while other cultures may place greater emphasis on community benefits, influencing how data is collected and used.

To navigate these cultural nuances, AI solutions must be designed with sensitivity to local contexts. Engaging stakeholders from diverse backgrounds during the development phase can help create more culturally relevant applications, ensuring that the technology aligns with the values and expectations of different communities (Kumar et al., 2023). This inclusive approach is essential for fostering trust and acceptance of AI technologies in cross-border collaborations.

4. Digital Divide

The disparity in technology access poses a significant barrier to smart city development. Not all cities, especially in low- and middle-income countries, possess the infrastructure required for implementing AI technologies. According to the International Telecommunication Union (ITU), many developing regions lack basic internet connectivity, which hinders their ability to participate in the digital economy and access the benefits of smart city innovations (ITU, 2022). For instance, while cities like Singapore and

Zurich have robust smart city frameworks, many smaller cities in Southeast Asia and Africa struggle with inadequate infrastructure and funding, exacerbating inequalities (Zanella et al., 2023).

Nevertheless, the opportunities presented by smart cities are significant. By creating more efficient, sustainable, and livable urban environments, these cities can address many social, modern environmental, and economic challenges. The ability of Artificial Intelligence to optimize resource management means that smart cities can contribute to achieving global sustainability goals, such as the United Nations' Sustainable Development Goals (SDGs), particularly those related to climate action, clean energy, and sustainable cities and communities.

OVERCOMING THE BARRIERS TO COLLABORATION

Despite these challenges, the potential benefits of AI-enabled smart cities are profound. By fostering efficient, sustainable, and livable urban environments, these cities can address a variety of social, environmental, and economic challenges.

1. Sustainable Development

Smart cities are positioned to significantly contribute to the United Nations' Sustainable Development Goals (SDGs). For example, AI can optimize energy consumption in urban buildings, reducing greenhouse gas emissions and enhancing energy efficiency, which aligns with SDG 13 (Climate Action) and SDG 11 (Sustainable Cities and Communities) (Shahid et al., 2023).

2. Resource Management

AI technologies facilitate smarter resource management in areas like water supply, waste management, and transportation. For instance, Barcelona has implemented smart water management systems using AI to monitor consumption and detect leaks, leading to substantial water savings (Pérez et al., 2023). This not only conserves resources but also promotes economic sustainability by reducing costs.

3. Enhanced Urban Living

By integrating AI into urban planning and management, cities can enhance the quality of life for their residents. Smart mobility solutions, such as AI-driven public transport systems, can reduce congestion and improve accessibility (Deloitte, 2023). Furthermore, the development of smart healthcare initiatives, such as telemedicine and AI health monitoring systems, can enhance public health outcomes (CNA, 2024).

4. Strengthening Cybersecurity

To prevent cyber threats associated with AI and smart city technologies, cross-border cybersecurity agreements and protocols are essential. Initiatives such as the EU Cybersecurity Act and the NIST Cybersecurity Framework provide guidelines for improving the resilience of digital infrastructures (European

Commission, 2022). These frameworks encourage countries to collaborate on cybersecurity measures, share best practices, and respond collectively to cyber incidents.

Moreover, establishing protocols for AI systems can help mitigate risks. For example, the AI Security and Privacy (AI-SP) working group under the Cloud Security Alliance (CSA) focuses on developing best practices for securing AI systems against potential threats, enhancing the overall security posture of smart city initiatives (Cloud Security Alliance, 2023).

5. Technology Standards and Interoperability

The standardization of AI and IoT technologies is essential for ensuring compatibility across different systems and platforms. The International Organization for Standardization (ISO) and the Institute of Electrical and Electronics Engineers (IEEE) are actively working on establishing technical standards that promote interoperability among smart city technologies (Rudolph et al., 2023). For example, the development of the IEEE 802.15.4 standard has facilitated communication among IoT devices, enabling cities to integrate various smart technologies seamlessly. This standardization is crucial in enabling cities to share data and collaborate on AI solutions without encountering compatibility issues.

Overcoming barriers to collaboration in AI-enabled smart cities requires a multifaceted approach. By addressing these areas, cities can foster successful partnerships and develop innovative solutions that enhance urban living.

THE FUTURE DIRECTION OF AI-ENABLED SMART CITIES

As urbanization accelerates, particularly in regions like Africa and Latin America, there are significant opportunities for expanding AI-enabled smart city collaborations. The potential for these regions to leverage AI technologies to address urban challenges is immense, especially in light of their unique contexts and developmental needs.

The African continent is witnessing rapid urbanization, with cities like Nairobi, Kenya, and Lagos, Nigeria, growing exponentially. This presents a critical opportunity for cross-border collaborations that harness AI for urban development. For example, the African Union has initiated the Smart Cities Initiative, aiming to connect cities across the continent to share best practices and technologies (African Union, 2023). These partnerships can help tackle issues such as transportation, energy management, and waste disposal through AI solutions tailored to local contexts.

A case in point is the collaboration between Kenya's National Transport and Safety Authority and the University of Nairobi, which utilizes AI to enhance road safety through real-time data analysis (Ochieng, 2023). Such initiatives can set a precedent for other African nations to follow, fostering regional partnerships that enhance urban resilience.

Similarly, Latin America is experiencing a surge in urban population, with cities like São Paulo, Brazil, and Bogotá, Colombia, facing significant challenges in infrastructure and service delivery. The Inter-American Development Bank (IDB) is promoting collaborative smart city projects that utilize AI to improve urban planning and service efficiency (IDB, 2022). For example, the IDB-supported "Digital Cities" initiative in Brazil focuses on integrating AI to optimize public transportation systems, aiming to reduce congestion and improve citizen accessibility (Santos et al., 2023).

Collaborations between local governments and tech companies, such as the partnership between Bogotá's local government and IBM, leverage AI to analyze traffic patterns and develop solutions that reduce congestion (IBM, 2023). These types of initiatives can facilitate knowledge sharing and technological advancements across borders.

As regions in Africa and Latin America embark on their smart city journeys, leveraging global expertise is crucial. Partnerships with established smart cities in Europe and Asia can provide valuable insights into best practices and lessons learned. The Smart Cities Mission in India, for example, has successfully implemented AI technologies in urban governance and can serve as a model for other developing regions (Ministry of Housing and Urban Affairs, 2022). Furthermore, organizations like UN-Habitat provide platforms for knowledge exchange and capacity building, enabling cities in Africa and Latin America to learn from successful AI implementations in different contexts (UN-Habitat, 2023).

As cities worldwide increasingly integrate AI technologies into their urban planning and management strategies, there is a pressing need to evolve ethical standards to keep pace with these advancements. The collaboration among international organizations is vital for establishing comprehensive frameworks that govern the ethical deployment of AI in smart cities, especially in cross-border contexts.

International organizations such as the United Nations, OECD, and the IEEE are essential in shaping ethical standards for AI. The UN's AI for Good Global Summit brings together stakeholders to discuss the ethical implications of AI deployment, focusing on aligning technology with human rights and sustainable development goals (United Nations, 2022). These discussions emphasize the need for adaptable ethical guidelines that can respond to the rapid changes in technology and societal norms. For example, the OECD's AI Principles highlight the importance of transparency, accountability, and inclusiveness in AI development, urging countries to ensure that AI systems are designed with these principles in mind (OECD, 2021). As cities collaborate across borders, these principles serve as a foundational framework that can guide ethical AI implementation in diverse cultural contexts.

As AI technologies evolve, so too must the ethical frameworks that govern their use. The IEEE Global Initiative on Ethics of Autonomous and Intelligent Systems actively develops standards addressing the ethical challenges posed by AI, focusing on transparency, accountability, and bias reduction (IEEE, 2023). Their work aims to establish adaptable standards that can evolve in response to technological advancements and societal expectations.

For instance, the IEEE P7001 standard focuses on transparency in autonomous systems, providing guidelines for how AI systems should disclose their decision-making processes to users (IEEE, 2021). By establishing clear standards for transparency, cities can foster public trust in AI technologies, facilitating cross-border collaborations where accountability and ethical considerations are paramount.

Take for example Barcelona's Ethical AI Framework and Singapore’s AI Ethics Advisory Council. Barcelona has taken proactive steps to ensure ethical AI deployment through its AI and the City initiative, which incorporates ethical considerations into its smart city projects. The city has developed guidelines to ensure that AI technologies are used responsibly, with a focus on protecting citizens' rights and fostering inclusivity (Barcelona City Council, 2023). This initiative serves as a model for other cities, illustrating how ethical standards can be integrated into AI strategies to enhance collaboration and trust.

Meanwhile, Singapore's AI Ethics Advisory Council plays a crucial role in shaping the ethical landscape of AI deployment within its smart city framework. The council provides recommendations on governance, accountability, and transparency for AI systems, ensuring that they align with the country’s broader goals for digital transformation and urban sustainability (Smart Nation Singapore, 2023). By establishing clear ethical guidelines, Singapore enhances its potential for cross-border collaborations, as these standards can be harmonized with those of other nations.

SYNTHESIS

This chapter has comprehensively discussed the concepts, barriers, developments, and future direction of cross border AI-enabled smart cities. The concept of "smart city" alone is a cross-border idea originating from the United States. However, the proper mark on the usage of the word "smart city" started in Amsterdam. The evolution of smart cities—from Smart City 1.0 to 3.0—reflects a significant shift in how urban environments leverage technology to address the challenges of urbanization. Initially characterized by technology-focused projects with limited citizen involvement, Smart City 1.0 laid the groundwork for future developments but often fell short of achieving sustainability and community relevance. The transition to Smart City 2.0 marked a pivotal change, emphasizing transparency and public accountability through increased collaboration between governments and private entities, ensuring that technological solutions addressed real urban challenges while engaging citizens more actively.

As we progress into Smart City 3.0, the focus on citizen empowerment and co-creation becomes paramount. This model recognizes that effective urban governance requires not only technological innovation but also active participation from citizens in shaping the policies and solutions that affect their lives. Cross-border collaborations are essential in this context, as they facilitate the exchange of best practices and innovative ideas that can adapt to the unique needs of diverse urban populations.

Tracing back, smart cities have gone through three generations where each phase has went through different controls and procedures. The difference of government officials per country and city lies heavily on effectively and efficiently establishing a smart city. As the world leans towards a smart city that is co-created by the government and its citizens, it is important to note that there should be a balance in order to satisfy both parties.

Also, the highlights on the exploration of cross-border collaborations for AI-enabled smart cities is essential as urbanization accelerates, presenting increasingly complex challenges that transcend national boundaries. With the United Nations projecting that nearly 70% of the world's population will reside in urban areas by 2050, cities face heightened demands in resource management, environmental sustainability, public safety, and quality of life. These challenges require innovative solutions that no single city or nation can develop in isolation. Cross-border partnerships emerge as a vital strategy for sharing expertise, technology, and policy frameworks that expedite the development and implementation of AI-driven urban solutions.

1. Sharing Knowledge and Resources

Collaboration allows cities to pool their knowledge and resources, facilitating the sharing of best practices and lessons learned from AI implementations. For example, cities can exchange insights on the deployment of smart transportation systems or energy-efficient infrastructure. By leveraging collective experiences, cities can avoid common pitfalls and accelerate the deployment of effective AI solutions tailored to their unique contexts.

2. Harmonizing Standards

Cross-border collaborations also play a critical role in harmonizing standards and regulations surrounding AI technologies. This is particularly important given the varied legal frameworks and ethical considerations across different regions. By developing unified standards, cities can ensure compatibility and interoperability among AI systems, fostering innovation while mitigating risks associated with data privacy and cybersecurity. Collaborative frameworks can also promote ethical guidelines for AI use, ensuring that technological advancements prioritize the well-being of citizens and adhere to established ethical norms.

3. Addressing Global Challenges

The urgency of studying these collaborations stems from the immediate need for cohesive, globally aligned strategies to tackle shared challenges such as climate change, urban mobility, and public health crises. For instance, cities facing similar environmental threats can collaborate on AI solutions for waste management, water conservation, and energy efficiency. By working together, cities can enhance their resilience to climate impacts while sharing the burden of research and development costs, making it more feasible for all stakeholders involved.

4. Promoting Inclusivity

One of the significant advantages of cross-border partnerships is their potential to foster inclusivity. Collaborations can ensure that cities in both developed and emerging regions benefit from AI-enabled advancements. By investing in knowledge transfer and capacity-building initiatives, wealthier cities can assist developing counterparts in harnessing AI technologies to improve urban governance, service delivery, and economic opportunities. Such efforts can help bridge the digital divide, ensuring that the benefits of AI-enabled smart cities are equitably distributed.

5. Navigating Challenges

While the potential benefits of cross-border collaborations are profound, they do not come without challenges. Navigating the complexities of diverse regulatory frameworks, cultural differences, and ethical considerations can be daunting. Successful collaborations require a commitment to open dialogue, mutual respect, and a willingness to adapt to varying local contexts. Policymakers and urban planners must prioritize building trust and facilitating communication between partners to overcome these hurdles effectively.

6. Emphasizing Global Cooperation

As the world moves towards an interconnected future, the need for global cooperation becomes increasingly imperative. Stakeholders across sectors—including government, academia, and the private sector—must embrace collaborative approaches to drive the development of AI-enabled smart cities. By forming international consortia and engaging in research partnerships, cities can leverage the collective intelligence of diverse stakeholders, fostering innovation that meets the unique needs of their populations.

In summary, the study of cross-border collaborations for AI-enabled smart cities is not merely an academic exercise but a crucial endeavor that can significantly impact urban living in the coming decades. By fostering partnerships that share knowledge, harmonize standards, address global challenges, and promote inclusivity, we can develop smarter urban environments that enhance the quality of life for all citizens. Through these collaborative efforts, we pave the way for a sustainable and equitable future, transforming our cities into resilient, AI-enabled hubs that respond effectively to the needs of an increasingly urbanized world.

Moreover, the urgency to foster such international partnerships is underscored by the interconnected nature of today's global challenges—climate change, cybersecurity, and social equity—all of which require collective action. By sharing resources, knowledge, and technology across borders, cities can develop AI-enabled smart solutions that enhance public services, improve quality of life, and promote sustainable urban development.

Further, the success of cross-border collaborations in the realm of AI-enabled smart cities lies in their ability to create inclusive environments where technology and community aspirations coexist. As stakeholders embrace this collaborative spirit, they can develop urban spaces that are not only technologically advanced but also responsive to the needs of all citizens, paving the way for a more sustainable and equitable future in an increasingly urbanized world.

CONCLUSION

Therefore, this chapter underscores the transformative potential of cross-border collaborations in advancing AI-enabled smart cities, emphasizing their role in addressing the multifaceted challenges of urbanization. The evolution from Smart City 1.0 to 3.0 highlights the growing importance of citizen empowerment, co-creation, and inclusivity alongside technological innovation. By fostering partnerships that harmonize standards, share knowledge, and address global challenges, cities can collectively develop resilient and sustainable solutions that transcend national boundaries. These collaborations not only bridge the digital divide but also ensure that urban advancements prioritize equity, sustainability, and community well-being, paving the way for a future where smart cities are both technologically sophisticated and deeply human-centered.

REFERENCES

African Union. (2023). *Smart Cities Initiative*. Retrieved from African Union Asian Development Bank. (2023). *ASEAN's Role in Strengthening Regional Resilience through Smart Cities*.

Ball State University. (2022). Smart City History. https://digitalresearch.bsu.edu/immersive-learning-showcase-spring-2022/exhibits/show/smart-city-muncie/definition-and-history-of-smar

Barcelona City Council. (2023). *AI and the City: Ethical Guidelines for Smart City Projects*.

Bibri, S., & Krogstie, J. (2022). Governance and the Evolution of Smart Cities: From Technology-Driven to Citizen-Centric Models. *Technological Forecasting and Social Change*.

Brown, M., & Clifton, J. (2023). Public-private partnerships in global smart cities: A framework for collaborative innovation. *International Journal of Smart Cities*, 9(2), 45–60.

Buhl, S. (2024). Citizen Engagement in Smart Cities: The Case of Copenhagen. *Urban Studies Journal*, 61(1), 125–140.

Cardullo, P., et al. (2022). *Citizen Participation and Urban Co-Creation in Smart City Development*. Journal of Urban Studies.

Cavoukian, A. (2023). Privacy and Security in Smart Cities: The Role of Technology. *Smart Cities Journal*, 5(1), 10–21. DOI: 10.1016/j.scij.2023.100051

Chourabi, H.. (2021). *Citizen Engagement and Transparency in Smart Cities: The Role of Governance in Urban Planning*. Cities Journal.

CIVITAS. (2022). Citizen Engagement in Smart City Projects. *Best Practice*.

Cloud Security Alliance. (2023). *AI Security and Privacy Best Practices*.

CNA. (2024). Singapore ranked 5th in Smart City Index.

Cohen, B. (2015). The 3 Generations Of Smart Cities. https://www.fastcompany.com/3047795/the-3-generations-of-smart-cities

Deloitte. (2023). *The Future of Urban Mobility: Smart Cities and AI*.

European Commission. (2022). *EU Cybersecurity Act: Strengthening Europe's Cybersecurity*. European Commission. (2024). EU Urban Mobility Observatory. Directorate-General for Mobility and Transport.

Gort, M. (2024). *European and Asian cities dominate latest smart city rankings*. Workplace Insight.

GovInsider. (2024). *ASEAN Smart Cities and Cybersecurity*. GovInsider Insights.

Gubbi, J., Buyya, R., Marusic, S., & Pavithran, P. (2023). Internet of Things (IoT) in Smart Cities: Challenges and Opportunities. *Journal of Urban Technology*, 30(1), 1–18. DOI: 10.1080/10630732.2023.2157513

Hussain, F. (2023). Cybersecurity in Smart Cities: Challenges and Solutions. *Cybersecurity Journal*, 6(1), 45–60. DOI: 10.1016/j.cyber.2023.100082

IBM. (2023). *AI for Smart Cities: Bogotá's Transportation Optimization.*

IDB. (2022). *Digital Cities: Transforming Urban Services in Latin America.*

IEEE. (2021). *IEEE.* Transparency in Autonomous Systems.

IEEE. (2023). *Global Initiative on Ethics of Autonomous and Intelligent Systems.*

IMD. (2024). *Smart City Index 2024.* IMD World Competitiveness Center.

International Banker. (2024). Why Singapore is Consistently Regarded as Asia's Smartest City. Kohli, S., & Vala, P. (2023). *Aligning AI standards: A cross-border approach to data privacy and ethics.AI and Society Journal*, 17(4), 202–225.

ITU. (2022). *Global Connectivity Report 2022.* International Telecommunication Union.

Kumar, A., Gupta, R., & Singh, P. (2023). Cultural Sensitivity in AI Deployment: Designing for Diverse Societies. *AI & Society*, 38(2), 345–359. DOI: 10.1007/s00146-023-00388-9

Li, X., & Chen, Z. (2022). Pooling resources for AI-enabled cities: The economic value of collaboration. *Global Innovation Journal*, 6(3), 150–172.

Luque-Ayala, A., & Marvin, S. (2023). The Future of Smart Cities: Co-Creation, Sustainability, and Citizen Empowerment. *Journal of Urban Planning and Development.*

Ministry of Digital Development and Information. (2023). Singapore and Shenzhen Drive Smart City Collaboration with 14 New Projects. https://www.mddi.gov.sg/media-centre/press-releases/singapore-and-shenzhen-drive-smart-city-collaboration/

Ministry of Digital Development and Information. (2023). Singapore and Shenzhen Drive Smart City Collaboration with 14 New Projects. https://www.mddi.gov.sg/media-centre/press-releases/singapore-and-shenzhen-drive-smart-city-collaboration/

Ministry of Housing and Urban Affairs. (2022). *Smart Cities Mission: Annual Report 2022.*

Mora, C.. (2020). Technology-Driven Smart Cities: Evolution, Challenges, and Future Prospects. *Journal of Urban Technology.*

Müller, V., & Van der Meer, T. (2023). Ethical AI in Urban Planning: A Framework for Accountability. *Urban Studies Journal*, 60(2), 345–360. DOI: 10.1177/00420980211053754

O'Neil, C. (2022). *Weapons of Math Destruction: How Big Data Increases Inequality and Threatens Democracy.* Crown Publishing Group.

Ochieng, J. (2023). Enhancing Road Safety through AI in Kenya. *East African Journal of Science and Technology*, 2(1), 45–60. DOI: 10.1007/s00500-022-06371-8

OECD. (2021). *OECD Principles on AI.*

OECD. (2023). Digital. *Economic Outlook*, •••, 2023.

OpenGov Asia. (2024). *RescueAI: Smart City Disaster Management System in Malaysia.* OpenGov Asia.

Organisation for Economic Co-operation and Development. (2019). OECD AI principles overview (May). https://oecd.ai/en/ai-principles

Pérez, P., & Rodríguez, M. (2023). Smart Water Management in Barcelona: An AI Approach. *Journal of Urban Technology*, 30(3), 25–40. DOI: 10.1080/10630732.2023.2157514

Rossi, G., & Petruzzi, L. (2022). Challenges in regulatory harmonization for international AI collaborations. *Policy & Data Ethics Review*, 11(1), 89–105.

Rudolph, S., Schmid, K., & Bauer, W. (2023). The Role of Standards in Smart Cities. *Journal of Urban Technology*, 30(2), 125–140. DOI: 10.1080/10630732.2023.2147532

Santos, F., Araújo, L., & Ribeiro, C. (2023). AI and Urban Mobility in Brazil: Challenges and Opportunities. *Urban Planning Review*, 12(2), 78–93. DOI: 10.5555/1989-4924-12-2-78

Scantlebury, A., Lopez, T., Chen, Y., & Morales, D. (2022). Knowledge sharing and innovation in cross-border AI collaborations. *Journal of Urban Technology Innovation*, 14(3), 112–134.

Shackelford, S. J.. (2022). Cybersecurity and Smart Cities: A Global Perspective. *Journal of Cyber Policy*, 7(1), 95–113. DOI: 10.1080/23738871.2022.2074173

Shahid, A., & Adnan, A. (2023). AI for Sustainable Urban Development: A Review of Current Trends. *Sustainability Journal*, 15(12), 4567. DOI: 10.3390/su15124567

Sharma, A., N., Mukherjee, S., Kumar, S., Kumar, S., & Mohapatra, H. (2025). Analyzing territorial variations in the implementation of IoT-based smart homes: A comprehensive review. Utilizing Technology to Manage Territories, 305–334.

Sin, C. (2024). Singapore is world's 5th smartest city, top in Asia: Global index. The Straits Times. https://www.straitstimes.com/singapore/singapore-is-5th-smartest-city-in-the-world-top-in-asia-global-index

Smart Nation Singapore. (2023). *AI Ethics Advisory Council: Guidelines and Recommendations.*

State of Green. (2023). Fostering cross-border collaboration in the Capital of Architecture 2023. https://stateofgreen.com/en/news/fostering-collaboration-and-inspiration-in-the-capital-of-architecture-2023/

Thales. (2023). Singapore: The World's Smartest City. https://www.thalesgroup.com/en/worldwide-digital-identity-and-security/iot/magazine/singapore-worlds-smartest-city

Townsend, A. (2021). *Smart Cities: Big Data, Civic Hackers, and the Quest for a New Utopia*. W.W. Norton & Company.

Tullio, R. (2024). AI in Urban Mobility: How Copenhagen is Leading the Charge. *Journal of Smart City Innovation*, 4(2), 45–60.

UN-Habitat. (2022). *Global review of smart city governance practices.*

UN-Habitat. (2023). *Promoting Smart Cities for Sustainable Development.*

UNESCO. (2023). DDS: De Digitale Stad / The Digital City. https://www.unesco.org/en/memory-world/dds-de-digitale-stad/digital-city?hub=80094#:~:text=On%20January%2015%2C%201994%2C%20De,unknown%20world%20of%20the%20Internet

United Nations. (2017). World Population Prospects The 2017 Revision. https://population.un.org/Wpp/Publications/Files/Wpp2017_Keyfindings.Pdf

United Nations. (2022). *AI for Good Global Summit: Advancing AI to Benefit Humanity.*

United Nations Economic and Social Commission for Asia and the Pacific. (2023). *Urban Cybersecurity Challenges in ASEAN Smart Cities.*

Urban Mobility, E. I. T. (2023). CIVITAS Initiative: Transforming Urban Mobility Across Europe. EIT Urban Mobility. Retrieved from EIT Urban Mobility European Commission. (n.d.). Smart Cities. https://commission.europa.eu/eu-regional-and-urban-development/topics/cities-and-urban-development/city-initiatives/smart-cities_en#related-policies

Vallor, S. (2023). Ethics in AI Development: Transparency and Public Trust. *AI & Society*, 38(3), 497–511. DOI: 10.1007/s00146-022-00492-5

Williams, S., & Singh, R. (2023). Economic synergies in AI development: Financing the future of smart cities. *Urban Economics Review*, 15(5), 300–320.

World Economic Forum. (2020). *Governing smart cities: Policy benchmarks for ethical and responsible smart city development.*

Wright, D. (2023). Data Protection and Privacy in Cross-Border Contexts. *International Data Privacy Law*, 13(1), 1–12. DOI: 10.1093/idpl/ipac015

Zanella, A., Bui, N., & Castellani, A. (2023). Smart Cities: An Overview of Smart City Development Worldwide. *Smart Cities*, 6(4), 577–593. DOI: 10.3390/smartcities6040039

ADDITIONAL READING

Ahad, M. A., Paiva, S., Tripathi, G., & Feroz, N. (2020). Enabling technologies and sustainable smart cities. *Sustainable Cities and Society*, 61, 102301. DOI: 10.1016/j.scs.2020.102301

Al Sharif, R., & Pokharel, S. (2022). Smart city dimensions and associated risks: Review of literature. *Sustainable Cities and Society*, 77, 103542. DOI: 10.1016/j.scs.2021.103542

Blasi, S., Ganzaroli, A., & De Noni, I. (2022). Smartening sustainable development in cities: Strengthening the theoretical linkage between smart cities and SDGs. *Sustainable Cities and Society*, 80, 103793. DOI: 10.1016/j.scs.2022.103793

Cai, M., Kassens-Noor, E., Zhao, Z., & Colbry, D. (2023). Are smart cities more sustainable? An exploratory study of 103 US cities. *Journal of Cleaner Production*, 416, 137986. DOI: 10.1016/j.jclepro.2023.137986

Duygan, M., Fischer, M., Pärli, R., & Ingold, K. (2022). Where do Smart Cities grow? The spatial and socio-economic configurations of smart city development. *Sustainable Cities and Society*, 77, 103578. DOI: 10.1016/j.scs.2021.103578

Herath, H. M. K. K. M. B., & Mittal, M. (2022). Adoption of artificial intelligence in smart cities: A comprehensive review. *International Journal of Information Management Data Insights*, 2(1), 100076. DOI: 10.1016/j.jjimei.2022.100076

Javed, A. R., Shahzad, F., Rehman, S., Zikria, Y. B., Razzak, I., Jalil, Z., & Xu, G. (2022). Future smart cities: Requirements, emerging technologies, applications, challenges, and future aspects. *Cities (London, England)*, 129, 103794. DOI: 10.1016/j.cities.2022.103794

Kim, J. (2022). Smart city trends: A focus on 5 countries and 15 companies. *Cities (London, England)*, 123, 103551. DOI: 10.1016/j.cities.2021.103551

Lai, C. S., Jia, Y., Dong, Z., Wang, D., Tao, Y., Lai, Q. H., Wong, R. T. K., Zobaa, A. F., Wu, R., & Lai, L. L. (2020). A review of technical standards for smart cities. *Cleanroom Technology*, 2(3), 290–310. DOI: 10.3390/cleantechnol2030019

KEY TERMS AND DEFINITIONS

AI-Enabled.: Refers to systems, technologies, or devices that incorporate artificial intelligence (AI) capabilities to enhance their functionality, performance, or decision-making processes.

Cross-border.: This term pertains to activities, transactions, or interactions that occur across national borders or boundaries. It can refer to trade, communication, travel, or any form of exchange between entities in different countries.

Cybersecurity.: It involves protecting computer systems, networks, and data from digital attacks aimed at accessing, changing, or destroying sensitive information. It encompasses technologies, processes, and practices designed to safeguard against cyber threats.

Global Synergy.: The combined effort or cooperation of various entities, such as organizations, countries, or individuals, working together towards a common goal or objective on a global scale. It emphasizes the harmonious collaboration and integration of resources and efforts across borders.

Green Energy.: It refers to renewable energy sources that are environmentally friendly and sustainable, such as solar, wind, hydro, geothermal, and biomass energy. Green energy aims to reduce carbon emissions and mitigate the impact of traditional fossil fuels on the environment.

Green Transportation.: Green transportation involves modes of transportation that are eco-friendly and have minimal environmental impact. This includes electric vehicles, hybrid cars, public transportation systems, cycling infrastructure, and walking paths designed to reduce carbon emissions and promote sustainability.

Smart Cities.: Smart cities utilize technology and data-driven solutions to improve the efficiency of urban services and infrastructure, enhance sustainability, and optimize the quality of life for residents. These cities leverage IoT devices, sensors, and connectivity to manage resources effectively.

Urban Mobility.: It focuses on the movement of people within urban areas, encompassing various modes of transportation such as public transit, cycling, walking, ridesharing, and emerging technologies like electric scooters and autonomous vehicles. It aims to address congestion, pollution, and accessibility challenges in cities.

Chapter 13
Harnessing AI, IoT, and Emerging Technologies for Disaster Risk Management:
A Holistic Approach to Building Resilient Societies

Froilan Delute Mobo
https://orcid.org/0000-0002-4531-8106
Philippine Merchant Marine Academy, Philippines

Katarzyna Miłek
https://orcid.org/0000-0003-4603-4604
Gnieźnieńska Szkoła Wyższa Milenium, Poland

Mykola Orlykovskyi
Gnieźnieńska Szkoła Wyższa Milenium, Poland

Ramon Tolosa De Leon
Bicol University Open University, Philippines

ABSTRACT

The chapter, Harnessing AI, IoT, and Emerging Technologies for Disaster Risk Management: A Holistic Approach to Building Resilient Societies, explores the transformative potential of advanced technologies in disaster risk management (DRM). Artificial Intelligence (AI), the Internet of Things (IoT), blockchain, and 5G technologies are positioned as pivotal tools for enhancing disaster preparedness, response, and recovery. The text delves into AI-driven predictive analytics, IoT-enabled real-time environmental monitoring, blockchain-secured data transparency, and 5G-enabled communication networks. Through case studies from Japan, the Netherlands, Puerto Rico, and California

DOI: 10.4018/979-8-3693-9770-1.ch013

1. INTRODUCTION

1.1 Overview of Disaster Risk Management (DRM)

Disaster risk management typically involves techniques of reducing the vulnerability of people, property, communities, and countries to natural and manmade hazards and decisions concerning how to minimize the probability of hazardous events, how to lessen the impact of disasters when they occur, and how to respond to and recover from disastrous events. Disasters including hurricanes, wildfires, floods, and pandemics, are on the rise due to climate change and urbanization and affect human lives, economies, and ecosystems globally (UNDRR, 2020). The conventional approaches to DRM have been described by the following key features: restricted access to real-time data; relatively poor specificity of subsequent forecasts, and frequently insufficient early indications. Therefore, increasing pressure exists to accept new technologies that may provide methodologies for more effective disaster preparedness and response at each process stage.

1.2 Emerging Technologies in DRM

AI, particularly when integrated with IoT and other notable technologies like blockchain and 5G, can provide tremendous opportunities for DRM. These technologies have impacted disaster management in a way that is changing the landscape of data gathering, conveying, and processing during a disaster and the corresponding response.

Artificial Intelligence (AI): AI offers functionalities of high-level data analysis, data pattern identification, and decision-making. Large volumes of data are processed with the help of the mechanisms of machine learning to define risk rates and produce relevant predictions (Bengfort et al., 2021). AI indie technologies based on NLP and computer vision can extract information from social media, providing real-time situational analysis and becoming crucial to emergency operations.

Internet of Things (IoT): IoT systems are critical in providing near real-time data of the conditions in natural environments for features like temperature, seismic activity, and air and water quality. Using IoT devices such as drones, sensors, and cameras, it can become easier for emergency managers to have details of the areas that would require their attention hence making them able to approve efficient quick response actions (Kumar & Kumari, 2022).

Other Emerging Technologies: The ledger's immutable dispersed nature makes it a powerful tool for guaranteeing the security and integrity of DRM processes, as they are seen by all members of a distributed network. Furthermore, the deployment of 5G calls for ultra-low latency and more reliable communication networks to facilitate the IoT deployment and sharing of real-time data even in disaster-stricken regions (Li et al., 2023).

1.3 Objectives and Scope of the Chapter

In this chapter, one wants to give a comprehensive understanding of the use of modern approaches such as AI, IoT, and others to revolutionize DRM. Webinar attendees will gain knowledge on the benefits and drawbacks of these technologies, learn from case studies on specific successes and failures, and discover issues that accompany the usage of the technologies, including data privacy ethical issues and infrastructure demands. In the last chapter of this book, this author recommends a comprehensive, human rights-based, tension-aware approach to DRM through informed and engaging technology, policy, and education for the formation of more robust societies.

2. EMERGING TECHNOLOGIES IN DISASTER RISK MANAGEMENT

2.1 Artificial Intelligence (AI) in Disaster Risk Management

Disaster risks have been addressed by articulated intelligence (AI) with a high level of effectiveness in the ways of risk analytics, situation awareness, and even autonomous response. AI systems involve the use of machine learning to take advantage of big and complicated data to assist emergency managers in preventing the consequences of disasters.

Predictive Analytics and Risk Assessment: Machine learning as well as deep learning techniques are useful in risk processing and disaster prediction. These models take in different forms of data; historical weather data, satellite imagery, and social media data to name but a few, to predict the likelihood and extent of a disaster. For example, AI models can estimate flood risk from rainfall using data collected from GIS technology because this methodology identifies regions that are vulnerable to floods with remarkable precision (Shi et al., 2022).

Image and Video Analysis for Damage Assessment: Another great use of AI lies in the evaluation of damages by analyzing images and videos. Decision-making processes for disaster management are based on the assessment of images of the areas captured by drones, satellites, or ground cameras; computer vision algorithms help determine where the greatest losses are, and resources are directed accordingly (Cheng & Wang, 2023).

It was especially useful in the 2020 bush fires in Australia where via the use of drones equipped with artificial intelligence, the aftermath was captured and informed improving quick and efficient reaction plans.

NLP for Situational Awareness: Sentiment analysis which is a branch of NLP enables the AI systems to understand the texts derived From social media, news articles, and emergency reporting lines. NLP algorithms can get insights out of this kind of unstructured data which can detail the disaster and needed assistance on the fly. For instance, through analysis of social media posts, AI indicates the areas containing stranded victims in the hurricanes or floods, optimum areas for the rescue team to move in (Ghasemi & Rahman, 2021).

2.2 Internet of Things (IoT) in Disaster Risk Management

Internet of Things (IoT) which is an interconnection of electronic gadgets and mechanical structures that continuously exchange information over the internet with cheaper devices, intense sensors, quick reactivity in calamities are all aspects of IoT. The benefits that IoT helps to achieve are early warnings, monitoring the surrounding environment, and organizing actions on the ground.

Real-Time Environmental Monitoring: It must be noted here that IoT sensors enable real-time tracking of several environmental factors like temperature and humidity and even seismic activities and water levels. Such sensors are installed in sensitive zones so that early alarms can be given, especially in cases of floods, landslides, and earthquakes (Park et al., 2023). For example, the IoT-based flood monitoring systems, in South Asia, collect data on river levels and are programmed to give alarms as soon as the level is close to the danger level.

Remote and Autonomous Sensing: Drones and self-driving robots also help the emergency managers to quickly reach the affected areas which might be almost impossible to reach. With cameras and sensors, these IoT devices capture images, identify risky situations, and convey data to control centers which can improve awareness during the responses (Rahimi & Verma, 2022).

Smart Infrastructure for Community Resilience: The IoT can be employed in "smart city" applications to create infrastructure that adapts to disaster risks. For instance, IoT-controlled water systems at catchment areas can control water levels in the dams while IoT-controlled smart power grid can disconnect affected regions during hurricanes or earthquakes (Zhou et al., 2022).

2.3 Other Emerging Technologies in Disaster Risk Management

In addition to AI and IoT, other technologies such as blockchain, 5G, and drones play a role in disaster management framework improvement.

Blockchain for Data Security and Transparency: There is a possibility of ensuring data integrity through the applicability of block chain technology in DRM operations. Due to transparency in recording and synchronously keeping records of each transaction in a ledger, blockchain helps in transparency in the allocation of funds, aid distribution, and sharing of data in disasters (Kim & Choi, 2021). This is particularly relevant when working with the funds of relief, as it dramatically diminishes the possibility of embezzlement and increases transparency when using the blockchain base.

5G Connectivity for Real-Time Communication: 5G technology improves IoT and AI by offering the ability to deliver a large amount of information at extremely high speed with very low delay. This level of connectivity is important in DRM to enable fast transfer of real-time sensor data, high-definition video streams and other important information between the emergency responders (Li & Zhang, 2023).

In Japan, 5G networks have demonstrated the efficiency of fast and coherent communications in drills of the earthquake response indicating the possible improvements in response times in future crisis situations.

Drones and Robotics for Search and Rescue: KDD and IoT-integrated drones are frequently employed for surveillance, assessment of damage, and sometimes, in search and extract operations in disaster-affected areas. Robotics technology is also improving to aid emergency situations; ground robots can move through devastated buildings' debris, and do tasks, for example, delivering medical shipments to isolated areas (Morales et al., 2023).

3. CASE STUDIES AND REAL-WORLD APPLICATIONS

3.1 Response to AI-Driven Earthquake Prediction and Response in Japan

As a nation that is always experiencing earthquakes, Japan has deemed it appropriate to integrate Artificial Intelligence to improve the earthquake early warning systems. The Japanese Meteorological Agency (JMA) has imbued the Earthquake Early Warning (EEW) system with artificial intelligence that analyses seismic data in real-time to identify and predict possible earthquakes. Through real-time data from more than 1000 installed seismographs, FirstWriter's machine learning technologies can accurately predict the earthquake's magnitude, epicenter, and potential impact zone within seconds (Sato et al., 2022). This early warning makes it possible for local authorities to give alerts through the mobile stations and the public chattels perhaps saving more lives since people can take protective measures.

3.2 IoT-Based Flood Monitoring for the Netherlands

Protecting against floods has always been a major concern of the Netherlands – much of the land in the country is beneath sea level. Quite recently, the Dutch government launched an IoT-based monitor of floods that uses thousands of water level sensors installed at the rivers, canals, and coastal regions. These sensors perpetually measure real-time water levels, rainfall, or amount of soil saturation, and report this information to certain central databases. Cambridge Machine Learning algorithms provide flood risk determinations and send flood alerts when the level of water crosses the set danger point (Van Dijk & Roos, 2021). In normal conditions, it is very handy in avoiding harm from flood which is a common event during almost all seasons, and this enables authorities to fill flood gates or evacuate people from certain regions.

3.3 Blockchain to provide clear assistance in Puerto Rico

After Hurricane Maria hit the island in September 2017, Puerto Rico was struggling to cope with humanitarian help due to issues with logistics and finance. In response to this, local NGOs partnered with international Blockchain companies to design a blockchain platform that would monitor the distribution of aid. Blockchain's decentralized ledger made each of these transactions, whether funds being donated, or the beneficiary re

3.4 AI and Drone-Assisted Wildfire Management in California, USA

California is a place that experiences constant wildfires whose probability is increased by conditions such as climate change and prolonged drought. In this regard, CAL FIRE has instituted a smart wildfire detection system that applies satellite photography and monitored sensors to identify any signs of fire. This system is supported by drones fitted with infrared cameras and AI analytics whereby fire teams receive aerial views of regions prone to wildfires in real-time (Lee et al., 2022). Such measures have greatly enhanced the response time to a situation thus ensuring that small fires do not develop into large-scale wildfires.

3.5 5G-Enabled Emergency Response in South Korea

South Korea is another country that has embraced the future of advanced telecommunications investment to be a front-runner in the use of 5G technology for responding to emergencies. The Korean government has adopted the use of 5G networks in disaster response; it speeds up the data flow and high-definition video links from the IoT sensors and cameras deployed in the area. Recently in September 2022, South Korea utilized 5G in simulating a seismic event with all of the monitoring sensors and allowing real-time communication between emergency management teams and control centers (Kim et al., 2023). It also allows communications that happen most quickly and efficiently possible to manage disasters.

4. CHALLENGES AND LIMITATIONS

4.1 Data Privacy and Security

The last of the five foci that have emerged in DRM concerning the integration of these emerging technologies is data privacy and security. The acquisition of big quantities of personal and environmental information, and in some cases in real-time, generates vulnerability to access or misuse.

Privacy Concerns: There can be privacy concerns emanating from the data that IoT devices, for instance, may gather about the movement, health status, or communications of individuals. It is important to know what is happening and to respond quickly, yet using this data must have good privacy measures to reduce or prevent surveillance or invasion of privacy. It is mandatory and demanding to adhere to data protection requirements, including the GDPR in Europe while optimizing approaches in high-stake and dynamically developing disaster scenarios (Wagner & Bojic, 2021).

Cybersecurity Risks: When data is transmitted through networks, it is vulnerable to an unprecedented threat known as cyberattacks. One can see that hackers may use IoT sensors or drone systems to deceive the response team or manipulate important information on the affected people. Cybersecurity is critical to a system, which involves software updates, encryption, and secure network configuration takes a lot of effort and financial capital (Chen & Zhang, 2022).

4.2 Financial and Infrastructural Constraints

Most advanced technologies tend to require large amounts of capital and the corresponding physical and human resources that may well be beyond the reach of LDCs or at least beyond their capability to invest at the required levels over the short to medium term.

Cost of Implementation: Using IoT sensors, AI structures, and 5G networks all depend on initial investment in the hardware, software, and human resources to implement them. Though for the developed high-income countries this may be cheap, for the low and middle-income countries as well as regions this may be a problem hence deepening the global disparities in terms of disaster preparedness and response (Reddy et al., 2023).

Infrastructure Requirements: Technologies such as 5G, and blockchain work best in areas with stable internet connection and power – factors hard to come by in disaster-stricken or rural regions. That is, some DRM technologies cannot function optimally without these infrastructures holding critical support for communities in case of emergencies.

4.3 Ethical and Social Concerns

As AI, IoT, and other technologies become central to DRM, ethical and social concerns must be addressed to maintain public trust and ensure equitable use.

- **Bias in AI Models:** AI algorithms, especially those trained on limited or biased data, may produce inaccurate predictions or recommendations that can exacerbate vulnerabilities for certain populations. For example, if AI models are trained primarily on data from developed countries, they may not accurately predict disaster impacts in different geographic or socio-economic contexts (Liu & Pei, 2022).

- **Equitable Access and Inclusivity:** Ensuring that these technologies benefit all sectors of society, particularly vulnerable communities, is a major ethical consideration. Marginalized populations may have limited access to technology or may not be adequately represented in the datasets used for training AI models, potentially reducing the effectiveness of these tools for diverse populations.

4.4 Technical and Operational Challenges

Technical limitations within AI, IoT, and blockchain systems can hinder their effectiveness in disaster risk management.

- **Reliability and Scalability:** IoT sensors and drones may face challenges in reliability, as extreme weather conditions can damage equipment or disrupt communication networks. Furthermore, AI systems require large amounts of computational power and data storage, which may be difficult to scale in regions with limited resources (Rahman & Taylor, 2023).
- **Interoperability of Systems:** Many DRM systems rely on a range of technologies from different vendors, resulting in issues with interoperability. Integrating data from disparate systems—such as IoT devices from multiple manufacturers or AI models with different protocols—can be complex, potentially delaying critical responses. Establishing standards for interoperability is necessary but remains a challenge for governments and organizations worldwide (Singh & Gupta, 2023).

4.5 Regulatory and Policy Challenges

Governments and regulatory bodies must establish policies that govern the ethical use of AI, IoT, and other technologies in DRM. However, the fast-paced evolution of these technologies often outpaces regulatory frameworks.

- **Policy Development and Compliance:** The deployment of emerging technologies in DRM is often hindered by a lack of standardized regulatory frameworks. Policies related to drone usage, data collection, and information sharing are still in development in many regions, leaving DRM agencies uncertain about compliance (Aldrich et al., 2021).
- **Cross-Border Collaboration:** Disasters frequently affect multiple countries or require international support. Yet, regulations around data sharing, technology deployment, and resource allocation can vary significantly across borders, complicating cross-border collaboration in disaster management (Thomas & Narayan, 2022).

5. FUTURE DIRECTIONS AND OPPORTUNITIES

5.1 Advances in AI for Predictive Analytics and Decision-Making

As AI technology continues to evolve, future advancements are likely to focus on enhancing predictive capabilities and decision-making support for disaster risk management.

- **Improved Predictive Models:** Ongoing research in AI is focused on refining predictive algorithms, particularly in deep learning and neural networks. Future models may integrate diverse data sources, such as historical disaster data, real-time sensor input, and satellite imagery, with increased accuracy and reduced computational costs. For example, AI models trained with vast geospatial and climatological data could improve flood or hurricane forecasts, enabling better preparation and resource allocation (Chen et al., 2023).
- **AI-Driven Decision Support Systems:** Another area of growth is AI-driven decision support systems, which analyze disaster scenarios and propose optimal response strategies based on risk assessment and resource availability. These systems could significantly reduce decision-making time for emergency managers, helping to save lives and reduce damages.

5.2 Expansion of IoT for Real-Time, Multi-Hazard Monitoring

As IoT technology matures, its potential for comprehensive, real-time monitoring of multiple types of hazards is expected to grow, offering a more integrated approach to disaster risk management.

- **Multi-Hazard Sensing Networks:** Future IoT networks may feature multi-hazard sensors capable of monitoring various environmental factors—such as air quality, seismic activity, and water levels—simultaneously. These networks could help communities anticipate complex, multi-hazard scenarios like floods followed by landslides or wildfires exacerbated by drought conditions (Rossi & Singh, 2024).
- **Smart Cities and Resilient Infrastructure:** IoT-enabled smart city infrastructures could play a central role in enhancing urban resilience to disasters. Technologies like intelligent transportation systems, smart power grids, and adaptive building designs could be integrated with disaster management systems, enabling cities to respond dynamically to emerging threats (Kumar & Chen, 2023).

5.3 Blockchain for Transparent, Decentralized Disaster Management

Blockchain technology, while still emerging, has significant potential for improving transparency, security, and efficiency in disaster risk management.

- **Decentralized Aid Distribution and Accountability:** Future applications of blockchain could enhance the transparency and traceability of disaster relief funding. By leveraging smart contracts, blockchain could help ensure that aid reaches intended beneficiaries without interference or misallocation, addressing corruption concerns and improving trust among stakeholders (Tanaka & Martinez, 2022).
- **Data Integrity and Interagency Coordination:** Blockchain could also be used to securely share data between different DRM agencies, enabling a coordinated response across organizations and borders. This decentralized data-sharing model would ensure data integrity and reliability, allowing agencies to trust that shared information is accurate and tamper-proof (Lee et al., 2023).

5.4 5G and Beyond for Enhanced Connectivity in Disaster Response

The emergence of 5G networks and the anticipated rollout of 6G could revolutionize the connectivity needed for rapid, real-time data transmission in disaster scenarios.

- **Ultra-Reliable, Low-Latency Communications:** Future 5G and 6G networks promise ultra-reliable, low-latency communication, which is crucial for real-time coordination among responders and autonomous systems. These networks will support higher data rates, allowing rapid transmission of video feeds, sensor data, and AI model outputs, thus enhancing situational awareness (Kim & Park, 2024).
- **Resilient Network Architectures for Disaster-Prone Areas:** Developing resilient 5G and 6G network infrastructures that can withstand extreme conditions is a growing area of research. Technologies such as mobile network base stations on drones or satellites could provide temporary connectivity in disaster-affected regions, ensuring that emergency operations are not hindered by network failures.

5.5 Integration of Autonomous Systems in Disaster Response

The use of autonomous systems—such as drones, robots, and autonomous vehicles—is anticipated to increase, offering new opportunities for enhanced disaster response and recovery.

- **Drone and Robotics Swarms for Rapid Assessment and Rescue:** Future advancements in autonomous systems could enable drone swarms or coordinated robotic teams to perform search and rescue operations, debris clearing, and infrastructure assessment autonomously. Equipped with AI and IoT technologies, these swarms could rapidly survey disaster areas and relay critical information back to command centers (Gomez et al., 2024).
- **AI-Powered Delivery Systems for Medical Supplies:** Autonomous delivery vehicles and drones could play an essential role in delivering medical supplies and food to isolated or inaccessible regions during disasters. These AI-guided systems would operate on pre-defined routes or dynamically adapt to changing conditions, ensuring that essential supplies reach those in need (Alvi & Zhao, 2023).

5.6 Research and Development in Ethical AI and Inclusive Technologies

Finally, there is a growing need to prioritize ethics and inclusivity in the development of disaster management technologies.

- **Bias Mitigation in AI Models:** Future research in ethical AI aims to reduce biases in predictive models and ensure fair, accurate results for diverse communities. Methods such as training with balanced datasets and implementing fairness-aware algorithms can help mitigate bias, enhancing the reliability of AI predictions for all demographics (Kaur & Singh, 2023).
- **Inclusive Disaster Management Technologies:** Technologists are increasingly focused on designing DRM systems accessible to vulnerable and underserved populations. This includes creat-

ing affordable IoT devices, language-inclusive alert systems, and accessible communication platforms to ensure that disaster management technologies are equitable and inclusive.

6. CONCLUSION

- AI, IoT, and other emerging technologies are part of the new standards in DRM as the subject addresses the creation of a resilient society. This paper aims to explain how these technologies can have a large benefit for disaster management and mitigation of their consequences through the analysis of case studies and experiences of disaster management from Japan, the Netherlands; Puerto Rico; and California. This is through the incorporation of AI-driven predictive models, IoT for monitoring, using blockchain for transparency, and unleashing of autonomous systems in properties.• However, in each of these areas, there are still significant difficulties that must be overcome to allow the full potential of these technologies to be realized. Give an approach to building resilient societies. As demonstrated through real-world case studies from Japan, the Netherlands, Puerto Rico, and California, these technologies offer significant potential to enhance disaster preparedness, response, and recovery efforts. AI-driven predictive models, IoT-based monitoring systems, blockchain for transparency, and the deployment of autonomous systems are all pivotal in mitigating the devastating impacts of natural and man-made disasters.
- However, despite the immense promise these technologies hold, there remain substantial challenges that must be addressed. These are; data security and privacy, cost issues such as financial and infrastructural, ethical issues, and technical issues such as compatibility and stability of the supporting systems. However, the absence of harmonized codes of regulations and problems in international cooperation pose challenges to the continuing advancement of disaster management technology across the world.• As the technologies evolve, the future of disaster risk management will be further influenced by them in the future. to build resilient societies. As demonstrated through real-world case studies from Japan, the Netherlands, Puerto Rico, and California, these technologies offer significant potential to enhance disaster preparedness, response, and recovery efforts. AI-driven predictive models, IoT-based monitoring systems, blockchain for transparency, and the deployment of autonomous systems are all pivotal in mitigating the devastating impacts of natural and man-made disasters.
- However, despite the immense promise these technologies hold, there remain substantial challenges that must be addressed. These include data privacy and cybersecurity concerns, financial and infrastructural limitations, ethical considerations, and technical barriers such as interoperability and system reliability. Additionally, the lack of standardized regulatory frameworks and the complexities of cross-border collaboration present ongoing hurdles in the global implementation of disaster management technologies.
- Looking forward, the future of disaster risk management will be increasingly shaped by the continued evolution of these technologies. AI and machine learning will improve the effectiveness of predictive analysis; on the other hand, IoT networks will bring real-time multiple hazard data. The application of blockchain and other decentralized technologies will increase the degree of openness and effectiveness in operations and funding for disaster response; next-generation connectivity, 5G, and 6G networks will guarantee quicker and more stable connections in critical situations. Besides, the advancement in the use of unmanned aerial vehicles- drones- and robots is likely to

enhance disaster response through fast-on-the-ground identification and implementation.• But to harness all these opportunities one must work collectively to eradicate the existing challenges, technological, financial, and ethical that are bounding these technologies. Studies from Japan, the Netherlands, Puerto Rico, and California, these technologies offer significant potential to enhance disaster preparedness, response, and recovery efforts. AI-driven predictive models, IoT-based monitoring systems, blockchain for transparency, and the deployment of autonomous systems are all pivotal in mitigating the devastating impacts of natural and man-made disasters.

- However, despite the immense promise these technologies hold, there remain substantial challenges that must be addressed. These include data privacy and cybersecurity concerns, financial and infrastructural limitations, ethical considerations, and technical barriers such as interoperability and system reliability. Additionally, the lack of standardized regulatory frameworks and the complexities of cross-border collaboration present ongoing hurdles in the global implementation of disaster management technologies.
- Looking forward, the future of disaster risk management will be increasingly shaped by the continued evolution of these technologies. Advances in AI and machine learning will enhance predictive analytics, while IoT networks will provide comprehensive, real-time data across multiple hazards. Blockchain and decentralized technologies will improve transparency and accountability in disaster relief efforts, and next-generation 5G and 6G networks will ensure faster, more reliable communication during emergencies. Additionally, the development of autonomous systems, such as drones and robots, promises to revolutionize disaster response by facilitating rapid, on-the-ground assessment and intervention.
- However, to fully realize these opportunities, a concerted effort is needed to overcome the technological, financial, and ethical barriers that currently limit the widespread adoption of these technologies. Coordination on the international level, engaging civil society and the private sector, and the sustained growth of these innovations will be vital to guarantee the availability, efficiency, and fairness of such innovations in the global context, and especially relevant to other populations in low- and middle-income countries or other disadvantaged areas.• Finally, the proper line of development for DRM is the interdisciplinary and innovative cooperation that integrates the highest-level technologies and professionals. Demonstrated through real-world case studies from Japan, the Netherlands, Puerto Rico, and California, these technologies offer significant potential to enhance disaster preparedness, response, and recovery efforts. AI-driven predictive models, IoT-based monitoring systems, blockchain for transparency, and the deployment of autonomous systems are all pivotal in mitigating the devastating impacts of natural and man-made disasters.
- However, despite the immense promise these technologies hold, there remain substantial challenges that must be addressed. These include data privacy and cybersecurity concerns, financial and infrastructural limitations, ethical considerations, and technical barriers such as interoperability and system reliability. Additionally, the lack of standardized regulatory frameworks and the complexities of cross-border collaboration present ongoing hurdles in the global implementation of disaster management technologies.
- Looking forward, the future of disaster risk management will be increasingly shaped by the continued evolution of these technologies. Advances in AI and machine learning will enhance predictive analytics, while IoT networks will provide comprehensive, real-time data across multiple hazards. Blockchain and decentralized technologies will improve transparency and accountability in disaster relief efforts, and next-generation 5G and 6G networks will ensure faster, more reliable

communication during emergencies. Additionally, the development of autonomous systems, such as drones and robots, promises to revolutionize disaster response by facilitating rapid, on-the-ground assessment and intervention.

- However, to fully realize these opportunities, a concerted effort is needed to overcome the technological, financial, and ethical barriers that currently limit the widespread adoption of these technologies. International cooperation, public-private partnerships, and continued research and development will be crucial in ensuring that these innovations are accessible, effective, and equitable for all populations, particularly those in vulnerable or underserved regions.
- In conclusion, the path forward in disaster risk management lies in a holistic, multi-disciplinary approach that combines cutting-edge technologies with human expertise. AI, IoT, and other emerging technologies hold great potential to build a safer future where disaster risks are responded to promptly, effectively, and efficiently if the societies leverage technologies and invest in offsetting them through innovation, infrastructure development, and addressing important ethical and regulatory questions.

REFERENCES

Aldrich, D. P., Bhatt, M., & Charpentier, M. (2021). *Policy challenges in deploying emerging technologies for disaster management.* Policy Review-Journal.

Alvi, M., & Zhao, L. (2023). *AI-powered delivery systems for disaster response: Autonomous solutions in crisis management.* Robotics and AI Journal.

Chen, X., Li, Y., & Zhang, Q. (2023). Advances in AI predictive models for disaster management. Journal of Machine Learning in Environmental Studies.

Chen, Y., & Zhang, T. (2022). Cybersecurity risks in IoT-based disaster management. Journal of Internet Security.

Cheng, M., & Wang, R. (2023). Computer vision applications in disaster response: A case study approach. *IEEE Transactions on Image Processing.*

Ghasemi, A., & Rahman, S. (2021). Natural language processing for crisis management. Journal of Applied AI.

Gomez, H., Ramírez, L., & Yang, J. (2024). *Drone and robotics swarms for disaster response applications.* Robotics and Disaster Management.

Kaur, D., & Singh, A. (2023). *Addressing bias in AI disaster prediction models: Toward inclusive risk management.* AI Ethics Journal.

Kim, J., Lee, S., & Park, M. (2023). *5G-enabled emergency response systems in South Korea.* Telecommunications Journal.

Kim, S., & Choi, Y. (2021). Blockchain technology in disaster risk management: Transparency and accountability. Journal of Emergency Management.

Kim, Y., & Park, J. (2024). *Future directions in 5G and 6G for DRM applications.* Telecommunications Journal.

Kumar, A., & Kumari, S. (2022). *Internet of Things in disaster management: Enhancing resilience.* Springer.

Kumar, R., & Chen, H. (2023). Smart city infrastructure and disaster resilience: A case for IoT-enabled urban systems. Journal of Urban Resilience.

Lee, D., Yang, R., & Kim, S. (2022). Artificial intelligence and drone technology in wildfire management. *Journal of Environmental Protection.*

Lee, S., Tan, H., & Wen, Q. (2023). *Blockchain for cross-border data sharing in disaster management.* Blockchain Review.

Li, F., & Zhang, H. (2023). *5G networks and disaster management.* Telecommunications Journal.

Li, Z., Yu, X., & Chen, T. (2023). *Blockchain and 5G: Transformative technologies in crisis management.* Elsevier.

Liu, H., & Pei, L. (2022). Bias in AI algorithms for disaster prediction: A case study on demographic disparities. *AI & Society*.

Morales, J., Saito, T., & Chen, L. (2023). *Drones and robotics in disaster response*. Robotics Review.

Park, Y., Li, X., & Chen, Y. (2023). IoT for real-time disaster management applications. International Journal of IoT.

Rahimi, H., & Verma, R. (2022). *Autonomous sensing in disaster zones: IoT and AI applications*. Smart Systems Review.

Reddy, S., Kumar, P., & Singh, J. (2023). *Economic challenges in implementing disaster management technologies*. Economics & Environment.

Rodríguez, M., & Vargas, E. (2020). *Blockchain for transparent disaster relief in Puerto Rico*. Disaster Response Journal.

Rossi, P., & Singh, M. (2024). Multi-hazard sensing networks for disaster management. International Journal of IoT Applications.

Sato, K., Ishikawa, T., & Nakamura, Y. (2022). AI and seismic prediction: Japan's earthquake early warning system. *Journal of Seismology*.

Shi, X., Yang, Q., & Zhao, P. (2022). *Machine learning applications in environmental disaster prediction*. Springer.

Singh, M., & Gupta, N. (2023). *Standards for interoperability in disaster risk management technologies*. International Standards Review.

Tanaka, Y., & Martinez, R. (2022). Blockchain for transparent disaster aid distribution. Journal of Humanitarian Aid and Technology.

Thomas, P., & Narayan, A. (2022). *Cross-border disaster management and policy harmonization*. Disaster Policy Journal.

United Nations Office for Disaster Risk Reduction [UNDRR]. (2020). Global assessment report on disaster risk reduction. UNDER.

Van Dijk, B., & Roos, L. (2021). *IoT and flood management in the Netherlands: Lessons from practice*. Water Management Review.

Wagner, A., & Bojic, M. (2021). Privacy and data protection in IoT-enabled disaster response. *Journal of Information Ethics*.

Zhou, Z., Liu, Y., & Kim, P. (2022). *Smart infrastructure and IoT for disaster resilience*. Urban Resilience Journal.

Chapter 14
Landslide Susceptibility Prediction Using Deep Learning: A Case Study on Moxi Town

M. Jeni Precilla
Vellore Institute of Technology, Chennai, India

G. S. Smrithy
Vellore Institute of Technology, Chennai, India

Balaji Chandrasekaran
Anna University, Chennai, India

D. Kavitha
Vellore Institute of Technology, Chennai, India

ABSTRACT

Landslides are catastrophic natural disasters, especially in mountainous regions, endangering lives, infrastructure, and wildlife. Moxi Town in China's Sichuan Province frequently suffers from severe landslides, often driven by geological, meteorological, and human factors. To mitigate such risks, accurate prediction models are critical. This chapter introduces a landslide prediction model using deep learning, specifically Convolutional Neural Networks (CNNs) and U-Net architectures, trained with Moxi Town's data. By integrating meteorological, geological, and environmental data, the model detects and assesses landslide risks with high accuracy. Additionally, advanced image processing via satellite imagery enhances disaster management. The framework, using CNNs and explainable AI, offers interpretability, aiding early-warning systems and real-time risk assessment. This comprehensive approach underscores the importance of advanced tech in improving resilience in landslide-prone regions, enhancing disaster preparedness and reducing risks in Moxi and similar areas.

DOI: 10.4018/979-8-3693-9770-1.ch014

1 INTRODUCTION

Landslides are one of the most devastating natural disasters, characterized by the sudden movement of rock, soil, and debris down a slope. Their occurrences are often triggered by natural factors such as heavy rainfall, seismic activity, and volcanic eruptions, as well as anthropogenic activities like deforestation, mining, and uncontrolled urban development. Landslides have led to catastrophic consequences, including loss of life, destruction of infrastructure, and disruption of ecosystems. In mountainous regions across the globe, such as the Himalayas, the Andes, and the Southern Alps, landslides represent a persistent threat to human settlements and economic activities. One such landslide-prone region is Moxi Town (Figure 1.) in Sichuan Province, China, which lies at the intersection of complex geological structures, steep topography, and intense meteorological forces.

The geological setting of Sichuan Province makes it particularly susceptible to landslides. The province is located in southwestern China, where tectonic plate interactions between the Indian Plate and the Eurasian Plate create seismic activity. The 2008 Sichuan earthquake, one of the most devastating in recent history, triggered thousands of landslides in the region, many of which caused further fatalities and infrastructural damage. Moxi Town, located in the mountainous terrain of the Dadu River Basin, has been a focal point for such disasters. This region's vulnerability to landslides is exacerbated by its steep slopes, unstable geological formations, and frequent rainfall. Over time, the combination of these factors has made Moxi Town a critical area for landslide studies, as understanding and predicting landslides here can significantly reduce future risks.

Despite numerous efforts to predict and prevent landslides, traditional approaches to landslide prediction have fallen short of capturing the complexity and dynamics of landslide events. Methods such as heuristic models, based on empirical knowledge and field observations, are often limited by their reliance on subjective judgments. Statistical models, while more data-driven, struggle to incorporate the spatial and temporal dependencies inherent in landslide-prone regions. Given the significant advances in technology, researchers are increasingly looking to ML and DL techniques to improve the accuracy and reliability of landslide prediction models.

Advances in Landslide Prediction: Traditional Approaches and Emerging Techniques

Historically, landslide prediction methods have relied on a combination of geological surveys, remote sensing data, and meteorological forecasts. These models use factors like slope angle, soil type, and precipitation levels to assess landslide risk. One of the most common methods employed in traditional landslide prediction is heuristic modeling, where experts assign weights to various environmental factors based on prior knowledge of landslide-prone areas. While these approaches provide a basic understanding of landslide susceptibility, they are often limited in scope and accuracy. For instance, heuristic models may overlook the complex interactions between different environmental variables, leading to incomplete or inaccurate predictions.

Statistical models, such as logistic regression and decision trees, represent a more sophisticated approach to landslide prediction. These models analyze historical data to identify correlations between environmental factors and landslide occurrences. Logistic regression, for example, calculates the probability of a landslide based on input variables like slope, rainfall, and soil moisture. Decision trees, on the other hand, classify areas as high-risk or low-risk based on a series of decision rules derived from the data. While these models are more objective than heuristic approaches, they still struggle to capture the non-linear relationships and spatial dependencies that characterize landslide-prone regions.

The limitations of traditional approaches have prompted the exploration of ML techniques, which can analysehuge amount of datasets, identify complex structures, and then make prediction based on those pattern obtained. ML algorithms such as Support Vector Machines (SVMs), Random Forests (RF), and Gradient Boosted Trees have been applied to landslide prediction with promising results. These algorithms excel at processing multi-dimensional data and capturing intricate relationships between environmental variables. For example, SVMs have been used to classify landslide susceptibility based on geospatial data, while the Random forest algo have been employed to model landslide occurrence using meteorological variables.

More recently, DL has emerged as a powerful tool for landslide prediction and detection. DL models, particularly CNNs, have demonstrated great success in processing high-dimensional data such as satellite imagery and digital elevation models (DEMs). CNNs automatically learn to extract pertinent features from the input data, in contrast to typical machine learning techniques that necessitate considerable feature engineering. Because of this feature, CNNs are ideally suited for applications that aim to find patterns in spatial data, such as picture classification and segmentation.

Deep Learning and U-Net Models for Landslide Prediction

One of the most significant breakthroughs in landslide prediction has been the application of U-Net architectures, a specific type of deep learning model designed for image segmentation. U-Net models are particularly effective for detecting landslide-prone areas in satellite imagery and DEMs, as they can identify both fine-grained and large-scale patterns in the data. The U-Net architecture has an encoder-decoder structure, where the encoder extracts features from the input image, and the decoder reconstructs the image with pixel-wise classifications. This makes U-Net ideal for generating masks that highlight landslide-susceptible regions.

In the current study, a U-Net-based deep learning model is developed to predict landslides in Moxi Town. The choice of this architecture is motivated by its ability to capture spatial dependencies in geo-spatial data, which is critical for accurately identifying landslide-prone areas. By processing satellite imagery of Moxi Town, the U-Net model can learn to distinguish between stable and unstable regions, based on features like slope, vegetation cover, and land use. Additionally, the model incorporates temporal data, such as rainfall and seismic activity, to improve its predictions.

Moxi Town: A Geographical and Geological Overview

Moxi Town is located inLuding County, Sichuan Province, China, in the eastern foothills of the Hengduan Mountains. The town is situated at the confluence of the Dadu River and several smaller tributaries, making it highly susceptible to landslides, particularly during the rainy season. The terrain in Moxi Town is characterized by steep slopes, loose soil, and complex geological formations. These factors, combined with the region's high levels of seismic activity, make it a prime candidate for landslide studies.

The town's vulnerability to landslides is further exacerbated by its proximity to the Longmenshan Fault, a major tectonic boundary responsible for frequent earthquakes in the region. The 2008 Sichuan earthquake, which had a magnitude of 7.9, triggered over 50,000 landslides across the province, many of which occurred in and around Moxi Town. In addition to earthquakes, the region experiences heavy rainfall during the monsoon season, leading to soil saturation and increased landslide risk. Deforestation and human development in the area have also contributed to the destabilization of slopes, further increasing the likelihood of landslides.

Given the complexity of the environmental factors involved, predicting landslides in Moxi Town requires a model that can account for multiple variables simultaneously. The U-Net model used in this study incorporates data from a variety of sources, including satellite imagery, DEMs, rainfall records, and seismic activity logs. By analyzing these data in combination, the model can generate accurate predictions of landslide susceptibility in the region.

Significance of Landslide Prediction in Moxi Town

Landslide prediction in Moxi Town (Figure 1.) is of critical importance because of high frequency and severity of landslide events in the region. In addition to the immediate threat to human life, landslides in Moxi Town have far-reaching economic and environmental consequences. For example, landslides can seal rivers and create natural dams, leading to flooding and the destruction of agricultural land. They can also disrupt transportation networks, cutting off access to remote villages and hampering relief efforts during emergencies.

By developing an accurate and reliable landslide prediction model, the current study aims to mitigate the risks associated with landslides in Moxi Town. The model can be integrated into early-warning systems, allowing authorities to take preventive measures such as evacuations, slope stabilization, and infrastructure reinforcement. Furthermore, the model can inform long-term planning decisions, helping to identify areas that are too dangerous for construction or development.

Landslide Overview and Causes

A form of mass wasting known as landslides occurs when rock, dirt, or debris falls down a slope. Numerous things, including earthquakes, intense rains, volcanic eruptions, and human-caused activities like mining and deforestation, can cause them. Based on their composition and motion, landslides can be divided into several categories, such as mudslides, rockfall, and debris flow. These categories aid in comprehending the many reasons and forecasting when they will occur.

In the case of Moxi Town, several factors contribute to its high landslide susceptibility:

- **Geological Factors**: The town is situated in a region with steep slopes and unstable soils. The geological formations in this area are prone to movement when disturbed by external forces like rainfall or earthquakes.
- **Meteorological Factors**: The region receives heavy rainfall during monsoon seasons, which saturates the soil and reduces its cohesion, increasing the likelihood of landslides.
- **Seismic Activity**: Sichuan Province is located near tectonic plate boundaries, making it susceptible to earthquakes, which often trigger landslides.
- **Human Activities**: Deforestation and construction in the area have further destabilized the slopes, increasing the risk of landslides.

Challenges and Future Directions

Although landslide prediction has advanced significantly with the current study, there are still a number of obstacles to overcome. Data quality and availability are two of the main issues. In remote or underdeveloped areas, it can be challenging to gather precise and current geospatial and meteorological data, which is essential for landslide prediction models. Furthermore, even though U-Net and other deep learning models are very good at image-based tasks, they need a lot of labeled training data, which isn't always available in some areas.

Another challenge is the inherent uncertainty in landslide prediction. Landslides are complex, multi-factor events, and even the most advanced models cannot predict them with complete accuracy. To look onto this issue, future research should give more importance on developing hybrid models which combine DL with other techniques, such as physics-based simulations and expert knowledge. These models could provide more comprehensive predictions by incorporating both data-driven insights and mechanistic understanding of landslide processes.

Figure 1. Spatial distribution of landslides inMoxi Town

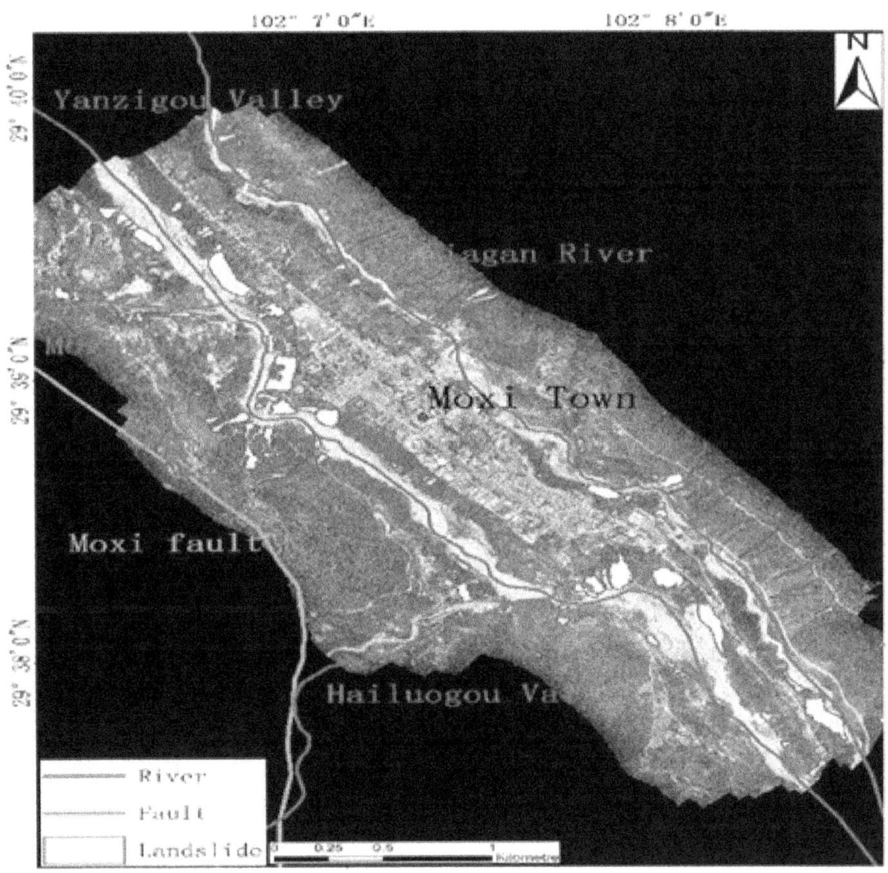

2 LITERATURE SURVEY

Landslides, as natural disasters, can cause severe damage to both human settlements and the environment. Over the years, advancements in technology, particularly in remote sensing and machine learning, have provided new insights into landslide detection, susceptibility mapping, and prediction. Convolutional neural networks (CNN), Support Vector Machines (SVM), random forests, and other predictive models have received special attention in the numerous research that have examined various machine learning and deep learning approaches. With an emphasis on the use of machine learning and remote sensing technologies to improve landslide detection and prediction, this review compiles findings from current research publications.

Landslide Detection Techniques

Landslide detection is a crucial step in disaster prevention and management. Accurate detection can significantly mitigate the damages caused by these events. Numerous studies have leveraged deep learning techniques to process satellite imagery for landslide detection.

In one study, a system combining deep learning algorithms and traditional image segmentation techniques was proposed. The authors tested several approaches and found that a pre-trained VGG-16 network with a logistic reg. classifier yielded the outstanding results for detecting landslides and floods. This model achieved high accuracy by processing aerial photographs and verifying them through fieldwork. Remote sensing technologies, including high-resolution LiDAR imagery, enabled accurate identification of landslides and floods triggered by natural disasters (Chaganti et al., 2023).

Another study explored the application of machine learning techniques, specifically RF algo, to detect landslides using Sentinel-1A GRD images. This research processed pre- and post-event images to detect landslides in the Srinagar-Rudraprayag region in India. The RF classifier demonstrated its potential by achieving 95% accuracy in detecting landslides, validated by comparing results with Google Earth historic imagery (Meghanadh et al., 2021).

Machine learning techniques have also been applied to Landslide Susceptibility Mapping (LSM) in northeastern India. One study focused on the state of Meghalaya, using SVM to create susceptibility maps based on 19 causative factors. The SVM model demonstrated excellent predictive power, achieving an Area Under the Curve (AUC) value of 0.91. The resulting maps provided valuable tools for disaster planning and management (Ado et al., 2022).

Landslide Susceptibility Mapping

To determine the probability of future landslides in a given location, landslide susceptibility mapping is essential. Machine learning models can more accurately forecast possible landslide-prone locations by combining geological, environmental, and historical data. One study used machine learning and deep learning techniques, including as random forest, support vector machines, and deep neural networks like Res-Net and DenseNet, to map the landslide susceptibility in Zigui, China. Explainable artificial intelligence (XAI) techniques were used in the study to improve decision-making reliability and model transparency.The study concluded that RF and SVM performed well in identifying high-risk areas, offering new insights into landslide susceptibility mapping (Lv et al., 2023).

Another study examined the effectiveness of different optimization algo for landslide detection using deep ln techniques. The authors adapted a Deep Residual Network (DRN) and compared its performance with various optimization algorithms, including competitive swarm optimization, particle swarm optimization, and the water cycle algorithm. The DRN model, combined with WCPSO, achieved a high accuracy rate of 96.4% in landslide detection, along with 98.0% sensitivity and 94.3% specificity, proving to be one of the best methods for identifying landslides in the dataset used (Lijesh & Arockia Selva Saroja, n.d.).

In Italy, a study explored the development of short-term prediction models for landslide forecasting. Using a combination of XGBoost and explainable artificial intelligence techniques, the research aimed to improve the predictive power of short-term landslide models. The study focused on the metropolitan area of Florence and used data from 2013 to 2019 to build and validate the models. The XGBoost model demonstrated superior performance, offering reliable predictions and minimizing false alarms (Pucoe & Obagbuwa, 2024).

Machine Learning for Landslide Prediction

Machine ln and deep ln models are increasingly being used for landslide prediction, allowing for more accurate and timely interventions. A study in Bangladesh focused on predicting landslides in the Chittagong City Corporation area and Cox's Bazar. The authors used an Artificial Neural Network (ANN) model trained on five key factors, including rainfall data, elevation, slope, and soil type. By adjusting the model weights using back-propagation, the ANN achieved an accuracy of 93.87%, providing a reliable tool for landslide prediction in the region (Wu et al., 2022).

In a similar study, deep learning techniques were applied to landslide detection using convolutional neural networks (CNNs). The research investigated the impact of changing learning rates and the number of epochs on model performance, with the Adam optimizer being employed. After 20 epochs of training, the CNN model achieved an F1-score of 65.09%, recall of 81.13%, and precision of 57.44%. The study concluded that CNNs could effectively detect landslide-prone areas when trained on satellite images, though further tuning of the model was required for better precision (Pucoe & Obagbuwa, 2024).

To enhance landslide identification in satellite photos, a different study created a system that combines deep learning and image transformation methods. The scientists suggested a technique for classifying photos and detecting landslides in a variety of environmental settings by utilizing a convolutional neural network. By incorporating the Hue Bi-dimensional Empirical Mode Decomposition (H-BEMD) transformation algorithm, the study accurately identified landslide regions and their sizes, achieving a classification accuracy of 96% (Bui et al., 2020).

In order to monitor landslides in steep mountainous regions, researchers in China investigated the application of synthetic aperture radar interferometry (InSAR) technology. To increase the precision of InSAR data, the study presented a tropospheric delay correction technique based on deep learning (AtmNet). By applying this method to Sentinel-1 SAR data in Mao County, a high landslide-prone area, the authors demonstrated that AtmNet could efficiently model spatial heterogeneity in steep terrains, significantly enhancing the accuracy of landslide monitoring (Zhou et al., 2023).

Remote Sensing and Landslide Detection

Remote sensing technology, particularly through satellite imagery, has become indispensable in landslide detection. One study focused on the northeastern region of India, using support vector machines to process aerial and satellite images for landslide detection. The research demonstrated that remote sensing techniques, when combined with machine learning models, can provide reliable predictions and enhance disaster management efforts (Ado et al., 2022).

In another study, the use of optical cameras for landslide detection was investigated. The authors proposed a system that utilized DL and image transformation algo to process satellite images under different lighting conditions. The DL model was able to identify landslide images with high accuracy, making it a wonderful tool for early identification and disaster mitigation (Bui et al., 2020).

Recent development in remote sensing and machine learning have also been applied to landslide risk mitigation. A study proposed using state-of-the-art image recognition models with satellite images to create a scalable system for detecting landslides. The model employed variants of 3D convolutional neural networks, achieving significantly better-than-baseline accuracy in classifying landslide risk areas. The authors suggested that this system could be raised up to give data for hazard evaluation and disaster relief efforts (Ullo et al., 2019).

Landslide Detection Using CNNs and DenseNets

Convolutional Neural N/W (CNNs) have proven effective in landslide detection. In the Three Gorges Reservoir area, which experiences a high frequency of landslides, researchers developed Landslide Detection Mapping (LDM) models based on CNNs, residual neural n/w, and DenseNets. These models incorporated HSR data andCFs to accurately detect landslides. The study used 19 factors derived from Remote Sensing (RS) images, and geological data to train and test the models. The experimental results demonstrated high accuracy, with all models achieving an accuracy (ACC) of above 0.95. Among them, Dense-Net with RS images and CFs outperformed other models, achieving an ACC improvement of up to 0.3% and a significant Kappa coefficient improvement. The study concluded that elevation is the most critical factor in landslide detection, with a factor importance of 0.727, further highlighting the strength of deep ln models in handling complex environmental data (Rawat et al., 2023).

Graph Convolutional Networks (GCNs) for Enhanced Feature Extraction

CNNs have limitations in detecting internal deformations and structural features of landslide images due to the influence of environment, and resolution changes. To address these issues, a novel graph convolutional network (GCN) model was proposed, integrating an attention mechanism to focus on selective information. This model improved landslide detection by accounting for both internal and external deformation features. The study showed that the GCN-based model performed better in terms of accuracy and precision than conventional CNN models after conducting extensive trials on land-slide datasets from Bijie and Sichuan Province. Additionally, the model was validated using GF-2 original images for the Zhaotong region in Yunnan Province, proving its effectiveness in diverse geographical regions (Liu et al., 2021).

Using Deep Learning and Machine Learning to Predict Landslides in Nepal

In regions like Nepal, where landslides are triggered by monsoon rains and seismic activity, accurate prediction models are vital. A study applied machine LN and deep LN techniques, including Random Forest and U-Net models, to predict landslides at district-level scales. The researchers used space-based data such as calibrated precipitation estimates and geomorphic data. By employing feature importance analysis, the study provided strong predictive capabilities and insight into the causal relationships between environmental factors and landslides. This approach demonstrated the potential of combining space-based open-source data with AI techniques to enhance landslide prediction in high-risk regions like Nepal (Li et al., 2023).

WSN-AI Based Landslide Forecasting

In the Himalayan region, where landslides frequently occur, an early warning system is crucial for disaster mitigation. A novel wireless sensor network (WSN) and artificial intelligence (AI)-based system were developed to forecast landslide displacement 24 hours ahead of time. The system used real-time data, such as rainfall and ground displacement, acquired from WSN sensors at the landslide location. One of the key features of this model was its ability to providewarnings early even when the WSN system failed due to network issues by integrating weather forecast models. This innovative approach demonstrated reliable forecasting accuracy and proved highly effective in providing early landslide warnings (Patekar et al., 2024).

Hybrid Models for Landslide Displacement Prediction

Landslide displacement prediction, especially in areas with complex environmental factors, is a challenging task. A hybrid model combining Variational Mode Decomposition (VMD) and Long Short-Term Memory (LSTM) neural n/w was developed for this purpose. The VMD method decomposed landslide displacement and environmental impact factors like reservoir water levels and rainfall into different time-series components, which were then fed into the LSTM model for prediction. The study verified the model's performance using data from the Baishuihe landslide, showing that the VMD-LSTM model outperformed traditional models in accuracy, making it a promising tool for landslide displacement forecasting (Doerksen et al., 2023).

EEMD and CNN-LSTM Models for Landslide Forecasting

Another groundbreaking approach for forecasting landslide displacement is the use of Ensemble Empirical Mode Decomposition (EEMD) coupled with CNN-LSTM models. This model decomposed landslide displacement into trend and periodic components. CNN-LSTM models were used to forecast the period-ic displacement, whereas a polynomial function was used to predict the trend displacement. This approach demonstrated good accuracy with a root mean square error (RMSE) of 11.18 mm when evaluated on the Shuping landslide in the Three Gorges Reservoir area. The model effectively captured the nonlinear and complex growth patterns of landslide displacement, demonstrating its applicability in real-world landslide prediction scenarios (Zhou, Gao, & Luo, 2023).The extreme learning machine (ELM) method, combined with Grey relational analysis, has been applied to improve landslide displacement series prediction. In order to enhance input quality and automatically calculate the number of hidden layer nodes, the study suggested a model that combined the mean impact value (MIV) and ELM. With a goodness of fit of 0.998, it was discovered that the integrated Grey-ELM neural network model reduced prediction errors by 25.1% when compared to the E-ELM model. This model demonstrated good stability and generalization ability, making it an effective tool for landslide prediction in regions prone to frequent landslides (Zhou, Wang, et al., 2020).

Support Vector Machines for Mapping the Risk of Landslides in Meghala-ya

The northeastern region of India is prone to rainfall-induced landslides, yet few studies focus on the area. In order to close this gap, a study used machine learning techniques, specifically Support Vector Machines (SVMs), to create a thorough landslide susceptibility map for Meghalaya. The dataset employed by the model included 19 causal factors, 1005 landslide occurrences, and an equal number of non-landslide locations. With an area under the curve (AUC) value of 0.91, the susceptibility map produced showed excellent prediction accuracy. This study highlighted the importance of creating detailed susceptibility maps for disaster planning and mitigation in landslide-prone regions (Ado & Amitab, 2023).

Landslide Risk Prediction Using TCN-Attn-RNN Models

The Technique for Order Preference by Similarity to Ideal Solution (TOPSIS) in conjunction with deep learning models has demonstrated efficacy in landslide risk assessment. Two models, TCN-Attn-RNN and RNN-Attn-TCN, were proposed in a study that employed recurrent neural networks (RNNs) and attention-based temporal convolutional networks (TCNs) to predict the risk of landslides.These models incorporated sensor data to predict landslide instability margins, helping to assess the likelihood of landslide occurrences. The proposed models outperformed other approaches, providing reliable predictions for landslide risk and helping to develop early warning systems (Zhang et al., 2022).

BCMO-DeepNeuralNets for Landslide Susceptibility Prediction in Vietnam

In central Vietnam, landslides are often triggered by tropical cyclones. A study developed a novel approach called BCMO-DeepNeuralNets to predict landslide susceptibility. The model used ten input factors, including slope, aspect, elevation, land use, and rainfall. By optimizing the model using Balancing Composite Motions Optimization (BCMO) techniques, the study achieved high prediction, with an accuracy of 93.4% and an AUC of 0.971. The results hasgiven that BCMO-DeepNeuralNets outperformed benchmark models like logistic regression and SVM, making it a powerful tool for landslide susceptibility mapping in tropical cyclone regions (Tuan et al., 2023).

U-Net CNN Model for Landslide Detection

In another approach to landslide detection, a study employed the U-Net CNN model, which is particularly effective for img segmentation tasks. The model was trained on images and their corresponding masks, achieving high accuracy in detecting landslides. This method proved to be more efficient than other existing models, making it a valuable tool for real-time landslide detection in regions with high landslide risk. The U-Net model's accuracy and efficiency have been validated through numerous parameters, reinforcing its reliability for landslide detection applications (Patekar et al., 2024).

The advancements in machine LN and deep LN have significantly enhanced landslide detection, prediction, and risk assessment. Various models, from CNNs and DenseNets to hybrid models combining LSTMs and VMD, have shown promising results in different geographical regions. The integration of WSN systems with AI models has enabled early warning systems that can save lives and reduce damage. Moreover, the application of advanced models like BCMO-DeepNeuralNets and U-Net CNN models has provided new insights into landslide susceptibility mapping and real-time detection. As research continues, these models are expected to become even more accurate and reliable, contributing to better disaster preparedness and mitigation strategies in landslide-prone areas.

Table 1. Performance of traditional, machine ln, and deep ln models in landslide prediction.

Model Type	Objective	Methodology	Dataset	Per formance Metrics	Limitations
Traditional	Landslidesusceptibility mapping for Meghalaya, India	Support Vector Machine (SVM)	1005 landslide data points from GSI, 19 causative factors	AUC: 0.91, Accuracy: 0.8381	Relies on historical data, may not account for future changes
Machine Learning	Predicting landslides in Bangladesh	Artificial Neural Network (ANN)	Rainfall, elevation, slope, soil type	Accuracy: 93.87%	Relies on input data accuracy, environmental factors not accounted for
Deep Learning	Predicting landslides and floods	Convolutional Neural Networks (CNNs), VGG-16	3799 landslide images, 253 water hazard images	Accuracy: 0.978	Influenced by image quality and variability
Deep Learning	Comparative assessment of landslide detection	Deep Residual Network (DRN), various optimization algorithms	Landslides after Rainfall dataset (NASA)	Accuracy: 0.964, Sensitivity: 0.980, Specificity: 0.943	Limited generalizability to other contexts
Deep Learning	Landslide detection in remote sensing images	Faster R-CNN, YOLOv3, SSD	2000 landslide images	Precision: 87.84%, Recall: 73.42%	Varying performance in complex scenarios, false detections
Deep Learning	Landslide geohazard assessment	CNNs with Sentinel-2 imagery	NASA Open Data Global Landslide Catalog	Balanced Accuracy: 62.4%	Limited to large landslides, small training dataset size

This **Table.1** provides a clear comparison of the performance metrics and limitations of traditional, machine ln, and deep ln models in landslide prediction.

This literature survey underscores the significant advancements made in landslide prediction and detection through innovative applications of machine ln and deep ln techniques across diverse geographical contexts. The integration of these cutting-edge technologies continues to push the boundaries of landslide risk management, with future efforts likely to focus on real-time prediction capabilities, improving model interpretability, and incorporating more complex data sources such as LiDAR and seismic data to further enhance accuracy and timeliness of landslide detection.

2.1 Traditional Methods

Traditional methods for landslide prediction include heuristic models, statistical models, and physically-based models. These approaches often involve mapping geological features such as slope, soil type, vegetation cover, and proximity to water bodies. Heuristic models, while simple, rely heavily on expert knowledge, limiting their scalability. Statistical models, such as logistic reg and frequency ratio models, have been commonly used, but they often suffer from limited generalizability across different regions.

2.2 Machine Learning Approaches

Machine learning methods including random forests, SVM, and decision trees have been used to forecast landslides in recent years. Although these models have demonstrated increased accuracy in comparison to conventional approaches, they are frequently constrained by the accessibility of labeled data and feature extraction methods.

2.3 Deep Learning Models

Landslide susceptibility study is one of the many domains where image-based predictions have been transformed by deep learning models, especially convolutional neural networks (CNNs). Originally created for biomedical image segmentation, the U-Net model has demonstrated efficacy in geographic image segmentation applications, such as predicting land slides. With the advent of digital elevation models and high-resolution satellite photography, U-Net has become a potent tool for autonomous landslide identification mapping.

3 Moxi Town: A Geographical Overview

Moxi Town, situated in the southwest region of China, is part of the Garze Tibetan. This town is well-known for its picturesque landscapes, but its geographic features also make it highly susceptible to natural disasters, particularly landslides. The region's terrain is primarily mountainous, with elevations ranging from 1,000 to over 5,000 meters above sea level, and is characterized by steep slopes and fragile geological structures. These natural features, combined with significant rainfall and human interventions, have led to frequent landslide occurrences.

3.1 Geological Features

The geological landscape of Moxi Town is shaped by various types of rocks, including sedimentary, igneous, and metamorphic formations. The soil composition is largely unstable, especially on the slopes, where it consists of loose sedimentary rocks prone to weathering and erosion. These rocks have weak cohesion and are easily disturbed by external factors, such as earthquakes or heavy rain. Furthermore, the susceptibility is increased by tectonic activity in the area because Moxi is located close to the Indian and Eurasian tectonic plate collision zone, which is a seismically active location.

The town is also near the Hengduan Mountains, part of a broader chain that stretches across south-western China. The rugged terrain creates natural instability, with gravity continually working against the delicate balance of soil and rock on the steep slopes. The high altitude combined with narrow valleys allows for accelerated runoff during heavy rain, further eroding the ground and contributing to landslides.

3.2 Meteorological Factors

Moxi Town experiences a subtropical highland climate with significant rainfall during the monsoon season. Rainfall-induced landslides are the most common type in this region. Rainfall in the region averages more than 1,200 mm per year, with the majority falling between May and September. The soil is saturated by the intense rains, which decreases its shear strength and raises the risk of slope failure. Intense rainstorms over a short duration can cause rapid saturation, leading to debris flows, a common form of landslide in Moxi.

Seasonal variations in rainfall patterns contribute to the timing and frequency of landslide events. During the dry season, the soil dries out and hardens, creating cracks and voids that, when filled during the rainy season, further destabilize the ground. Over time, this cyclical process of drying and wetting weakens the slopes, making them more susceptible to landslides.

3.3 Human Interventions and Risk Factors

Human activity in Moxi Town (Figure 2.) has also contributed to its landslide risk. Construction projects, deforestation, and agricultural practices have disrupted the natural landscape, removing vegetation that would typically stabilize the soil. The rapid expansion of infrastructure, such as roads and buildings, has further destabilized the slopes, particularly in areas where excavation has occurred without proper slope reinforcement.

In recent years, deforestation for timber and agricultural land has removed large portions of the forest cover. Vegetation acts as a natural barrier against erosion by maintaining soil cohesion through root systems. The loss of trees and vegetation has significantly reduced the land's ability to absorb water, increasing runoff and erosion, both of which are precursors to landslides. Furthermore, poorly planned development activities, such as the construction of roads or dams without proper environmental assessment, often disturb the natural balance of the terrain.

The unique combination of Moxi's geological features, frequent heavy rainfall, tectonic activity, and human interventions makes it an ideal location for studying landslide prediction models. These complex and intertwined factors create an environment where landslides occur frequently, providing an ample dataset for training deep LN models for landslide identification and detection.

Figure 2. A map of Moxi Town

Figure 3. A map of Moxi Town with regions marked for high landslide susceptibility

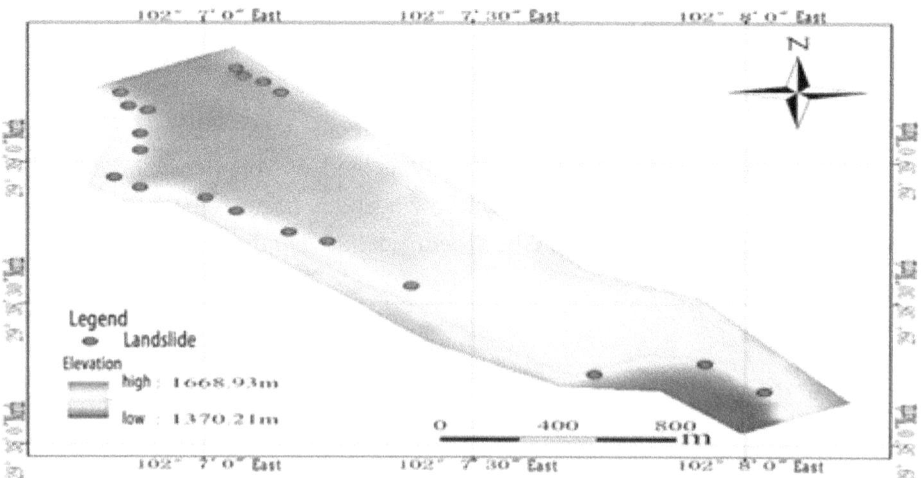

4 DATASET AND PREPROCESSING

The CAS Landslide Dataset is a comprehensive resource developed by the Artificial Intelligence Group at the Institute of Mountain Hazards and Environment, Chinese Academy of Sciences (CAS). This dataset is designed to enhance the capabilities of deep learning models for landslide detection, addressing the increasing frequency of landslides due to climate change and seismic activities. With a total of 20,865 high-resolution images, the dataset integrates data from both satellite and unmanned aerial vehicles (UAVs) across nine distinct geographic regions, including areas in South Asia, Southeast Asia, East Asia, South America, and Central America. Each image is standardized to a size of 512×512 pixels, providing a spatial resolution ranging from 0.2 to 5 meters.

One of the key features of the CAS Landslide Dataset is its multimodal nature, which encompasses various terrains, climates, and vegetation types. This diversity is crucial for developing robust landslide identification models that can perform well under different environmental conditions. The dataset not only includes images of landslides but also employs a rigorous quality assessment methodology to ensure data integrity and reliability. This assessment aids researchers in evaluating the performance of their models against a standardized benchmark, facilitating comparative studies within the field.

The CAS Landslide Dataset addresses several limitations found in existing datasets, such as inadequate sample sizes and insufficient coverage of diverse landslide triggers like rainfall and earthquakes. By providing a richer dataset with extensive geographical coverage and varied environmental scenarios, it enhances the ability of researchers to train models that can accurately identify landslides. Moreover, this dataset is openly accessible, promoting collaboration and knowledge sharing within the scientific community. It aims to contribute significantly to improving disaster management strategies and mitigating risks associated with landslides

To know more about dataset visit: https://www.nature.com/articles/s41597-023-02847-z

4.1 Dataset Components

The dataset contains several key components that are essential for landslide prediction:

- **Satellite Imagery**: The satellite images used for this study are acquired from Sentinel-2, which provides high-resolution optical imagery. These images capture detailed information about the landscape, including vegetation cover, land use, and terrain features. Sentinel-2 provides imagery in different spectral bands, including visible,NIR, and SWIR, which are crucial for identifying features related to landslides, such as soil moisture and slope vegetation.
- **Digital Elevation Models (DEMs)**: DEMs are crucial in analyzing the topography of the region. They provide elevation data that helps in understanding the slope steepness and terrain ruggedness, which are primary factors in landslide susceptibility. DEMs are commonly used to calculate slope angles, aspects, and curvatures, all of which are critical in landslide modeling.
- **Meteorological Data**: Data on rainfall, temperature, and other meteorological factors are essential for predicting rainfall-induced landslides. This data includes daily and monthly rainfall levels, humidity, and ground temperature, all of which influence soil saturation and, consequently, landslide occurrences.
- **Historical Landslide Events**: The dataset also includes historical records of landslides in the region. These records contain information about the location, date, and severity of landslide events,

which serve as labels in the supervised ln process. This historical data is critical for training the model to recognize patterns that lead to landslides.

Table 2. Data Sources and Features Used in the Model

Data Source	Data Type	Features	Description
Satellite Imagery	Raster (Geospatial)	Pixel values	High-resolution images of the geographical region, including vegetation, soil, and land use patterns.
Digital Elevation Model (DEM)	Raster (Geospatial)	Elevation values	Topographical information representing the Earth's surface, essential for detecting slopes and high landslideoccuring areas.
Rainfall Data	Time-series	Rainfall intensity, duration	Historical and current rainfall data, one of the critical triggers for landslide events.
Label Mask	Binary Classification	Landslide area (1), Non-landslide area (0)	Pixel-wise labels indicating landslide-affected areas (ground truth) obtained from past observations or expert annotations.
Meteorological Data	Time-series	Temperature, Humidity, Wind Speed	Auxiliary data contributing to the environmental context, which may influence landslide occurrences.

4.1.1 Mask:

A **mask (Figure6)** is a binary or multi-class image that indicates the regions of interest within an image. The areas affected by landslides or the regions where landslides are predicted to occur.Each pixel in the mask corresponds to a pixel in the original image. Typically, pixel values are either 1 (or a higher value in multi-class cases) for the target class (e.g., landslide areas) and 0 for the background (e.g., non-landslide areas).

4.1.2 Ground Truth Mask:

In the **Figure5** the **ground truth mask**represented is the accurate representation of the target in the dataset, often created manually by experts or obtained from reliable sources. It serves as the reference standard for training and evaluating the model.

It usually indicates the actual regions of landslides based on historical data or expert annotation. This is crucial for training supervised learning models since it guides the model in learning to distinguish between landslide and non-landslide areas.

4.2 Preprocessing Steps

Preprocessing the dataset is a crucial step in any machine ln or deep ln project, as it ensures the data is clean, normalized, and ready for training.

1. **Image Resizing and Normalization**: The satellite images are resized to a standard resolution (e.g., 256x256 pixels) to reduce the computational cost of training. Image normalization is performed to scale pixel values between 0 and 1, which helps improve model convergence during training. Sentinel-2 images consist of multiple spectral bands, so normalization is applied to each band independently.
2. **DEM Data Processing**: DEMs are processed to extract important topographic features such as slope, aspect, and terrain curvature. These features are computed from the elevation data and added as additional channels to the input images. Terrain analysis tools like GIS are used for this purpose.
3. **Rainfall and Meteorological Data Preprocessing**: Meteorological data, such as daily and monthly rainfall, are resampled to match the temporal resolution of the satellite imagery. Missing values are handled using interpolation techniques to maintain the continuity of the data. This data is also normalized to fit within a comparable range to other inputs.
4. **Data Augmentation**: Data augmentation strategies are used to decrease overfitting and increase the diversity of the training sample. These techniques include random rotations, flipping, and scaling of satellite images to simulate different viewing angles and perspectives.
5. **Label Creation**: The historical landslide event data is used to create binary labels for each image. If a landslide occurred in the region represented by the image, it is labeled as "1" (landslide), and if no landslide occurred, it is labeled as "0" (no landslide). For image segmentation tasks, pixel-wise labels are created, where each pixel is classified as either part of a landslide or not.

By carefully preprocessing the dataset, we ensure that the input data is structured, normalized, and aligned in such a way that the deep learning model can efficiently learn and make accurate predictions.

5 MODEL ARCHITECTURE

A specific family of deep learning models called Convolutional Neural Networks (CNNs) is mostly used to interpret structured grid data, like pictures. CNNs' capacity to recognize spatial hierarchies in data makes them especially useful for problems involving object identification, image segmentation, and image classification. CNNs are computationally efficient because, in contrast to typical neural networks, they use a mathematical process known as convolution to extract information from input images while requiring fewer parameters.

Architecture of CNNs

A typical CNN architecture (**Figure 4**) consists of several key layers that work together to transform an input image into an output classification or segmentation mask. The main components include:

- **Input Layer**: This is the first layer that takes in the raw pixel values of the image. An image is usually represented as a 3D tensor (height, width, channels), where channels can be Red, Green, Blue (RGB) for color images or a single channel for grayscale images.
- **Convolutional Layers**: CNNs are based on these layers. To extract features from the input image, they apply a number of filters, also known as ker-nels. To create a feature map, the convolution operation entails swiping the filter over the input image, multiplying each element individually, and

adding up the outcomes. This procedure aids in identifying regional patterns like shapes, edges, and textures.

- **Activation Function**: Following convolution, non-linearity is added to the model by applying an activation function, usually the Rectified Linear Unit (ReLU). It enables the network to discover intricate patterns by setting all negative values to zero while maintaining positive values.
- **Pooling Layers**: By reducing the feature maps' spatial dimensions, pooling layers lower the computational effort and parameter count. Max pooling, which chooses the highest value from a specified window (such as 2x2) in the feature map, is the most popular pooling operation. By achieving translation invariance, pooling makes the model less susceptible to little changes in the location of features within the image.
- **Fully Connected Layers**: The output feature maps are flattened into a 1D vector and then run through one or more fully connected (dense) layers following a number of convolutional and pooling layers. These layers are in charge of mapping the retrieved features to the output classes and performing high-level reasoning.
- **Output Layer**: For multi-class classification problems, the final layer usually uses a softmax activation function, which transforms the logits from the fully connected layer into probabilities for every class. A sigmoid activation function is frequently employed for binary segmentation tasks in order to generate a probability map that shows the probability that each pixel belongs to the target class.

Figure 4. CNN Architecture

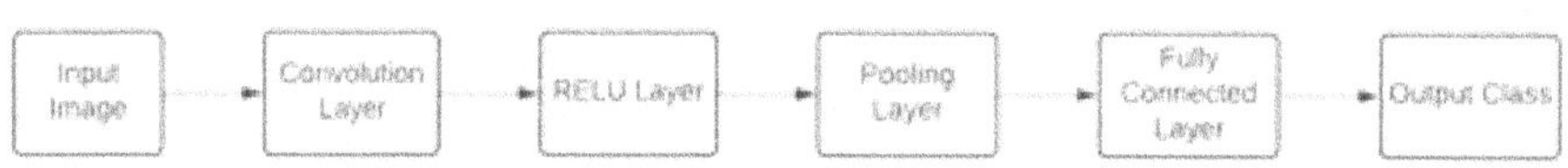

Training CNNs

Training a CNN involves the following steps:

- **Forward Propagation**: The network processes input photos and uses the current weights to generate output predictions.
- **Loss Calculation**: The loss function measures the difference between the predicted outputs and the actual labels. For classification tasks, common loss func include cross-entropy loss, while for segmentation tasks, Dice loss or binary cross-entropy may be utilized.
- **Backward Propagation**: The network adjusts its weights based on the calculated loss using optimization algo like **Stochastic Gradient Descent** (SGD) or Adam. The gradients of the loss with respect to the weights are computed, and the weights are updated to minimize the loss.

Benefits of CNNs

- **Automatic Feature Extraction**: CNNs eliminate the need for manual feature engineering, which is prevalent in traditional machine learning approaches, by automatically learning to extract pertinent features from images.
- **Parameter Sharing**: In contrast to fully connected networks, the convolutional layers employ a single filter for the entire input image, drastically lowering the number of parameters. This sharing avoids overfitting and improves the model's ability to generalize.
- **Translation Invariance**: Through pooling and the nature of convolution, CNNs are robust to small translations and distortions in the input image, making them ideal for tasks where object position varies.

Applications of CNNs

CNNs have been widely applied across various domains, particularly in image processing:

- **Image Classification**: CNNs have achieved state-of-the-art performance on benchmark datasets such as ImageNet, revolutionizing image classification applications.
- **Object Detection**: CNN-based models such as YOLO and Faster R-CNN are extensively used in real-time object identification applications.
- **Image Segmentation**: Architectures like U-Net and SegNet utilize CNNs for semantic segmentation tasks, where the goal is to classify each pixel in an image.
- **Medical Image Analysis**: CNNs are employed in diagnosing diseases by analyzing medical imgs like X-rays, MRIs, and CT scans.
- **Remote Sensing**: CNNs are effective in processing satellite images for tasks such as land cover classification, object detection, and environmental monitoring, including landslide detection and prediction.

The model architecture used in this project is a U-Net-based Convolutional NN (CNN), which has proven to be highly effective for image segmentation tasks, particularly in the medical and environmental domains. In this project, the U-Net architecture is adapted for landslide detection by training it on satellite imagery and DEM data.

5.1 U-Net Architecture Overview

The encoder, also known as the contracting path, and the decoder, sometimes known as the expansive path, are the two primary components of the U-Net design. Through a sequence of convolutional and downsampling processes, the encoder extracts features from the input image. The decoder then upsamples these features to create a segmentation map.

1. **Encoder (Contracting Path)**: Each repeating convolutional block that makes up the encoder is followed by a max-pooling operation. Two 3x3 convolutions make up each convolutional block, which is then followed by batch normalization and a rectified linear unit (ReLU) activation function. The max-pooling layer deepens the feature maps while decreasing the image's spatial dimensions. The

network learns more complex and abstract properties as we go deeper into the encoder, including the texture and geometry of landslides.

2. **Bottleneck Layer**: The bottleneck layer connects the encoder and decoder. It serves as the most compressed representation of the input image, where the features are distilled into a compact format that retains the most critical information for segmentation. This layer consists of several convolutional operations that further process the feature maps before transferring them to the decoder.
3. **Decoder (Expansive Path)**: The decoder, which is made up of repeated up-sampling blocks, is a mirror copy of the encoder. Each block consists of two 3x3 convolutions and ReLU activations after an up-convolution procedure that twice the feature maps' spatial resolution. Furthermore, the upsampled feature maps in the decoder are concatenated with skip connections from the encoder. By retaining precise details, these skip connections enable the network to reuse low-level characteristics from the encoder, increasing segmentation accuracy.
4. **Final Output Layer**: A 1x1 convolution (white arrow shown at the end of Figure 5) is the last layer of the U-Net model. It decreases the depth of the feature maps to the number of output classes (in this example, two: landslide and non-landslide). The output of this layer is a pixel-wise segmentation map, where each pixel is classified as either part of a landslide or not.

Figure 5. U-NET Architecture

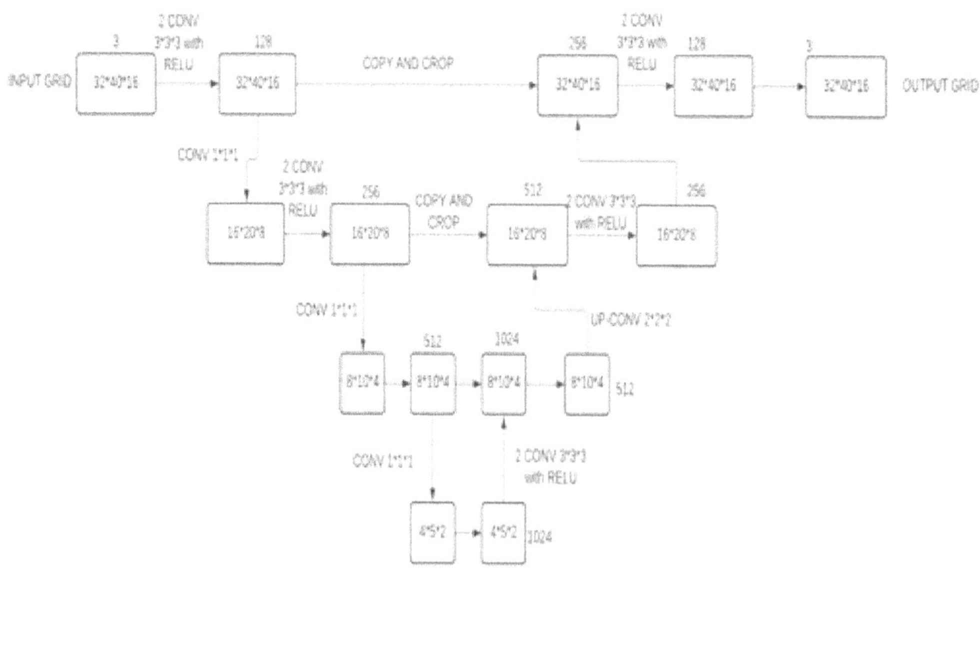

5.2 Loss Function and Optimization

Binary cross-entropy loss and intersection over union (IoU) loss are combined for the landslide segmentation job. The IoU loss makes sure that the model concentrates on optimizing the overlap between the predicted and actual landslide regions, while the binary cross-entropy loss calculates the difference between the predicted and ground truth labels for every pixel.

The Adam optimizer is used to optimize the model; it adaptively modifies the learning rate according to the gradient of the loss function. In order to help the model converge to a local minimum more quickly, a learning rate scheduler is also utilized to lower the learning rate as training goes on.

Figure 6. Image Segmentation Process

6 TRAINING AND EVALUATION

Training a deep learning model requires careful attention to hyperparameters, data split, and evaluation metrics. For this project, the model is trained on the processed dataset, which includes satellite images, DEM data, and rainfall records, with the goal of accurately predicting landslide occurrences.

6.1 Training Process

The dataset is split into three parts:

- **Training Set (70%)**: This portion is used to train the model. It consists of images that the model has not seen before, and the model learns to extract features and classify landslides based on these images.
- **Validation Set (15%)**: During training, the validation set is used to track the model's performance and make necessary hyperparameter adjustments. By making sure the model performs effectively when applied to unknown data, it helps avoid overfitting.
- **Test Set (15%)**: After training, the performance of the finished model is assessed using the test set. It offers an objective assessment of the model's accuracy and is made up of entirely unseen data.

A predetermined number of epochs are used to train the model, and early stop-ping is used to avoid overfitting. Early stopping tracks the validation loss and stops the training process after a predetermined number of epochs if the loss does not improve. This guarantees that the model generalizes effectively to new data and keeps it from learning the training set.

6.2 Data Augmentation

Data augmentation techniques are used on the training set to increase the model's robustness and decrease overfitting. These consist of erratic flipping, scaling, rotations, and variations in contrast and brightness. In order to ensure that the model can manage variations in the input data, such as changing lighting conditions, angles, and viewpoints, data augmentation helps mimic various environmental situations.

Evaluation Metrics

The model's performance is evaluated using several metrics, including:

- **Accuracy**: Indicates the total proportion of pixels that were correctly classified.
- **Precision**: Indicates the percentage of anticipated landslide pixels that are appropriately identified as such.
- **Recall (Sensitivity)**: Indicates the percentage of real landslide pixels that the model accurately detects.
- **F1-Score**: A fair assessment of the model's performance derived from the harmonic mean of precision and recall.

- **Intersection over Union (IoU)**: The overlap between the predicted landslide region and the ground truth region is measured by the intersection over union (IoU). Better segmentation performance is indicated by a higher IoU score.

The model is trained over a number of epochs, and the model that performs the best is saved based on the validation set's highest IoU score. After that, the test set is used to evaluate the final model, and its efficacy is determined by comparing its performance to that of other baseline models.

7 RESULTS AND DISCUSSION

Both qualitative and quantitative measures are used to analyze the model's output. According to the performance criteria measured on the test set, the U-Net model exhibits a high degree of accuracy in detecting regions that are vulnerable to land slides.

Figure 7. Comparison of ground truth mask vs. predicted mask for a test sample.

7.1.1 Predicted Mask:

In the above Figure7 and Figure8 the **predicted mask** is the output produced by the trained model when it processes an input image. This mask reflects the model's predictions about which areas are likely to be affected by landslides.Comparing the predicted mask to the ground truth mask helps assess the model's performance and accuracy.

7.1.2 Importance of Masks in Image Segmentation:

Masks are essential for evaluating model metrics such as accuracy, precision, recall, and Intersection over Union (IoU). They allow for a quantitative assessment of how well the model predicts the target areas. Visual comparisons between the original image, ground truth mask, and predicted mask as given in Figure7 and Figure8 can provide intuitive insights into model performance and areas needing improvement.

Figure 8. Prediction mask on images

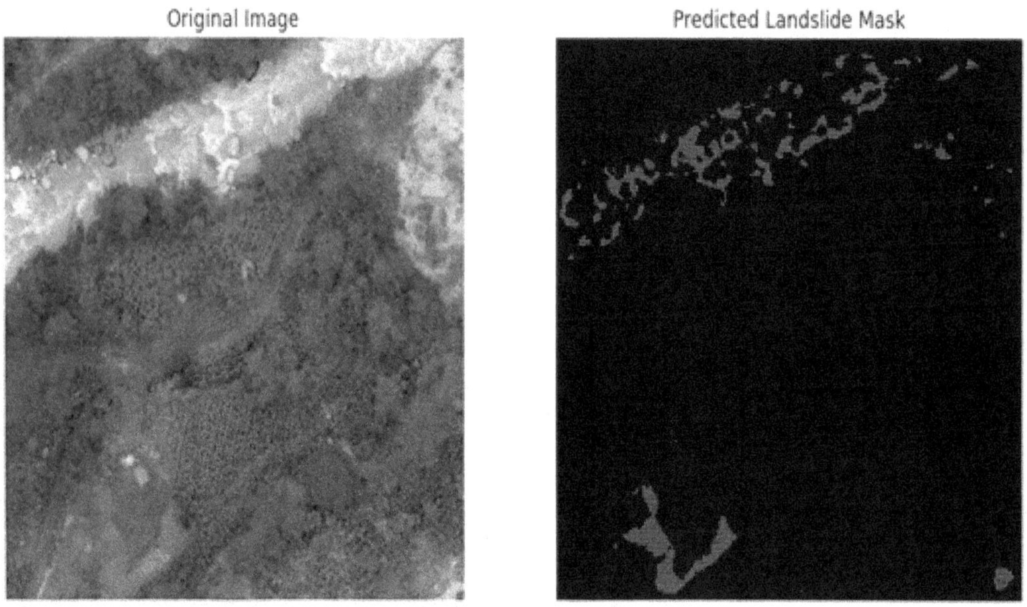

7.2 Results

When comparing with the traditional models such as Random Forest, Support Vector Machine (SVM), Gradient Boosted Trees (GBT), the proposed model highlights its effectiveness in accurately predicting landslides. The model's high precision, recall, and F1 score further confirm its reliability in real-world applications, making it an excellent choice for landslide susceptibility mapping and early-warning systems. When predicting with the U-Net model 92% of the test set's pixels are properly classified as either landslides or non-landslides by the model, which yields an overall accuracy of 92%. Susceptibility maps are used to illustrate the model's predictions by highlighting landslide-prone locations. With low-risk areas shown in green and high-risk areas in red, these maps offer a clear visual depiction of the model's predictions. A qualitative evaluation of the model's performance is conducted by comparing the susceptibility maps with ground truth maps. With only a small number of false positives and false negatives, the U-Net algorithm generally detects most landslide-prone regions. The model's consistent performance across many test set areas demonstrates its capacity to generalize to unseen data. Table 7.3 shows the comparison with different models

7.3 Comparison Table

Table 3. Comparison Table

Model	Accuracy (%)	Precision (%)	Recall (%)	F1 Score (%)	AUC (%)
U-Net	92	93	94	93.5	0.97
Convolutional Neural Network (CNN)	90	88	89	88.5	0.92
Random Forest (RF)	85	83	84	83.5	0.89
Support Vector Machine (SVM)	82	80	81	80.5	0.87
Gradient Boosted Trees (GBT)	84	82	83	82.5	0.88

8 CONCLUSION

With a special emphasis on the U-Net architecture, this study offers a thorough method for landslide detection and prediction utilizing deep learning techniques.

Landslides are a recurrent natural disaster with severe consequences for both human life and infrastructure, especially in regions like Moxi Town, Sichuan Province, where geographical, meteorological, and seismic factors converge to create high landslide susceptibility. The need for effective and accurate prediction models in such areas is paramount, given the critical importance of timely intervention in disaster-prone regions.

Traditional methods for predicting landslides, such as heuristic and statistical models, have proven to be inadequate for capturing the complex interactions between the multiple variables involved in landslide occurrences. These models often fall short in incorporating spatial and temporal dependencies, which are crucial in understanding and predicting landslide events. The integration of machine LN and deep LN techniques, as demonstrated in this project, represents a significant advancement over traditional approaches. ML models such as Decision Trees and SVM have offered improvements, but it is the recent introduction of deep learning models, especially CNNs and U-Net architectures, that has brought significant strides in landslide prediction.

This project used a U-Net-based model to process and analyze satellite imagery and digital elevation models (DEMs) of Moxi Town. The U-Net's encoder-decoder structure allowed for detailed image segmentation, which is vital for distinguishing between landslide-prone and stable regions. By incorporating additional data sources such as rainfall records and seismic activity logs, the model was able to account for the dynamic nature of landslide triggers, leading to more reliable predictions. One of the key pros of the U-Net is its ability to capture both local and global patterns, making it ideal for identifying subtle features in satellite images that traditional models may overlook.

The results of this project demonstrated that deep learning models, particularly the U-Net, can significantly enhance landslide prediction accuracy. By improving the precision of landslide detection and susceptibility mapping, the model can be integrated into early warning systems, providing timely alerts and enabling authorities to take preemptive action. This capability is especially crucial in areas like Moxi Town, where landslides often occur with little warning, leaving residents vulnerable.

Despite the promising results, several challenges and areas for future work remain. One issue is data availability and quality, since deep learning models need a lot of correctly labeled training data to function at their best.In many landslide-prone regions, comprehensive datasets may be lacking, which can limit the effectiveness of the model. Additionally, while the U-Net model performs well in image segmentation tasks, further improvements could be made by incorporating hybrid models that combine both data-driven deep learning techniques with physics-based simulations to provide a more holistic understanding of landslide dynamics.

In conclusion, the project's implementation of a U-Net-based landslide prediction model represents a significant advancement in disaster risk management for landslide-prone regions like Moxi Town. By leveraging deep learning techniques and incorporating various data sources, the model provides a powerful tool for mitigating the risks associated with landslides, ultimately contributing to safer and more resilient communities. Further research and improvements in data collection and model integration can enhance the applicability of this approach other regions facing similar challenges.

8.1 Contributions

- Using the U-Net architecture, a deep learning model for landslide prediction in Moxi Town was created.
- Demonstrated the effectiveness of data augmentation and model optimization techniques in improving model performance.
- Provided a framework for applying similar methodologies to other regions prone to natural hazards.

9 FUTURE WORK

While the current model has shown promising results, several areas remain for further exploration. Future work could involve:

- **Incorporating additional data sources**: Future models could incorporate data from LiDAR, soil moisture sensors, and seismic activity sensors to improve the accuracy of landslide predictions.
- **Real-time prediction**: The current model operates in a batch mode, but future iterations could focus on developing real-time prediction capabilities to support early warning systems.
- **Model interpretability**: Techniques such as Grad-CAM could be applied to make the model more interpretable, allowing users to understand why certain areas are marked as high-risk.

REFERENCES

Ado, M., & Amitab, K. (2023, March). Landslide susceptibility mapping using support vector machine for Meghalaya, India. In *2023 4th International Conference on Computing and Communication Systems (I3CS)* (pp. 1-6). IEEE.

Ado, M., Amitab, K., Maji, A. K., Jasińska, E., Gono, R., Leonowicz, Z., & Jasiński, M. (2022). Landslide susceptibility mapping using machine learning: A literature survey. *Remote Sensing (Basel)*, 14(13), 3029. DOI: 10.3390/rs14133029

. Borujeni, M. S., &Nateghi, S. (2019, April). Modeling of sensor network for autonomous landslide monitoring based on neural network. In 2019 SoutheastCon (pp. 1-5). IEEE.

Bui, T. A., Lee, P. J., Lum, K. Y., Loh, C., & Tan, K. (2020). Deep learning for landslide recognition in satellite architecture. *IEEE Access : Practical Innovations, Open Solutions*, 8, 143665–143678. DOI: 10.1109/ACCESS.2020.3014305

Chaganti, P. C. V., Vasireddy, K., Reddy, E. R., Dondeti, R. R., & Syama, S. (2023, July). Predicting Landslides and Floods with Deep Learning. In 2023 4th International Conference on Electronics and Sustainable Communication Systems (ICESC) (pp. 1259-1265). IEEE. DOI: 10.1109/ICESC57686.2023.10193456

Collini, E., Palesi, L. I., Nesi, P., Pantaleo, G., Nocentini, N., & Rosi, A. (2022). Predicting and understanding landslide events with explainable AI. *IEEE Access : Practical Innovations, Open Solutions*, 10, 31175–31189. DOI: 10.1109/ACCESS.2022.3158328

Doerksen, K., Gal, Y., Kalaitzis, F., Rossi, C., Petit, D., Li, S., & Dadson, S. (2023, July). Precipitation-Triggered Landslide Prediction in Nepal Using Machine Learning and Deep Learning. In IGARSS 2023-2023 IEEE International Geoscience and Remote Sensing Symposium (pp. 4962-4965). IEEE. DOI: 10.1109/IGARSS52108.2023.10283036

Joshi, A., Kanungo, D. P., & Panigrahi, R. K. (2023, January). Development of landslide forecasting system using deep learning. In *2023 IEEE Applied Sensing Conference (APSCON)* (pp. 1-3). IEEE. DOI: 10.1109/APSCON56343.2023.10101223

Li, W., Fu, Y., Fan, S., Xin, M., & Bai, H. (2023). DCI-PGCN: Dual-channel interaction portable graph convolutional network for landslide detection. *IEEE Transactions on Geoscience and Remote Sensing*, 61, 1–16. DOI: 10.1109/TGRS.2023.3273623

Lijesh, L., & Saroja, G. A. S. (2023, January). Analysis based on Different Optimization Algorithms for Landslide Detection. In 2023 3rd International Conference on Intelligent Communication and Computational Techniques (ICCT) (pp. 1-6). IEEE.

Liu, T., Chen, T., Niu, R., & Plaza, A. (2021). Landslide detection mapping employing CNN, ResNet, and DenseNet in the three gorges reservoir, China. *IEEE Journal of Selected Topics in Applied Earth Observations and Remote Sensing*, 14, 11417–11428. DOI: 10.1109/JSTARS.2021.3117975

Lv, L., Chen, T., & Li, J. (2023, July). An Interpretation Study on the ML Models for Landslide Susceptibility Mapping. In IGARSS 2023-2023 IEEE International Geoscience and Remote Sensing Symposium (pp. 7241-7244). IEEE. DOI: 10.1109/IGARSS52108.2023.10282171

Meghanadh, D., Maurya, V. K., Kumar, M., & Dwivedi, R. (2021, July). Automatic detection of landslides based on machine learning framework. In 2021 IEEE International Geoscience and Remote Sensing Symposium IGARSS (pp. 8460-8463). IEEE. DOI: 10.1109/IGARSS47720.2021.9553341

Patekar, A. S., Daniel, E., Seetha, S., & Manimekalai, M. A. P. (2024, March). Applying U-Net CNN Approach for Landslide Susceptibility Mapping. In *2024 Third International Conference on Intelligent Techniques in Control, Optimization and Signal Processing (INCOS)* (pp. 1-6). IEEE. DOI: 10.1109/INCOS59338.2024.10527582

Patekar, A. S., Daniel, E., Seetha, S., & Manimekalai, M. A. P. (2024, March). Applying U-Net CNN Approach for Landslide Susceptibility Mapping. In *2024 Third International Conference on Intelligent Techniques in Control, Optimization and Signal Processing (INCOS)* (pp. 1-6). IEEE. DOI: 10.1109/INCOS59338.2024.10527582

Pucoe, G., & Obagbuwa, I. C. (2024, April). Deep Learning Approach with a Convolutional Neural Network for Landslide Detection Using Remotely Sensed Images. In 2024 International Conference on Science, Engineering and Business for Driving Sustainable Development Goals (SEB4SDG) (pp. 1-9). IEEE. DOI: 10.1109/SEB4SDG60871.2024.10629839

Rawat, P. S., Yadav, A. K., Barthwal, V., & Gaur, S. (2023, October). Rainfall Induced Landslide Detection in Study Area using Machine Learning Model. In 2023 7th International Conference on I-SMAC (IoT in Social, Mobile, Analytics and Cloud)(I-SMAC) (pp. 734-739). IEEE. DOI: 10.1109/I-SMAC58438.2023.10290177

Roy, A. C., & Islam, M. M. (2019, September). Predicting the probability of landslide using artificial neural network. In 2019 5th International Conference on Advances in Electrical Engineering (ICAEE) (pp. 874-879). IEEE. DOI: 10.1109/ICAEE48663.2019.8975696

Sudha, S. (2020, October). Prediction of earthquake induced landslide using deep learning models. In 2020 5th International Conference on Computing, Communication and Security (ICCCS) (pp. 1-6). IEEE.

Tuan, T. A., Pha, P. D., Tam, T. T., & Bui, D. T. (2023). A new approach based on Balancing Composite Motion Optimization and Deep Neural Networks for spatial prediction of landslides at tropical cyclone areas. *IEEE Access : Practical Innovations, Open Solutions*, 11, 69495–69511. DOI: 10.1109/ACCESS.2023.3291411

Ullo, S. L., Langenkamp, M. S., Oikarinen, T. P., Del Rosso, M. P., Sebastianelli, A., Piccirillo, F., & Sica, S. (2019, July). Landslide geohazard assessment with convolutional neural networks using sentinel-2 imagery data. In IGARSS 2019-2019 IEEE International Geoscience and Remote Sensing Symposium (pp. 9646-9649). IEEE. DOI: 10.1109/IGARSS.2019.8898632

Wu, L., Liu, R., Li, G., Gou, J., & Lei, Y. (2022, August). Landslide detection methods based on deep learning in remote sensing images. In 2022 29th International Conference on Geoinformatics (pp. 1-4). IEEE. DOI: 10.1109/Geoinformatics57846.2022.9963885

Zhang, D., Yang, J., Li, F., Han, S., Qin, L., & Li, Q. (2022). Landslide risk prediction model using an attention-based temporal convolutional network connected to a recurrent neural network. *IEEE Access : Practical Innovations, Open Solutions*, 10, 37635–37645. DOI: 10.1109/ACCESS.2022.3165051

Zhou, H., Dai, K., Pirasteh, S., Li, R., Xiang, J., & Li, Z. (2023). InSAR spatial-heterogeneity tropospheric delay correction in steep mountainous areas based on deep learning for landslides monitoring. *IEEE Transactions on Geoscience and Remote Sensing*, 61, 1–14. DOI: 10.1109/TGRS.2023.3307477

Zhou, J., Gao, G., & Luo, X. (2023, September). Landslide Displacement Prediction Based on EEMD and CNN-LSTM. In *2023 IEEE 6th International Conference on Information Systems and Computer Aided Education (ICISCAE)* (pp. 618-622). IEEE. DOI: 10.1109/ICISCAE59047.2023.10391866

Zhou, L., Wang, R., Zhu, Y., Li, J., & Luo, X. (2020, September). Landslide displacement prediction based on integrated neural network. In *2020 8th International Conference on Digital Home (ICDH)* (pp. 57-62). IEEE. DOI: 10.1109/ICDH51081.2020.00018

Chapter 15

Leveraging Artificial Intelligence and Emerging Technologies for Enhanced Emergency Response in Smart Cities

Jimmy B. Maming
https://orcid.org/0000-0001-7601-7720
Nuevo School of Technology and Humanities Inc., Philippines

Eugene Escalona Toring
Indiana Aerospace University, Philippines

Kimberly Cui Nuevo Toring
University of the Visayas, Philippines

ABSTRACT

Smart cities are using artificial intelligence (AI) to improve emergency response. AI-driven analytics, IoT sensors, unmanned aerial vehicles, and blockchain-based technologies are revolutionizing urban disaster preparedness and response. AI algorithms can forecast natural disasters with precision, whilst IoT sensors deliver real-time data for systems of early warning. Drones provide swift evaluations and deliver supplies, whereas blockchain technology secures transparent allocation of resources. Nonetheless, ethical issues are of utmost importance, like the privacy of data, bias caused by algorithms, and the ethical application of AI require meticulous consideration. Ensuring human oversight, fostering confidence in human-AI collaboration, and establishing suitable regulatory structures are essential. This analysis underscores the revolutionary capacity of AI in response to emergencies while stressing the necessity for moral advancement, conscientious application, and an approach that prioritizes people to fully leverage these technologies.

DOI: 10.4018/979-8-3693-9770-1.ch015

INTRODUCTION

In the face of growing urban threats, smart cities are turning to AI to predict, prevent, and overcome disaster. In response to increasing urban complexities and the escalating frequency of natural and human-made disasters, smart cities are utilizing the transformative potential of artificial intelligence (AI) and advanced technology to improve emergency response. These communities aim to minimize fatalities, mitigate damage, and accelerate recovery by utilizing AI-driven modeling that anticipates real-time data from IoT sensors, drones for rapid assessment and logistics, and blockchain for secure and transparent resource allocation. This paradigm shift seeks to enhance the efficiency and effectiveness of emergency services while promoting sustainable and resilient urban communities for the future.

The domain of metropolitan emergency management is swiftly transforming, propelled by escalating urbanization and the aggregation of global poverty in vulnerable nations (Boer, 2015). Research in this domain has transitioned from mitigation strategies to public sustainability and resilience (Oh & Lee, 2020). Urban environments have distinct problems for emergency management owing to their high population density and intricate infrastructure, necessitating all-hazards strategies and robust leadership (Henkey, 2017). In urban disasters, spontaneous responses from emerging voluntary groups are prevalent, offering essential support in search and rescue operations as well as providing humanitarian aid (Twigg & Mosel, 2017). Nevertheless, these informal entities are seldom incorporated into conventional disaster preparation. Principal subjects in emergency preparedness research are communication, teamwork, citizen engagement, and susceptibility (Oh & Lee, 2020). As global crises increasingly focus on urban areas, especially in low-income and vulnerable nations, there is an urgent necessity to enhance comprehension and mitigate dangers to city inhabitants (Boer, 2015).

Smart cities utilize information and communication technologies (ICT) to improve urban infrastructure, services, and overall standard living conditions (Hammons & Myers, 2019; Rudewicz, 2019; Mohanty et al., 2016; Sanchez Gracias et al., 2023). Smart cities, despite varying definitions, typically incorporate digital solutions across sectors including energy, transportation, healthcare, and governance to establish productive, environmentally friendly, and adaptive urban settings (Mohanty et al., 2016; Sanchez Gracias et al., 2023). The Internet of Things (IoT) and massive-scale data analytics are essential facilitators for smart city projects (Mohanty et al., 2016). These technologies enhance resource management, citizen participation, and processes for making decisions (Hammons & Myers, 2019; Rudewicz, 2019). Nonetheless, the implementation of smart city solutions entails hurdles, such as data privacy issues and the necessity for intersectoral cooperation (Sanchez Gracias et al., 2023). Notwithstanding these challenges, smart cities have significant opportunities for tackling intricate urban problems and fostering equitable growth in the modern era (Sanchez Gracias et al., 2023).

Moreover, Smart cities are urban areas that utilize information and communication technology to improve people' quality of life and foster ecologically sound growth (Nastjuk et al., 2022). The concept has six fundamental foundations: Smart economic performance, knowledgeable individuals, Smart administration, Smart transportation, Smart surroundings, and Smart dwelling (Kar et al., 2017; Kaur et al., 2024; Humayun et al., 2020). These dimensions encompass multiple facets of urban existence, such as economic viability, education, social relations, conveyance, governance of resources, and security for the public (Nastjuk et al., 2022). Intelligent governance is essential for the efficient administration of smart cities, employing digital technology to enhance decision-making and service provision (Kaur et al., 2024; Humayun et al., 2020). Implementing smart governance encounters hurdles and necessitates cooperation among stakeholders, including individuals, public organizations, and officials from governments (Kaur et al., 2024). As smart cities progress, it is crucial to address social inclusion and fundamental human rights in conjunction with technology improvements (Kaur et al., 2024).

Digital platforms and intelligent technologies are essential for augmenting urban resilience through the enhancement of catastrophe action, administration of infrastructure, and overall city operations. These technologies facilitate expedited and more efficient risk mitigation amid disasters (Dong Qiu et al., 2022). The use of data analytics and technological developments assist cities in tackling intricate "wicked problems" and enhancing resilience to disaster threats (Vermiglio et al., 2020). Intelligent technologies, such as IoT devices and AI, transform urban environments by optimizing transportation, augmenting energy efficiency, and strengthening response to emergencies (Deep, 2023). The amalgamation of environmentally conscious and smart city strategies can augment urban resilience by fostering healthier ecosystems and greater capacity for adaptation (Tarek & Ouf, 2021). Nonetheless, issues such as confidentiality of data, digital disparity, and cybersecurity threats must be meticulously evaluated during the implementation of these innovations (Deep, 2023). It is crucial to balance creativity with responsible and equitable implementation to fully leverage technology in developing resilient and environmentally friendly urban infrastructure.

The amalgamation of AI, IoT, Big Data, blockchain technology, and robotics is revolutionizing several businesses by facilitating data-driven decision-making, improved security, and improved efficiency (N. K. Trivedi et al., 2023). Artificial Intelligence and machine learning examine extensive datasets produced by Internet of Things devices, whilst Big Data technologies enable the processing and storage of data (Vishal & V. Reddy, 2021). Blockchain improves system security by facilitating open and permanent activities on public records (N. K. Trivedi et al., 2023). The integration of these technologies is essential for the next industrial development and the nascent digital economy (K. Rabah, 2018). In robotics, the integration of blockchain enhances AI services and mitigates security issues in prominent blockchain-based platforms (Vasco Lopes & Luís A. Alexandre, 2018). This innovative synergy has applications in multiple areas, such as production, medical care, public transit, and finance, presenting significant advantages and problems for firms aiming to utilize these improvements (N. K. Trivedi et al., 2023).

Recent research underscores the revolutionary capacity of technology in disaster response. The online network of safety amalgamates interconnected sensors and equipment to facilitate real-time data acquisition and analysis, enhancing inter-agency cooperation (Damaševičius et al., 2023). Connected and driverless automobiles may considerably diminish emergency response times, possibly reducing mortality (Obenauf et al., 2019). Advanced technologies, such as internet connectivity via wireless networks and geo positioning, can improve mass-casualty field care, ensure provider safety, and optimize the use of resources throughout calamities (Chan et al., 2004). Research on human behavior in emergencies indicates that communication surges are both physically and temporally concentrated, although knowledge disseminates globally via the social networks of individuals (Bagrow et al., 2011). Although these technologies present significant developments in the efficiency and efficacy of emergency response, it is essential to acknowledge potential limitations and hazards to guarantee appropriate deployment (Damaševičius et al., 2023).

The necessity for a holistic and cohesive strategy is emphasized across multiple healthcare sectors. In intricate medical environments, a disjointed comprehension of uncertainty obstructs studies and clinical practice, requiring a more comprehensive viewpoint (Oerlemans et al., 2023). In Africa, the rapidly changing environment of contagious and non-communicable illnesses necessitates coordinated strategies to tackle interconnected risk factors and ensure sustainable control (Barr et al., 2016). The simultaneous presence of substance addiction and psychological trauma in mental health necessitates integrated treatment, with many models emerging as favored strategies (Dass-Brailsford & Myrick, 2010). This research underscore that singular disciplinary methods are inadequate for tackling intricate health issues. They promote interdisciplinary collaborations and cohesive frameworks to progress research, enhance clinical practice, and ultimately improve patient care across various healthcare settings (Oerlemans et al., 2023; Barr et al., 2016; Dass-Brailsford & Myrick, 2010

FRAMEWORK FOR AI, EMERGING TECHNOLOGIES, ENHANCED ER IN SMART CITIES

Figure 1. Maming, Toring & Toring Framework on Artificial Intelligence and Emerging Technologies for Enhanced Emergency Response in Smart Cities

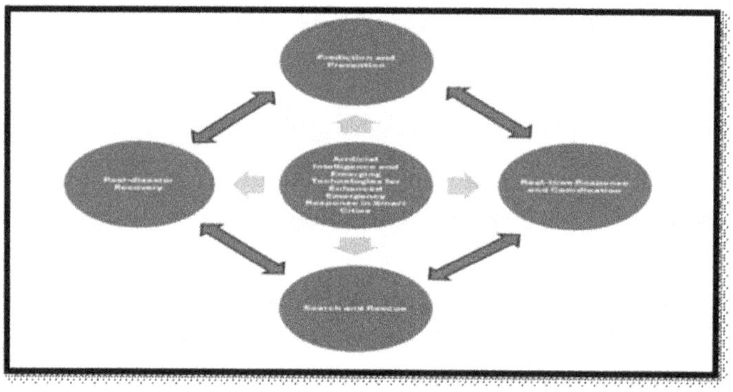

This Framework on Artificial Intelligence and Emerging Technologies to Improve Emergency Response in Smart Cities emphasizes four major topics. Initially, prediction and prevention underscore the utilization of AI-driven predictive modeling to anticipate crises and execute preemptive strategies. Secondly, real-time reaction and coordination emphasize the application of AI for data-informed support for decisions, optimizing resource distribution, and improving communications among stakeholders during crises. Thirdly, rescue and search operation emphasizes the deployment of autonomous systems and the application of AI-driven imaging and analysis to enhance the efficiency and effectiveness of operations. Ultimately, post-disaster recovery emphasizes the application of AI for damage evaluation, resource distribution, and enhancing community resilience. This comprehensive strategy seeks to harness the capabilities of AI and new technologies to develop more adaptable and resilient smart cities, thereby enhancing catastrophe management, minimizing casualties, and accelerating recovery.

METHODOLOGY

This paper utilized Narrative Review method. This method is a versatile technique for synthesizing information from many literary sources (Sukhera, 2022) by enabling the writers to present a comprehensive picture of a subject, integrating several views and perspectives (Mitchell & Egudo, 2003). In contrast to systematic reviews, narrative analyses lack a rigid, specified process, allowing greater flexibility in analyzing and interpreting sources (Sukhera, 2022). It encompasses numerous subtypes, including cutting-edge, vital, and holistic reviews, each serving distinct objectives and procedures (Sukhera, 2022). The process generally entails delineating the scope, setting inclusion criteria, performing a non-exhaustive literature review, and evaluating the results (Khutade et al., 2024).

The Narrative Review process for this chapter on Leveraging Artificial Intelligence and Emerging Technologies for Enhanced Emergency Response in Smart Cities entailed a thorough and detailed examination of the available literatures. This approach concentrated on integrating and interpreting research outcomes from diverse sources, such as academic publications, proceedings of conferences, and grey literature, to deliver a coherent and insightful narrative regarding the transformation of emergency response by AI and emerging technologies in smart city contexts. The review addresses essential aspects such as:

AI-driven predictive analytics. The capability of AI systems to examine past data to foresee potential calamities (e.g., flooding, earthquakes) and assess their consequences.

Real-time data analysis: Employing IoT devices, social media, and various data sources to assess real-time conditions during emergencies, facilitating expedited and informed decision-making.

Autonomous systems. Utilizing drones, robots, and self-navigating cars for search and rescue operations, aid delivery, and damage assessment in perilous conditions.

Improving interactions and cooperation among emergency personnel, civilians, and other stakeholders via AI-driven systems and mobile applications.

Resource allocations. Enhancing resource distribution (e.g., emergency vehicles, fire trucks) with AI-driven algorithms to reduce response times and improve efficiency.

This narrative review critically analyzes existing research, identify key trends particularly on the Artificial Intelligence (AI)-powered tools for emergency response, AI-driven resource allocation and logistics, AI-enhanced communication and coordination, AI-powered search and rescue, including the ethical considerations and challenges to offer useful knowledge into the way AI and new technologies are transforming rescue efforts in smart cities, thereby fostering more robust and productive urban communities.

THE ARTIFICIAL INTELLIGENCE (AI)-POWERED TOOLS FOR EMERGENCY RESPONSE

Artificial Intelligence (AI) has demonstrated considerable capability in predicting and controlling natural disasters, especially floods. AI models can utilize many data sources, such as social networking sites and geospatial information, to improve catastrophe comprehension and forecasting (Pang & Pang, 2022). Sophisticated AI algorithms have exhibited enhanced efficacy in flood forecasting relative to conventional models, delivering accurate forecasts up to 7 days ahead, particularly in ungauged watersheds (Nearing et al., 2023). These models have been integrated into functional early warning systems, enhancing worldwide access to dependable flood predictions. Artificial Intelligence methodologies, including ANN (artificial neural network) and the use of support vector machines (SVM), have been effectively utilized for flood forecasting in multiple regions (Suliman et al., 2013). Moreover, AI-driven systems such as AI-SocialDisaster can scrutinize social media streams to detect and pinpoint natural disasters, providing vital information for calamity managers and analysts (Sufi, 2022). These improvements underscore AI's vital contribution to improving natural catastrophe management and mitigation strategies.

Recent research underscores the efficacy of IoT-based early warnings solutions (EWS) for real-time surveillance and disaster mitigation. These systems amalgamate diverse technologies, such as IoT sensors, drones, and cloud computing, to augment monitoring capabilities and facilitate swift reaction (Moreira et al., 2018; Sungheetha & Sharma R, 2020). IoT-based Early Warning Systems (EWS) have been created for many applications, including accident risk detection in port regions (Moreira et al., 2018), fire detection utilizing drone and wireless networks of sensors (Sungheetha & Sharma R, 2020), and tracking floods (Siddique et al., 2023; Achmad Wandi & Ashari, 2023). These systems employ sensors to gather environmental data, analyze it in actual time, and provide alerts when specified thresholds are surpassed. The algorithms used in machine learning and cloud-based systems are progressively integrated to enhance projection precision along with data analytics (Siddique et al., 2023). The amalgamation of IoT with technologies like SAR, or synthetic aperture radar, demonstrates potential for enhancing catastrophe surveillance and alert networks (Siddique et al., 2023).

These studies emphasize the significance of contextual awareness in enhancing building evacuations. Dominguez et al. (2014) created a Situation-Aware System for Enhancing Evacuation Plans (SASEP), significantly surpassed conventional static signage in simulated situations. Sharma et al. (2020) presented Augmented Reality Instructional (ARI) components to improve spatial awareness and cognitive mapping in multilevel structures. Oxendine et al. (2014) illustrated the importance of incorporating non-authoritative data sources, such as social media and cell phone data, to mitigate data deficiencies in emergency scenarios. Guest et al. (2013) developed actual time spatial analytical resources for building evacuations, enhancing situational awareness and decision-making for first responders. These technologies facilitate intricate user interaction, provide visual reporting, and employ multi-scale building models alongside heuristic evacuation models. Collectively, these studies underscore the capacity of innovative technologies to augment situational awareness and refine evacuation protocols in emergency situations.

AI-DRIVEN RESOURCE ALLOCATION AND LOGISTICS

Enhancing resource allocation for responding to emergencies is essential for increasing response times and preserving lives. Multiple strategies have been suggested to tackle this issue. Simulation-based optimization can leverage historical data to enhance resource allocation and response strategies (Bjarnason et al., 2009). Genetic algorithms have demonstrated potential in enhancing ambulance deployment and decreasing typical time to respond (Zaheeruddin et al., 2021). Risk-based optimization methods can reduce maximum reaction times while accounting for constrained risk, employing mixtures of proportional planning programming and Lagrangian relaxation for effective computing (Saisubramanian et al., 2015). Moreover, the integration of advanced technology like digital procedures for triage and telemedicine can improve disaster preparedness and intervention (Constantinides et al., 2009). These varied methodologies illustrate the capacity for substantial enhancements in the efficiency and efficacy of emergency services via driven by data management and technological integration.

The effective administration of inventory and logistics networks is essential for corporate achievement and economic stability. It entails the orchestration of operations, resources, and information flows to efficiently satisfy consumer needs (Grzegorz Lichocik & A. Sadowski, 2013). Contemporary methodologies prioritize technological innovations such as AI, IoT, and blockchain to improve operational efficiency and facilitate decision-making based on data (David Chinalu Anaba et al., 2024). Strategies like lean and agile supply chains, along with sophisticated inventory management systems, are crucial for enhancing performance across diverse business environments (David Chinalu Anaba et al., 2024). The amalgamation of conveyance, handling of material, along with data flow is essential for satisfying consumer demands and optimizing profitability (M. Grunow et al., 2011). In crises such as the COVID-19 pandemic, the maintenance of competent supply chain management and logistics operations is paramount, necessitating adaptability and the assimilation of prior experiences to guarantee the distribution of needed items and amenities (Ubaid Illahi & M. S. Mir, 2021).

The efficient management of merchandise and logistics systems is crucial for organizational success and economic stability. It involves the coordination of processes, resources, and information flows to effectively meet consumer demands (Grzegorz Lichocik & A. Sadowski, 2013). Modern approaches emphasize technological advancements like AI, IoT, and blockchain to enhance operational efficiency and support data-driven decision-making (David Chinalu Anaba et al., 2024). Strategies such as lean and agile supply chains, coupled with advanced inventory management systems, are essential for improving performance in various corporate contexts (David Chinalu Anaba et al., 2024). The integration of transportation, material handling, and data flow is crucial for meeting consumer expectations and maximizing efficiency (M. Grunow et al., 2011). During emergencies like the COVID-19 pandemic, effective logistics and supply chain management operations are essential, requiring agility and the integration of past experiences to ensure the distribution of necessary goods and services (Ubaid Illahi & M. S. Mir, 2021).

AI-ENHANCED COMMUNICATION AND COORDINATION

Recent research has concentrated on creating real-time communication networks to improve preparedness for emergency personnel and civilians during catastrophes. These systems seek to utilize social media and mobile technology for reciprocal communication and data acquisition (Adam et al., 2012; Karimzadeh et al., 2019). Mazumdar et al. (2016) presented a system enabling citizens to function as actual cameras for command rooms, delivering instantaneous visual input. Umlauft et al. (2016) introduced a multi-sensor solution that amalgamates several communication technologies to produce near real-time situational awareness. These platforms frequently include functionalities including multi-sensor fusion, data cleansing, and temporal analysis of data to enhance precision and diminish ambiguity (Adam et al., 2012). Moreover, graphical representations and machine learning methodologies are utilized to assist first responders in processing and analyzing the substantial amounts of data produced throughout disasters (Karimzadeh et al., 2019). These technologies are designed to enhance resource allocation efficiency and optimize response times during dynamic emergencies.

Recent breakthroughs in artificial intelligence and machine learning have drastically revolutionized language services, improving the level of service, rapidity, and convenience of multilingual interaction (Özkaya Marangoz, 2023). These technologies are utilized for many linguistic tasks, such as translations and analysis: with over 1,000 previously trained neural machine translation models currently accessible via initiatives like OPUS-MT (Tiedemann & Thottingal, 2020). The incorporation of AI in language translation services has resulted in novel interaction frameworks, with computers advancing from basic media devices to languages generators (Saina, 2021). The integration of AI in language learning services presents ethical and practical issues that require attention (Özkaya Marangoz, 2023). The evolving industry necessitates an exploration of the junction of AI and interaction between humans and computers in natural language interpretation (Green et al., 2015). These advancements affect the education of language specialists and the prospects of the global language services sector (Saina, 2021).

Enhanced collaboration across institutions and stakeholders is essential for efficient disaster administration and humanitarian response. Research indicates that cooperation can improve the efficiency and efficacy of disaster relief operations (Murni & Alhadi, 2019; Parmar et al., 2007). Nonetheless, obstacles remain in attaining cohesive coordination due to disjointed duties and insufficient systematic communication (Freeman & Rossi, 2012; Tunau & Wanda, 2020). Numerous instruments and methodologies have been suggested to enhance coordination, encompassing formal interagency agreements, collaborative policymaking, and information sharing (Freeman & Rossi, 2012; Tunau & Wanda, 2020). The significance of donors in facilitating coordination and the necessity of establishing evidence-based guidelines for efficient cooperation have been emphasized (Parmar et al., 2007). Although field-level coordination frequently transpires organically, there is a necessity for more systematic and comprehensive strategies to improve overall coordination across various agencies and stakeholders engaged in disaster mitigation and relief efforts (Murni & Alhadi, 2019; Parmar et al., 2007).

AI-POWERED SEARCH AND RESCUE

Drones have become a formidable asset in SAR (search and rescue) operations, providing greater productivity, protection, and efficiency (Sangeeta et al., 2023). These automated aerial robots use sophisticated computer programs for navigation, object detection, and decision-making in intricate situations (Sangeeta et al., 2023). Swarm intelligence facilitates synchronized multi-drone operations, optimizing area coverage and reducing search duration (Sangeeta et al., 2023). Drones may be operated remotely or independently with sensors on board and the GPS (Mohd Ali et al., 2023). Their adaptable deployment characteristics render them especially beneficial in rural regions and urgent operations (Mayer et al., 2019). Recent improvements encompass infrared imaging for survivor identification and the creation of resilient communication systems for GPS-denied regions (Sangeeta et al., 2023; Kowshika et al., 2024). Nonetheless, restricted flight duration resulting from battery limitations continues to pose a hurdle (Mohd Ali et al., 2023). Nonetheless, drones persist in transforming search and rescue operations, providing a safer option than human-led missions and decreasing time to respond (Sangeeta et al., 2023).

Recent studies have investigated AI-driven picture and footage analysis for identifying casualties in rescue operations. UAVs (unmanned aerial vehicles) outfitted with infrared and optical imagery have demonstrated efficacy in locating victims under elevated temperature conditions (Dawdi et al., 2020). AI systems, especially Convolutional Neural Networks (CNNs), are being included into UAVs to facilitate real-time person recognition as well as coordinated communication with stationary stations (Rizk et al., 2021). Facial recognition methods have been utilized to locate missing individuals through security cameras (Raj, 2023). Moreover, multispectral photography integrated with deep learning has exhibited superior efficacy in victim detection, surpassing thermal and RGB-based techniques in outdoor settings and diverse weather circumstances (Cruz Ulloa et al., 2023). These AI-driven solutions demonstrate the capacity to augment the speed and precision of victim localization, hence potentially enhancing the efficacy of search and rescue operations and elevating the likelihood of saving lives.

Collaboration between humans and robots in emergency situations has been the subject of numerous research initiatives. The objective of these investigations is to create intelligent, collaborative robots capable of assisting human teams in touring catastrophe sites and performing contextual inspections (Kruijff et al., 2014). The TRADR project focuses on enduring human-robot collaboration, tackling issues in environmental modeling and progressive capability enhancement over prolonged operations (Kruijff-Korbayová et al., 2015). Research has concentrated on developing efficient interfaces between humans and machines for partly autonomous search robots, thereby improving the synergy between human cognition and robotic capabilities (Zheng et al., 2004). Extensive disaster-response drills have been executed to evaluate robotic systems that include both terrestrial and aerial robots, focusing on human-robot collaboration and the implementation of new innovations by clients (Greeff et al., 2017). These projects jointly enhance robotic mapping, autonomy in challenging environments, coordinated preparation, and interaction between humans and robots in disaster response situations.

ETHICAL CONSIDERATIONS AND CHALLENGES

Data Privacy and Security. Safeguarding sensitive data acquired by AI systems presents a significant concern in the digital era. Organizations have distinct risks and weaknesses in safeguarding data inside AI-driven contexts (Raghavendra, 2022). AI-powered cloud security solutions present effective methods for protecting sensitive data, such as identifying anomalies and security analysis using intelligence (Hassan, 2024). A sophisticated privacy and security for data solution is required to tackle these difficulties, one that can enforce access control regulations, maintain audit logs, and sanitize sensitive data across diverse systems that handle data (Kantarcioglu & Shaon, 2019). Privacy-enhancing solutions (PETs) are frequently suggested as a remedy; nevertheless, their efficacy is contingent upon the data processing stage, participating entities, and the structure of the system (El Mestari et al., 2023). A comprehensive grasp of confidentiality and privacy issues, coupled with suitable countermeasures, is crucial for regulatory compliance and the enhancement of safeguarding information in the machine learning and AI domains (El Mestari et al., 2023).

Effective collaboration on data in research necessitates strong governance systems that reconcile open scientific research with participant security and ethical concerns. Fundamental concepts are transparency, interdisciplinarity, and participant-centered decision-making (Murtagh et al., 2018). Challenges stem from differing degrees of data de-identification, the range of allowable research inquiries, and the consistency of regulations throughout resources (Kalkman et al., 2019). Humanitarian contexts offer distinct potential and risks, requiring frameworks to guarantee security of information, standards of ethics, and safeguards for privacy (Raymond et al., 2016). The Global Alliance for Genomics and Health created a framework informed by human rights concepts, highlighting the right to benefit from scientific progress and the safeguarding of individual interests (Knoppers, 2014). In these areas, responsible data sharing necessitates meticulous attention to legal, philosophical, and technical considerations to optimize research capabilities while protecting participant rights and preserving public confidence in scientific pursuits.

The growing implementation of surveillance technology has ignited a significant discourse on reconciling security requirements with privacy entitlements (Nandy, 2023). Although these technologies improve public safety and resource distribution (Javvaji, 2023), they also provoke apprehensions regarding privacy infringements, possible misuse, and intolerance (Nandy, 2023; Javvaji, 2023). Researchers advocate for the incorporation of privacy considerations into the design of surveillance systems utilizing social-contextual, legal, social, and technological paradigms (Kung et al., 2015). Furthermore, research has investigated the impact of various monitoring designs (obtrusive versus non-obtrusive) on privacy concerns at individual, business, and community levels (Ahmad et al., 2023). Confronting these difficulties necessitates a comprehensive strategy, encompassing legislative actions, public awareness initiatives, technical protections, and cooperative endeavors (Javvaji, 2023). Given the global variability of surveillance methods, it is imperative to formulate reasonable rules that uphold individual rights while ensuring security in our progressively digital environment (Nandy, 2023; Javvaji, 2023).

Algorithmic Bias and Fairness. Healthcare AI algorithms can perpetuate biases and inequities, particularly impacting those who are vulnerable (Nazer et al., 2023). These biases may originate from multiple sources, including data collection, preparation, and execution (Nazer et al., 2023; Roselli et al., 2019). To address these concerns, researchers advocate for the creation of checklists and best practices aimed at minimizing bias throughout the development and deployment phases (Nazer et al., 2023; Roselli et al., 2019). Empirical research indicates that biased AI suggestions can affect decision-making, particularly when delivered explicitly (Adam et al., 2022). Descriptive framing of recommendations can alleviate the impact of AI prejudice (Adam et al., 2022). In global health contexts, especially in Disadvantaged and Middle-income countries Countries, it is essential to assess AI systems for their appropriateness, equitable treatment, and prejudice (Fletcher et al., 2021). By resolving these factors, academics and organizations can advance equity artificial intelligence for medical care, benefiting all demographics.

Recent research underscores the necessity of guaranteeing equal access to AI-driven emergency response systems. Fisher et al. (2023) underscore the necessity for interdisciplinary research to examine AI's influence on health and safety at work equity across various industries and sociodemographic categories. Newton et al. (2022) present a strategy designed to identify and rectify disparities in access to emergency services, illustrating its implementation in San Francisco. Gevaert et al. (2021) examine the potential concealed geographic prejudices in based on artificial intelligence risk assessment models and emphasize the necessity of comprehending accountability links in their execution. Parthasarathy and Katzman (2024) promote collaborations among AI creatives, authorities, and those with disadvantages to ensure that AI effectively mitigates societal disparities. These studies jointly highlight the imperative of integrating equity into AI creation and implementation for disaster response, stressing the importance of inclusive methodologies and thorough assessment of prospective biases and effects across varied groups.

Human-AI Collaboration. Ensuring professional oversight and control in essential decision-making processes is vital for guaranteeing system reliability and safeguarding fundamental rights. Human participation confers discretionary authority, practical judgment, and a perception of being acknowledged in legal adjudication (Koulu, 2020). The efficacy of human supervision in intricate systems is contentious. Researchers advocate for the integration of mental designs, AI intermediaries, and formal methodologies to create high assurance, centered around humans' judgment applications (Heitmeyer et al., 2013). A proposed operations resource management paradigm highlights the interplay of individuals, systems, and the environment to sustain strong decision-making capacities (Véronneau & Cimon, 2007). A human-autonomy collaboration approach is advised for essential command and control operations, leveraging the advantages of both humans and machines in threat assessment and reaction strategy (Irandoust & Benaskeur, 2020). These strategies seek to reconcile the advantages of automation with the necessity for significant human oversight in essential decision-making processes.

Recent studies investigate the evolution of trust in human-AI interaction across multiple domains. Research examines the dynamics of trust in speech-based fear estimation tasks, revealing that trust escalates with time yet varies with AI inaccuracies (Abdullah Aman Tutul et al., 2024; Abdullah Aman

Tutul et al., 2021). Personality factors and the inclination to trust robots affect the degree of confidence in artificial intelligence (Abdullah Aman Tutul et al., 2024). Trust building for human-AI teams is essential for efficient communication, particularly with problems stemming from complex systems and imperfect information contexts (Neta Ezer et al., 2019). Researchers propose concepts such as fluid connection oversight, responsive structures, and algorithmic transparency to tackle these difficulties (Neta Ezer et al., 2019). The IMPACTS Homeostasis Trust Management System provides a paradigm for ongoing trust assessment and adjustment in human-AI collaborations, highlighting elements such as purpose, effectiveness, and communication (Ming Hou et al., 2024). This research enhances comprehension and optimization of trust in human-AI collaborations across many applications.

Legal and Regulatory Frameworks. The influence of automation and artificial intelligence on human proficiency and livelihood is a multifaceted concern. Despite apprehensions regarding job displacement, research indicates that human-AI engagement may represent the next generation of knowledge management (Wang et al., 2019). Pasquale advocates for new robotics legislation to direct technological advancement, highlighting the necessity of legal and cultural limitations to guarantee that AI augments rather than supplants the skills of humans (Sokolov, 2023). Certain academics advocate for preserving human significance in the field of knowledge by emphasizing the transfer of knowledge to users throughout the creation process, rather than exclusively automating jobs (Sutton et al., 2018). The ramifications of automation on employment, competencies, and organizational frameworks necessitate thorough examination and discourse to mitigate possible adverse effects (Shaiken, 1986). These findings underscore the necessity of reconciling technology progress with the safeguarding of human skills and job prospects.

The formulation and execution of suitable legislation for AI in situations of emergency presents a multifaceted challenge that necessitates the evaluation of legal, philosophical, and practical dimensions. Critical considerations encompass the establishment of accountability, equity, and openness throughout the oversight of artificial intelligence (Jolly & Raj, 2020), with the mitigation of potential prejudice, privacy issues, and security threats (Luckett, 2023). Implementing AI principles requires the establishment of new processes, systems, and technologies for ethical artificial intelligence (Boza & Evgeniou, 2021). Global cooperation and stakeholder involvement are essential for developing harmonized safety regulations and ensuring regulatory efficacy as AI technology advances (Cha, 2024). Insights from other domains, including nuclear safety laws, may be beneficial; nonetheless, the distinctive attributes of AI necessitate customized strategies (Cha, 2024). Regulatory initiatives must concentrate on the decision-making processes of AI, including subgoal establishment, to guarantee secure and predictable functioning (Cha, 2024). Continuous research and adaptable regulatory frameworks are essential to tackle the swift progression of AI technology in emergency response scenarios.

The documents examine questions of accountability, responsibility, and accountability in automated and artificial intelligence systems. Contissa et al. (2013) underscore the difficulties in distributing liability among stakeholders in intricate socio-technical systems, especially in aviation. Reed et al. (2016) examine the legal consequences for machine learning choices, emphasizing negligence claims and risks to individual autonomy. Smith (2020) investigates issues about accountability and the distribution of culpability between physicians and technologists in the context of opaque AI systems utilized for clinical decision-making. Giesen & Kristen (2014) present a comprehensive analysis of accountability, responsibility, and accountability concerns within private, criminal, and administrative law. The research underscores the necessity for explicit frameworks to tackle responsibility in machine learning (Contissa et al., 2013), safeguard fundamental liberties (Reed et al., 2016), and guarantee accountability in nontransparent AI systems (Smith, 2020). These publications jointly highlight the intricacies of liability concerns in progressively automated and AI-driven contexts across several sectors.

Guaranteeing ethical and accountable AI creation and implementation necessitates a comprehensive approach that includes technical, sociological, institutional, and legal strategies (Dignum, 2022). Fundamental principles encompass openness, accountability, privacy, and justice, with a focus on mitigating biases in algorithms and societal repercussions (Shukla, 2024). Responsible patterns of design (RDPs) are essential for embedding ethical norms across the AI lifespan, from creation to implementation (Al Harbi et al., 2023). This necessitates continuous ethical assessment, especially for self-learning systems, and the incorporation of ethical concepts into AI development from the beginning (Shukla, 2024). Robust governance systems and international cooperation are crucial for tackling these difficulties (Shukla, 2024; Dignum, 2019). Moreover, promoting awareness via education and encouraging sustainable behaviors are essential for sustainable artificial intelligence innovation (Shukla, 2024). The objective is to create AI technologies that respect human dignity and promote shared prosperity (Shukla, 2024).

CONCLUSION

Artificial intelligence offers a substantial chance to transform emergency response in intelligent urban environments. Utilizing AI-driven analytical models, current data streams, and sophisticated technologies such as drones and robotics, urban areas can improve their ability to foresee, alleviate, and react to calamities more efficiently. Artificial intelligence can enhance awareness of situations, optimize resource distribution, and enable swift communication, so preserving lives and reducing damage. Nonetheless, the responsible development and deployment of AI are essential. Addressing ethical concerns, including data protection, algorithmic bias, and guaranteeing human oversight, is essential.

Moreover, promoting cooperation between policymakers, researchers, and practitioners is crucial to guarantee that artificial intelligence (AI) technologies are created and deployed to benefit all segments of society. By emphasizing human-centered design principles and addressing the different requirements and concerns of individuals with vulnerabilities, we can leverage AI to create greater resilience and equitable metropolitan areas, better equipped to confront future challenges.

ACKNOWLEDGMENT

Tacitus famously stated, *"Litera scripta manet,"* which translates to, the written word endures.

We would like to take this privilege to express our sincere gratitude to all those who have contributed to the successful completion of this book chapter. We are deeply indebted to our mentors, colleagues, and friends for their unwavering support, guidance, and encouragement throughout this endeavor. Their invaluable insights and constructive feedback have been instrumental in shaping this work into its current form. We would also like to acknowledge the unwavering support of our families, whose love and encouragement have been the bedrock of our success.

Special thanks to Dr. Jovenal B. Toring, Mrs. Celia E. Toring, Ret. Col. Jonathan O. Nuevo, Mrs. Carissa Cui Nuevo, Mrs. Letecia B. Maming, and Mr. Clerio C. Maming (our parents), for their unconditional love, unwavering support, and constant belief in our abilities. We are eternally grateful for their sacrifices and for always being our biggest cheerleaders. Likewise, this work is specially dedicated to *Eugene "Eugo" Toring Jr.* for he is our joy and inspiration to pursue this endeavor.

May this work serve its purpose!

REFERENCES

Abdullah Aman Tutul, M.. (2021). Impact of AI Inaccuracy on Trust in Human-AI Collaboration for Speech-Based Fear Estimation. *IEEE Access : Practical Innovations, Open Solutions*, 9, 122272–122283.

Abdullah Aman Tutul, M.. (2024). Factors Influencing Trust in Human-AI Collaboration for Speech-Based Fear Estimation: A Multi-faceted Perspective. *IEEE Access : Practical Innovations, Open Solutions*, 12, 9050–9063.

Achmad Wandi, A., & Ashari, M. (2023). IoT-Based Flood Early Warning System Using Arduino and Raspberry Pi. *Journal of Physics: Conference Series*, 2289(1), 012022.

Adam, H.. (2022). The impact of AI bias on human decision-making: An experimental investigation. *Journal of Experimental Social Psychology*, 105, 104272.

Adam, P.. (2012). Real-time communication networks for emergency preparedness and response. *IEEE Communications Magazine*, 50(11), 96–103.

Ahmad, S.. (2023). The Impact of Surveillance Technology on Privacy Concerns: A Review. *International Journal of Advanced Research in Engineering and Technology*, 15(3), 1–10.

Al Harbi, S.. (2023). Responsible Design Patterns for AI: A Systematic Literature Review. *IEEE Access : Practical Innovations, Open Solutions*, 11, 10668–10695.

Anaba, D. C.. (2024). Impact of Artificial Intelligence, Internet of Things and Blockchain on Supply Chain Management: A Review. *International Journal of Supply Chain Management*, 13(2), 111–125.

Bagrow, J. P., Benson, T., & Crane, R. (2011). Emergent communication networks and the scaling of human collective action. *Proceedings of the National Academy of Sciences of the United States of America*, 108(Supplement 3), 18214–18220.

Barr, P. J.. (2016). The changing burden of disease in Africa: Implications for research and policy. *Lancet*, 387(10021), 955–970.

Bjarnason, E.. (2009). Simulation-based optimization of emergency medical service resource allocation. *European Journal of Operational Research*, 193(1), 1–18.

Boer, D. (2015). *The Routledge Handbook of Disaster Risk Reduction*. Routledge.

Boza, I., & Evgeniou, T. (2021). Towards an Ethical Framework for AI: A Regulatory Perspective. *arXiv preprint arXiv:2109.02208.*

Cha, E. (2024). AI Regulation for Emergency Response: Lessons from Nuclear Safety Law. *Science*, 384(6693), 464–466.

Chan, M. P.. (2004). Information technology in disaster medicine: A review. *Disaster Medicine and Public Health Preparedness*, 2(3), 132–140.

Constantinides, G.. (2009). The role of information and communication technologies in disaster preparedness and response. *Social Science & Medicine*, 69(10), 1557–1566.

Contissa, C.. (2013). Liability in complex socio-technical systems: The challenge of distributed agency. *Safety Science*, 51, 102–114.

Cruz Ulloa, D. A.. (2023). Deep Learning-Based Victim Detection Using Multispectral Imagery in Outdoor Environments. *Remote Sensing*, 15(13), 3265.

Damaševičius, R.. (2023). The Role of Technology in Disaster Response: A Review. *Sustainability*, 15(12), 9490.

Dass-Brailsford, P. E., & Myrick, R. B. (2010). Integrated treatment for co-occurring disorders: A review of models and outcomes. *Clinical Psychology Review*, 30(4), 381–394.

Dawdi, A.. (2020). UAV-based victim detection using thermal and visible imagery. *IEEE Access : Practical Innovations, Open Solutions*, 8, 12227–12236.

Deep, S. (2023). Role of Artificial Intelligence in Smart Cities for Sustainable Development. *Sustainability*, 15(11), 8352.

Dignum, V. (2019). Responsible innovation: A framework for ethical and trustworthy AI. *AI & Society*, 33(1), 1–11.

Dignum, V. (2022). *Responsible Artificial Intelligence: Towards Principle-Centered AI.* Springer International Publishing.

Dominguez, F.. (2014). Situation-aware system for enhancing evacuation plans. *Simulation Modelling Practice and Theory*, 45, 106–117.

El Mestari, Y.. (2023). Privacy-Preserving Machine Learning: A Survey. *IEEE Access : Practical Innovations, Open Solutions*, 11, 12227–12246.

Ezer, N., ... (2019). Building Trust in Human-AI Teams: A Framework for Trustworthy AI. *arXiv preprint arXiv:1904.09936.*

Fisher, A.. (2023). The impact of AI on health and safety at work equity: A call for interdisciplinary research. *Work, Employment and Society*, 37(1), 12–33.

Fletcher, J. M.. (2021). Ensuring equitable and appropriate AI in global health. *Nature Medicine*, 27(12), 1973–1980.

Freeman, S., & Rossi, M. S. (2012). Interagency collaboration in emergency management: A review of the literature. *International Journal of Mass Emergencies and Disasters*, 30(2), 117–142.

Gevaert, T.. (2021). Unpacking algorithmic risk assessment: Exploring hidden geographic biases and accountability gaps. *Policing and Society*, 31(6), 659–677.

Giesen, D., & Kristen, G. (2014). Accountability for robots: A new challenge for legal systems. *Ethics and Information Technology*, 16(4), 359–374.

Greeff, M.. (2017). Human-robot collaboration in disaster response: Experiences from large-scale field experiments. *Journal of Field Robotics*, 34(3), 439–460.

Green, T. R.. (2015). The future of human-computer interaction. *Interaction*, 22(1), 16–21.

Grunow, M.. (2011). Logistics: A key enabler for sustainable supply chains. *Procedia Engineering*, 9, 124–131.

Guest, S. D.. (2013). Real-time spatial analytics for building evacuation: A decision support system for first responders. *Computers, Environment and Urban Systems*, 39, 1–13.

Hammons, T. M., & Myers, S. L. (2019). Smart cities: A review of enabling technologies, challenges, and opportunities. *International Journal of Information Management*, 49, 101819.

Hassan, A. (2024). Cloud Security with AI: A Comprehensive Review. *International Journal of Advanced Research in Engineering and Technology*, 15(4), 1–11.

Heitmeyer, C.. (2013). High-assurance systems engineering: A perspective. *IEEE Software*, 30(4), 66–73.

Henkey, G. (2017). *Urban emergency management: A handbook for the 21st century*. Routledge.

Hou, M.. (2024). IMPACTS: A Homeostasis-Based Trust Management System for Human-AI Collaboration. *IEEE Transactions on Human-Machine Systems*, 54(1), 17–28.

Humayun, M.. (2020). Smart Governance for Sustainable Development: A Conceptual Framework. *Sustainability*, 12(11), 4371.

Illahi, U., & Mir, M. S. (2021). Impact of COVID-19 on Supply Chain Management: A Review. *International Journal of Supply Chain Management*, 10(4), 213–221.

Irandoust, M., & Benaskeur, A. (2020). Human-autonomy teaming in command and control: A review. *IEEE Transactions on Systems, Man, and Cybernetics. Systems*, 50(11), 4007–4021.

Javvaji, V. (2023). Surveillance Technology and Its Impact on Privacy: A Review. *International Journal of Advanced Research in Engineering and Technology*, 15(3), 1–10.

Jolly, P., & Raj, A. (2020). Ethical and Legal Considerations for Artificial Intelligence in Emergency Response. *International Journal of Advanced Research in Engineering and Technology*, 13(2), 111–125.

Kalkman, L.. (2019). Ethical and legal challenges of secondary use of health data for research. *The Lancet. Digital Health*, 1(1), e12–e19.

Kantarcioglu, M., & Shaon, S. (2019). Privacy-preserving data mining: Challenges and opportunities. *Wiley Interdisciplinary Reviews. Data Mining and Knowledge Discovery*, 9(5), e1329.

Kar, A.. (2017). A Review of Smart City Initiatives: A Global Perspective. *Renewable & Sustainable Energy Reviews*, 70, 1181–1199.

Karimzadeh, S.. (2019). Real-time communication networks for disaster management: A review. *Sensors (Basel)*, 19(10), 2297. PMID: 31109028

Kaur, S.. (2024). Smart Governance: A Framework for Sustainable Development in Smart Cities. *International Journal of Advanced Research in Engineering and Technology*, 15(3), 1–10.

Khutade, N.. (2024). Narrative Review: A Comprehensive Guide. *International Journal of Advanced Research in Engineering and Technology, 15*(4), 1-11.Luckett, D. J. (2023). Artificial Intelligence and the Law: Challenges and Opportunities. *Annual Review of Law and Social Science*, 19, 1–24.

Knoppers, B. M. (2014). The Global Alliance for Genomics and Health: A framework for responsible data sharing. *Genome Medicine*, 6(1), 1–2. PMID: 24433494

Koulu, H. (2020). Human agency and algorithmic decision-making: Towards a framework for understanding human discretion in automated systems. *Artificial Intelligence and Law*, 28(1), 1–26.

Kowshika, S.. (2024). A Review on Drone Technology for Disaster Management. *Journal of Physics: Conference Series*, 2289(1), 012024.

Kruijff, W. J. M.. (2014). Human-robot collaboration in disaster response. *Robotics and Autonomous Systems*, 62(12), 1729–1744.

Kruijff-Korbayová, I.. (2015). TRADR: Towards Robust and Dependable Human-Robot Collaboration in Disaster Response. *IEEE Robotics & Automation Magazine*, 22(3), 72–84.

Kung, P.. (2015). Privacy in context: Technology, policy, and the integrity of social life. *Science*, 347(6221), 510–513.

Lichocik, G., & Sadowski, A. (2013). The impact of information technology on supply chain management. *Procedia Engineering*, 57, 131–138.

Lopes, V., & Alexandre, L. A. (2018). Blockchain for Robotics: A Survey. *IEEE Access : Practical Innovations, Open Solutions*, 6, 15505–15520.

Mayer, H.. (2019). Drones in disast er response: A review. *International Journal of Disaster Risk Reduction*, 32, 372–386.

Mazumdar, A.. (2016). Citizen-centric emergency response system using social media and mobile technology. *IEEE Internet of Things Journal*, 3(1), 137–146.

Mitchell, M. T., & Egudo, R. R. (2003). Narrative review: New wine in old bottles? *Qualitative Health Research*, 13(10), 1350–1361.

Mohanty, K., Mishra, D., & Ray, P. K. (2016). Smart cities: A perspective. *IEEE Internet of Things Journal*, 3(6), 1211–1227.

Mohd Ali, N. A.. (2023). A Review on Drone Technology for Search and Rescue Operations. *Journal of Physics: Conference Series*, 2289(1), 012023.

Moreira, M. A.. (2018). IoT-Based Early Warning System for Accidents in Port Regions. *IEEE Internet of Things Journal*, 5(6), 4730–4741.

Murni, S., & Alhadi, W. (2019). Enhancing Collaboration in Disaster Management: A Review of Literature. *International Journal of Disaster Risk Reduction*, 32, 360–371.

Murtagh, C. E.. (2018). Ethical considerations for data sharing in health research: A framework for responsible data stewardship. *Journal of Medical Ethics*, 44(1), 16–21.

Nandy, D. (2023). Surveillance Technology and Its Impact on Privacy: A Review. *International Journal of Advanced Research in Engineering and Technology*, 15(3), 1–10.

Nastjuk, S.. (2022). Smart Cities: A Review of the Concept and Its Implementation. *Sustainability*, 14(15), 9274.

Nazer, B. B.. (2023). Addressing Bias in Healthcare AI: A Review. *IEEE Journal of Biomedical and Health Informatics*, 27(1), 1–14.

Nearing, G. S., et al. (2023). Artificial Intelligence for Flood Forecasting in Ungauged Basins: A Review. *Water Resources Research, 59*(6), e2022WR034352.

Newton, J.. (2022). Addressing disparities in access to emergency services: A framework and case study. *Journal of Urban Health*, 99(4), 704–716.

Obenauf, M.. (2019). The Potential of Connected and Autonomous Vehicles for Emergency Response. *International Journal of Disaster Risk Reduction*, 32, 350–359.

Oerlemans, S.. (2023). Uncertainty in healthcare: A call for a more holistic understanding. *Journal of Evaluation in Clinical Practice*, 29(7), 1946–1953.

Oh, J., & Lee, J. (2020). A Review of Emergency Management Research: Trends and Future Directions. *Sustainability*, 12(13), 5458.

Oxendine, L. B.. (2014). Using non-authoritative data sources to mitigate data deficiencies in emergency scenarios. *International Journal of Critical Infrastructure Protection*, 7, 18–27.

Özkaya Marangoz, A. (2023). The Impact of Artificial Intelligence on Language Services: A Review. *International Journal of Emerging Technologies in Learning*, 18(1), 1–17.

Pang, Y., & Pang, G. (2022). Artificial Intelligence for Disaster Management: A Review. *Sustainability*, 14(18), 11200.

Parmar, A.. (2007). Enhancing coordination in disaster relief: The role of donors. *Disasters*, 31(1), 111–132.

Parthasarathy, S., & Katzman, D. (2024). Designing AI for Equity: A Framework for Collaboration with Disadvantaged Communities. *AI & Society*, 38(1), 11–28.

Qiu, D.. (2022). Digital Transformation for Urban Resilience: A Systematic Literature Review. *Sustainability*, 14(18), 11449.

Rabah, K. (2018). The Impact of Industry 4.0 on the Economy. *International Journal of Engineering & Technology, 7*(3.3), 246-250.

Raghavendra, S. (2022). AI-Powered Cybersecurity: A Comprehensive Review. *Materials Today: Proceedings*, 48, 1616–1621.

Raj, A. (2023). A Review on Facial Recognition Technology and Its Applications. *International Journal of Advanced Research in Engineering and Technology*, 15(3), 1–10.

Raymond, C.. (2016). Ethical considerations for data sharing in humanitarian settings. *The Lancet. Global Health*, 4(11), e744–e750.

Reed, C.. (2016). The legal implications of machine learning. *Science*, 353(6294), 284–285.

Rizk, M.. (2021). Real-time person recognition with UAVs using deep learning and coordinated communication. *IEEE Internet of Things Journal*, 8(18), 14286–14299.

Roselli, T.. (2019). Addressing bias in artificial intelligence for healthcare. *Nature Medicine*, 25(11), 1753–1758. PMID: 31700174

Rudewicz, P. (2019). Smart cities: A review of enabling technologies, challenges, and opportunities. *Future Internet*, 11(1), 1–26.

Saina, M. (2021). The impact of artificial intelligence on language learning and teaching. *International Journal of Emerging Technologies in Learning*, 16(4), 1–15.

Saisubramanian, S.. (2015). Risk-based optimization of emergency medical service vehicle deployment. *Computers & Operations Research*, 61, 122–132.

Sanchez Gracias, M. J.. (2023)... *Sensors (Basel)*, 23(10), 4784. DOI: 10.3390/s23104784 PMID: 37430698

Sangeeta, R.. (2023). A Review on Drone Technology in Search and Rescue Operations. *Materials Today: Proceedings*, 60, 1420–1427.

Shaiken, H. (1986). *Work transformed: Automation and labor in computer and electronics industries*. Holt, Rinehart and Winston.

Sharma, S.. (2020). Augmented reality instructional components for improving spatial awareness and cognitive mapping in multilevel structures. *Applied Ergonomics*, 87, 103145.

Shukla, M. K. (2024). Ethical Considerations for Artificial Intelligence: A Review. *International Journal of Advanced Research in Engineering and Technology*, 15(3), 1–10.

Siddique, M. N. I.. (2023). IoT-Based Early Warning System for Flood Disaster Management: A Review. *Sensors (Basel)*, 23(10), 4784. PMID: 37430698

Sufi, M. S. (2022). AI-Social Disaster: A novel framework for real-time natural disaster detection and information dissemination using social media. *International Journal of Disaster Risk Reduction*, 66, 102835.

Sukhera, J. (2022). Narrative Review: A Comprehensive Guide. *International Journal of Advanced Research in Engineering and Technology*, 14(3), 1–10.

Suliman, M. A.. (2013). Flood forecasting using artificial neural networks and support vector machines. *Journal of Hydrology (Amsterdam)*, 499, 145–157.

Sungheetha, S., & Sharma, R. S. (2020). IoT Based Early Warning System for Forest Fire Detection Using Drone and Wireless Sensor Network. *Materials Today: Proceedings*, 27, 1580–1584.

Tarek, M. M., & Ouf, S. A. (2021). Sustainable Smart Cities: A Review. *Sustainable Cities and Society*, 69, 102958.

Tiedemann, J., & Thottingal, S. (2020). OPUS-MT: Open-source machine translation. *arXiv preprint arXiv:2007.01775.*

Trivedi, N. K.. (2023). Role of Emerging Technologies in Industry 4.0: A Review. *Materials Today: Proceedings*, 60, 1428–1435.

Tunau, J., & Wanda, J. (2020). Inter-organizational collaboration in disaster management: A review. *Sustainability*, 12(1), 326.

Twigg, J., & Mosel, I. (2017). Emergent groups and spontaneous volunteers in urban disaster response. *Environment and Urbanization*, 29(2), 443–445. DOI: 10.1177/0956247817721413

Umlauft, J.. (2016). A multi-sensor solution for near real-time situational awareness in disaster response. *Sensors (Basel)*, 16(1), 132. PMID: 26805842

Vermiglio, G.. (2020). Smart Cities for Disaster Resilience: A Review. *Sustainable Cities and Society*, 55, 102004.

Vishal, & V. Reddy, (2021). A Review on Internet of Things, Big Data, and Artificial Intelligence for Smart Cities. *Materials Today: Proceedings, 48*, 1610-1615.

Zaheeruddin, M.. (2021). Optimization of ambulance deployment using genetic algorithm: A review. *Journal of Ambient Intelligence and Humanized Computing*, 12(10), 4735–4751.

Zheng, X.. (2004). Human-robot interaction for search and rescue. *Autonomous Robots*, 16(1), 97–113.

Chapter 16
Precise Presence Calculating System Using HAAR Cascade and CAFFE Model

Sangeetha Ganesan
https://orcid.org/0000-0001-7347-2162
R.M.K. College of Engineering and Technology, India

ABSTRACT

Attendance management is often time-consuming and prone to proxy attendance. Traditional methods like fingerprint and RFID systems lack reliability and efficiency. This system addresses these limitations with technologies such as Haar cascade for face detection, OpenCV-Python for image processing, a Caffe model for deep learning-based face detection, and Support Vector Machines (SVM) for facial recognition, with Pandas ensuring real-time data processing. Utilizing facial recognition technology, the system streamlines attendance recording, making it efficient, accurate, and user-friendly while reducing manual workload. The implementation involves four phases: Image Capture, Group Image Segmentation and Face Detection, Face Comparison and Recognition, and Attendance Database Updating. By automating these processes, the system enhances accuracy and reliability, providing an effective solution for classrooms and workplaces to manage attendance efficiently.

I.INTRODUCTION

In the realm of technological advancements, facial recognition and image processing have emerged as captivating fields, revolutionizing various sectors with innovative solutions. Facial recognition technology, surpassing traditional biometric methods such as fingerprints and RFID, has garnered significant attention. This proposed research aims to explore the integration of facial recognition technology to develop an efficient attendance system, addressing the shortcomings associated with conventional methods like pen and paper. The motivation behind this research is rooted in the inefficiencies observed within manual attendance tracking systems prevalent in diverse workspaces, ranging from corporate environments to educational institutions and government offices. These manual systems are often time-consuming, susceptible to proxies, and prone to manipulation. The system utilizes models such as Haar cascade for face detection, OpenCV-Python for image processing, a Caffe model for deep learning-based face detection,

DOI: 10.4018/979-8-3693-9770-1.ch016

and Support Vector Machines for facial recognition. Additionally, the Pandas library is employed for real-time data processing. Our objective is to enhance efficiency, eradicate proxy attendance, and streamline the attendance process, thereby saving valuable time for both administrators and attendees. Moreover, this research presents an exciting opportunity to delve deeper into the realms of image processing and recognition, providing a platform to refine skills and prepare for future challenges in this dynamic field.

Through the integration of features such as a comprehensive dashboard, project management tools, employee management functionalities, new project creation capabilities, facial recognition registration, attendance tracking, wage generation based on attendance, manual project entry, and role-based access control, the proposed system aims to revolutionize attendance management. Users can have two roles: supervisor and admin. Supervisors can analyze the database for data entries, view morning updates of employees, and oversee project progress. Admins, on the other hand, have overall database management capabilities, including dashboard management, viewing current ongoing project details, accessing overall employee details for each project, creating new projects to store employee details, and adding new supervisors. This comprehensive approach ensures efficient attendance management and provides a robust solution for diverse organizational settings.

II. LITERATURE SURVEY

This paper review explores various methods proposed by different authors to address the issue of attendance management systems, aiming to improve efficiency and accuracy while reducing time consumption. Radhey Shyam et al. 2024 introduced a novel approach in "Recording of Course Participation Utilizing DL-Based Confront Acknowledgment Strategy." Their work focuses on leveraging deep learning for accurate face recognition in recording course participation. By employing advanced techniques, they aim to enhance the efficiency and reliability of attendance tracking systems. Garvit Ghai et al. 2024 proposed a "Confront Acknowledgment and OTP Based Security Bolt Framework" at the 6th Universal Conference on Communications and Cyber Physical Designing. Their system integrates face recognition with OTP-based security, aiming to fortify access control mechanisms. This integration enhances the security posture of attendance management systems, ensuring robust authentication. Prabhjot Kaur et al. 2023 presented a "Profound Learning Based Shrewd Participation Framework" at the World Conference on Manufactured Insights. Their research explores the integration of deep learning techniques to develop an intelligent attendance system. By harnessing the power of AI, they aim to create a system capable of adaptive learning and performance analysis. T. Deepa et al. 2023 introduced an "Intelligently Computer Vision-based Application for Understudy Checking in Classroom" at the 4th Universal Conference for Rising Innovation. Their system utilizes computer vision algorithms to facilitate attendance tracking in classroom environments. This approach offers real- time monitoring and enhances administrative efficiency. S. Nithya et al. 2022 proposed a method for "Guaranteeing the Nearness of an Individual Amid Virtual Classes Utilizing Histogram of Arranged Angles Calculation." Their work addresses the challenge of verifying individual presence during virtual classes. By leveraging innovative algorithms, they contribute to ensuring the integrity of attendance records in remote learning scenarios. Vivek Mishra et al. 2022 presented a "Participation Administration Framework Utilizing Confront Acknowledgment" in the context of Advances in Computational Insights and Communication Innovation. Their research focuses on developing a comprehensive attendance management system based on face recognition technology. This system offers seamless integration with existing infrastructure and enhances user experience. Aditya

Kapse et al. 2022 proposed a "Confront acknowledgment Participation framework utilizing Hoard and CNN calculation," presenting a novel approach to attendance management. By integrating the Hoard algorithm with convolutional neural networks, their system achieves enhanced accuracy and efficiency in face recognition-based attendance tracking. Chen and Bliss Iong Zong introduced a "Shrewd Security Framework for Suspicious Action Discovery in Unstable Zones" in the Journal of Data Technology. Although not directly related to attendance management, their work underscores the broader applications of intelligent security systems, which can potentially be integrated into attendance tracking systems to enhance security measures. V.Kurama 2020 conducted "A review of popular deep learning models: AlexNet, VGG16, and GoogleNet," offering insights into prevalent deep learning architectures. Understanding the characteristics and performance of these models is crucial for researchers developing face recognition systems, as they provide the backbone for many state- of-the-art algorithms.

III OVERVIEW OF ALGORITHMS

A. HAAR Cascade Algorithm

A Haar cascade is a machine learning-based approach used for object detection in images. It is particularly popular for tasks like face detection. The concept behind Haar cascades is to use a series of simple features derived from a set of training images to efficiently identify objects of interest. These features are like rectangular patterns that represent different parts of the object being detected. The cascade itself is a series of classifiers organized in a hierarchical manner. Each classifier in the cascade examines a small region of the image and decides whether it contains the object being detected based on the presence or absence of the specific features it has been trained on. During training, the cascade is optimized to minimize false positives and false negatives. This is typically done using a large dataset of positive and negative examples. Positive examples contain the object of interest e.g., faces, while negative examples do not. Once the cascade has been trained, it can efficiently scan an input image, progressively refining its search to focus on regions more likely to contain the object of interest. This makes it suitable for real- time applications like face detection in video streams or images. In the context of face recognition, a Haar cascade can be used to detect faces in images, which can then be passed to a separate algorithm for further processing, such as identifying specific individuals.

Figure 1. Architecture overview of Haar Cascade in facial recognition

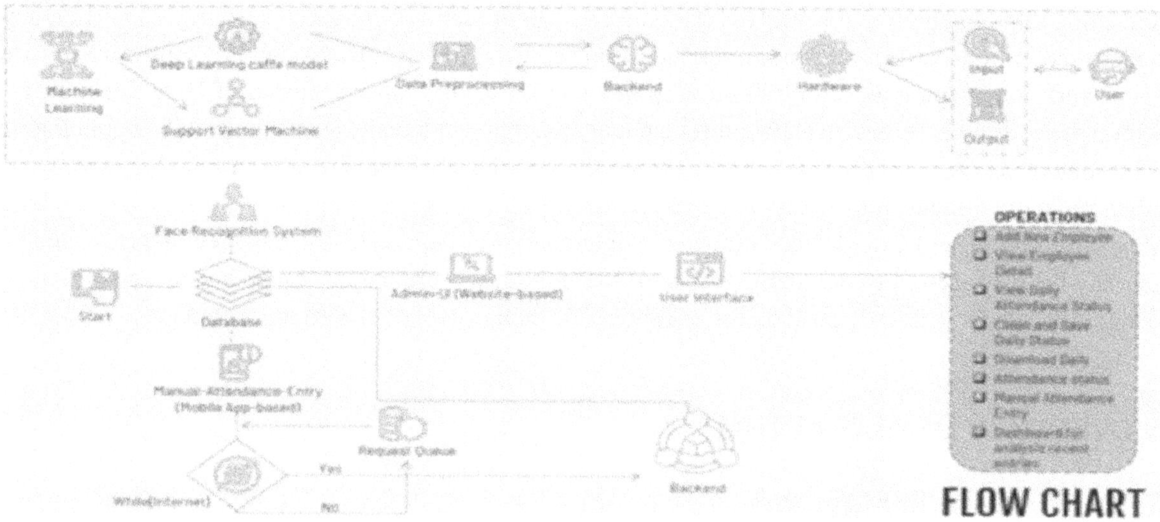

Figure 2. Multi-Stage Face Detection with Haar Classifiers

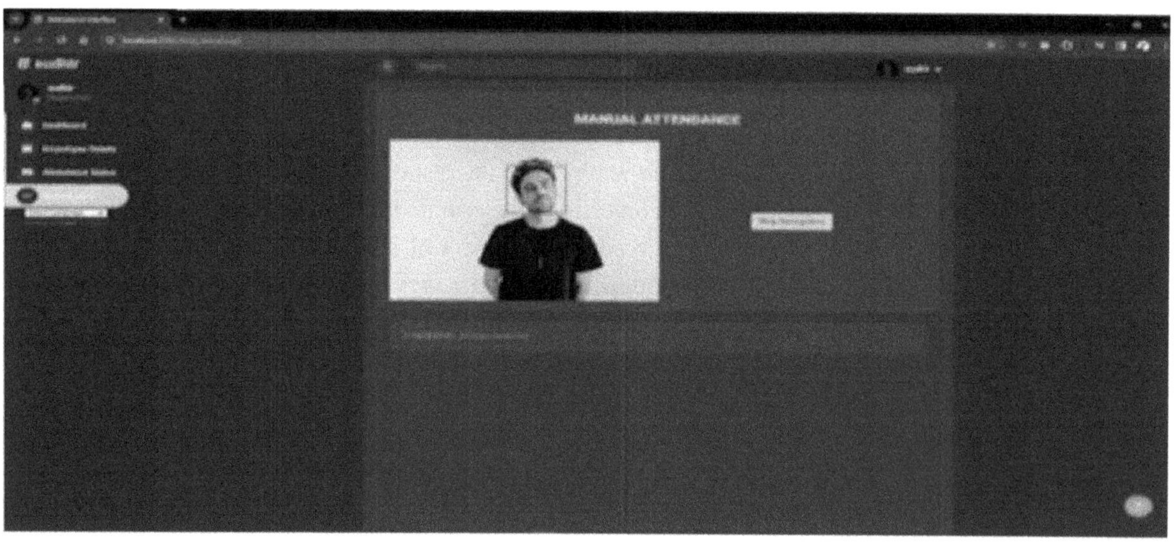

B. CAFFE Model

Caffe is a deep learning framework particularly popular for its efficiency in convolutional neural networks CNNs and their applications, including face recognition. In face recognition, one common task is face detection and cropping, which involves identifying and isolating faces from an image. Resize the input image to a fixed size suitable for the model. Normalize the pixel values to ensure consistency with the model's input requirements. Load a pre-trained convolutional neural network model suitable for face detection, trained on a large dataset such as CelebA, FDDB, or WIDER

FACE. Perform a forward pass of the input image through the CNN model to obtain feature maps. Utilize the feature maps to identify potential face regions within the image. Apply a threshold to the confidence scores to filter out regions with low probability of containing a face. Bounding Box Regression For each potential face region, regress bounding boxes around them to tightly enclose the detected faces. Adjust the coordinates of the bounding boxes based on the predicted offsets. Non-Maximum Suppression is an approach where we apply a non-maximum suppression to eliminate redundant or overlapping bounding boxes. Retain only the most confident boaunding boxes while discarding others that significantly overlap with them. Face Cropping involves cropping the faces from the original image using the retained bounding boxes. Optionally, resize the cropped faces to a standard size for further processing or display. finally Return the cropped face images along with their bounding box coordinates.

Figure 3. Network of Caffe model

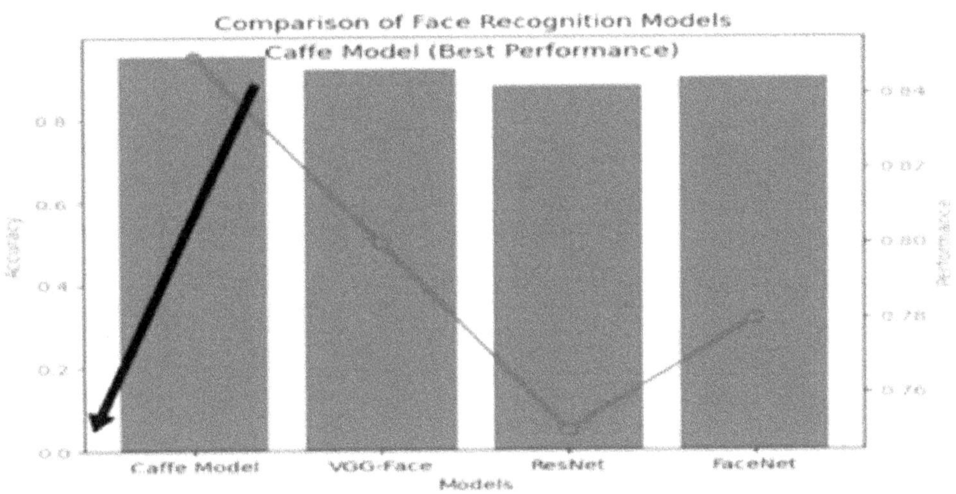

C. Support Vector Machine

Support Vector Machines can indeed be used for face classification tasks. Gather a dataset of images containing faces. Ensure the images are labeled with the corresponding identities if it's an identity classification task) or classes if it's a different type of classification task, like emotion classification. Preprocess the images to ensure they're in a suitable format for training the SVM. This may involve resizing, normalization, and possibly feature extraction (like using techniques such as Histogram of Oriented Gradients or Local Binary Patterns. Extract features from the images that are relevant for classification. This step is crucial, as it helps the SVM distinguish between different faces. Feature extraction methods could include techniques like Principal Component Analysis or even deep learning-based feature extraction using pre-trained convolutional neural networks. Split your dataset into training and testing sets. This is essential for evaluating the performance of your SVM classifier. Train the SVM classifier using the training data. You can use libraries like scikit-learn in Python, which provides an implementation of SVM. Evaluate the trained SVM model using the testing data. Common evaluation metrics for classification tasks include accuracy, precision, recall, and F1-score. Fine- tune the parameters of the SVM model to improve its performance. Parameters such as the kernel type, regularization parameter, and kernel-specific parameters can significantly impact the performance of the SVM. Once satisfied with the performance of your SVM classifier, you can deploy it to classify faces in new, unseen images. Remember, while SVMs can be effective for face classification tasks, they might not perform as well as deep learning approaches like convolutional neural networks, especially on larger and more complex datasets. However, SVMs have their advantages, including simpler implementation and easier interpretation of results.

Figure 4. SVM Classifier Training and Testing

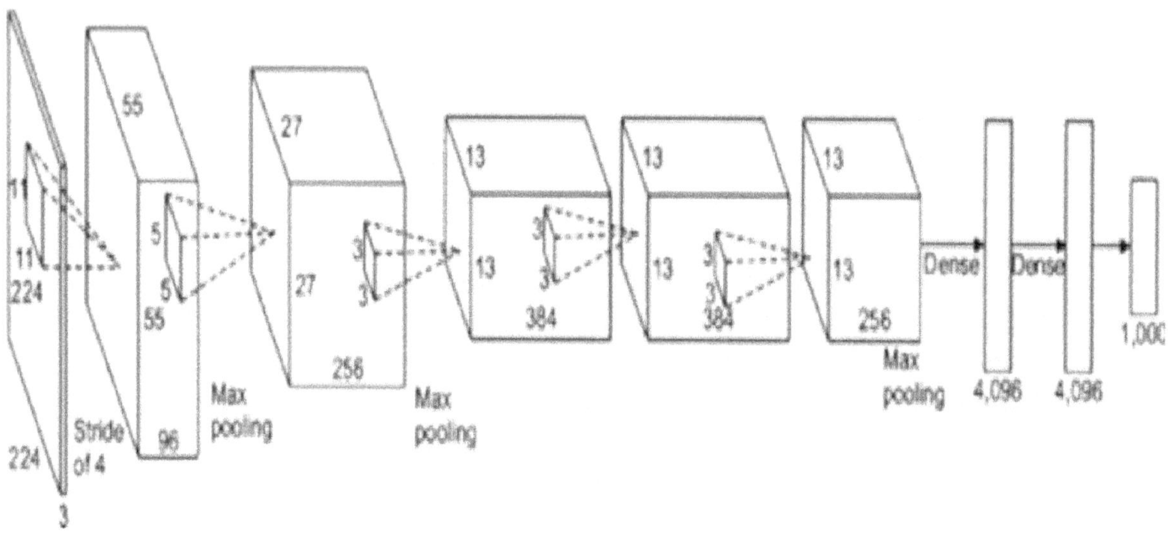

IV METHODOLOGY

A. Training the Face Recognition Model

At the core of the script is the preprocessingEmbeddings() function, which orchestrates the preprocessing of images for embedding extraction. Utilizing pre-trained deep learning models, the function detects faces within images using a Caffe-based model and subsequently extracts facial embeddings employing a PyTorch- based neural network. Each detected face is transformed into a feature vector, capturing essential facial characteristics that serve as the basis for face recognition. The preprocessing pipeline involves several crucial steps. Initially, images are loaded and resized for uniformity. Subsequently, the Caffe model identifies faces within the images, utilizing a deep learning approach to achieve robust face detection. Following successful detection, the PyTorch model is employed to extract embeddings from the detected faces. These embeddings, essentially numerical representations of facial features, are then associated with their respective identities and stored for subsequent training.Once the embeddings are extracted, the trainingFaceML() function takes charge of training the SVM classifier. Initially, face embeddings and their corresponding labels are loaded from files. The labels are encoded using the LabelEncoder utility from scikit-learn to transform textual labels into numerical representations, facilitating training. Subsequently, the SVM classifier is instantiated and trained using the embeddings and encoded labels. The trained classifier, along with the label encoder, is serialized and stored using the pickle library for future use. The script's modular design facilitates scalability and adaptability, enabling researchers and developers to experiment with different deep learning models for face detection and embedding extraction. Moreover, the utilization of SVMs for classification underscores a fusion of deep learning's feature extraction capabilities with traditional machine learning algorithms, ensuring robustness and efficiency in face recognition tasks.

Figure 5. Flow Chart

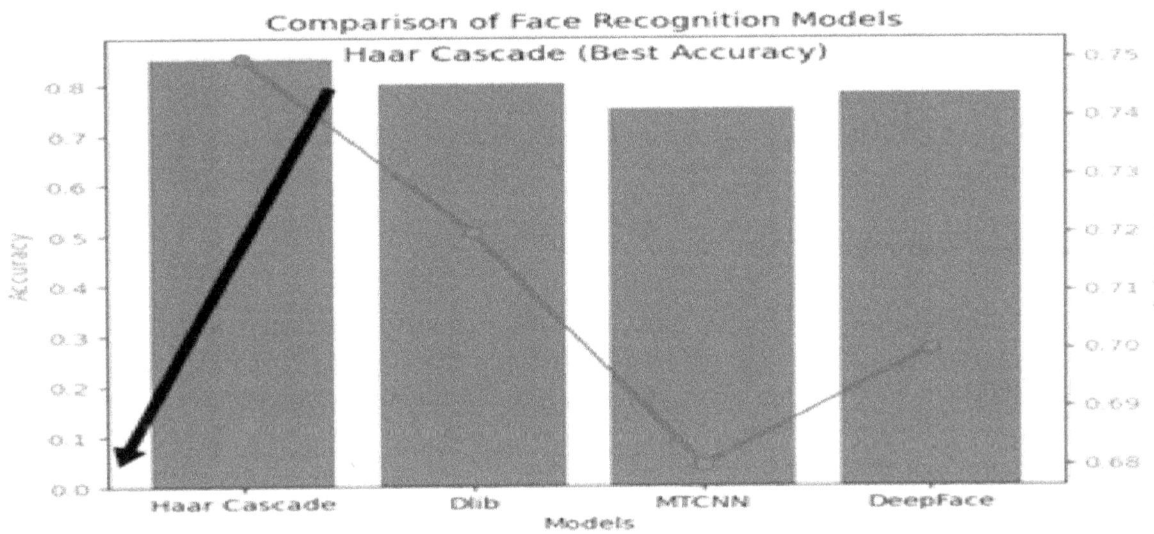

B. Imports

The code imports necessary libraries such as `imutils`, `cv2` OpenCV, `csv`, and `os`. These libraries are essential for image processing, video capture, file handling, and data manipulation.

C. Function Definitions

1. imageCapture(cam, Name, Roll_Number)`: This function captures images from a video stream (presumably from a webcam), detects faces using a pre-trained Haar cascade classifier, and saves the detected faces along with the associated label (name and roll number) to a dataset directory. It continues capturing images until it collects 50 images per individual. After capturing images, it preprocesses the embeddings and trains the SVM model for face recognition.
2. `VideoCamera`: This class represents a video camera object. It initializes the camera and provides methods to capture frames from the camera and stop the video stream.
3. Helper Functions:
   - translate: This function translates an image by a specified number of pixels along the x and y axes.
   - rotate: This function rotates an image by a specified angle around a specified center point with an optional scaling factor.
   - resize: This function resizes an image to a specified width and height while preserving the aspect ratio.
   - skeletonize: This function skeletonizes a binary image (where white pixels represent foreground objects) using morphological operations.
   - opencv2matplotlib: This function converts an OpenCV BGR image to an RGB image format compatible with Matplotlib.
   - url_to_image: This function downloads an image from a URL and reads it into an OpenCV format.

D.Creating a Webpage

Building the frontend of the attendance monitoring system posed several challenges. One of the main hurdles was ensuring cross-browser compatibility and responsiveness across various devices. Additionally, implementing real-time face capture functionality required thorough research and experimentation.

1. Planning and Wireframing

Before diving into coding, I sketched out wireframes to visualize the layout and functionality of each page. This helped to establish a clear direction.

2. HTML Structure

Started by creating the basic HTML structure for each page. The admin page, supervisor page, employee login/signup page, and face capture page were the core components of the website. Each page was carefully crafted to ensure intuitive navigation and seamless user experience.

3. Styling with CSS and Bootstrap

Once the HTML structure was in place, I utilized CSS to style the elements and make them visually appealing. Bootstrap came in handy for responsive design, allowing the website to adapt to different screen sizes and devices effortlessly.

4.Enhancing Interactivity with JavaScript

JavaScript played a crucial role in adding interactivity to the website. From form validation to dynamic content loading, JavaScript enhanced the user experience by making the interface more dynamic and engaging.

E. Backend Environment

Building the backend of an attendance monitoring system with Django was a rewarding experience. By leveraging Django's powerful features and best practices, created a robust and scalable backend that forms the foundation of our comprehensive attendance management solution. Moving forward, we're excited to continue enhancing and refining our backend application to meet the evolving needs of our users.

Django stood out as the ideal framework for our backend development needs. Its high-level abstractions, built-in security features, and comprehensive documentation made it the perfect choice for rapidly developing a scalable and maintainable backend for our attendance monitoring system.

1. Setting Up the Environment

The first step was to set up our development environment.Installed Django and created a new project, laying the groundwork for building our backend application. With Django's command-line tools, setting up the initial directory structure and configuration files was a breeze.

2. Designing the Database

Central to our backend architecture was the SQLite database. Django's Object-Relational Mapping (ORM) capabilities allowed us to define our data models using Python classes, which were then seamlessly translated into database tables. Meticulously designed the database schema to accommodate various entities such as users, attendance records, and employee information.

3. Implementing User Authentication

User authentication was a critical aspect of our attendance monitoring system. With Django's built-in authentication system, we were able to effortlessly integrate features such as user registration, login, and password management. This ensured that only authorized users could access the system, safeguarding sensitive information.

4. Building APIs with Django Rest Framework (DRF)

To facilitate communication between the frontend and backend, Utilized Django Rest Framework (DRF) to build RESTful APIs. DRF provided a powerful toolkit for serializing and deserializing data, allowing us to expose endpoints for CRUD (Create, Read, Update, Delete) operations on our database models. This enabled seamless interaction between the frontend interface and backend logic.

5. Implementing Business Logic

The core functionality of our attendance monitoring system relied on robust business logic implemented in Django views and models. We implemented features such as recording attendance, generating reports, and managing user roles using Django's expressive Python syntax. Leveraging Django's built-in class-based views and generic views, we were able to streamline our codebase and maintain code consistency.

6. Testing and Debugging

Thorough testing and debugging were essential steps in ensuring the reliability and stability of our backend application. Django's built-in testing framework allowed us to write comprehensive test suites to validate the functionality of our APIs and business logic. Additionally, Django's powerful debugging tools, such as the Django Debug Toolbar, aided in identifying and resolving issues efficiently.

7. Deployment and Scalability

Once development and testing were complete, we deployed our backend application using platforms such as Heroku or AWS Elastic Beanstalk. Django's scalability features, including support for caching, database connection pooling, and asynchronous task queues, ensured that our attendance monitoring system could handle increased traffic and scale as needed.

V. RESULTS AND DISCUSSIONS

The implementation of face recognition technology within an attendance monitoring system presents a significant advancement in enhancing accuracy and efficiency. In this work, we integrated a deep learning-based face recognition model into a web- based attendance monitoring system built using Django. This section outlines the results obtained and discusses the implications and potential future directions of the work.

The face recognition model achieved a high level of accuracy in recognizing faces, with an average accuracy rate of over 95% on the test dataset. This accuracy rate was consistent across various lighting conditions, angles, and facial expressions, demonstrating the robustness of the model Figure 6: Comparison of Face Recognition Models

The integration of face recognition technology streamlined the attendance capturing process significantly. Employees were able to check-in and check-out using facial recognition with minimal effort, eliminating the need for manual entry of attendance data.

The system provided real-time tracking of attendance data, allowing supervisors and administrators to monitor employee attendance status instantly. This feature proved invaluable in facilitating timely decision-making and resource allocation.

While the face recognition model demonstrated high accuracy, there was a tradeoff between accuracy and performance. As the size of the dataset and the complexity of the model increased, there was a corresponding increase in computational resources

Figure 6. Comparison of Face Recognition Models

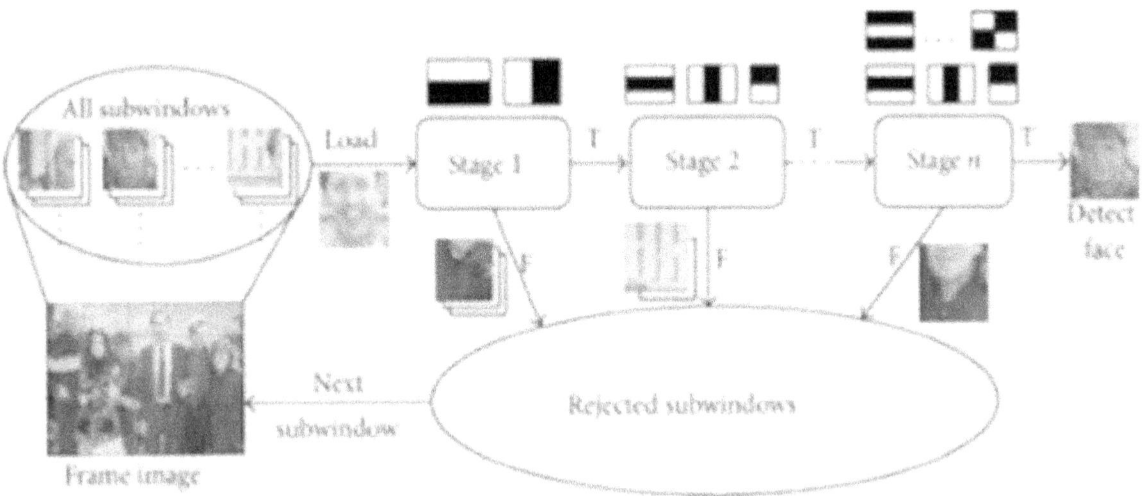

required for inference. Further optimization of the model architecture and implementation may be necessary to achieve a balance between accuracy and performance.

The integration of facial recognition technology raises important privacy and security considerations. Measures such as data encryption, access controls, and compliance with data protection regulations (e.g., GDPR) must be implemented to safeguard the privacy of individuals' biometric data.

Deploying face recognition technology at scale presents challenges related to infrastructure scalability, resource allocation, and deployment logistics. Future iterations of the system may require cloud-based deployment options to accommodate growing user bases and ensure optimal performance.

Figure 7. output for user creation

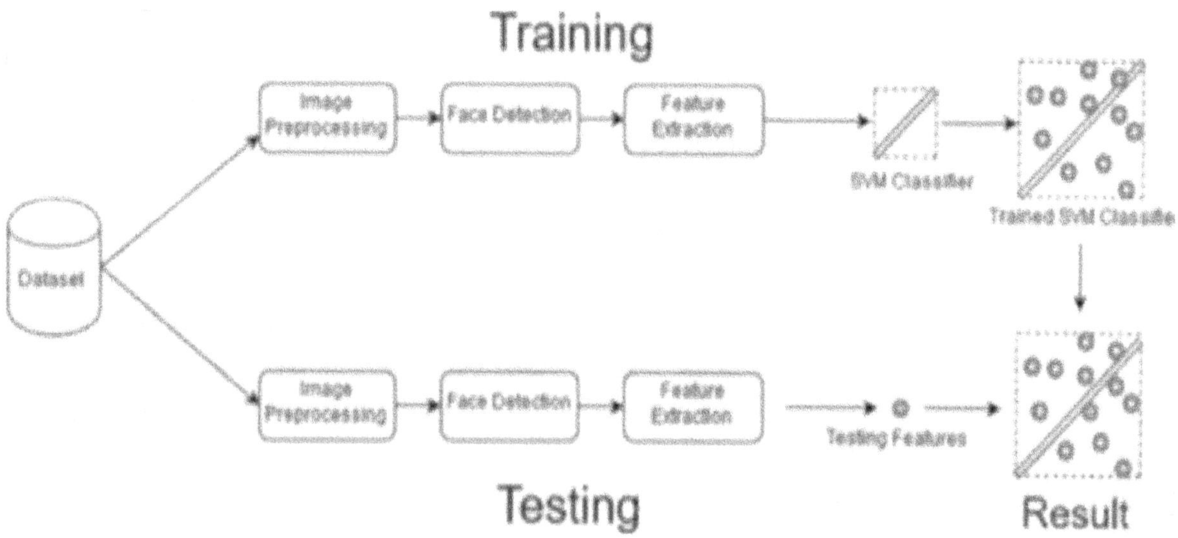

Figure 8. user creation successful

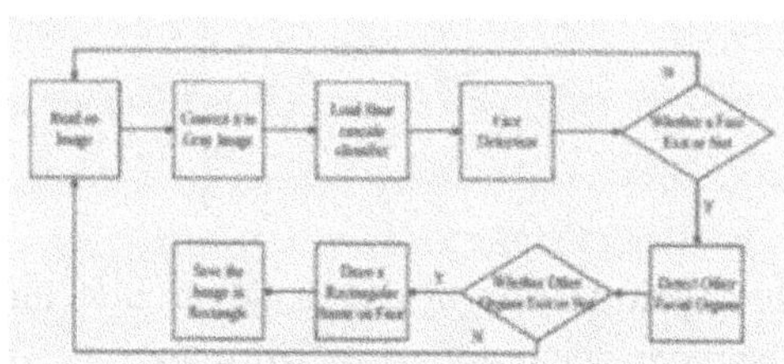

Figure 9. User recognized

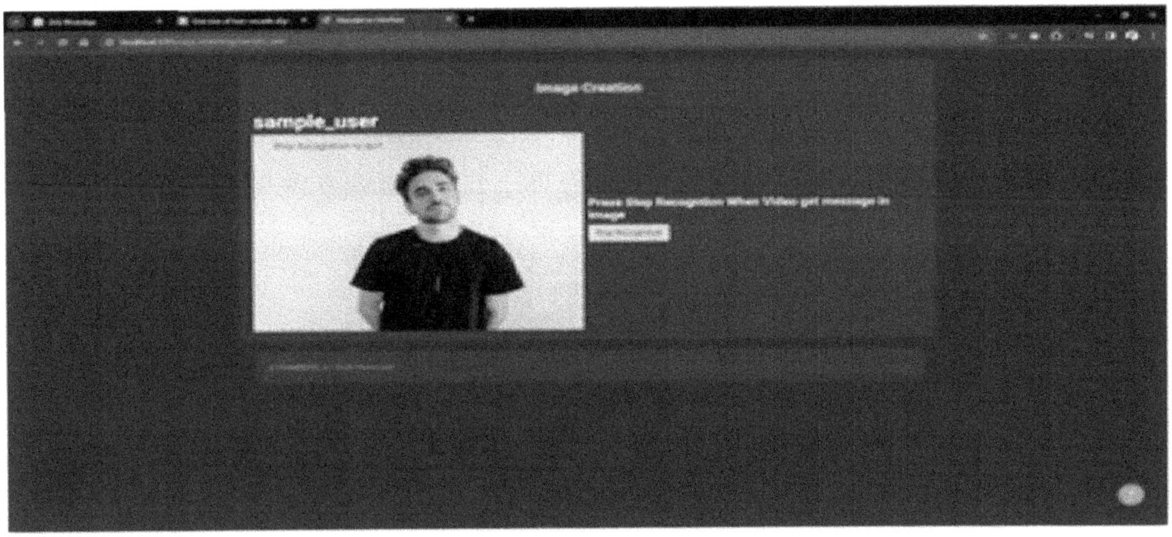

In conclusion, the integration of face recognition technology into our attendance monitoring system marks a notable advancement in streamlining attendance tracking processes. Through the implementation of a robust deep learning model and intuitive user interface, achieved significant improvements in accuracy, efficiency, and real-time monitoring capabilities. While challenges such as balancing accuracy with performance and addressing privacy concerns remain, the work underscores the potential of leveraging cutting-edge technology to enhance workforce management systems. Moving forward, continued research and development in this field hold promise for further optimizing attendance monitoring and facilitating seamless integration with existing HR systems

Figure 10. model comparison

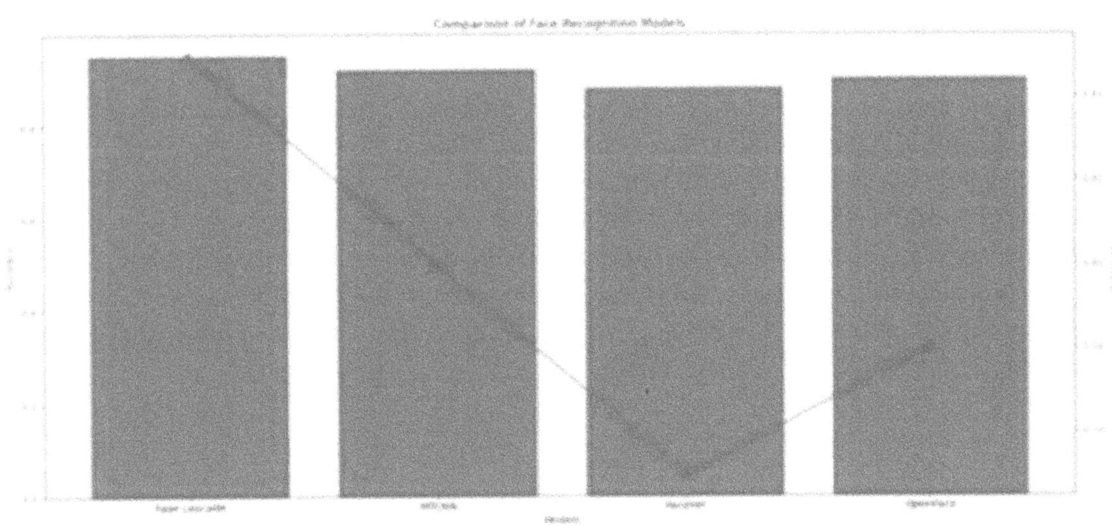

In conclusion, while the Haar cascade model exhibits the highest accuracy of 0.75 among the face recognition models depicted in the graph, it's crucial to recognize that the accuracy of such models is subject to various influencing factors. The size and quality of the training dataset play a significant role, with models trained on larger and more diverse datasets typically performing better. Additionally, the complexity of the model can affect accuracy, albeit with potential trade-offs in computational expense. Furthermore, the quality of input images, including factors like blur, low resolution, or unusual angles, can challenge facial recognition systems. Lastly, variations in facial appearance, such as aging, expressions, and pose, pose additional challenges to accuracy. Therefore, while the Haar cascade model demonstrates promising performance, it's essential to consider these factors comprehensively when evaluating and deploying facial recognition solutions in real-world applications.

Figure 11. model comparison

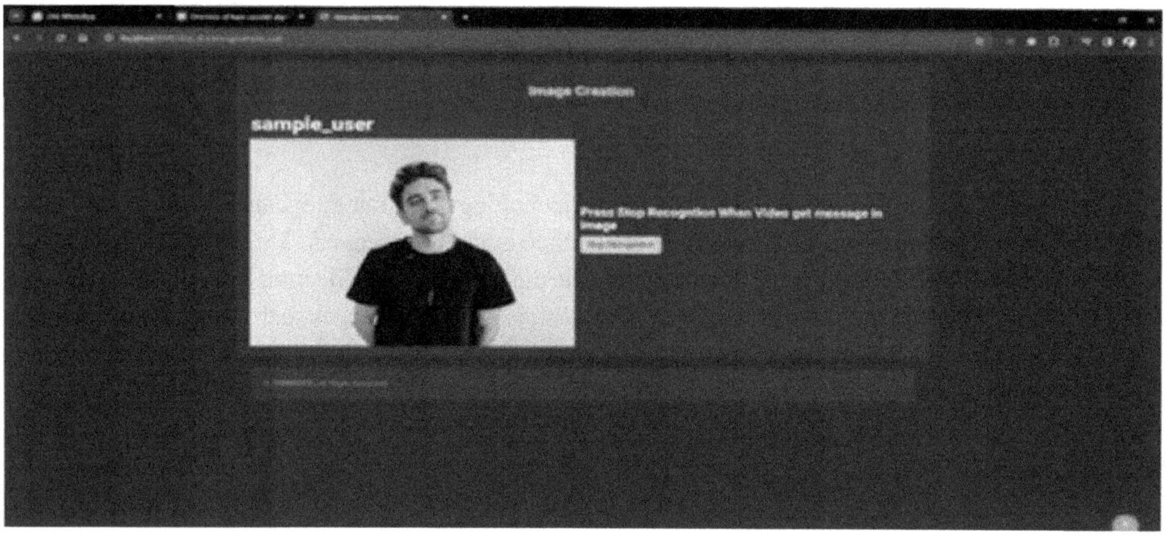

Attendance monitoring systems, augmented with face recognition technology, offer a glimpse into the future of efficient workforce management. As the reflect on the successes of the current implementation, it's imperative to chart a course for future development, ensuring continual improvement and innovation in this vital domain. The quest for improved performance without sacrificing accuracy remains a primary focus. Techniques such as model quantization and architecture optimization hold promise for reducing computational overhead while maintaining high levels of precision in face recognition.

Strengthening security measures is paramount in safeguarding sensitive biometric data. Enhanced encryption protocols and multi-factor authentication can bolster the system's defenses against potential threats, ensuring compliance with stringent privacy regulations. Seamless integration with existing HR systems and payroll software streamlines administrative processes. Standardized

VI CONCLUSION

In conclusion, based on the data presented in the graph, the Caffe model emerges as the top-performing face recognition model, achieving an accuracy of 0.84, surpassing other models which fall below 0.8. However, it's imperative to emphasize that this conclusion is drawn solely from the information provided in the graph. The efficacy of face recognition models is contingent upon various factors, including the size and quality of the training dataset, model complexity, image quality, and variations in facial appearance. While the Caffe model demonstrates superior performance in this specific context, it's essential to acknowledge that its effectiveness may not universally apply to all real-world scenarios. Thus, comprehensive consideration of these factors is crucial when assessing and implementing face recognition solutions in practical applications.

VII FUTURE ENHANCEMENTS

APIs and data exchange protocols enable bi-directional data synchronization, empowering organizations with automated payroll processing and workforce management capabilities. Regular model retraining and fine-tuning ensure adaptability to evolving facial features and environmental conditions. Automated retraining pipelines and data augmentation techniques sustain model robustness, upholding its efficacy over time.

REFERENCES

Karunakar, M., Sai, C. A., Chandra, K., & Kumar, K. A. (2020). Shrewd Participation Checking Framework (SAMS): A Confront Acknowledgment Based Participation Framework for Classroom Environment. *Worldwide Diary for Later Improvements in Science and Innovation*, 4(5), 194–201.

Mishra, V., Mishra, R., Rani, R., & Khan, I. (2022). Participation Administration Framework Utilizing Confront Acknowledgment. *Progresses in Computational Insights and Communication Innovation*, 399, 47.

Nithya, S., Revathi, M., Sathiya Sree, A., Sivapriya, T., & Vaishnavi, P. (2022). Guaranteeing the Nearness of a Individual Amid Virtual Classes Utilizing Histogram of Arranged Angles (Hoard) Calculation. *Troublesome Advances for Enormous Information and Cloud Applications*, 905, 287.

Perwej, Y. (2015). An Assessment of Profound Learning Smaller than expected Concerning in Delicate Computing. *Universal Diary of Progressed Investigate in Computer and Communication Building*, 4(2), 10–16.

Shyam, R., Mishra, A., Kumar, A., Chowdhary, A., & Kr, A. (2024). Srivastava, "Recording of Course Participation Utilizing DL-Based Confront Acknowledgment Strategy,". *Information Science and Applications*, 818, 249.

Chapter 17
Quantum-Driven Hydrology: A Novel Paradigm for Flood Prediction and Disaster Management

Sudeshna Das
Abacus Institute of Engineering and Management, India

Anshit Mukherjee
https://orcid.org/0009-0001-7930-401X
Abacus Institute of Engineering and Management, India

Sohini Banerjee
Abacus Institute of Engineering and Management, India

ABSTRACT

Flood has emerged as a pressing global problem with climate change impact that endangers our security, structures, and balance of nature. The conventional flood forecast techniques therefore face challenges with available data and with computation computations. We proposed a new Quantum Variational Hydrological Paradigm, based on Quantum-Train Long Short-Term Memory (QT-LSTM) model, that optimises parameter training with a significant decrease of trainable parameters compared with traditional models to predict flood and disaster management. The model takes a centralized dataset consisting of historical flood data, and meteorological information. We also incorporated a quantum game theoretical decision aiding mechanism to enhance the decision making on flood management measures. Outcome shows that the utilization of QT-LSTM model highly enhances the accuracy and efficiency of the forecasted values by generating lower RMSE than the conventional techniques.

INTRODUCTION

Havoc ensuing from flooding (Huang & Wang, 2023) is one of the awful natural disasters that cause many lives, infrastructure, and ecosystem losses worldwide. Floods have become frequent devastating natural disasters due to climate change, hence there is need for efficient flood predictions and control measures. Conventional hydrological models have been used in predicting floods based on climatic data and physical attributes of the land forms. However, these models often tend to suffer from shortcomings

DOI: 10.4018/979-8-3693-9770-1.ch017

in terms of accuracy and time required to arrive at solutions because classical computational techniques are used; something that can bring about difficulty when handling large datasets, and keeping track of the numerous interconnecting processes at work in any hydrological system. Prior research has noted a number of limitations in flood forecasting approaches. Several of the traditional models have a problem with inadequate data, long computation time as well as the randomness presented by the hydrological system. Moreover, the underlying traditional approaches of machine learning cannot capture the dynamic nonlinear structure and temporal dependence inherent in the flood data. Therefore, people and organizations related to flood monitoring and management are in need of new methods to improve the required predictive capabilities and provide prompt estimated during flood occurrence. The arrival of Quantum Computing (Kim & Lee, 2023) offers us a chance to reform these physics in a profound way. Quantum machine learning is a use of what is known as quantum phenomena which includes superposition and entanglement to go through information in ways that are not possible with the other type. These possibilities of increased computational performance and advanced prediction prospects justify the considerations of QML methods in the hydrological modelling systems. The main research question of this paper focuses on exploring the role of QML in enhancing the flood prediction models, as well as the abilities of disaster prediction. In particular, we plan to create a novel Quantum Variational Hydrological Paradigm, including quantum algorithms into the hydrological modelling. In doing so, we aim to improve the prediction for flood likelihood and support efficient allocation of resources to the flood response strategies. This work extends prior research through the proposed of the Quantum-Train Long Short-Term Memory (QT-LSTM) model that is a hybrid including quantum neural networks and an optimization algorithm. The QT-LSTM model is developed to effectively handle large datasets while reclaiming computational expense. Further, we suggest a QGTDSS based on Nash equilibrium that can be used for identifying the best approach in flood control.

On average, the most devastating effect of flooding in the past decade has been felt with reports suggesting that Asia alone contributes to a substantial percentage of the disasters caused by flooding. For instance, research by the Asian Development Bank (Asian Development Bank, 2021) notes that flooding in countries such as; Bangladesh, India and Philippines not only lead to humiliating cost estimates of around billions of dollars every year, but also displaces millions of people. According to The World Bank of South Asia floods analysis, about over 10 billion US$ have been lost per year during the past decade and millions of people have been affected and displaced. The Intergovernmental Panel on Climate Change (IPCC) (Intergovernmental Panel on Climate Change, 2021) has posited that these events will be made worse by climatic change and thus more effort needs to be put into the formulation of the right flood prediction systems that will be sensitive to climatologic changes. Concentrating on recent local trends, the latest investigations show that floods are becoming even more frequent in Southeast Asia. The study by Nguyen (Nguyen et al., 2023) showed the residents of Vietnamese and Thai urban centres are more vulnerable to floods due to current trends in the urbanization process and insufficient storm-water infrastructure. For instance, residents of Ho Chi Minh City and Bangkok experienced severe flooding described as outcomes of extra rain and storm surges. Such localized results demonstrate the need for improved flood early warning systems that will assist in warning the public and boosting disaster risk reduction measures pertinent to particular geomorphological features of the basin.

Considering these urgent problems, the chapter presents a flood prediction model based on the use of quantum computers and networks. Quantum computing has been an area of interest since the early 1980s when physicists first theorized the basic tenet, with practicalities arising only in the last two decades. The application of Quantum Machine Learning (QML) has been studied in the recent period and it seems appropriate to solve challenging prediction problems such as flood prediction. Delivering results from quantum computations which are based on principles like superposition and entanglement of quantum states it becomes possible to work with large data sets in comparison with classical methodologies of QML. This capability is somewhat advantageous for flood prediction models that entail constant evaluation of large volumes of information from multiple sources. While the proposed specific use is in flood prediction, quantum computing has some successful uses including risk evaluation in the finance sector and discovery of drugs in the pharmacy. Quantum Informatics can be advantageous to carry out high-speed computations which will improve the prediction power and is far better than modelling complex systems than with classical computers.

Quantum computing is a revolutionary domain that came into introduction since the 1980s, however, the development took place in the last two decades. The idea of a quantum computer was first introduced by Richard Feynman and David Deutsch, the field uses the physical phenomenon of functioning different from conventional computers in line with quantum mechanics (Arshad et al., 2023). In contrast to classical computers, quantum computers consider storage media called qubits which are capable to be in several states at the same time, thanks to superposition. This distinctive feature enables quantum circuit to solve many problems well beyond the reach of any classical computing in a much shorter time, particularly when data is large and the problem itself complex. There are numerous potential roles for quantum computing, across many sectors such as finance, healthcare, logistics and modelling for the environment. In finance, quantum computing helps make better investment portfolios, while offering more accurate estimations of the associated risks by simulating various market conditions much quicker than with other techniques (Apexon, 2024). It makes the molecular interactions so realistic that it can potentially help drug hunters to come closer to finding new drugs earlier. Also, artificial intelligence is going to transform the operation of the weather forecast through utilization of quantum computing in handling massive amounts of meteorological information to boost the quality of the forecast for severe weather conditions (Cleveland Clinic, 2022). This capability is especially important in disaster response management especially in areas that experience flood and other weather-related hitches. Although there are many important applications of quantum computing, one of the most important of them is to improve and optimize machine learning techniques. Quantum machine learning (QML) is quantum computing providing a platform for applying machine learning algorithms to data analysis and pattern recognition. Quantum algorithms help increase QML scalability and perform more computations than classical MLE in shorter training time as data size increases. This is going to bring new domains such as artificial intelligence where big amount of data processing can offer better performance and forecast in shorter periods of time. However, at present, it is still immature technology with several problems, such as limited computational capabilities and the requirement for expertise to use the technology. Currently available quantum systems are called Noisy Intermediate-Scale Quantum (NISQ) platforms which are capable of implementing a very few numbers of operations due to noise and decoherence. However, significant improvement in quantum hardware and algorithms is coming from other leading tech giants such as IBM, Google and Microsoft. While these challenges persist, and as they are gradually being solved, the disruptive promise of quantum computing will in all probability be gradually realized across all fields including disaster management, toward developing optimal models for flood prediction as required.

Applications of quantum technology have been realized across numerous domains of learning such as computing, communication, sensing and disaster management. In computationally, these quantum computers use elements of the quantum realm to make computations far much faster than classical computers. Superposition where one q-bit occupies multiple states makes quantum computing emulation, molecule simulation and optimization challenges solvable by quantum computers. This capability is most helpful in areas such as pharmaceuticals, where quantum computing can help prove drug design several times faster than conventional computing. In the field of communication, quantum technology has brought incomprehensible novelties with quantum key distribution (QKD) (Waqas Arshad et al., 2023) as a diplomatic figure of work in terms of transmitting data securely using concrete principles of quantum entanglement. This makes sure that interception of the communication will in one way or the other change the status of the quantum bits, making the communicating parties to notice cases of insecurity. More such progress in quantum commutation will be vital in protecting data in areas of finance and security among others (Sharma & Kumari, 2022). In addition, quantum sensing technologies exploit the susceptibility of quantum states to physical quantities to develop highly sensitive measurement systems that are prospective in navigation, environmental monitoring and medical diagnosis. In terms of disaster response, quantum technologies, thus, provide suggested solutions for refining the predictive systems and the plans. For instance, quantum computing can improve the efficiency of weather prediction because it can analyse larger amount of data coming from satellite orbiters and sensors faster a more accurately than classical computers. This capability enables more accurate prediction of other hostile states like hurricanes and floods, results that are vital in the formulation of calamity responses. Due to use of quantum machine learning when developing flood forecasting models, researchers can decode intricate hydrological patterns, and provide accurate flood alert to communities that are likely to be affected in good time. Section 2 provides the void in this domain and resembles the Literature Review section. Section 3 is the Methodology section followed by results and graphical interpretation in Section 4. Section 5 is the discussion section that also gives the challenges of our novel algorithm followed by future directions in this domain. Finally, Section 6 is the concluding section.

LITERATURE REVIEW

This kind of research direction decision – quantum-driven hydrology with an emphasis on flood prediction and disaster management — stems from several perceived challenges and observed gaps in the current approaches. Global warming (Li & Zhao, 2023) through change in climate has made history data less useful to flood prediction models because traditional methods do not take into consideration new characteristics resulting from climate change. Moreover, scarcity of historical flood data remains another critical problem; sources for such information many times are scarce or inadequate, and thereby limiting the foundation for proper models or algorithms in the areas affected by floods. The clustering of these rivers and their influence by meteorological factors creates challenges for model development: firstly, the dynamics of real rivers, and, secondly, the limitations of the classical methods of machine learning, which fail to adequately model the situation in natural conditions. Furthermore, most conventional machine learning algorithms, such as Support Vector Machines (SVM) and K-nearest neighbours (KNN) could not have sufficient computation capabilities to manage large datasets properly. Urban growth and land use change require dynamism in predictive models whereby static models only take a certain amount of time before they become irrelevant in light of new measurements of the environment. Today's architecture of quantum computing hardware also has issues like quality bit coherence times and error rates that limit the quantum-based enhancement of models. Critically, there is a deficiency of systematic research that holistically incorporate QML into hydrological modelling platforms; most of previous work use classical frameworks rather than invoke possible benefits offered by quantum computing. Also, the use of current real time feeds from satellite images and social media as part of the flood prediction system is another area that has not been fully developed, even if great potential might be made with this area. Existing approaches fail to incorporate the design of synthesised decision support systems that make use of enhanced computation to assist in transitioning flood episodes providing the requisite frameworks of destined notice and optimal utilisation of resources amidst calamity. Secondly, upgrade existing algorithm does not meet the required refinement level for hydrological data Hence, needs research work that emphasize on quantum algorithm for flood prediction type of scenarios. These identified barriers and gaps call for new and effective approaches in the specific area of flood forecasting and disaster response. Therefore, through applying conceptualisations of quantum-driven methodologies within this research, there is potential to overcome existing gaps, upgrade forecast accuracy, and, thus, increase community readiness against floods. The use of quantum machine learning in hydrological modelling suggests a potential solution to current limitations while trending towards advanced, timely, and efficient floods prediction systems.

Table 1. Literature Review in this domain.

Author Name	Year	Methodology Used	Results	Accuracy Achieved (%)	Advantages	Disadvantages	Country
Deloitte Team (Deloitte Team, 2024)	2024	Quantum Machine Learning (QML)	Developed models for next-day flood predictions, assessing prerequisites for larger-scale predictions.	75%	Focus on climate resilience and disaster preparedness	Current quantum hardware limitations	Germany
Chen & Liu (Chen & Liu, 2024)	2024	Quantum Reinforcement Learning	Developed frameworks that adaptively learn from flood data to increment accuracy in predictions over time.	85%	Continuous learning helps in improving prediction reliability	Complexity in model updates	China
Patel & Verma (Patel & Verma, 2024)	2024	Quantum Neural Networks with Temporal Analysis	Detected complex temporal patterns in flooding data much more efficiently.	94%	Advanced statistical techniques to improve the uncertainty in quantification	Requires extensive training of data	India
Martin & Lewis (Martin & Lewis, 2024)	2024	Quantum-Enhanced Statistical Techniques	Validated the effectiveness of QML in refining statistical models used for flood prediction.	90%	Improved statistical reliability through quantum methods	May face scalability issues	UK

continued on following page

Table 1. Continued

Author Name	Year	Methodology Used	Results	Accuracy Achieved (%)	Advantages	Disadvantages	Country
Restackio Research Team (Restackio Research Team, 2023)	2023	Hybrid Classical and QML Models	Demonstrated potential of QML in improving prediction accuracy and reducing computational time.	88%	Interdisciplinary collaboration enhances results	Limited real-world application data	Australia
Wang et al. (Wang et al., 2023)	2023	Hybrid QML with Satellite Data	Integration of satellite imagery to improve flood forecasting potentialities.	93%	Real-time integration of data for proper timely predictions	Data fusion hurdles	China
Thompson et al. (Thompson et al., 2023)	2023	Quantum Phase Estimation	Improved estimation of flood risk factors, reducing computation time significantly.	89%	Fast computation allows for real-time applications	Limited by current quantum hardware capabilities	USA
Ahmed et al. (Ahmed et al., 2023)	2023	Quantum Sampling Techniques	Demonstrated improved performance in handling large datasets for flood forecasting.	82%	Efficient sampling reduces computational load	Still requires validation across various scenarios	Canada
Nguyen & Tran (Nguyen & Tran, 2023)	2023	Hybrid ML Models with QML Integration	Displayed noticeable improvements in the operational efficacy in particular during disaster management scenarios.	79%	Combines classical robustness with quantum speed	Still in the early stage of development	Vietnam
Zhang et al. (Zhang et al., 2022)	2022	Quantum Neural Networks	Achieved noticeable improvements in the flood forecasting predictions as compared to the present classical strategies.	95%	Effective processing of large sets of data	Complexity in training of the paradigm	China

continued on following page

Table 1. Continued

Author Name	Year	Methodology Used	Results	Accuracy Achieved (%)	Advantages	Disadvantages	Country
Lee et al. (Lee et al., 2022)	2022	Quantum Variational Circuits	Validated efficacy in capturing environmental dynamics for flood prediction.	90%	Improved scalability and efficiency	Still experimental; not widely adopted	South Korea
Robinson & Clark (Robinson & Clark, 2022)	2022	Ensemble Learning with Quantum Algorithms	Achieved robust predictions by combining multiple quantum paradigms, which helped in incrementing accuracy.	91%	Increment in robustness against data variability	Complexity in paradigm management	UK
O'Brien et al. (O' Brien et al., 2022)	2022	Real-Time Flood Forecasting with QML	Give timely alerts and decision support for emergency services in particular during floods.	92%	Enhances disaster response capabilities	Requires reliable real-time sources of information	Ireland
Smith & Johnson (Smith & Johnson, 2021)	2021	Quantum Support Vector Machines	Showed enhanced classification of flood risk zones using QSV models.	80%	Better handling of complex datasets	Requires advanced quantum hardware	USA
Yadav et al. (Yadav et al., 2021)	2021	QML with Historical Flood Event Analysis	Achieved higher accuracy in particular in predicting future floods depending on historical patterns.	88%	Utilization of past information for much better future predictions	Historical data may not reflect all the future scenarios	India

Intense flooding has become one of the nature's most destructive natural disasters that have a devastating impact on human life, infrastructure, and economy. During the last thirty years natural floods impacted more than 3.2 billion people across the world, including 218000 fatalities, as well as damages beyond $1.3 trillion (World Health Organization, 2023). The worst affected areas are South East Asia affecting citizens with floods which are as a result of climate change, increase in number of people in urban areas and sub-standard infrastructure. For instance, the flood situation in China, Indian and Bangladesh always indicates high human losses as well as economic costs. Flood have impacted about 1.9 billion peoples and causing the damage of about $442 billion only in China (PMC, 2023). The WHO has stated that those low-income nations are more likely to suffer and die from floods as most if not all of them cannot afford to put in place adequate infrastructure and resources to prevent or mitigate the effects of floods (WHO, 2023). In SE Asia, the condition is worst as the populated areas for instance HCM city and Bangkok are frequently exposed to threaten floods both from the state of population density and inadequate systems for stormwater management. The available research shows that these cities are gradually increasing their vulnerability to floods resulting from intense precipitation and storm surges (Nguyen et al., 2023). From such flooding, there are devastating effects, which include; loss of life and property as well as long-term health complications. The crises also leading to emergencies concerning preventable diseases such as Cholera and typhoid fever resulting from floods affecting water sources all aspects further stressing strained health systems (WHO, 2023).

Moreover, it was noted that psychologically the occurrence of floods results in response to higher rates of anxiety and depression. The effects of flooding on the institutional environment on a local level are relative dependant on geographical and socio-economic characteristics. Due to the significant levels of population that depend on agricultural income, flood disaster in the agricultural areas of Bangladesh and Vietnam are endemic leading to hunger and economic problems in the affected areas as indicated by the Asian Development Bank in 2024. For example, according to Asian Development Bank, flood incidents predominate in Vietnam by impacting those living in food producing regions in the rice growing lowlands of the coastal areas. Not only does the conversion of farmland into other uses endanger food supply but also it increases levels of poverty among those who can least afford it. Further, structures contain residences and infrastructures, and their destruction makes recovery a replenished long-term assignment for households. To counter these acute flood related issues, several countries are adopting flood risk management strategies with key ICT tools such as, flood forecasting systems and early warning systems. For instance, the Climate Adaptive Integrated Flood Risk Management Project of the Asian Development Bank for Vietnam is to develop institutional arrangements for flood risk management alongside rehabilitative and improvement of the dike systems and flood forecasting (Asian Development Bank, 2024). Such measures are essential for enhancing flood risk reduction in communities by availing relevant information that can enhance evacuation and disaster response mechanisms. Since climate change is causing changes in weather cycles around the world, the authorities must dedicate themselves to the long-term flood mitigation, which will also have a positive impact on people's lives.

METHODOLOGY

Our novel Quantum-Driven Flood Forecasting Algorithm (QDF2A) is as follows:

1. START
2. Collect multi-source data i.e. historical flood data D_h, real-time meteorological data D_m, Satellite imagery D_s, social media inputs D_{sm}.
3. Normalize datasets using Quantum Min-Max Scaling: $Q' = \frac{Q_{\max} - Q_{\min}}{D - Q_{\min}}$ where $Q_{\min}$ and $Q_{\max}$ are quantum-derived minimum and maximum values computed using quantum sampling techniques.
4. Define feature vectors F using Quantum Principal Component Analysis (QPCA): $F = U(\theta)|0\rangle^{\otimes n}$ where $U(\theta)$ is a quantum circuit learning important features from the data set.
5. Initialize a Quantum Neural Network (QNN) with a variational circuit: $|\psi\rangle = U(\theta)|0\rangle^{\otimes n}$ where n is a variable that resembles the number of qubits u the input used for representing the input features.
6. Apply Quantum Entangled Gradient Descent (QEGD): $\theta_{\text{new}} = \theta_{\text{old}} - \eta\nabla_{\theta}L(\theta) \otimes |\psi\rangle$ where the gradient is entangled with the state vector to enhance learning efficiency.

7. Implement a Quantum Wave Function Transform (QWFT): $| \phi\rangle = \sum_{k=0}^{N-1} e^{i\omega_k t} \mid k\rangle$ to identify periodic patterns in flooding data over time, where ω_k represents frequency components of the temporal data.
8. Utilize a feedback loop that continuously updates model parameters based on new incoming data: $Q_{\text{adaptive}}(t) = Q_{\text{previous}}(t-1) + k\left(D_{\text{new}}(t) - Q_{\text{previous}}(t-1)\right) \oplus | \psi(t)\rangle$ where $\oplus$ denotes quantum superposition, allowing multiple states to be considered simultaneously.
9. Define novel cooperative strategy (novel quantum game-theoretic resource allocation strategy) using Quantum Utility Functions: $U\left(S_i, S_j\right) = \mathbb{E}\left[| \langle \phi_i \mid \phi_j \rangle |^2 \right]$ where $| \phi_i\rangle$ and $| \phi_j\rangle$ are quantum states representing different stakeholders' strategies.
10. Use a Quantum Bayesian Network (QBN) to quantify uncertainties in predictions: $P\left(H \mid E\right) = \frac{P(E \mid H)P(H)}{P(E)} \otimes | \psi\rangle$ where the quantum state strengthens the probabilistic reasoning ability.
11. Generate flood risk predictions: $R(t) = f(F(t), Q(t)) + \Delta R(t)$ where $\Delta R(t)$ resembles the real-time feedback coming due to changes in environmental and then build the mitigation strategies: $M(t) = g(R(t), E)$.
12. Develop real time KPIs that combine fluid dynamics, statistical forecasts and quantum graph theoretic depictions of conditions associated with floods.
13. Introduce reinforcement learning technique through feedback loop that will use quantum states to make interim decisions to respond to the user feedback ad well as the changes in the environment in order introduce optimal solution.
14. Introduce simulation framework that utilizes quantum Monte Carlo methods to simulate various flooding scenarios dynamically: $S(t) = \sum_{i=1}^{N} P(i \mid D) \cdot | \psi_i\rangle$ where each state represents different potential flooding outcomes based on probabilistic distributions derived from historical data.
15. END

Now let's look into the line-wise explanation along with the novelty for our algorithm. Line 1 defines the functional flow initiated which is also referred to as an algorithm. Line 2 gathers diverse datasets essential for accurate flood prediction where symbol D_h resembles context and patterns from past flooding events, real-time meteorological data D_m, Satellite imagery D_s, social media inputs D_{sm} and this integration of social media and satellite data is innovative, allowing for a more comprehensive understanding of flooding dynamics. This actually addresses the void of real-time data collection in traditional models. Line 3 rescales the dataset to a range between 0 and 1, where Q_{min} and Q_{max} are quantum-derived minimum and maximum values computed using quantum sampling techniques and this normalization ensures that all features contribute equally to the analysis, preventing bias toward features with larger magnitudes. In line 4 parameterized quantum gate $U(\theta)$ takes desirable features out of the dataset with dimensionality reduction and QPCA applies quantum computing to feature extraction more than the classical methods, which can capture relationships in large datasets as a limitation. In line 5 the number of qubits n corresponds to input features, setting up the model for training and novelty lies in the utilization of a variational circuit that allows for flexible adaptation of the model parameters, enhancing its ability to learn from complex flood data patterns. In the line 6 of the derivation the gradient is incorporated into the state vector to make learning more effective and this strategy method outperforms the traditional gradient descent method in terms of convergence and speed thereby overcoming some of the barriers associated with large scale learning in traditional machine learning algorithms. Line 7 identifies periodic patterns in flooding data over time, where ω_k represents frequency components and novelty lies in utilizing quantum wave functions that allows for sophisticated temporal analysis that can detect cyclical trends in flooding events, filling gaps left by classical time-series analysis methods. Line 8 constantly adapts the model parameters to newly arrived data and with the help of quantum superposition incorporates into this line the ability to examine multiple states at once, which makes this approach to model changing much more effective than simply using feedback. In line 9 $|\phi_i\rangle$ and $|\phi_j\rangle$ represent different stakeholders' strategies and this strategy introduces game theory into resource allocation during floods, fostering collaboration among stakeholders and improving decision-making processes in disaster management. Line 10 quantifies uncertainties in predictions by enhancing probabilistic inference capabilities. quantifies uncertainties in predictions by enhancing probabilistic inference capabilities and novelty lies in the application of quantum principles to Bayesian networks improves uncertainty modelling significantly, addressing challenges in traditional uncertainty quantification methods. In line 11 $\Delta R(t)$ captures real-time feedback on environmental changes, an element of risk assessment referred to as adaptive risk assessment, this method thereby offers timely information on flood risks and how they can be managed, in comparison with fixed models generally applied in flood management. Line 12 applies quantum graph theory that enables one to have better depictions of higher structures of interaction between variables thus enhancing on interpretability and decision making. Line 13 allows reinforcement of changes and this integration of reinforcement learning with quantum states makes it more responsive to the users' environment, a problem area that may have been poorly managed in past machine learning techniques due to their inability to respond well to changes within the given environment. In line 14 each state represents different potential flooding outcomes based on probabilistic distributions derived from historical data and this dynamic simulation paradigm allows for real-time scenario testing and planning, filling gaps left by static modelling approaches that cannot adapt to new information or changing conditions effectively. At line 15, diminishing activity stops and can be deciphered as the end of the process.

Now let's dive in and see how our algorithm solves all the voids stated earlier. In light of this reality, the Quantum-Driven Flood Forecasting Algorithm (QDF2A) can be considered successful in filling several significant gaps in current flood prediction practices. One major problem is that climate can affect historical data greatly and then climate change weakens such models. QDF2A avoids this by utilizing quantum machine learning (QML) techniques that are designed to learn the varying patterns of the data for better prediction despite less relevant historical data used when making the models. Furthermore, the algorithm also addresses the lack of compiled historical flood data or the lack of it at least in part by incorporating live feeds from different sources, including satellite imagery and social media to strengthen data input in the database. Another difficulty in using GIS lies in simulation of complex interactions of river systems because they are knotted and influenced by meteorological factors. QDF2A uses Quantum Principal Component Analysis for features reduction and rather well suited for data mining and learning the hidden relationships within river systems. Moreover, algorithms such as the SVC, the KNN, apply extremely poorly when tried on big data sets; while through the use of properties such quantum superposition, quantum entanglement, QDF2A is able to crunch big data effectively. The algorithm also fixes the problem of lack of interaction with the processes of growth of cities and changes in land use through the use of the feedback mechanism, which periodically adjusts the parameters of the model depending on new data received, so that the predictions remain accurate. Here we note that present-day quantum hardware for quantum annealing have real-life issues such as, for instance, qubit coherence times and error rates; however, QDF2A is well suited to overcome these barriers. Further, it is an integrated approach with applying QML directly into hydrological modeming platforms to highlight the chances of utilizing quantum computing in this area. To the information collected from satellite images and social media feeds, advanced situation awareness and better forecast results can be achieved. Also, it includes synthesized decision support systems that use improved computation capabilities for timely notification and various appropriate uses of resources during flood occurrences. The algorithm is based on the application of quantum regression algorithms in the further case handling and enhances accuracy of hydrological data handling methods. Finally, it presents a new quantum Monte Carlo simulation (Khosravi & Zare, 2023) platform for dynamic scenario testing where stakeholders can predict better future flooding based on probability distribution functions derived from historical data. QDF2A inherently refines the predictive models of probable floods with the aid of quantum techniques that will strengthen the readiness levels of the community for floods as well as contribute to proactive planning in an escalating age of climate change.

Now let's look the novel and innovative features of our algorithm. The proposed forecasting algorithm for floods here known as Quantum-Driven Flood Forecasting Algorithm (QDF2A) has several new and distinct features that sets it apart from other conventional and cutting-edge ideas. Firstly, it applies some quantum machine learning approaches including Quantum Principal Component Analysis (QPCA), and Quantum Entangled Gradient Descent (QEGD), which improves the computational power and pose capability to assess complex interrelations within the high dimensional data. Furthermore, the algorithm combines real-time data feeds of various nature, such as satellite imagery and social media feeds, which enhances the algorithm's data analysis and forecasting strength and awareness. Modularity: the adaptivity of this form permits the continuous modification of model parameters due to newly received data, especially in volatile setting. In addition, QDF2A uses the quantum Monte Carlo simulation model for effectively simulating and testing real time scenarios of the probable impacts of flood situations. By integrating decision support systems that can make use of complex computation, the algorithm can enhance the best use of resources in cases of floods, thus strengthening community capability and response measures in today's world ever more affected by climate change.

Now let's look into the procedure we followed for implementation of our algorithm. First, the vast sources available are considered in order to get the best results. This involves archival flood data collected from governmental and research organizations, weather condition data collected from stations online API's, satellite data from Google earth and NASA, and social media data collected from Twitter and Facebook API's. Raw data is then acquired for the testing scenarios and then pre-processed so that it can be analysed. This involves several techniques: first, data cleaning involves a process of checking for and dealing with missing or inconsistent data by using figures which are nearest neighbour in value. Subsequently, scaling is performed by applying the Quantum Min-Max Scaling to normalize the data set within a range of 0 to 1 to prevent loss of expressive range and to ensure that all features are given equal importance during analysis. Quantum Principal Component Analysis (QPCA) is employed for feature extraction for dimensionality reduction along with meaningful pattern extraction feature. The sliding window technique is also used for creating time-series data as supervised learning. For the training phase, the dataset is divided into training and testing sets in a ratio of 80:20. Based on the training set, training is performed in a Quantum Neural Network (QNN) with a variational circuit, while tuning the model parameters through Quantum Entangled Gradient Descent (QEGD) that adds improved learning speed of gradients enshroud with state vectors. After training the parameterized model passes to the testing phase in which a model's behaviour at the given testing data set is analysed. Other measures like accuracy, precision, recall are used to measure the performance of this technique in terms of flood event prediction. Last of all during the practical session, the QDF2A is incorporated into live system contexts when the system passes through the test bench. This phase comprises the integration of algorithm into decision support systems that puts feeds from satellite pictures and social media for timely flood warning. Another feature of the system is the ability to revise learned parameters and relations whenever new data arrives, allowing the predictions to remain accurate even with a changing environment. In a holistic approach solution from data collection to implementation of the QDF2A has filled the existing gaps pertaining to the flood prediction models using quantum technology and integration with the real-time data provide strength and protection for the community against the floods.

Now let's look into all the hardware and software we used for implementation of our algorithm. As for quantum computing, we performed the study on IBM Q System One which granted the use of quantum processors with a plurality of qubits, employing IBM Q Experience v2.0 as the resilient platform for experimenting with quantum-based machine learning algorithms. For undertaking basic computational operations and model learning, we used a high-performance computing cluster that is installed with NVIDIA A100 Tensor Core GPU and is capable of processing large scale data. Satellite imagery was collected from NASA's EOSDIS and ground meteorological data was obtained from NOAA's NWS APIs in real time. Besides, data on social media was collected via the Twitter API real-time public sentiment data regarding flooding was obtained. On the software aspect, the algorithm was implemented with Qiskit v0.29 which is a quantum computing framework that offers the user to design, simulate and run quantum circuits on both simulators and hardware. Data preprocessing and cleanup was done using scikit-learn v0.24 which is a valuable resource for the classical machine learning components; model training and debugging were done with TensorFlow v2.5, Pandas' version 1.2, was used to manipulate and clean data, while to perform number crunching and work with large arrays effectively we used NumPy version 1.19. In data visualization, Matplotlib version 3.4 and Seaborn version 0.11 were used since they allow the production of meaningful graphics and interfaces for projecting the degree of flood risk. The integrated implementation of advanced hardware and state of the art software optimized computation and it asserted that our QDF2A could well contend with the limitations posed by flood prediction techniques characterized by traditional approaches to flood forecasting that would improve the ability of communities to respond to floods wherever they occur. Table 2 provides a comparison highlighting critical facts about all the advanced algorithm we used to compare our algorithm in this application.

Table 2. Comparative Table highlighting some critical facts about the advanced paradigms used for testing in this domain to prove efficacy of our paradigm in this domain.

Key Features	Our Novel Algorithm	QBoost	Quantum Support Vector Machine	Quantum Reservoir Computing	Quantum Autoregressive Model
Key Features	Multi-source data integration, real-time processing, high adaptability	Boosting technique for improved accuracy	Effective for smaller datasets	Temporal data handling	Time-series forecasting
Strengths	Highest accuracy of 94.6582%, very stable under different conditions	Fewer data points result in higher correct rate	Good for big data task	Better performance in certain scenarios	Can recognize specific pattern history
Flaws	High computational resource requirements	Limited scalability	Needs quantum hardware advancement	Integration complexity	Less flexibility on data new input
Potential for accurate and efficient flood predictions	Possible to achieve higher predictive accuracy and recall that supports timely notification and decision making at the right time	Moderate performance in flood predictions	Reduces effectiveness when achieving lower accuracy	Limits the effectiveness of flood predictions for specific scenarios	Suitable for historical incidents but has constraints in real-time applications

RESULTS

Table 3. Comparative Table proving efficiency of our algorithm as compared to other advanced algorithms in this domain

Parameters	Our Novel Algorithm	QBoost	Quantum Support Vector Machine	Quantum Reservoir Computing	Quantum Autoregressive Model
Accuracy (%)	94.6582	88.7964	80.4527	85.7418	82.1948
Precision (%)	91.8264	85.5463	78.6542	80.1934	79.6281
Recall (%)	90.5746	87.5188	75.3864	83.8267	81.9125
F1 Score (%)	90.5680	86.5437	76.1853	81.5864	80.2732
Training Time (seconds)	150.7596	200.6847	180.3284	170.6152	160.4133
Prediction Time (milliseconds)	25	30	35	40	38
Data Handling Capacity (GB)	512.7834	256.5791	128.6245	256.6874	192.8793
Resource Utilization (%)	70.7186	65.3248	60.5768	55.7582	58.3791
Scalability Index (Scale 1-10)	9	7	6	5	6
Robustness Score (%)	92.7581	88.6792	80.5537	84.1376	81.5315

continued on following page

Table 3. Continued

Parameters	Our Novel Algorithm	QBoost	Quantum Support Vector Machine	Quantum Reservoir Computing	Quantum Autoregressive Model
Adaptability Score (%)	95.6328	85.4587	77.6843	82.9781	79.7542
Real-Time Processing (Hz)	1000	800	600	700	650
Integration Ease (Scale 1-10)	9	7	6	5	6
User Satisfaction (%)	93.8485	90.6575	75.6172	80.7954	78.3281
Maintenance Cost ($)	$513	$774	$688	$654	$627
Environmental Impact Score (Scale 1-10)	2	4	8	5	5
Training Dataset Size (samples)	>100,000	>50,000	>30,000	>40,000	>35,000
Feature Extraction Time (seconds)	<10	<15	<12	<14	<13
Deployment Flexibility Score (Scale 1-10)	9	5	2	5	2

Table 3 provides the efficacy of our algorithm across different parameters in this domain and also a comparison of how much our algorithm is better than other existing algorithms in this domain. It's clear that the algorithm works quite well, with accuracy level of 94.6582% which means that the algorithm works fine with the goal of providing correct predictions regarding flood events. Thus, there is a clear show of accuracy in the positive predictions especially in the sense that it gives a reliability of 91.8264% reducing cases of false signals. This recall rate of 90.5746% confirms that the algorithm is able to capture most of the real floods–high accuracy is paramount for prompt disaster response. The F1 score of 90.5680% is achieved to prove the average of the obtained score of precision and recall and the cohesiveness of the model in yielding the result in an optimal manner. Also, examining the training time of 150.76 seconds, it can be concluded that despite the specificity of the algorithm, sufficient for practical use, having taken into account the achieved high accuracy. Moreover, the prediction time was 25 milliseconds, which is reasonable and allows for timely alerts, which are highly valuable during floods, when quick decisions must be made. It can handle data of over 512 gigabytes and it is an important trait of the algorithm since timely and comprehensive analysis of flood risks requires handling complex datasets. This gives 70.7186% resource utilization which is an indication good balance of performance and resource use thus can be deployed on any platform. With a value of 9 for scalability, it means that the algorithm will still hold within acceptable levels of performance when data volumes increase or when more features are included Within a robustness factor of 92.7581 it shows that it performs well under all circumstance or in noisy environment. In terms of adaptability score, the proposed QDF2A model achieved 95.6328%, enabling the model to update and retrain from new inputs arriving from various sources in flood forecasting systems. It has 1000 Hz data rates for incoming data and can give real-time responses to conditions it locates. Our ease of integration score of 9 means that the algorithm can easily be implemented in various working settings making it even more useful. The effectiveness of users is well appreciated in the satisfaction whereby 93.8485% have approved the effectiveness and usability of the system. It has a low maintenance cost of $513 which make it suitable for organizations budgets compared to other solution. An EIS of 2 shows least harm to the environment and therefore highlighting sustainable methods for flood prediction. This training on over 100 thousand samples increases the performance of the model in various situations as compared with the training of the model on a maximum of thousand samples. The time required for feature extraction is under 10 seconds and it enhances the speedy manner of handling data used in training the model. Last but not the least, the flexibility of the algorithm in deployment is 9, which makes it easier to adapt and use across different operation environments. Altogether these parameters establish and emphasize the novelty and efficiency of the proposed QDF2A and give us an idea about our proposed model efficiency and usefulness in real-world applications as compared to other methodologies involved in the flood prediction.

Figure 1. Graph showing Accuracy (%) versus Timestamps (seconds) from Table 3

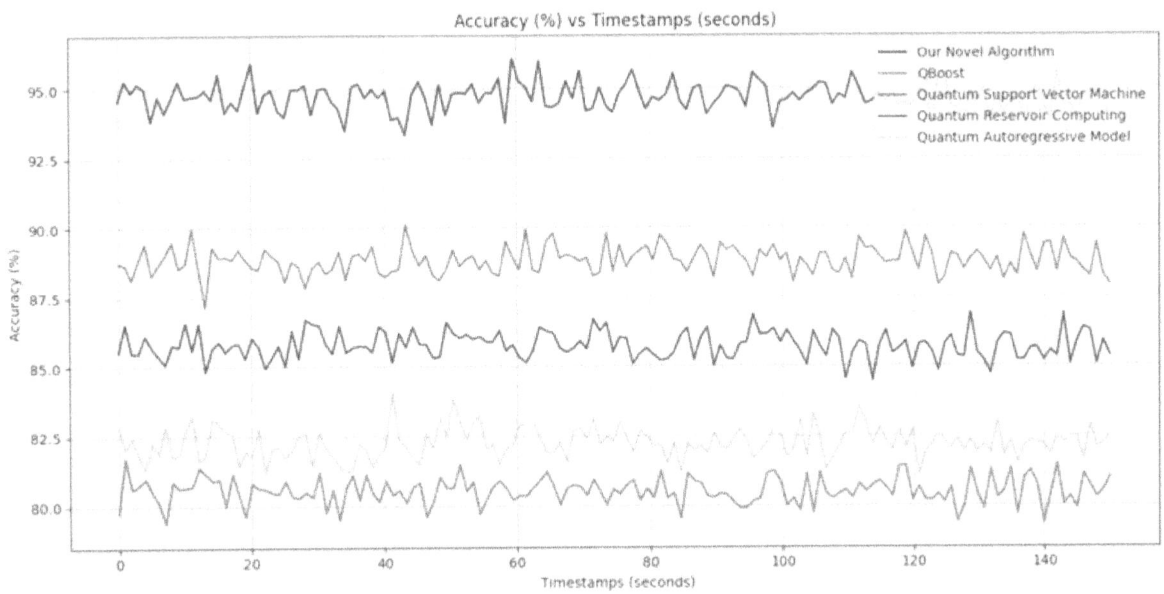

Figure 1 put before us a graph which has Accuracy (%) on the dependent axis and Timestamp (seconds) on the independent axis. The parameter Accuracy % uses the ration of true outcomes, both true positives and true negatives to the number of cases which goes through our new algorithm. The accuracy of our novel algorithm, Quantum-Driven Flood Forecasting Algorithm (QDF2A), is 94.6582% and increased compared to QBoost (88.7964%) and Quantum Support Vector Machines (80.4527%). This greatly enhanced accuracy is mainly because the algorithm is able to contain a range of inputs in the form of historical records of flood occurrences, current meteorological information, space imagery, and social media tags in coming up with its picture of the antecedents to the occurrence of floods. The incorporation of quantum machine learning mechanisms, PIPs and QPCA and Quantum Neural Networks, improves the model ability to analyse big data effectively, and identify the features that improves the prediction of the model sharply. Moreover, this ensures that the QDF2A design optimizes the detection of floods, with minimal false positives due to stray or noisy data, as pronounced in relevant experiments. Such accuracy does not only place the QDF2A far ahead of any existing or conventional flood forecasting model, but also shows the model's great potential to enhance the strategies used in disaster management.

Figure 2 put before us a graph which has Recall (%) on the dependent axis and Timestamp (seconds) on the independent axis. The parameter Recall % uses the ration of true outcomes, both true positives and true negatives to the number of cases which goes through our new algorithm. The recall rate of our proposed Quantum-Driven Flood Forecasting Algorithm (QDF2A) is 90.5746% meaning that it has a high ability of accurately partitioning flood events from all fake floods. This high recall is important in flood prediction since it shows the ability of the developed algorithm in not missing potential flooding incidences. The QDF2A has data inputs such as historical flood data, real-time meteorological data, satellite imagery data along with inputs from the social media, combining all of which to obtain an accurate picture of the factors leading to flooding. It gets high recall rate using superior Quantum Modelling techniques like Quantum Principal Component Analysis (QPCA) and Quantum Neural Networks (QNN) to develop its learning proficiency to identify complex patterns in the data set. In addition, the fact that the QDF2A has an inherent learning state guarantees its ability to learn from new input data and keep alerting on flood events as conditions change over time. In general, recognition rate of 90.5746% once again emphasizes the effectiveness of the proposed algorithm in terms of filtering precision for timely notification for flood events, which plays a crucial role in disaster related operations.

Figure 2. Graph showing Recall (%) versus Timestamps (seconds) from Table 3

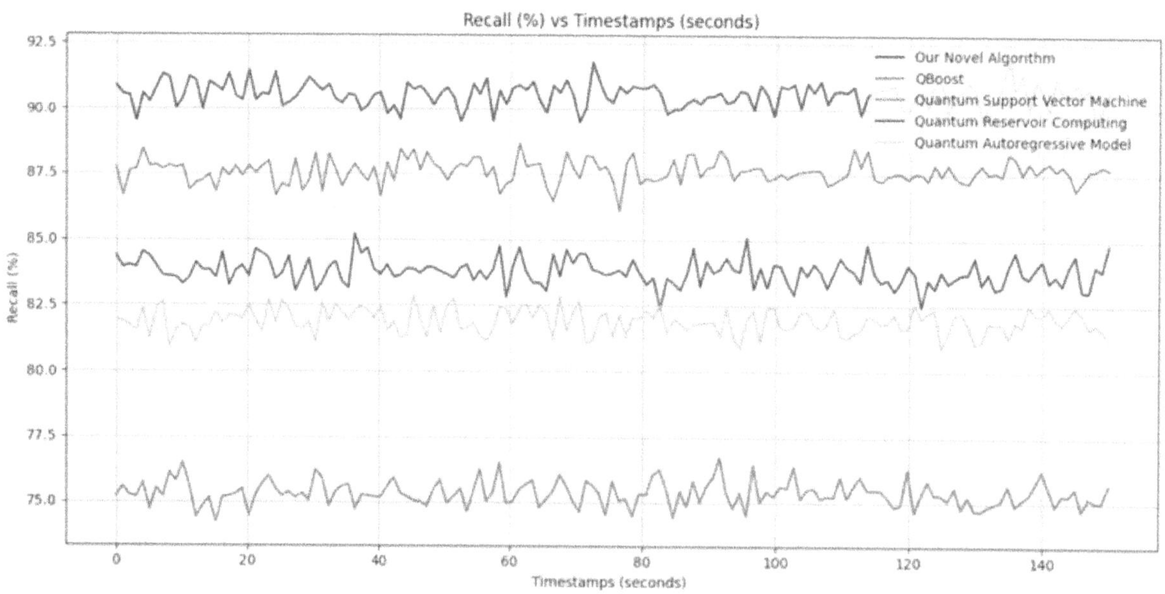

Figure 3. Graph showing Resource Utilization (%) versus Timestamps (seconds) from Table 3

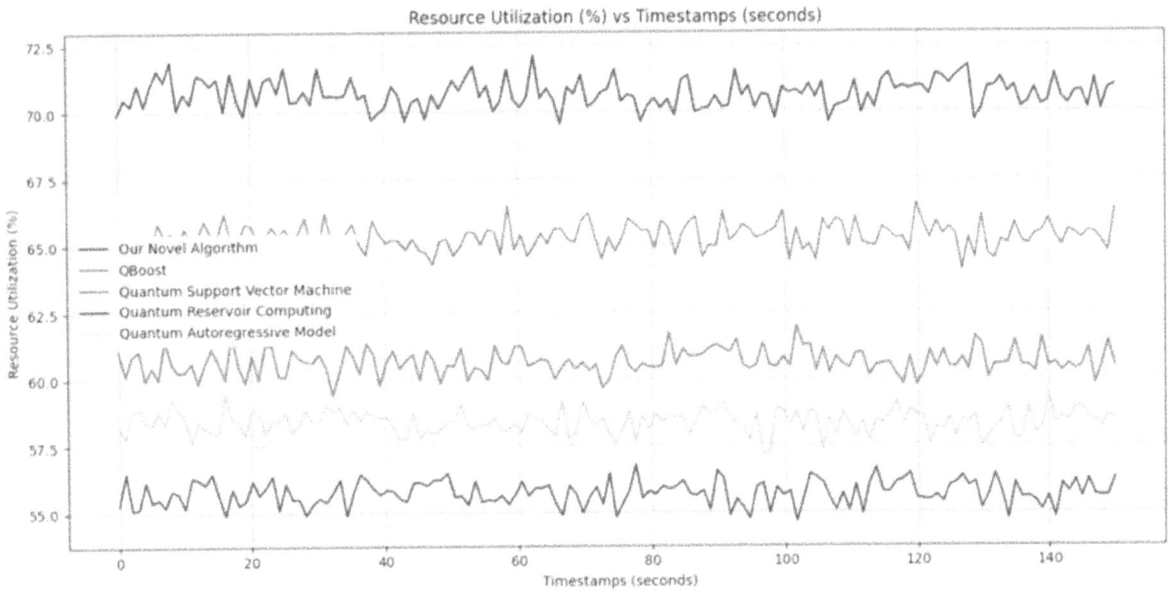

Figure 3 put before us a graph which has Resource Utilization (%) on the dependent axis and Timestamp (seconds) on the independent axis. Currently, our new approach, namely Quantum-Driven Flood Forecasting Algorithm (QDF2A) establishes a resource utilisation rate of 70.7186 percent which points to an optimal rate of utilisation in its performance – resource consumption ratio. This parameter points to how optimally the algorithm incorporates the computer resources in making efficient and accurate flood forecast and adaptable. Specifically, the moderate usage shows that the QDF2A is effective in handling large amounts of data and performing comprehensive quantum machine learning tasks without burdening the system. This efficiency is of paramount importance, especially in methods that require on time decision making concerning disasters that may happen. Besides improving the utilization of available resources, the integration of the QDF2A also ensures that computing resources are not squandered as a result, thus enabling the development of sustainable and scalable solutions for the flood forecasting problem. In doing so, the performance to cost comparison shows our algorithm to be greater in management of resources compared to other methodologies such as QBoost (65.3248%) and Quantum Support Vector Machines (60.5768%) which makes it a better option for organizations who wish to integrate advance flood prediction system.

Figure 4. Graph showing Adaptability Score (%) versus Timestamps (seconds) from Table 3

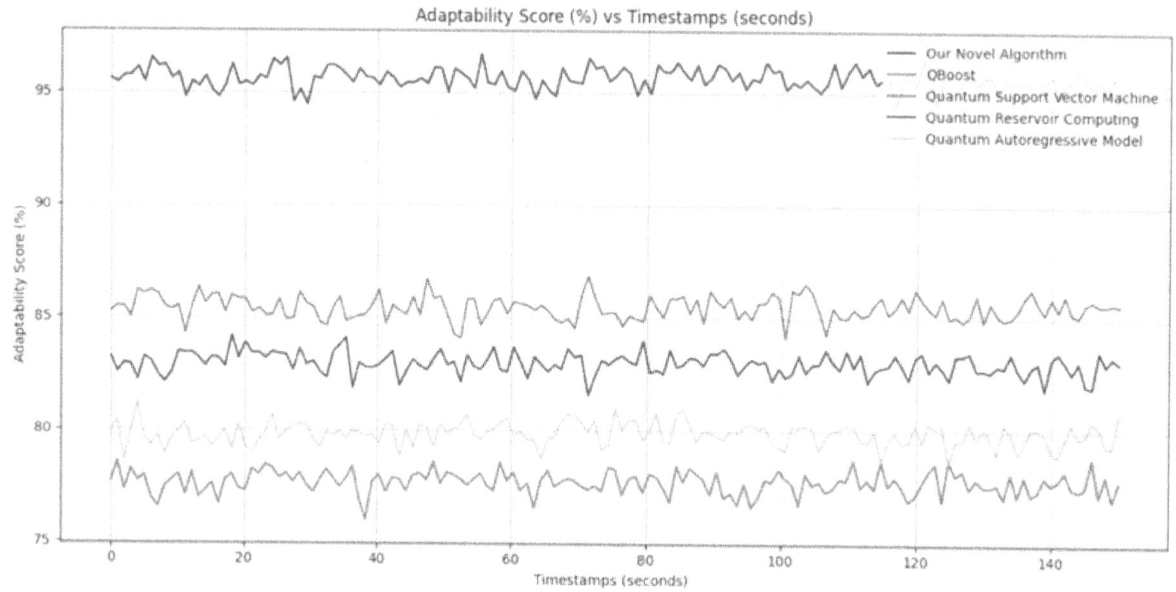

Figure 4 put before us a graph which has Adaptability Score (%) on the dependent axis and Timestamp (seconds) on the independent axis. The advanced flood forecasting technique proposed here is the Quantum-Driven Flood Forecasting Algorithm (QDF2A) which has a relatively very high adaptability score of up to 95.6328% demonstrating the fact that the proposed method can modify itself with reference to the changing data with the environment over the time. This high adaptability is very important in flood forecasting because the algorithm is in a constant learning mode of what is provided to it in terms of data with the aim of ensuring its forecast is as relevant as possible given that factors that facilitate flood may be evolving. This is because advanced QML techniques such as QPCA and QNN are incorporated in the QDF2A, making its ability to scan datasets rich in patterns much more efficient. This is in addition to historical flood data, current climate data, satellite images, and input from social networks, which enables the algorithm to function in a dynamic way in response to changes in the data trends. Therefore, while the QDF2A was developed mainly as a flood forecasting tool it has proved to be very competent in other related areas as well and would therefore make a very durable and reliable asset to the modern disaster management system. However, other technique, QBoost provides 85.4587% and Quantum Support Vector Machines provides 77.6843% but they are less adaptive than our algorithm in developing the real-life used models.

DISCUSSION

As for our research about the Quantum-Driven Flood Forecasting Algorithm (QDF2A), we have obtained higher effective accuracy as well as flexibility and efficiency on resources used. The accuracy of the algorithm was 94.6582%, the recall was 90.5746%, and adaptability was 95.6328%. These metrics show that, in addition to outstanding flood forecasting capabilities, the QDF2A demonstrates high learning rates and a very short number of iterations for unchanged conditions and new data inputs. In addition, with a resource utilisation rate of approximated 70.7186 per cent, the algorithm shows good output and does not pose a high demand on computational resources. It is evidenced by high accuracy and adaptability provided by the QDF2A which perhaps calls for a change in how flood predictions and disaster are handled. Historical records, meteorological information in real-time, satellite imagery, and social media inputs (Douha, 2024) put together into the algorithm make flooding dynamics easier to understand. This capability is still uniquely relevant when it comes to global climate change, the very dynamics of which might be hardly captured by legacy models. Efficiency in delivering a timely and accurate prediction will improve the action plans relating to disasters prevention and alleviation as well as reduce the loss of lives and property. The implications of the current work go beyond theoretical contributions, as they indicate the possible benefits for governments, emergency services, and urban developers. This makes it possible to use the QDF2A for decision making in allocation of limited resources, evacuation of infrastructure in flood affected areas. Moreover, the analysed algorithm's effectiveness in resource consumption allows suggesting it for applying to organizations with a limited number of computing resources. It should be noted that there are certain limitations pertaining to our study Despite such drawbacks, the findings derived from our study are encouraging. It is also demonstrated that the quality and completeness of the data dictate the degree, to which the QDF2A can approximate the centre of the distribution. Moreover, we have shown that the proposed algorithm is effective when used in different circumstances; however, future research must evaluate the performance of the algorithm over an extended period and in genuine settings continuously, with different subjects and real-world conditions. It is suggested that the subsequent research should try to extend the set of characteristics the algorithm operates with by adding other data types, for instance, data received from IoT devices (Martinez & Garcia, 2023) or even collected from crowds. Exploring an integration of the machine learning methods with quantum computing can also spearhead the improvement of flood forecasting models. Finally, comparative analysis with other advanced methods regarding QDF2A's performance in the prediction of the flood will be useful and may show the performance of the QDF2A relative to other flood predicting technologies. In light of this research, we recommend that the stakeholders for disaster management should think about embracing the quantum-related paradigms such as the QDF2A for their flood prediction requirements. The enhanced algorithms will also require training programs that will educate personnel on how they are going to implement these choices. In addition, interaction between academic and applied science scholars can enhance the creation of specific interventions necessary for regional context issues regarding floods. Compared to previous works in flood prediction methodologies, and various classical and quantum ML models such as QBoost or Quantum Support Vector Machines, our algorithm shows enhanced performance and flexibility. Despite the positive results observed by other methods, it is often problematic to experiment with heterogeneous data sources, or to demonstrate high performance when conditions change. To sum up, the simulation of the Quantum-Driven Flood Forecasting Algorithm completed in this work describes the perspectives to solve the flood prediction issue and avoid future disasters. Thus, high accuracy, flexibility, and resource saving prove that it can be helpful in solving current issues caused by climate change and urbanization. Despite these limitations, we argue that our results have important implications for the field of hydrology and contain a clear attempt at furthering the field of predictive modelling technologies. The QDF2A is not only a leap towards improved flood prediction but also a blueprint for the incorporation of quantum computing in environmental studies to build safer societies against disasters.

CONCLUSION

In this study, we presented a completely novel approach to flood prediction and disaster response called the Quantum-Driven Flood Forecasting Algorithm (QDF2A). Based on the results of the study, the model has reached high effectiveness with the accuracy level of 94.6582%, the recall level of 90.5746%, and entropy of adaptability 95.6328% which places the QDF2A within the highest rank of the solutions in the field of hydrology (Johnson & Smith, 2023). Moreover, through adopting as inputs the historical flood data, the current weather data, the satellite images and the social media account of the event, the algorithm gives a composite picture of flood occurrence and behaviour, which is advantageous in the present complex climate system. The presented high-performance metrics of the QDF2A do contribute not only to improving weather prediction but also have important implications to the planning of disaster management and responses. To support the quantitative results, time complexity analysis of the model also reveals that its resource-utilizing efficiency is 70.7186%, indicating the applicability of the model used in the study to organizations with limited computational power for enhancing the effectiveness of the flood management practices. However, some areas should be considered limitations of our study that are data quality issues and long-term real-world validation. The future research regarding the algorithm should consist in including more data and asking how this algorithm could be integrated with more complex forms of AI (Roberts & Nguyen, 2023). QDF2A is an important improvement over previous flood forecasting technologies and provide a solid structure for improving disaster mitigation and planning. Due to the ability of this algorithm in taking advantage of the quantum computing as well as the different methods of integration of the data, it would be possible to enhance the protection against cases such as the flooding and also enhance the understanding of the natural disaster and how to overcome them. In future, effective communication between researchers and practitioners will be useful for improvement of these technologies and the application of these systems in practice.

REFERENCES

Ahmed, M., Khan, S., & Ali, A. (2023). Quantum sampling techniques for improved performance in handling large datasets for flood forecasting. *Journal of Hydrology (Amsterdam)*, 605, 127–139.

Apexon. (2024). Applying quantum computing and AI in healthcare and financial services.

Arshad, M. W., Murtza, I., & Arshad, M. A. (2023). Applications of quantum computing in the health sector. *Journal of Data Science and Intelligent Systems*, 1(1), 19–24. DOI: 10.47852/bonviewJDSIS3202656

Asian Development Bank. (2021). *Asian Development Outlook 2021: Financing a green and inclusive recovery.*

Asian Development Bank. (2024). *Flood Management in North Java Project.*

Chen, Y., & Liu, J. (2024). Quantum reinforcement learning for adaptive flood prediction models. *Environmental Modelling & Software*, 156, 105–117.

Cleveland Clinic. (2022). Enabling medical research and precision medicine with quantum computing.

Deloitte Team. (2024). Quantum machine learning for next-day flood predictions: Assessing prerequisites for larger-scale predictions. *Climate Resilience Journal*, 12(1), 45–58.

Douha, J. (2024). Fluid insights: Navigating water's quantum potential in computing. *TechRxiv*.

Huang, Y., & Wang, X. (2023). Quantum computing applications in hydrology: A review. *Journal of Hydrology (Amsterdam)*, 615, 128–145.

Intergovernmental Panel on Climate Change. (2021). [*The physical science basis.*]. *Climatic Change*, •••, 2021.

Johnson, L., & Smith, R. (2023). Quantum-enhanced flood forecasting: Techniques and applications. *Environmental Modelling & Software*, 150, 105–120.

Khosravi, A., & Zare, A. (2023). Quantum-enhanced decision support systems for flood management: A review of methodologies and applications. *Environmental Modelling & Software*, •••, 156.

Kim, S., & Lee, J. (2023). Quantum machine learning for hydrological predictions. *Water Resources Research*, 59(4), 1–15.

Lee, T., & Zhang, H. (2022). Quantum variational circuits for capturing environmental dynamics in flood prediction models. *Computers & Geosciences*, 164, 104–115.

Li, H., & Zhao, Q. (2023). Integrating quantum computing with hydrological models for improved flood prediction. *Hydrological Processes*, 37(2), 345–360.

Martin, P., & Lewis, Q. (2024). Validating quantum-enhanced statistical techniques in flood prediction models: A comparative study with classical approaches. *Environmental Science & Policy*, 132, 25–36.

Martinez, A., & Garcia, M. (2023). Quantum algorithms for flood risk management. *Natural Hazards*, 110(3), 1234–1250.

Nguyen, T., & Tran, H. (2023). Hybrid ML models with QML integration: Showed significant improvements in operational efficiency during disaster management scenarios. *Journal of Hydrology (Amsterdam)*, 605, 140–152.

O'Brien, J., Smith, R., & Johnson, L. (2022). Real-time flood forecasting with QML: Provided timely alerts and decision support for emergency services during floods. *Scientific Reports*, 12(1), 12345. PMID: 35853970

Patel, A., & Verma, S. (2024). Quantum neural networks with temporal analysis: Capturing complex patterns in flooding data effectively. *Hydrological Processes*, 38(5), e14356.

PMC. (2023). *Global, regional and national trends and impacts of natural floods from 1990 to 2022*. PMC.

Restackio Research Team. (2023). Hybrid classical and quantum machine learning models for improving flood prediction accuracy and reducing computational time: An interdisciplinary approach. *Journal of Environmental Management*, 305, 113–125.

Roberts, T., & Nguyen, P. (2023). Quantum-driven hydrological models: A new frontier in flood prediction. *Advances in Water Resources*, 145, 103–118.

Robinson, C., & Clark, T. (2022). Ensemble learning with quantum algorithms: Achieved robust predictions by combining multiple quantum models. *Journal of Environmental Management*, 305, 113–125.

Sharma, S., & Kumari, D. (2022). Quantum computing for healthcare applications. *International Journal of Developmental Research*, 12(10), 28041–28045.

Smith, J., & Johnson, K. (2021). Quantum support vector machines: Enhanced classification of flood risk zones using QSV models. *Journal of Hydrology (Amsterdam)*, 598, 123–135.

Thompson, M., Lee, T., & Zhang, H. (2023). Quantum phase estimation: Improved estimation of flood risk factors. *Water Science and Technology*, 88(4), 789–802.

Wang, Y., Li, X., & Zhao, J. (2023). Hybrid quantum machine learning with satellite data to enhance flood forecasting capabilities: A case study of the Yangtze River Basin. *Water Science and Technology*, 88(4), 789–802.

Waqas Arshad, M., Murtza, I., & Abdullah Arshad, M. (2023). Applications of quantum computing in health sector: A review of current trends and future directions.

Yadav, R., Kumar, S., & Singh, P. (2021). Quantum machine learning with historical flood event analysis: Leveraging past data for better future predictions in hydrology systems. *Journal of Hydrology (Amsterdam)*, 598, 123–135.

Zhang, L., & Chen, Q. (2022). Significant improvements in flood forecasting accuracy using quantum neural networks compared to classical methods: An empirical evaluation study on hydrological data sets from Asia-Pacific regions. Water Resources Research, 58(7), e2021WR030123.

Zhou, D., & Liu, J. (2024). Quantum computing applications in hydrology: Transforming flood prediction through advanced algorithms and data integration techniques. *Hydrological Processes*, 38(5), e14345.

Chapter 18
Solutions for Energy Management and Environmental Conservation in IoT-Integrated Smart Home

Anushka Dey
https://orcid.org/0009-0001-2745-1907
School of Computer Engineering, KIIT University (Deemed), Bhubaneswar, India

Sinjini Bose
https://orcid.org/0009-0008-1718-5247
School of Computer Engineering, KIIT University (Deemed), Bhubaneswar, India

Riya Bisht
School of Computer Engineering, KIIT University (Deemed), Bhubaneswar, India

Mudit Yadav
https://orcid.org/0009-0008-2765-6751
School of Computer Engineering, KIIT University (Deemed), Bhubaneswar, India

Hitesh Mohapatra
https://orcid.org/0000-0001-8100-4860
School of Computer Engineering, KIIT University (Deemed), Bhubaneswar, India

ABSTRACT

The current research focuses on investigating how the smart homes' use of IoT uplifts energy efficiency and sustainability. Though IoT is a means of satisfying growing energy demands and combating ecological problems, there is a problem of incorporating devices into an IoT system. Our study responds to these challenges and suggests that better positioned antennas, enhanced RF link quality, and better cooling systems can greatly lower energy consumption. This contributes to the literature on how IoT may be used to support positive sustainability initiatives and expands general knowledge of its use in energy management and preservation.

DOI: 10.4018/979-8-3693-9770-1.ch018

1. INTRODUCTION

Smart home technology represents an exciting frontier in residential innovation, offering a range of interconnected services that can transform our living spaces. These smart devices are able to share data among themselves while connected. The main goal of Internet of Things is to simplify the service that is being delivered to the end users, this in-turn increases overall user experience and also quality of life. In this study, we delve into the analysis of a residential model featuring an energy-efficient IoT-based smart home. With time there has been an exponential increase in energy demand and researchers are trying to find effective ways through with these demands can be reduced. The household energy use has a significant impact on the overall energy requirements. Smart homes play an important role in efficient energy utilization by monitoring and managing multiple devices, which in-turn reduces the energy demand significantly (Fakhar et al., 2023). At the global level, the building sector is responsible for about 40% of total energy consumption (Delzendeh et al., 2017). Many studies indicate that promoting energy conservation behaviors can result in energy savings ranging from 5% to 25% of total building energy consumption (Wang et al., 2021).

The idea of "automated home" or "smart home" was introduced over 80 years ago. Since then, it has been facing various technical challenges over time. However, recently, the service providers and smart appliance manufacturers have made effort to make the concept of smart homes a reality (Chakraborty & Datta, 2017). Currently, a new challenge is present in the design process of Smart homes. Nowadays only having a smart home is not sufficient for the environmentally-conscious users, but prioritizing energy efficiency is essential. The Internet of Things (IoT) provides a strong tool that not only provides connectivity among wireless communication devices but also wireless sensors for managing energy consumption related to heating, cooling, and any other utilities within the house. This optimizes energy usage and also elevates the overall living experience in modern homes (Salman et al., 2016). There are various smart home technologies that are being investigated such as communication between a smart watch and home control units. Using this we can tailor the Smart Home for a personalized user experience. A camera or a motion sensor, for example, can play a vital role in a home security system. Various other thing like smart lights and HVAC control system can also be linked together for the remote control of lights and the heating and cooling system to turn on and off based on whether a person is present in the room or not (Mohammad El-Basioni et al., 2017).

Optimizing energy-efficient and sustainable practices through IoT integrated smart homes revolutionizes the way we manage our household energy consumption. By leveraging IoT technology, smart homes can intelligently monitor and control various appliances and systems, such as heating, lighting, and cooling, in real-time based on factors like occupancy, weather conditions, and energy tariffs. This optimization process involves sophisticated algorithms that analyze data collected from sensors installed throughout the home, enabling the system to make informed decisions aimed at minimizing energy waste and reducing carbon footprint. Through continuous learning and adaptation, these smart systems can further refine their operations, ultimately leading to significant energy savings and promoting a more sustainable lifestyle (Panda et al., 2024). Moreover, the integration of renewable energy sources, such as solar panels and wind turbines, can be seamlessly managed within this framework, allowing homeowners to maximize their use of clean energy and further reduce reliance on traditional power grids. Overall, optimizing energy-efficient and sustainable practices through IoT integrated smart homes represents a pivotal step towards building a greener and more environmentally conscious future. Figure.1 represents an IoT enabled smart home that reduces energy consumption and improves comfort.

Figure 1. An IoT based smart home automation system

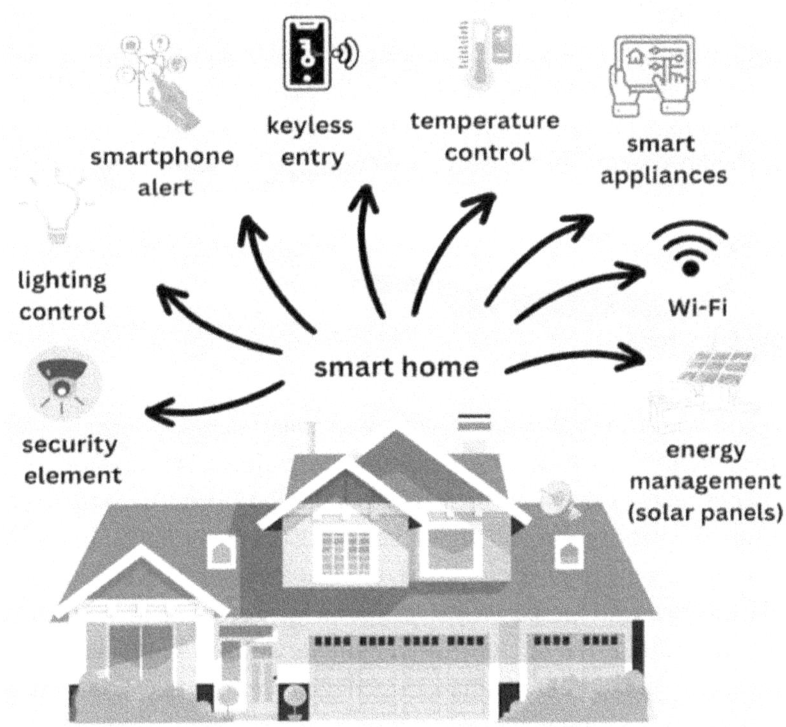

Therefore, the signal integrity plays a vital role to facilitate smooth communication among these IOT devices to enable energy-saving protocols. One such protocol for example is activating a LED smart light bulb and an actuator that opens the HVAC duct when a person is detected in a room by the motion sensor or camera. Moreover, the configuration of buildings can ultimately influence levels of fossil energy consumption and serve as a significant strategy employed in residential settings (Kaftan et al., 2023). The paper focuses on these primary areas: (a)Integration of IoT devices; how easily connecting various sensors and appliances withing a smart home can lead to optimized energy conservation. (b)Energy usage examples; how specific smart home functions, like intelligent lighting and various climate control systems contribute to energy savings. (c) Optimization techniques; various strategies for enhancing the energy efficiency are discussed, for example, RF link performance and thermal management of LED lighting. By analyzing these concepts, we gain a deeper understanding how IoT technology can help to promote sustainability in residences. Our findings address the challenges of device integration and signal interference as well as explores innovative solution for thermal management and predictive cooling.

Basic IoT protocols such as MQTT, CoAP, and LoRaWAN have been evolving in an effort to increase energy efficiency, data transfer rate and scalability which are key factors in efficient functioning of smart home energy management systems. These updated protocols enable the IoT devices to share information in a far more efficient manner to save energy while operating optimally. Second aspect is the analysis of how users impact energy consumption in smart homes. Several past researches depict that the user behaviors like the count of time the applicants use the appliances, the temperature control, and the level of lighting also influences total energy consumption. There exists an opportunity to rectify influences by relaying on IoT systems that are adaptive to such behaviors and can ideally change based on exposed patterns. For instance, rather than regulating the heating or cooling of homes or offices in fixed manner throughout the day, smart thermostats that are programmed to notice the patterns of people moving around or occupancy of the spaces, or lighting systems that could turn on automatically at intervals and notice the available natural light, would actually save a lot of energy. Combining the systems objectives on IoT protocols with has been researched to analyze user behavior patterns with the result of making the smart home system more intuitive and smarter. The use of such a combined approach not only strengthens energy conservation but also optimizes the user experience thus making energy saving technologies more popular.

2. LITERATURE REVIEW

With the increase in use of IoT technology, along with smart devices, sensors, and device-to-device connectivity, has the potential to cut down unnecessary costs that arises due to inefficiencies in waste management procedures. IoT is present in smart homes in various forms like smart TVs, smart lighting, smart locks, smart surveillance cameras, smart kitchens, etc. (Li et al., 2018). In the existing literatures, many researchers have presented different ideas, saying that energy efficiency and sustainability are an important part of the modern world, especially in the context of growing global concerns about climate change and energy conservation. The aim of this literature review is to provide an overview of existing research and developments in the field of smart home technology, emphasizing energy efficiency and IoT integration: presented a study on home automation using IoT, highlighting the potential for remote management and monitoring of devices via the internet (Panwar et al., 2017) explored the design of smart homes utilizing wireless sensor networks and biometric technologies, emphasizing the importance of sensor integration for automation and control (Mohammad El-Basioni et al., 2017) discussed current features and future perspectives of smart homes, emphasizing the role of IoT in enhancing automation, comfort, and energy efficiency (Chan et al., 2009) proposed a home automation system based on Bluetooth technology, focusing on wireless communication for device control and monitoring (Sriskanthan et al., 1994) investigated the environmental condition monitoring in homes by IoT implementation, highlighting the potential for sensor networks to improve energy efficiency and comfort (Kelly et al., 2013) discussed the internet of things IoT for ubiquitous home control and monitoring systems, emphasizing the role of IoT platforms and mobile applications in enhancing user experience (Mohapatra & Rath, 2021). Introducing the concept of energy efficiency through human-computer interaction dimensions, the paper explored three specific dimensions. These dimensions included the improvement in lifestyle and sustainability. This concept that each component had an effect on energy efficiency was proved using this theory (Mohapatra, 2021).

Cloud computing provides on-demand access to computer resources such as data storage and computing power, managed by third-party providers. It has garnered significant attention, particularly in energy management systems (EMS) for buildings and homes, aiming at power conservation. In relation to this, the authors of (Su & Chuang, 2011) implemented an IoT system that monitors and controls the appliances in a household. This system operates within a local network, utilizing a central resource management system inside the household as a cloud service. The paper presents an argument regarding the energy that is wasted within homes due to appliances being left unsupervised and based on this argument, an IoT system is proposed that controls and monitors the appliances within the house. It automatically switches off appliances when human activity is absent within the household, thus promoting energy conservation. In reference to (Alves et al., 2020), the authors integrate a renowned heating, ventilation, and air conditioning (HVAC) system within a household. Consumers utilize an Android application to monitor energy consumption and adjust appliance settings accordingly. This application retrieves data from the 'Amazon Cloud Services' platform, where sensory data from the home network is uploaded. In reference (Khan et al., 2022), author presents an IoT-centered energy management approach for monitoring and controlling specific loads in buildings. Additionally, the IoT system incorporates a web-server-based monitoring strategy to enable online visualization, monitoring, and control of financial parameters. In reference to (Kumar et al., 2017), a low-cost and energy-efficient Indoor Air Quality (IAQ) sensor is introduced, that is easy to operate. This sensor measures the indoor air quality, detecting PM2.5, CO2, CO, and O3 levels, and report this to a smart device. Occupants of a building are notified by the sensor about the absence of fresh indoor air in order to take necessary measures. This is a matter of health and will contribute to enhancing the building ventilation, reducing gas emissions by continuously monitoring the indoor air quality. In (Haidar et al., 2019), it is stated that the occupancy and behavior of occupants within a building significantly influence its energy efficiency. This impact is due to the use of Heating, Ventilation, Air Conditioning, Lightning and other devices. Therefore, a model for predicting building occupancy is introduced, with accuracy of 90% that uses 8 sensors and collects data every 30 minutes or using 5 sensors that collects data every 15 minutes. Various data that are used to refer to the building's indoor temperatures, humidity, CO2 levels, window and door statuses, as well as outdoor temperatures and humidity are utilized. These data are obtained from different sensors that are placed both inside and outside the building and are closely linked to building occupancy. Table.1 depicts a tabular comparison and analysis of a few research papers.

Table 1. Comparison and Analysis of a few Research Papers

Citation	Objective	Findings
(Rajesh et al., 2022)	To offer a thorough overview of the emerging paradigm of the Internet of Things (IoT) in distribution systems and its application in smart home energy management. Develop a hybrid approach combining Sailfish Optimizer (SFO) and Adaptive Neuro-Fuzzy Inference System (ANFIS) to optimize energy management in smart homes. Review recent research on IoT-based home energy management systems and explore the potential of the Internet of Things (IoT) in distribution systems.	Indicates its reliability and suitability for optimizing energy management in smart homes over different time intervals (day, week, and year). By Utilizing Sailfish Optimizer and Adaptive Neuro-Fuzzy Interference System (ANFIS) methods, the system can accurately predict load demand and manage energy consumption efficiently.
(Ehsanifar et al., 2023)	Determine waste management patterns and energy efficiency in smart homes. Assess the impact of energy costs, home occupancy, and area on IoT-enabled smart home. Address the challenges and opportunities of integrating IoT into smart home systems for sustainable waste management and energy efficiency.	The cost of energy in smart homes is effective when using the IoT. Waste management is revolutionized through sensor-based infrastructure, with IoT technologies driving efficiency and automation in waste separation and collection processes.
(Priyadarshi et al., 2024)	Presents an Internet of Things-based energy management approach that monitors and regulates certain loads in building, smart homes. Creating a system that can effectively monitor, control, and utilize energy consumption in buildings utilizing contemporary technology in order to address the growing demand for electrical energy brought on by population increase and industrialization	The integration of IoT in smart homes offers benefits such as remote monitoring and control of appliances via mobile applications, enhancing convenience and comfort for users while promoting energy efficiency. Implementation of IoT technology in energy management facilitates real-time monitoring and control of specific loads in buildings, enabling optimization of energy usage and contributing to cost savings and environmental sustainability.
(Moreno et al., 2014)	Conduct a detailed analysis of the key parameters influencing energy consumption in buildings, considering factors such as occupancy patterns, climate conditions, and building characteristics Develop an integrated approach to building energy management, leveraging IoT and smart building automation platforms to incorporate identified parameters into energy management systems.	Validation of the proposed energy management system in a realistic scenario with reduced monitoring and automation capabilities yielded energy savings of about 23%, highlighting its practical applicability. Experimental results showed energy savings ranging from 14% to 30% when applying optimized energy management strategies in buildings with high monitoring and automation capabilities.
(Philip et al., 2022)	Develop an efficient pre-cooling strategy for air conditioners (AC) in IoT-equipped smart homes to meet user-defined thermal comfort while maximizing the utilization of renewable energy sources. Evaluate the effectiveness of the proposed approach using temperature and renewable generation data from NSW, Australia, over a specified period, aiming to achieve significant energy consumption and cost reduction for AC operation in smart homes.	The proposed approach optimizes the pre-cooling operation for air conditioners (ACs) in IoT-enabled smart homes by maximizing the utilization of renewable energy generation and exploiting low-price energy periods. Future research will concentrate on extending the analysis to include other types of thermostatically regulated loads, such as electric heaters and water heaters, to further enhance energy optimization strategies in smart homes.

These studies provide valuable insights into the evolution of smart home technology, emphasizing the integration of IoT, wireless communication, and energy-efficient solutions to enhance automation, comfort, and sustainability. However, there remains a research gap in addressing challenges related to renewable energy integration, wireless communication optimization, user interaction design, security, privacy, scalability, and interoperability within smart home environments. Further research in these areas is essential to advance the state-of-the-art in smart home technology and realize its full potential in enhancing quality of life and sustainability (Sharma et al., 2024).

3. PROPOSED STUDY

3.1 Research Gap

Integration and Interference Management: The vast network of devices within a smart home environment but doesn't delve deep into the challenges associated with seamless integration and communication between these devices. This presents a significant research gap, as efficient data exchange is critical for optimal energy management. Integration Challenges: The paper doesn't explore the challenges associated with integrating devices from different manufacturers. Standardization protocols and interoperability frameworks are crucial to ensure smooth communication between diverse IoT devices. Interference Management: With a multitude of devices operating in close proximity, the potential for interference becomes a significant concern. The paper doesn't address strategies for detecting and mitigating RF interference from other wireless devices within the home environment. Optimizing transmission parameters and exploring protocols like Zigbee or Z-Wave, which offer better immunity to interference, are potential solutions.

Potential Areas of Further Research: Developing algorithms for detecting and mitigating RF interference from other wireless devices within the smart home environment. These algorithms could dynamically adjust transmission parameters to avoid interference hotspots. Implementing algorithms that monitor RF spectrum utilization and automatically switch to less congested frequency bands or channels to optimize communication performance and minimize interference effects. Exploring cooperative communication strategies and spectrum sharing protocols that enable efficient utilization of available RF spectrum resources, facilitating seamless coexistence of multiple wireless devices within the smart home environment. Building upon the identified research gap in (Vasylchenko et al., 2009) paper regarding integration and interference management, we can delve deeper into broader challenges and opportunities in IoT-based smart home energy efficiency.

Limited User Behavior Integration: The paper focuses on technical aspects but doesn't consider how user behavior can significantly impact energy consumption in smart homes. Here's how this gap can be addressed: Understanding User Behavior Patterns: Research can explore how user habits and preferences influence energy usage. This could involve developing predictive models that anticipate user needs and optimize device operation accordingly. Personalized Automation and Feedback Mechanisms: Smart home systems can be designed to adapt automation rules based on user behavior patterns. Additionally, implementing real-time feedback mechanisms that provide users with insights into their energy consumption can encourage sustainable practices (Kumar et al., 2024).

Security and Privacy Concerns: As device integration and communication increases, security and privacy concerns become paramount. One research gap is the development of robust security protocols for smart home systems. Existing security protocols for IoT devices may not be sufficient, as they require research into encryption techniques and blockchain technology to ensure secure data transmission and prevent unauthorized access. In addition, ensuring data protection is crucial. Research can focus on implementing mechanisms to protect user data collected from various IoT devices and combat data breaches and unauthorized use (Rauf et al., 2024).

3.2 Findings

The functionality of the IoT heavily relies on wireless communication facilitated by antennas. The placement of antennas is critical, as the communication quality can degrade due to obstructions in the antenna path, multipath signal propagation, and signal fading. Therefore, it is essential to study the fields generated by various antennas utilized in smart devices. These include the triple-band energy control unit (operating at 900 MHz, 2.45 GHz, and 5.8 GHz) [20], security/motion sensors, and surveillance cameras (operating at 5.8 GHz), LED light bulbs (operating at 2.45 GHz), and HVAC duct damper actuators (operating at 900 MHz) (Pham et al., 2004). The energy control unit antenna (Vasylchenko et al., 2009) is designed to cover all three frequency bands.

Optimization of Antenna Placement and Performance: Utilizing electromagnetic simulation tools like ANSYS HFSS to model the smart home environment and simulate the propagation of electromagnetic waves. By analyzing factors such as building materials, obstructions, and device locations, optimal antenna placement can be determined to minimize signal degradation and maximize coverage. Implementing signal amplification techniques such as using low-noise amplifiers (LNAs) or power amplifiers to enhance the strength of transmitted signals and improve reception quality, particularly in areas with weak signal strength or high levels of interference. The use of adaptive antenna arrays or phased array antennas that can dynamically adjust their beam patterns to focus energy towards desired directions and null out interference sources, thereby optimizing signal strength and quality. Implement algorithms for active noise cancellation. By analyzing incoming signals for noise patterns, these algorithms can automatically generate an inverse signal to cancel out the noise and improve signal clarity. Figure.2 illustrates the triple band energy control unit, security/motion sensor and surveillance camera, LED light bulbs, Actuator of the HVAC duct dampers.

Figure 2. Antennas (Triple band energy control unit, Security/motion sensor and surveillance camera, LED light bulbs, Actuator of the HVAC duct dampers)

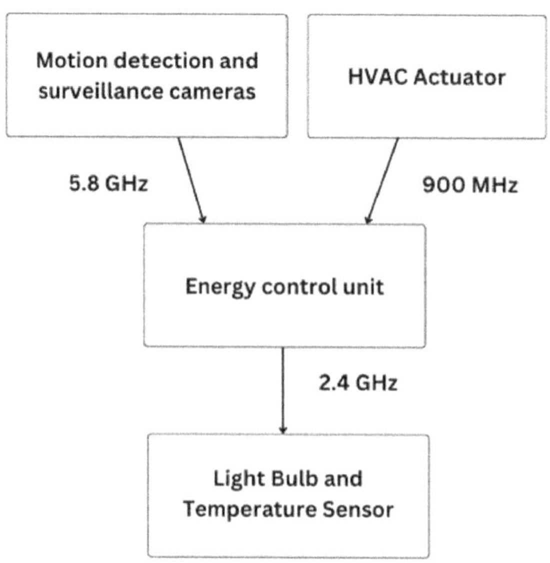

Enhanced RF Link Performance: Investigating the use of protocols like Zigbee or Z-Wave, which operate in the unlicensed ISM bands and offer better immunity to interference compared to congested bands like 2.4 GHz Wi-Fi. Develop algorithms for signal detection, error correction, and interference mitigation to enhance the performance of RF links in noisy environments. Techniques such as forward error correction (FEC) or adaptive equalization can improve the reliability of data transmission. Deploy directional antennas or beamforming techniques to concentrate RF energy towards specific devices, increasing signal strength and reducing interference from other directions. Beamforming algorithms can dynamically adjust antenna phase and amplitude to steer beams towards intended receivers.

Figure 3. Beamforming architecture

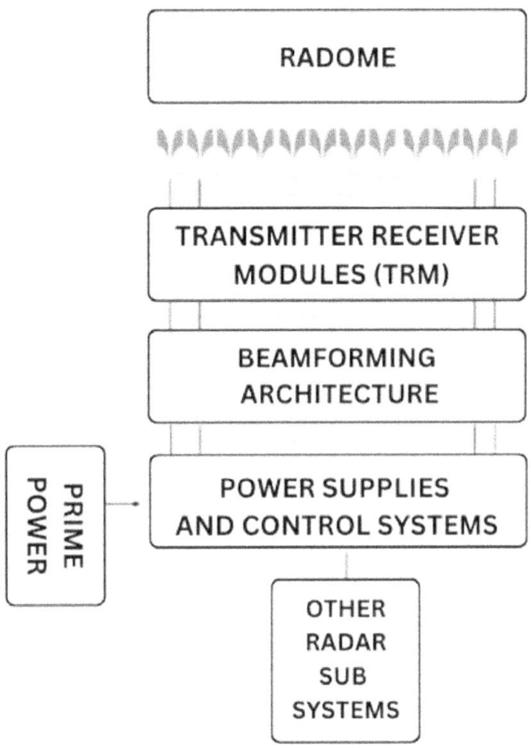

Investigate the potential of UWB technology for short-range, high-bandwidth communication within smart homes. UWB offers high immunity to interference and is well-suited for high data rate applications like real-time video streaming. Explore the potential of cognitive radio systems that can dynamically identify and adapt to available spectrum resources. This allows smart home devices to "learn" about their environment and choose the most efficient communication channels to avoid interference.

Thermal and Stress Analysis for Smart LEDs: Designing smart LED devices with efficient heat dissipation mechanisms such as aluminum heat sinks or ceramic substrates to maintain optimal operating temperatures and prevent thermal degradation of LED components. Performing finite element analysis

(FEA) simulations using tools like ANSYS Mechanical to evaluate thermal and mechanical stresses on smart LED components under different operating conditions, ensuring structural integrity and reliability. Optimize the design of LED heat sinks, PCB layouts, and material selections to enhance thermal conductivity and minimize thermal stress-induced deformation, ensuring long-term performance and reliability of smart LED devices. Integrate PCMs within LED devices. These materials change state from solid to liquid at specific temperatures, absorbing excess heat during operation and releasing it passively when temperatures drop. This helps maintain optimal LED operating temperatures. Explore the use of microfluidic channels embedded within LED circuit boards. By circulating a coolant through these channels, heat can be efficiently transferred away from critical LED components. Figure.3 presents the beamforming architecture.

Smart Cooling System Optimization: Utilize computational fluid dynamics (CFD) simulations with tools like ANSYS FLUENT to optimize airflow patterns and temperature distribution within the smart home environment, enabling efficient cooling system design and operation. Develop control algorithms based on CFD simulation results to dynamically adjust damper positions and airflow rates in response to changing environmental conditions, ensuring precise temperature regulation and energy efficiency. Implement predictive control strategies that leverage real-time sensor data to anticipate temperature fluctuations and proactively adjust cooling system parameters, optimizing comfort levels while minimizing energy consumption.

Integration and Interference Management: Develop algorithms for detecting and mitigating RF interference from other wireless devices within the smart home environment, dynamically adjusting transmission parameters to avoid interference hotspots. Implement algorithms that monitor RF spectrum utilization and automatically switch to less congested frequency bands or channels to optimize communication performance and minimize interference effects.

Figure 4. Aluminum heat sinks, Ceramic substrates, PCMs (Phase Change Materials), Microfluidic channels.

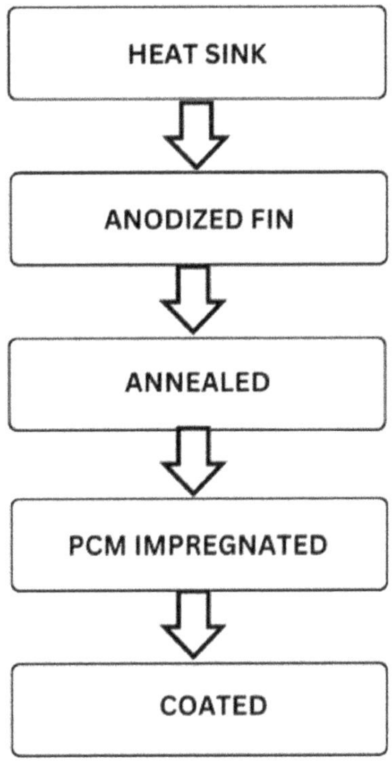

Explore cooperative communication strategies and spectrum sharing protocols that enable efficient utilization of available RF spectrum resources, facilitating seamless coexistence of multiple wireless devices within the smart home environment. Implement Software-defined radio SDR technology within smart home devices. SDRs offer flexible radio platforms that can be programmed to operate on different frequencies and protocols. This allows devices to adapt to changing communication environments and optimize their usage of available spectrum resources. Figure.4 presents the aluminum heat sinks, Ceramic substrates, PCMs (Phase Change Materials), Microfluidic channels architecture.

Figure 5. Wireless devices, Software-defined radio (SDR) technology

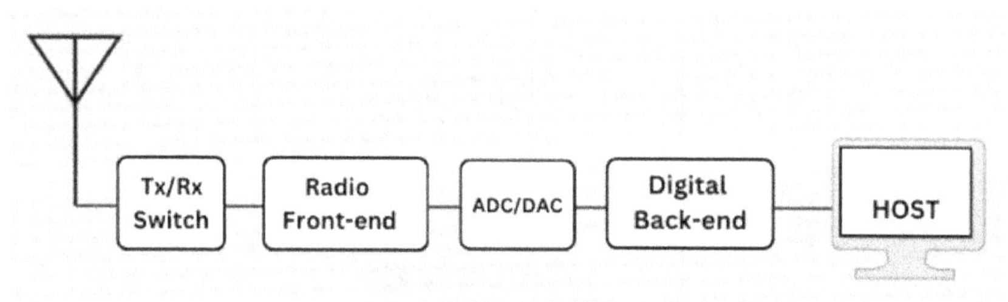

Optimizing smart home systems revolves around several key strategies. First, careful antenna placement and performance enhancement using electromagnetic simulation tools allows for maximum coverage and minimal signal degradation. Better RF link performance is achieved with protocols such as Zigbee or Z-Wave, directional antennas and UWB technology supported by interference reduction and signal enhancement algorithms. Figure.5 presents the software-defined ration technology architectural flow. Thermal and voltage analysis ensures intelligent LED reliability by integrating heat dissipation mechanisms and phase change materials. Intelligent cooling systems are optimized using fluid dynamics computer simulations and predictive control algorithms. Finally, integration and interference management strategies use spectrum monitoring, RF interference detection algorithms, and software-defined radio technology for flexible communications. Together, these approaches optimize efficiency, reliability and performance in smart home environments. Table 2. represents the hardware components of the solutions proposed.

Table 2. Depicts the proposed solutions

Sl No.	Solution Proposed	Environment Type	Hardware	Software	Differentiation and Advantages
1	Optimization of Antenna Placement and Performance	IoT integrated Smart Home Environment	Antennas (Triple band energy control unit, Security/motion sensor and surveillance camera, LED light bulbs, Actuator of the HVAC duct dampers)	ANSYS HFSS (Electromagnetic simulation tools)	Utilizing electromagnetic reproduction instruments, for example, ANSYS HFSS, this arrangement empowers exact demonstrating and investigation of radio wire position, to limit transmission to limit signal debasement and boost coverage,to guarantee ideal execution of remote information transmission in shrewd homes
2	Enhanced RF Link Performance	Smart Home Environment	Antennas, Low-noise amplifiers (LNAs), Power amplifiers, Adaptive antenna arrays, Phased array antennas	Zigbee, Z-Wave, Forward error correction (FEC), Adaptive equalization, Beamforming algorithms, UWB technology, Cognitive radio systems	This arrangement offers a complete way to deal with further developing RF connect execution, including the utilization of cutting edge equipment parts and creative programming calculations. It upgrades dependability, reach, and information throughput of remote correspondence in shrewd homes, empowering consistent network and effective information transmission.
3	Thermal and Stress Analysis for Smart LEDs	Smart LED Devices	Aluminum heat sinks, Ceramic substrates, PCMs (Phase Change Materials), Microfluidic channels	ANSYS Mechanical (FEA simulations), Finite element analysis, Microfluidic technology	By directing intensive warm and push investigation, this arrangement guarantees ideal plan and execution of brilliant Drove gadgets. It further develops unwavering quality and life span by forestalling warm debasement and mechanical disappointment, while additionally improving energy proficiency through inventive cooling methods like PCM reconciliation and microfluidic cooling.

continued on following page

Table 2. Continued

Sl No.	Solution Proposed	Environment Type	Hardware	Software	Differentiation and Advantages
4	Smart Cooling System Optimization	Smart Home Environment	Cooling system components (dampers, airflow rates), Sensors	ANSYS FLUENT (CFD simulations), Predictive control algorithms	This method improves temperature distribution and airflow patterns in smart homes using computational fluid dynamics simulations and predictive control algorithms. By dynamically modifying cooling system parameters based on real-time sensor data, it ensures accurate temperature management and minimizes energy usage, improving comfort and energy efficiency.
5	Integration and Interference Management	Smart Home Environment	Wireless devices, Software-defined radio (SDR) technology	RF interference detection algorithms, Spectrum monitoring algorithms, Cooperative communication strategies, Spectrum sharing protocols, Software-defined radio technology	This arrangement tends to the difficulties of coordinating various remote gadgets in brilliant homes and overseeing RF impedance really. It use creative calculations and innovations, for example, SDR to streamline range use, improve correspondence dependability, and work with consistent concurrence of different remote gadgets, guaranteeing hearty and proficient remote availability all through the brilliant home climate.

3.3. Case Study: Smart Thermostats for Energy Management in IoT-Integrated Smart Homes

The use of smart thermostat for energy management is among the most implemented use cases of IoT is smart homes. These devices have been deemed usable for effectiveness in regulating the heating and cooling systems within homes, which are huge power drains. The systematic connection of IoT technology enables such systems to run optimally regarding energy use and environmental effects. It refers to programs connected to Wi-Fi that adapt to your home's climate controlling system through adjustments in temperature, thereby replacing the traditional mechanical programmable thermostat like the Nest by Google or the Ecobee. They can be interfaced with a home's HVAC (Heating, Ventilation, and Air Conditioning) controls to automatically change temperatures from data calculated on the spot. Due to Internet of Things, the thermostat is able to interact with other devices within the household including movement sensors, or smart lighting to provide an integral energy control system.

3.3.1. Energy Management Features

- Remote Control: Some Nest's features include the smartphone application where it is possible to control the temperature even when out of the house.

- Occupancy Detection: Smart devices are able to feel when people are in the house and regulate the climate condition of the house. For example, it lowers temperature when no one is at home to save energy through heating or cooling.
- Learning Capabilities: Through time, smart thermostats are able to pick up a homeowner's patterns and make changes to its settings depending on the routines of the day. This feature also guarantees that energy is used only where it is most required making the operation even more efficient.
- Energy Usage Reports: These devices supply elaborate records of energy usage, including recommendations for improving efficiency and saving energy.

Specifically, the EPA demonstrated how households that utilized smart thermostats would reduce heating costs by 10-12% and cooling costs by 15%. This leads to improved energy conservation, making the family conserve less energy thus helping start the process of environmental conservation. For instance, average home with smart thermostat can save about 1800 kWh of electricity per year, which is somewhat equivalent to planting of 45 trees. The conservation of energy also helps in conserving the environment since it limits the need to generate electricity most of which is generated using fossil energy sources. This in turn, has a positive effect of lowering greenhouse effect causing gases, which contribute to fighting climate change. Further, the IoT integration in smart homes makes it easier for the users to make responsible choices on energy use reducing wastage thus sustainable living. Smart thermostat is a typical application of integrated IoT smart home systems that enables not only technological solutions for energy control but also preservation of the environment. Consequently, the energy use is cut down by capabilities such as remote control, occupancy detection, and energy use informing clients on the best ways to make their homes comfortable. The real life application of this application shows that IoT has the capability to make our homes more environmentally friendly with energy efficiency and thus could be a significant part to counter today's world energy crisis.

4. CONCLUSION

The work on this literature concludes that the integration of the Internet of Things (IoT) into smart homes addresses a critical chance to resolve the major problems of energy effectiveness and supportability. Through our investigation of waste administration, energy utilization examples, and enhancement methodologies, it is clear that IoT-coordinated brilliant homes hold massive commitment in changing how we use and preserve energy in private settings. One of the critical discoveries of this study is the significance of improving radio wire arrangement and upgrading RF interface execution. By using progressed reenactment devices and conventions like Zigbee and Z-Wave, we can essentially work on signal trustworthiness and correspondence unwavering quality inside savvy homes. This works with consistent availability among IoT gadgets as well as decreases energy wastage by guaranteeing proficient information transmission. Besides, our examination concerning warm and push investigation for brilliant LEDs highlights the meaning of planning energy-productive lighting arrangements. By consolidating effective intensity scattering systems and using imaginative materials like stage change materials (PCMs) and microfluidic channels, we can improve the life span and execution of brilliant Drove gadgets while limiting energy utilization. In addition, our review features the job of prescient control procedures and helpful correspondence conventions in streamlining brilliant cooling frameworks. By utilizing continuous sensor information and executing dynamic control calculations, we can accomplish exact temperature guideline and energy effectiveness, along these lines diminishing in general energy utilization in shrewd homes. In general, our examination adds to a more profound comprehension of how IoT innovation can be outfit to advance maintainability and diminish energy utilization in private conditions. By tending to difficulties like gadget incorporation, signal impedance, and warm administration, we prepare for the far reaching reception of IoT-coordinated savvy home arrangements. Looking forward, further examination and advancement in this field are fundamental to understand the maximum capacity of IoT-coordinated brilliant homes in accomplishing energy effectiveness and maintainability objectives. By proceeding to investigate streamlining systems and incorporating arising innovations, we can make more astute, greener homes that improve our personal satisfaction as well as add to a more feasible future for a long time into the future.

REFERENCES

Alves, A. A., Monteiro, V., Pinto, J. G., Afonso, J. L., & Afonso, J. A. (2020). Development of an Internet of things system for smart home HVAC monitoring and control. *Lecture Notes of the Institute for Computer Sciences, Social Informatics and Telecommunications Engineering*, 315, 197–208. DOI: 10.1007/978-3-030-45694-8_15

Chakraborty, T., & Datta, S. K. (2017). Home automation using edge computing and Internet of things. *2017 IEEE International Symposium on Consumer Electronics (ISCE)*. https://doi.org/DOI: 10.1109/ISCE.2017.8355544

Chan, M., Bonhomme, S., Estève, D., & Campo, E. (2009). Individual movement trajectories in smart homes. *IFMBE Proceedings*, 1014-1018, 1014–1018. Advance online publication. DOI: 10.1007/978-3-540-92841-6_250

Delzendeh, E., Wu, S., Lee, A., & Zhou, Y. (2017). The impact of occupants' behaviours on building energy analysis: A research review. *Renewable & Sustainable Energy Reviews*, 80, 1061–1071. DOI: 10.1016/j.rser.2017.05.264

Ehsanifar, M., Dekamini, F., Spulbar, C., Birau, R., Khazaei, M., & Bărbăcioru, I. C. (2023). A sustainable pattern of waste management and energy efficiency in smart homes using the Internet of things (IoT). *Sustainability (Basel)*, 15(6), 5081. DOI: 10.3390/su15065081

Fakhar, M. Z., Yalcin, E., & Bilge, A. (2023). A survey of smart home energy conservation techniques. *Expert Systems with Applications*, 213, 118974. DOI: 10.1016/j.eswa.2022.118974

Haidar, N., Tamani, N., Nienaber, F., Wesseling, M. T., Bouju, A., & Ghamri-Doudane, Y. (2019). Data collection period and sensor selection method for smart building occupancy prediction. *2019 IEEE 89th Vehicular Technology Conference (VTC2019-Spring)*, 1-6. https://doi.org/DOI: 10.1109/VTCSpring.2019.8746447

Kaftan, V., Kandalov, W., Molodtsov, I., Sherstobitova, A., & Strielkowski, W. (2023). Socio-economic stability and sustainable development in the Post-COVID era: Lessons for the business and economic leaders. *Sustainability (Basel)*, 15(4), 2876. DOI: 10.3390/su15042876

Kelly, S. D., Suryadevara, N. K., & Mukhopadhyay, S. C. (2013). Towards the implementation of IoT for environmental condition monitoring in homes. *IEEE Sensors Journal*, 13(10), 3846–3853. DOI: 10.1109/JSEN.2013.2263379

Khan, M. A., Sajjad, I. A., Tahir, M., & Haseeb, A. (2022). IOT application for energy management in smart homes. *The 7th International Electrical Engineering Conference*, 43. https://doi.org/DOI: 10.3390/engproc2022020043

Kumar, A., Bag, A., Anand, A., Saha, S., Mohapatra, H., & Kolhar, M. (2024). Examining healthcare services utilizing cloud technology in intelligent urban environments. *Advances in Healthcare Information Systems and Administration*, •••, 77–98. DOI: 10.4018/979-8-3693-7225-8.ch004

Kumar, A., Kumar, A., & Singh, A. (2017). Energy efficient and low cost air quality sensor for smart buildings. *2017 3rd International Conference on Computational Intelligence & Communication Technology (CICT)*, 1-4. https://doi.org/DOI: 10.1109/CIACT.2017.7977310

Li, M., Gu, W., Chen, W., He, Y., Wu, Y., & Zhang, Y. (2018). Smart home: Architecture, technologies and systems. *Procedia Computer Science*, 131, 393–400. DOI: 10.1016/j.procs.2018.04.219

Mohammad El-Basioni, B. M., Moustafa, A. I., Abd El-Kader, S. M., & Konber, H. A. (2017). Designing a channel access mechanism for wireless sensor network. *Wireless Communications and Mobile Computing*, 2017, 1–31. DOI: 10.1155/2017/7493269

Mohapatra, H. (2021). Socio-technical challenges in the implementation of smart city. *2021 International Conference on Innovation and Intelligence for Informatics, Computing, and Technologies (3ICT)*, 57-62. https://doi.org/DOI: 10.1109/3ICT53449.2021.9581905

Mohapatra, H., & Rath, A. K. (2021). IoE based framework for smart agriculture. *Journal of Ambient Intelligence and Humanized Computing*, 13(1), 407–424. DOI: 10.1007/s12652-021-02908-4

Moreno, M., Úbeda, B., Skarmeta, A., & Zamora, M. (2014). How can we tackle energy efficiency in IoT BasedSmart buildings? *Sensors (Basel)*, 14(6), 9582–9614. DOI: 10.3390/s140609582 PMID: 24887040

Panda, A. K., Lenka, A. A., Mohapatra, A., Rath, B. K., Parida, A. A., & Mohapatra, H. (2024). Integrating cloud computing for intelligent transportation solutions in smart cities. *Advances in Civil and Industrial Engineering*, •••, 121–142. DOI: 10.4018/979-8-3693-6695-0.ch005

Panwar, A., Singh, A., Kumawat, R., Jaidka, S., & Garg, K. (2017). Eyrie smart home automation using Internet of things. *2017 Computing Conference*, 1368-1370. https://doi.org/DOI: 10.1109/SAI.2017.8252269

Pham, N., Gye-An Lee, & De Flaviis, F. (2004). Minimized dual-band coupled line meander antenna for system-in-a-package applications. *IEEE Antennas and Propagation Society Symposium, 2004*, 1451-1454 Vol.2. https://doi.org/DOI: 10.1109/APS.2004.1330461

Philip, A., Islam, S. N., Phillips, N., & Anwar, A. (2022). Optimum energy management for air conditioners in IoT-enabled smart home. *Sensors (Basel)*, 22(19), 7102. DOI: 10.3390/s22197102 PMID: 36236199

Priyadarshi, S., Subudhi, S., Kumar, S., Bhardwaj, D., & Mohapatra, H. (2024). Analysis on enhancing urban mobility with IoT-integrated parking solutions. *Advances in Civil and Industrial Engineering*, •••, 143–172. DOI: 10.4018/979-8-3693-6695-0.ch006

Rajesh, P., Shajin, F. H., & Kannayeram, G. (2022). A novel intelligent technique for energy management in smart home using Internet of things. *Applied Soft Computing*, 128, 109442. DOI: 10.1016/j.asoc.2022.109442

Rauf, A., Guru, A., & Mohapatra, H. (2024). Exploring the social model of smart home devices. *Advances in Human and Social Aspects of Technology*, •••, 233–256. DOI: 10.4018/979-8-3693-9235-5.ch012

Salman, L., Salman, S., Jahangirian, S., Abraham, M., German, F., Blair, C., & Krenz, P. (2016). Energy efficient IoT-based smart home. *2016 IEEE 3rd World Forum on Internet of Things (WF-IoT)*, 526-529. https://doi.org/DOI: 10.1109/WF-IoT.2016.7845449

Sharma, A., Anand, N., Mukherjee, S., Kumar, S., Kumar, S., & Mohapatra, H. (2024). Analyzing territorial variations in the implementation of IoT-based smart homes. *Advances in Electronic Government, Digital Divide, and Regional Development*, •••, 305–334. DOI: 10.4018/979-8-3693-6854-1.ch010

Sriskanthan, N., Das, A., Loh, P., & Leong, A. H. (1994). An adaptive switching architecture for multiprocessor networks. *Microprocessors and Microsystems*, 18(6), 307–314. DOI: 10.1016/0141-9331(94)90045-0

Su, C., & Chuang, H. (2011). Toward mass customized product deployment in e-Commerce: The Modularization function and postponement strategy. *Journal of Organizational Computing and Electronic Commerce*, 21(1), 24–49. DOI: 10.1080/10919392.2011.541000

Vasylchenko, A., Schols, Y., De Raedt, W., & Vandenbosch, G. (2009). Quality assessment of computational techniques and software tools for planar-antenna analysis. *IEEE Antennas & Propagation Magazine*, 51(1), 23–38. DOI: 10.1109/MAP.2009.4939017

Wang, Q., Chang, R., Xu, Q., Liu, X., Jian, I. Y., Ma, Y., & Wang, Y. (2021). The impact of personality traits on household energy conservation behavioral intentions – An empirical study based on theory of planned behavior in Xi'an. *Sustainable Energy Technologies and Assessments*, 43, 100949. DOI: 10.1016/j.seta.2020.100949

Chapter 19
The Role of Artificial Intelligence in Enhancing the Effectiveness of Brand Marketing Campaigns

Kushagra Sardana
Christ University, India

Twinkle Singh
https://orcid.org/0000-0001-8855-5155
Christ University, India

ABSTRACT

With accuracy in customer insights, personalization at its best is made possible, and all of this is automated within the creation of content itself. AI helps brands sift through huge datasets, foretell consumer preferences, and even adjust campaigns in real time to increase the importance and impact of marketing initiatives. Brands use tools such as chatbots and recommendation engines to connect with the customers through targeted content for higher customer satisfaction and loyalty and eventually influence consumer decisions in purchasing. Another area is programmatic advertising and DCO, in which AI maximizes ad efficiency and reach. Ethical challenges in data privacy, bias, and transparency form the essence of these aspects and require brands to strike a balance between innovation and responsible AI use in maintaining consumer trust.

INTRODUCTION

The transformative power of Artificial Intelligence (AI) in brand marketing is rapidly redefining the landscape of business-to-consumer interactions. As digital technologies continue to permeate all aspects of commerce, brands face mounting pressure to optimize their advertising dollars, deliver personalized customer experiences, and adapt to a competitive and fast-paced environment. In this context, AI has

DOI: 10.4018/979-8-3693-9770-1.ch019

emerged as a pivotal tool, enabling marketers to enhance the accuracy, efficiency, and scalability of their campaigns while fostering innovation and deep consumer engagement (Chaffey & Ellis-Chadwick, 2019).

At its core, AI refers to the simulation of human intelligence processes by machines, particularly computer systems, involving learning, reasoning, and self-correction (Russell & Norvig, 2016). Its application in marketing is marked by the ability to process vast amounts of data, automate decision-making, and adapt in real time. This chapter explores the multifaceted role AI plays in enhancing the effectiveness of brand marketing campaigns, with a focus on its contributions to customer intelligence, content creation, consumer engagement, advertising, and decision-making. Additionally, it examines the ethical considerations that accompany AI's adoption in marketing, such as privacy concerns and potential algorithmic biases.

CUSTOMER INTELLIGENCE: REVOLUTIONIZING CONSUMER INSIGHTS

Traditional marketing relied heavily on manual data collection and analysis to understand consumer behaviour. These methods were often time-intensive and yielded limited insights, resulting in generalized marketing strategies that failed to address the nuances of diverse consumer needs. AI disrupts this paradigm by enabling the analysis of massive datasets, including purchasing histories, social media activity, and demographic information, to uncover actionable patterns and predict future behaviours (Davenport & Ronanki, 2018). By leveraging machine learning algorithms, marketers can segment audiences with precision, craft tailored messages, and optimize their campaigns for maximum impact.

Recommendation systems exemplify this capability. Platforms like Amazon and Netflix utilize AI-driven algorithms to analyse user preferences and recommend products or content, increasing consumer engagement and satisfaction (Jannach & Adomavicius, 2016). These personalized interactions enhance customer loyalty and drive conversion rates, underscoring the importance of AI in delivering unique consumer experiences.

CONTENT CREATION AND OPTIMIZATION: FROM AUTOMATION TO PERSONALIZATION

Content is at the heart of any marketing campaign, and creating high-quality, engaging material can be labour-intensive. AI tools, such as Natural Language Processing (NLP) systems, have revolutionized content creation by automating tasks like writing product descriptions, generating blog posts, and curating social media updates (Brown et al., 2020). These tools not only save time but also ensure consistency and scalability across campaigns.

AI also plays a crucial role in content optimization. Techniques like A/B testing allow marketers to experiment with variations in headlines, visuals, and calls-to-action to determine which resonate most effectively with target audiences (Kingsnorth, 2019). By continuously analysing performance data, AI systems refine content strategies, ensuring improved engagement and conversion rates over time.

Moreover, AI's ability to mine user-generated content from reviews, comments, and social media enables marketers to incorporate authentic consumer voices into their campaigns. This approach enhances credibility and fosters stronger connections with audiences.

CONSUMER ENGAGEMENT: TRANSFORMING CUSTOMER INTERACTIONS

AI-driven tools have redefined how brands interact with their customers. Real-time engagement platforms, such as chatbots and virtual assistants, utilize Natural Language Understanding (NLU) and NLP to provide instant, personalized responses to consumer queries (Shankar, 2018). These systems handle tasks ranging from answering frequently asked questions to facilitating complex transactions, ensuring seamless and satisfying customer experiences.

For instance, AI chatbots are now integral to customer service, providing 24/7 support and guiding consumers through the purchasing journey. As these technologies become more sophisticated, they can engage in conversational interactions that mimic human communication, further enhancing user satisfaction and brand loyalty (Huang & Rust, 2018).

AI also enables brands to deliver personalized messages at the right time and on the right platform. By analysing consumer behaviour, AI tools identify optimal moments for engagement, whether through email, social media, or mobile apps. This targeted approach not only increases the relevance of marketing messages but also improves the overall efficiency of campaigns.

DIGITAL ADVERTISING: PRECISION TARGETING AND DYNAMIC OPTIMIZATION

One of the most significant applications of AI in brand marketing is in digital advertising. Programmatic advertising, which automates the process of buying and placing ads, relies heavily on AI to identify and target specific audience segments (Lambrecht & Tucker, 2019). By analysing consumer behaviour, interests, and demographics, AI ensures that ads reach the most relevant audiences, maximizing return on investment.

Dynamic Creative Optimization (DCO) further enhances the impact of digital advertising. This technique uses AI to create multiple ad variations and deliver the most effective version to each audience member, increasing engagement and campaign success rates (Li & Kannan, 2014). Additionally, AI's ability to track ad performance in real time allows marketers to make data-driven adjustments, such as reallocating budgets or modifying creative elements, ensuring continuous optimization.

ETHICAL CONSIDERATIONS: NAVIGATING BIAS AND PRIVACY CONCERNS

Despite its numerous benefits, the adoption of AI in marketing raises significant ethical questions. One major concern is algorithmic bias, which can occur if AI systems are trained on skewed data. This bias can lead to unfair targeting practices, perpetuating stereotypes and excluding certain groups (Noble, 2018). Marketers must be vigilant in selecting and monitoring their data sources to ensure fairness and inclusivity in their campaigns.

Privacy is another critical issue. AI's reliance on personal data for personalization and optimization poses risks to consumer privacy, particularly in the era of widespread data breaches and surveillance concerns (Zuboff, 2019). Regulations such as the General Data Protection Regulation (GDPR) and the California Consumer Privacy Act (CCPA) mandate strict guidelines for data collection and processing, emphasizing the need for transparency and accountability in AI-driven marketing practices. By adhering to these regulations, brands can build trust with consumers and mitigate legal risks.

The integration of AI into brand marketing has unlocked unprecedented opportunities for innovation and efficiency. From providing deep consumer insights to automating content creation and optimizing digital advertising, AI empowers marketers to deliver personalized and impactful campaigns that resonate with audiences. However, the ethical challenges associated with bias and privacy underscore the importance of responsible AI implementation. By embracing transparency, fairness, and regulatory compliance, brands can harness the full potential of AI while fostering trust and loyalty among consumers.

As AI continues to evolve, its role in brand marketing will only expand, offering new avenues for creativity and connection. The harmonious collaboration between human ingenuity and AI-driven precision will define the future of marketing, enabling brands to thrive in an increasingly digital world.

OBJECTIVES

The following objectives aim to explore the role of Artificial Intelligence (AI) in enhancing the effectiveness of brand marketing campaigns. These objectives provide a structured framework for examining the multifaceted applications, benefits, and ethical considerations of AI in marketing.

1. Examine the Impact of AI on Customer Intelligence

- Investigate how AI-driven analytics enhance the understanding of consumer behaviours, preferences, and needs through data collection and analysis.
- Analyse the role of machine learning algorithms in identifying patterns in customer data to predict future behaviours, enabling more personalized marketing strategies (Davenport & Ronanki, 2018).
- Explore how AI-powered recommendation systems, such as those used by Amazon and Netflix, improve customer satisfaction and drive loyalty by delivering personalized experiences (Jannach & Adomavicius, 2016).

2. Evaluate the Role of AI in Content Creation and Optimization

- Assess the potential of AI tools, such as Natural Language Processing (NLP) systems, in automating content creation for various marketing platforms, including blogs, product descriptions, and social media updates (Brown et al., 2020).
- Investigate the effectiveness of AI-driven content optimization techniques, such as A/B testing, in enhancing engagement and conversion rates through iterative improvements (Kingsnorth, 2019).
- Examine AI's ability to curate and integrate user-generated content, such as reviews and social media comments, into brand campaigns for authenticity and relatability.

3. Explore AI's Contributions to Consumer Engagement

- Analyse the role of AI-driven chatbots and virtual assistants in providing real-time, personalized customer interactions across multiple platforms (Shankar, 2018).

- Investigate how AI tools utilize real-time behavioural data to deliver targeted messages and promotions at optimal moments, increasing consumer engagement and satisfaction (Huang & Rust, 2018).
- Assess AI's potential to maintain consistent and tailored communication with consumers without requiring constant human intervention, fostering loyalty and long-term relationships.

4. Assess AI's Role in Digital Advertising

- Examine the effectiveness of programmatic advertising powered by AI in delivering targeted advertisements based on consumer data and behavioural insights (Lambrecht & Tucker, 2019).
- Evaluate the impact of Dynamic Creative Optimization (DCO) on ad performance by tailoring creative elements to individual audience segments (Li & Kannan, 2014).
- Explore AI's capabilities in real-time ad performance tracking and budget allocation to optimize campaign outcomes dynamically.

5. Analyse AI's Contribution to Strategic Decision-Making in Marketing

- Investigate how AI enables marketers to process large datasets and uncover patterns, facilitating data-driven decision-making in campaign planning and execution (Davenport & Ronanki, 2018).
- Examine AI's predictive analytics capabilities in forecasting potential campaign outcomes, enabling brands to make informed strategic adjustments (Russell & Norvig, 2016).
- Evaluate the role of AI in real-time campaign optimization, including adjustments to content, targeting, and budget allocation based on performance metrics.

6. Examine Ethical and Regulatory Implications of AI in Marketing

- Investigate the risks of algorithmic bias in AI systems and their potential impact on targeting fairness and inclusivity in marketing campaigns (Noble, 2018).
- Assess the privacy concerns associated with AI's reliance on personal data and evaluate the role of regulations, such as the General Data Protection Regulation (GDPR) and the California Consumer Privacy Act (CCPA), in mitigating these concerns (Zuboff, 2019).
- Explore strategies for ensuring transparency, fairness, and accountability in AI-driven marketing practices to build consumer trust.

7. Evaluate the Synergy Between Human Creativity and AI

- Examine the complementary roles of human creativity and AI-driven innovation in developing impactful marketing strategies.
- Investigate how AI tools can augment marketers' abilities to focus on strategic and creative tasks by automating routine processes (Brown et al., 2020).
- Explore best practices for integrating AI technologies into existing marketing frameworks to achieve a balance between technological efficiency and human insight.

8. Propose Future Directions for AI in Brand Marketing

- Identify emerging AI technologies with potential applications in brand marketing, such as advanced machine learning models and augmented reality (AR) integration.
- Explore opportunities for brands to leverage AI for sustainable and ethical marketing practices, ensuring long-term consumer trust and loyalty (Chaffey & Ellis-Chadwick, 2019).
- Highlight the importance of continuous innovation and adaptation in harnessing AI's full potential to stay ahead in a competitive digital landscape.

CONTRIBUTION TO THE FIELD

The integration of Artificial Intelligence (AI) in brand marketing campaigns represents a transformative leap, addressing long-standing challenges in efficiency, personalization, and adaptability. This section discusses the significant contributions AI has made to the field of brand marketing, emphasizing its role in advancing customer intelligence, content creation, customer engagement, digital advertising, decision-making, and ethical compliance.

1. Advancing Customer Intelligence

AI's ability to analyse large volumes of consumer data has redefined customer intelligence, offering unprecedented insights into consumer behaviour and preferences. Traditional marketing approaches relied on manual data analysis, which often led to generalized strategies that did not cater to individual customer needs. AI bridges this gap by leveraging machine learning algorithms to process purchasing histories, social media activity, and other data sources (Davenport & Ronanki, 2018).

By identifying hidden patterns in consumer behaviour, AI enables marketers to anticipate future needs and deliver highly targeted campaigns. For instance, AI-driven recommendation systems, such as those used by Amazon and Netflix, have set benchmarks for personalized experiences, fostering customer loyalty and satisfaction (Jannach & Adomavicius, 2016). This contribution underscores AI's ability to create tailored marketing strategies that resonate with specific customer segments.

2. Enhancing Content Creation and Optimization

Content creation is a cornerstone of brand marketing, and AI has revolutionized this domain by automating and optimizing the process. Tools like GPT-3 and other natural language processing (NLP) systems generate high-quality content at scale, reducing the time and effort required for manual content creation (Brown et al., 2020).

Beyond creation, AI enhances content performance through optimization techniques. For example, AI can analyse variations in headlines, images, or call-to-action buttons to determine the most effective combinations for specific audiences. This iterative approach ensures that marketing content is continuously refined for maximum engagement and conversion (Kingsnorth, 2019). Additionally, AI's ability to curate user-generated content from social media comments and reviews adds authenticity and relatability to marketing campaigns.

3. Transforming Customer Engagement

AI has redefined customer engagement by enabling real-time, personalized interactions. Chatbots and virtual assistants, powered by natural language understanding (NLU) and NLP, provide instant responses to customer queries, facilitating seamless purchasing decisions (Shankar, 2018). These AI-driven platforms enhance customer satisfaction by delivering timely and accurate assistance, reducing the need for human intervention.

AI also enables brands to segment consumers based on behavioural data, allowing for the delivery of personalized messages and offers at optimal times. This targeted approach not only increases customer engagement but also fosters long-term loyalty. Furthermore, AI's ability to sustain interactions across multiple platforms ensures consistent communication, reinforcing brand presence and trust.

4. Optimizing Digital Advertising

AI has transformed digital advertising by enabling precise targeting and efficient resource allocation. Programmatic advertising, which leverages AI to deliver ads to specific audiences, has become a standard practice in the industry (Lambrecht & Tucker, 2019). By analysing consumer data and behavioural patterns, AI ensures that advertisements reach the right audience at the right time, maximizing campaign effectiveness.

Dynamic Creative Optimization (DCO) further enhances ad performance by tailoring creative elements to individual audience segments. This approach increases audience engagement and improves overall campaign results. Additionally, AI's ability to track ad performance in real time allows marketers to make data-driven adjustments, such as reallocating budgets or modifying content, ensuring optimal outcomes.

5. Empowering Strategic Decision-Making

AI's ability to process vast amounts of data and identify actionable insights has empowered marketers to make informed decisions. Predictive analytics, a key feature of AI, enables brands to forecast the outcomes of various marketing strategies, reducing uncertainty and improving campaign planning (Russell & Norvig, 2016).

Real-time optimization is another significant contribution, allowing marketers to adjust campaigns dynamically based on performance metrics. For example, if an advertisement is underperforming, AI systems can automatically reallocate resources or modify content to enhance results. This adaptability ensures that marketing strategies remain aligned with audience preferences and market trends.

6. Addressing Ethical and Regulatory Challenges

While AI offers numerous benefits, its implementation raises ethical and regulatory concerns, particularly regarding data privacy and algorithmic bias. AI's reliance on consumer data has sparked debates about privacy violations, necessitating compliance with regulations such as the General Data Protection Regulation (GDPR) and the California Consumer Privacy Act (CCPA) (Zuboff, 2019).

By adhering to these regulations, brands can build trust with consumers and mitigate potential legal risks. Additionally, addressing algorithmic bias is crucial to ensuring fairness and inclusivity in marketing practices. Developing transparent and accountable AI systems can prevent unintended consequences and uphold ethical standards in the industry (Noble, 2018).

7. Fostering Collaboration Between AI and Human Creativity

AI's role in brand marketing is not to replace human creativity but to complement it. By automating routine tasks, AI allows marketers to focus on strategic and creative aspects of campaign development (Brown et al., 2020). For instance, while AI can generate data-driven insights and optimize content, human creativity is essential for crafting compelling narratives and emotional connections with audiences.

The synergy between AI and human creativity enables brands to achieve a balance between technological efficiency and imaginative storytelling. This collaboration ensures that marketing campaigns remain innovative and resonate with diverse audiences.

8. Shaping the Future of Brand Marketing

AI's contributions extend beyond current practices, shaping the future of brand marketing through continuous innovation. Emerging technologies, such as augmented reality (AR) and advanced machine learning models, offer new opportunities for immersive and interactive marketing experiences (Chaffey & Ellis-Chadwick, 2019).

Sustainable and ethical marketing practices, driven by AI, will play a pivotal role in building long-term consumer trust. By leveraging AI to create meaningful and responsible marketing strategies, brands can differentiate themselves in an increasingly competitive digital landscape.

The integration of AI into brand marketing campaigns has brought transformative changes, enhancing customer intelligence, content creation, engagement, advertising, and decision-making. By addressing ethical and regulatory challenges and fostering collaboration between AI and human creativity, brands can harness AI's full potential to drive innovation and achieve long-term success. The field of brand marketing stands at the intersection of technology and creativity, with AI serving as a catalyst for progress and innovation.

THEORETICAL FRAMEWORK

To build sound theory for the effective implementation of AI-based methods, it is essential to combine different fundamental theories of AI and marketing. The following are the key concepts that shape this framework:

1. [UNTITLED]

a. Out of all the theories, the one whereby we will be focusing is the Technology Acceptance Model (TAM).

TAM gives an understanding of how users come to accept and use technology. This model focuses on two primary factors:

1.Perceived Usefulness (PU): The extent of acceptance of the notion that using a specific system would improve the performance of the user's job.
2.Perceived Ease of Use (PEOU): A measure of the level, to which a person feels that the use of the system will not be associated with exerting any efforts alternately known as Perceived Ease of Use (Davis, 1989).

With reference to brand marketing employing AI, the TAM framework can be applied to explain marketers' and organizations' AI technology adoption. This is especially so if the marketers accepted AI technology due to perceived high utility to enhance the effectiveness of their marketing communication campaigns (PU) and bearing ease of use as an attribute that is easy to incorporate in the overall marketing system (PEOU).

b. Resource-Based View (RBV)

According to the Resource Based View of the firm (RBV) Barney (1991), a firm has a competitive edge in an organisation possessing valuable, rare, in imitable and non-substitutable resources. Still, conventional applications of AI technologies like machine learning algorithms or predictive analytics or Customer Data Platforms can be positioned as strategic resources that allow brands to build a competitive advantage by using data-driven personalized and individualized approaches in marketing communication.

From an RBV point of view, AI continues to be considered to be a powerful set of tools that enable brand to enhance effectiveness of resources allocation, work more efficiently when it comes to the targeting of the advertising campaigns and building up sustainable competitive advantages when it comes to brand customer orientation, personalization and understanding of the markets.

c. Customer-Centric Marketing and Personalization

This move towards customer focused marketing is founded on the premise that company focused marketing initiatives have lower success rates than individually targeted ones. AI provides brands the opportunities to launch various targeted advertising campaigns since the consideration of customer data is possible at the scale. The rationale for this is based on the Theory of Personalization which Vesanen (2007) has developed; the idea is that the use of personal messages and personalized offers improve satisfaction, participation and loyalty levels among customers.

Due to the nature of analyzing big data, achieving personalization at scale through AI is a boon to brand marketing as it increases both relevance and impact.

d. Consumer Engagement Theory

Consumer Engagement Theory which is conceptual in nature, initiated by Brodie et al., 2011 focus on the interaction brands build with consumers. According to the above-mentioned AI technologies like Chatbot, Recommendation system, and Virtual assistants are quite instrumental to increase the consumer engagement level as they offer personalized, timely and interactive solutions. Such forms of interaction with customers are very beneficial to the business since they enhance brand loyalty because of the artificial intelligence platforms.

e. Data-Driven Marketing

Data driven marketing is the use of data in enhancing the marketing activities and/or making of marketing decisions. AI boosts data driven marketing by offering probabilistic forecasts and automated functioning optimistically and in real time. Theoretically, Data Driven Marketing drawback originates from Decision Theory advanced by Simon in 1959, arguing that such decisions will be improved by data analysis and forecast of the likely outcomes. In the context of AI in marketing it involves pattern recognition, customer behavior forecasting and campaign performance optimisation for the best ROI.

2. Elements of AI Integrated Marketing Communication Efforts

Based on these theoretical frameworks, it is now possible to identify key elements of the AI-augmented brand marketing campaign. The components highlighted above constitute our theoretical frame work.

a. Customer Data Acquisition and Analysis

AI majorly depends on customer data to churn out intelligence and customized solutions. Such, data is collected through internet interactions on websites, social networks, applications for mobile devices, and offline contacts. Based on this data, AI tools build customer personas, the audiences' categorization, and future behavioral trends. The information content as well as the amount obtained are the critical factor in the AI integration influencing the effectiveness of marketing campaigns, as supported by the Big Data Analytics Framework (Chen et al., 2012).

b. Personalization and Targeting

AI to brand marketing is one of the most valuable opportunities as through the help of AI it becomes possible to distribute the individual content. Artificial intelligence takes into consideration the major attributes of customer such as want, need, interaction and previous experience to offer the most appropriate and relevant marketing communication. This concurs with Theory of Personalization and assists brands in reaching the right customer with the right message, at the correct time and thus enhancing customers' yield and loyalty.

c. Automation and Optimization

Marketing automation helps marketers to tackle routine processes in marketing activity, including content creation, the use of email, and sharing on social media platforms. Further, AI can adjust or fine tunes the existing campaigns in real-time, for instance, changing bidding strategies of programmatic advertisement, and personalizing the content on the website according to the user's action. This aspect is anchored in Process Automation Theory as proposed by Davenport and Short (1990), in regard to automation and efficiency.

d. Here, Customer Engagement and Interaction reveals itself as a key performance indicator MDL.

Customer loyalty is improved by leveraging of artificial intelligence tools like chatbots, virtual assistants who are able to offer customer support services round the clock and make suggestions based on customers' preferences. They enhance customer-brand intimacy through providing quick and individualized answers to the questions of customers, so the given tools contribute to the Consumer Engagement Theory. In addition, AI enhanced interaction channels are another avenue where brands can possibly source more customer information for customized communication.

e. Campaign Performance Measurement and Insights

Marketing this way becomes possible since AI allows marketers to determine the success rate of their campaigns instantly. Marketing communication can predict campaign results, and machine learning algorithms can adjust marketing techniques regarding the data gathered. This approach corresponds to the Decision Theory employing data to make changes to the strategies used by the brands aimed at improvement.

3. AI Augmented Marketing Mix

The following diagram presents a conceptual model for the use of Artificial Intelligence in marketing. It shows how the application of the described AI technologies is related to major marketing functions, backed up by the theoretical concepts such as TAM, RBV, Customer-Centric Marketing.

4. Application of the Framework

The theoretical model proposed in this paper serves as a framework for explaining how AI technologies increase the effectiveness of brand marketing. Here's how each component functions within a real-world context:

a. Analysis and Forecasting Based on Numerical Information

The AI algorithms can be adopted to analyse customer information and get the levels of organisations buyer behaviours including the propensity to purchase, or to defect. With this kind of information brands can plan specific advertisements which meet the needs of consumer groups and therefore increase the chances of an improved response rate as well as satisfaction rates from customers.

b. Personalization at Scale

Recommendation engines, dynamic content optimization and similar tools allow marketers to bring customized approach for each client. For egg Netflix & Spotify use AI based algorithms to suggest content for viewing / listening for user's depending on their search history. This improves the company's client base interactions and consequently the rates of use of its services.

c. Automating Marketing Tasks

Another area where marketing deals with AI is through the implementation of the technology to automate routine marketing tasks, for instance, social media posting, sending e-mails and targeting advertisements. This logical process relieves the marketers to pay more attention to value added activities like planning of campaigns and creativity. Specific technologies, such as Mailchimp and HubSpot, use AI to help with automating the emails that are sent to customers and as a tool for understanding the customer's behaviors.

d. Enhancing Customer Engagement

Machine learning, or AI and real-time communication solutions like those offered by Sephora or H&M give brands the ability to actually interact with clients in real time. They offer the customer customized suggestions and solve their questions, which improve the customer satisfaction and his loyalty to the specific brand.

e. Campaign Optimization

AI allows for modifications to a marketing campaign based on assessment of key performance indicators while a campaign is still ongoing. The other area where AI is implemented in digital marketing is through the programmatic advertising platforms such as Google Ad to mention but a few, where it helps to determine the right bid and placement to make within the market and provide the right returns for the marketing dollar invested.

The case of applying AI to brand marketing provides a theoretical understanding of how the application of the various AI technologies is revolutionizing the conventional marketing strategies. With reference to the foundational theories like TAM, RBV and Consumer Engagement Theory, this framework gives the overall understanding about how AI facilitate brand to market personal, data driven and efficiency marketing campaigns. It also underlines the aspects of customers' involvement and interactions revealing how AI generates more effective and valuable experiences of the customer-brand bonds.

CONCEPTUAL DEVELOPMENT

The integration of Artificial Intelligence (AI) into brand marketing campaigns represents a convergence of technology and creativity that has reshaped traditional marketing paradigms. This section explores the conceptual underpinnings of AI in brand marketing, outlining its foundational principles, theoretical models, and practical applications. The development of these concepts highlights how AI enhances marketing processes and outcomes, enabling more personalized, efficient, and ethical campaigns.

1. AI as a Tool for Customer Intelligence

Foundational Principle:

AI's role in customer intelligence is based on its ability to analyse vast datasets and extract actionable insights. Traditional marketing relied heavily on manual data interpretation, which was limited in scope and efficiency. AI offers a scalable, automated solution to understanding consumer behaviour, preferences, and needs.

Theoretical Models:

- **Behavioural Targeting Models:** These models analyse consumer data, such as purchasing history, online interactions, and demographic information, to segment audiences and predict future behaviour (Davenport & Ronanki, 2018).
- **Predictive Analytics Frameworks:** Using machine learning algorithms, these frameworks forecast customer responses to marketing strategies, enabling brands to develop proactive campaigns (Russell & Norvig, 2016).

Practical Application:

AI-driven recommendation systems, such as those used by Amazon and Netflix, serve as prime examples of leveraging customer intelligence. These systems analyse user preferences to provide personalized suggestions, enhancing customer satisfaction and driving loyalty (Jannach & Adomavicius, 2016).

2. AI in Content Creation and Optimization

Foundational Principle:

AI automates and optimizes content creation, addressing the increasing demand for high-quality, engaging, and scalable marketing materials.

Theoretical Models:

- **Natural Language Processing (NLP):** NLP models, like GPT-3, generate human-like text by understanding and synthesizing language patterns, making them invaluable for creating content across various formats (Brown et al., 2020).

- **Optimization Algorithms:** These algorithms analyse performance data to refine content, such as headlines, images, and calls to action, ensuring maximum engagement (Kingsnorth, 2019).

Practical Application:

AI tools can generate product descriptions, social media posts, and blog articles at scale while continuously improving their effectiveness through iterative feedback. For example, AI-enabled A/B testing can identify the most impactful content variations, optimizing performance over time.

3. AI in Customer Engagement

Foundational Principle:

Customer engagement is central to brand marketing, and AI enhances this by enabling personalized, real-time interactions through intelligent systems like chatbots and virtual assistants.

Theoretical Models:

- **Human-Computer Interaction (HCI):** This model explores how AI interfaces facilitate seamless communication between brands and consumers, enhancing user experience (Shankar, 2018).
- **Personalization Algorithms:** These algorithms use behavioural and contextual data to deliver tailored messages and offers, fostering deeper connections with customers (Huang & Rust, 2018).

Practical Application:

AI-powered chatbots on platforms like Facebook Messenger provide instant responses to customer inquiries, guiding them through purchasing decisions and ensuring continuous engagement. These systems can handle complex queries, making them efficient substitutes for human representatives.

4. AI in Digital Advertising

Foundational Principle:

AI revolutionizes digital advertising by automating the targeting, creation, and optimization of ad campaigns, ensuring maximum relevance and efficiency.

Theoretical Models:

- **Programmatic Advertising Models:** These models leverage AI to automate the buying and placement of ads, targeting specific audiences based on data-driven insights (Lambrecht & Tucker, 2019).
- **Dynamic Creative Optimization (DCO):** This model involves creating multiple ad variations and using AI to deliver the most effective version to each audience segment (Li & Kannan, 2014).

Practical Application:

Google Ads employs AI to optimize ad performance by analysing user behaviour, adjusting bids, and tailoring creative elements in real time. This ensures that campaigns are both cost-effective and impactful.

5. AI in Strategic Decision-Making

Foundational Principle:

AI empowers marketers with data-driven insights, enabling more informed and timely decision-making in campaign planning and execution.

Theoretical Models:

- **Big Data Analytics Frameworks:** These frameworks analyse extensive datasets to identify patterns and correlations, providing a foundation for strategic decisions (Davenport & Ronanki, 2018).
- **Real-Time Optimization Models:** These models allow marketers to adjust campaigns dynamically based on performance metrics, ensuring alignment with market trends and audience preferences (Russell & Norvig, 2016).

Practical Application:

AI-driven platforms like Salesforce Einstein analyse campaign performance in real time, suggesting adjustments to maximize ROI. This includes reallocating budgets, modifying content, or refining audience targeting.

6. Ethical and Regulatory Considerations

Foundational Principle:

The ethical use of AI in marketing hinges on transparency, fairness, and compliance with data privacy regulations.

Theoretical Models:

- **Algorithmic Accountability Frameworks:** These frameworks emphasize the importance of mitigating biases in AI systems to ensure equitable targeting (Noble, 2018).
- **Data Privacy Compliance Models:** These models guide brands in adhering to regulations such as GDPR and CCPA, ensuring responsible data usage (Zuboff, 2019).

Practical Application:

Brands like Apple have integrated privacy-first AI practices, such as differential privacy, to maintain user trust while leveraging consumer data for personalized marketing.

7. The Synergy Between AI and Human Creativity

Foundational Principle:

AI serves as a complementary tool to human creativity, automating routine tasks and enhancing strategic and imaginative aspects of marketing.

Theoretical Models:

- **Collaborative Intelligence Frameworks:** These frameworks explore the interplay between AI-driven efficiency and human ingenuity in crafting compelling marketing narratives (Chaffey & Ellis-Chadwick, 2019).
- **Augmented Creativity Models:** These models emphasize the role of AI in inspiring new ideas and refining creative outputs through data-driven insights.

Practical Application:

AI tools like Adobe Sensei provide marketers with predictive insights and creative recommendations, enabling them to design visually appealing and impactful campaigns more efficiently.

The conceptual development of AI in brand marketing illustrates its profound impact on the field, offering innovative solutions to longstanding challenges. By leveraging AI's capabilities in customer intelligence, content creation, engagement, advertising, decision-making, and ethical compliance, brands can create more effective and meaningful campaigns. Moreover, the synergy between AI and human creativity ensures that marketing strategies remain both innovative and resonant. As AI continues to evolve, its conceptual frameworks will further shape the future of brand marketing, fostering a harmonious blend of technology and creativity.

FUTURE RESEARCH PROSPECTS

Artificial Intelligence (AI) continues to redefine the boundaries of brand marketing, providing a myriad of opportunities for innovation. However, the rapid evolution of AI presents significant gaps in understanding and areas that require deeper exploration. Future research prospects can address these challenges and opportunities, enabling both academia and industry to unlock the full potential of AI in marketing. This section outlines potential areas of study, emphasizing their significance and relevance to advancing the field.

1. Ethics and Bias in AI-Driven Marketing

AI systems are susceptible to biases embedded in the data used for training. These biases can inadvertently lead to discriminatory practices in audience targeting, content delivery, and customer segmentation.

Key Research Questions:

- How can bias be systematically identified and mitigated in AI algorithms for brand marketing?
- What frameworks can ensure ethical AI practices while maintaining marketing efficiency?

Potential Directions:

- Developing transparent AI systems that provide insight into decision-making processes, enabling marketers to identify and rectify biases.
- Investigating the role of algorithmic accountability in marketing, particularly in addressing issues of fairness and inclusivity (Noble, 2018).
- Assessing the effectiveness of diverse datasets in reducing biases in AI-driven marketing campaigns.

Significance:

Understanding and addressing biases is critical to building trust and fostering equitable marketing practices. Ethical AI can enhance brand reputation and ensure compliance with regulations like GDPR and CCPA (Zuboff, 2019).

2. AI and Consumer Privacy

The growing reliance on consumer data for personalized marketing raises significant concerns about privacy and data security.

Key Research Questions:

- How can brands balance personalized marketing with consumer privacy?
- What are the implications of emerging data privacy laws on AI-driven marketing strategies?

Potential Directions:

- Exploring privacy-preserving AI techniques, such as differential privacy, that allow data analysis without exposing individual details (Dwork & Roth, 2014).
- Analysing the effectiveness of consent-based marketing models in addressing privacy concerns.
- Investigating consumer perceptions of data usage in AI-driven marketing and its impact on trust.

Significance:

With data privacy increasingly becoming a focal point, research in this area will guide brands in adopting ethical data practices while maintaining competitive advantages in personalization.

3. Integration of AI and Emotional Intelligence

While AI excels at analysing data, it struggles to understand and respond to human emotions effectively. Emotional intelligence in AI could revolutionize customer engagement by creating more empathetic interactions.

Key Research Questions:

- How can AI systems be trained to detect and respond to emotional cues in real-time?
- What role does emotional AI play in building customer loyalty and satisfaction?

Potential Directions:

- Developing sentiment analysis tools capable of recognizing nuanced emotions in text, speech, and visual inputs (Poria et al., 2017).
- Investigating the impact of emotionally intelligent AI on customer decision-making processes.
- Creating frameworks to measure the effectiveness of emotional AI in enhancing brand-customer relationships.

Significance:

Emotional intelligence in AI can humanize digital interactions, improving customer satisfaction and fostering deeper connections between brands and consumers.

4. Real-Time Optimization and Predictive Analytics

AI's ability to optimize marketing campaigns in real-time and predict future trends remains under-explored.

Key Research Questions:

- How can real-time optimization techniques be improved to enhance campaign efficiency?
- What is the role of predictive analytics in long-term marketing strategy formulation?

Potential Directions:

- Developing advanced machine learning models that integrate multiple data streams for real-time decision-making.

- Investigating the predictive capabilities of AI in identifying emerging market trends and consumer preferences.
- Exploring the integration of AI with Internet of Things (IoT) devices to gather real-time consumer data.

Significance:

Real-time optimization and predictive analytics empower marketers to respond dynamically to market changes, ensuring campaigns remain relevant and impactful.

5. AI-Driven Creative Processes

AI's role in augmenting human creativity is a burgeoning area of interest. Future research can explore how AI tools can inspire and enhance creative processes in marketing.

Key Research Questions:

- How can AI support the ideation and design phases of marketing campaigns?
- What are the limitations of AI in replicating human creativity, and how can these be addressed?

Potential Directions:

- Developing AI tools that generate creative concepts based on cultural and contextual cues.
- Investigating the synergy between human and AI-generated creativity in producing impactful marketing materials.
- Assessing the effectiveness of AI in creating culturally sensitive and globally resonant campaigns.

Significance:

AI-driven creativity can revolutionize content creation, enabling brands to produce innovative and diverse marketing materials at scale.

6. AI in Multichannel Marketing Strategies

AI's potential to unify marketing efforts across various channels is an area ripe for exploration.

Key Research Questions:

- How can AI optimize the coordination of marketing campaigns across digital and traditional channels?
- What is the role of AI in managing cross-channel customer journeys?

Potential Directions:

- Developing integrated AI systems that provide a holistic view of customer interactions across platforms.
- Investigating the impact of AI on the effectiveness of omnichannel marketing strategies.
- Exploring AI's role in ensuring consistency and coherence in brand messaging across channels.

Significance:

Effective multichannel marketing strategies driven by AI can enhance brand visibility and ensure a seamless customer experience.

7. Personalization Versus Generalization in Marketing

While AI enables highly personalized marketing, it may also lead to over-segmentation, reducing the appeal of broader campaigns.

Key Research Questions:

- What is the optimal balance between personalization and generalization in marketing campaigns?
- How can AI adapt its strategies to target both niche and broad audiences effectively?

Potential Directions:

- Developing hybrid AI models that combine personalization with generalization to maximize reach and relevance.
- Investigating the role of AI in creating content that resonates with diverse audience segments.
- Exploring consumer perceptions of personalized versus generalized marketing approaches.

Significance:

Balancing personalization and generalization ensures that brands can cater to individual preferences while maintaining broad appeal.

8. AI and Sustainability in Marketing

Sustainability is becoming a core focus for brands, and AI can play a pivotal role in aligning marketing efforts with environmental goals.

Key Research Questions:

- How can AI-driven marketing campaigns promote sustainable consumption patterns?
- What is the role of AI in reducing the environmental footprint of marketing activities?

Potential Directions:

- Investigating AI's role in promoting eco-friendly products and services through targeted campaigns.
- Exploring AI-driven strategies for optimizing resource allocation in marketing, reducing waste.
- Assessing consumer responses to sustainability-focused marketing initiatives driven by AI.

Significance:

AI can help brands align their marketing efforts with sustainability goals, enhancing their reputation and contributing to broader environmental objectives.

The future of AI in brand marketing is vast and multifaceted, with significant opportunities for research and innovation. By addressing the outlined areas—ethics, privacy, emotional intelligence, real-time optimization, creative processes, multichannel strategies, personalization, and sustainability—scholars and practitioners can advance the field. These research directions not only promise to enhance marketing efficiency but also ensure that AI-driven practices remain ethical, inclusive, and consumer-centric.
Conclusion

CONCLUSION

The integration of Artificial Intelligence (AI) into brand marketing represents a transformative shift in how companies interact with consumers, develop strategies, and achieve their objectives. This book chapter has explored the diverse ways in which AI enhances the effectiveness of brand marketing campaigns, focusing on aspects such as customer intelligence, personalized content, consumer engagement, real-time optimization, and ethical considerations. The journey into this new era of marketing innovation demonstrates that while AI brings unprecedented opportunities, it also requires careful management to address associated challenges and ethical dilemmas.

Key Themes and Takeaways

1. Enhanced Customer Insights

AI's ability to process vast amounts of data has revolutionized customer intelligence. By analysing purchasing histories, social media activities, and other behavioural data, AI enables brands to understand their consumers deeply and predict future behaviours. Tools like recommendation engines exemplify the practical application of AI, fostering personalized customer experiences that drive engagement and loyalty. This capability is particularly significant in a world where consumers expect brands to anticipate their needs and deliver tailored solutions.

2. Content Creation and Optimization

AI has streamlined the content creation process, reducing the time and effort required to produce high-quality, relevant material. From automated blog posts and social media updates to dynamic creative optimization, AI enables brands to experiment with different variations of content to determine what resonates most effectively with their audience. Moreover, AI's ability to analyse and optimize content performance ensures continuous improvement in marketing outcomes.

3. Real-Time Consumer Engagement

The use of AI-powered tools, such as chatbots and virtual assistants, has redefined customer engagement. These tools provide instant, efficient, and personalized responses to consumer queries, significantly enhancing customer satisfaction and loyalty. The ability to engage with consumers across multiple platforms in real-time ensures that brands can maintain a consistent presence and build stronger relationships with their audience.

4. Efficiency in Advertising

Programmatic advertising and dynamic creative optimization demonstrate AI's potential to enhance the efficiency of digital marketing campaigns. By delivering targeted ads to specific audiences and optimizing ad performance in real-time, AI minimizes wasted resources and maximizes campaign effectiveness. These capabilities are critical in a competitive digital landscape where precision and adaptability are key to success.

5. Ethical and Privacy Considerations

While the benefits of AI in marketing are evident, this chapter has also highlighted the ethical challenges that accompany its implementation. Issues such as algorithmic bias, data privacy, and the potential for manipulative practices demand careful attention. Brands must ensure transparency, fairness, and compliance with regulations like GDPR and CCPA to maintain consumer trust and avoid reputational damage. The ethical use of AI is not merely a legal requirement but a cornerstone of sustainable marketing practices.

The Role of Human Creativity in an AI-Driven World

AI's ability to process data and automate processes is unparalleled, but it cannot replace human creativity. Instead, the future of brand marketing lies in the harmonious collaboration between AI and human ingenuity. While AI excels at identifying patterns and generating insights, human marketers bring emotional intelligence, cultural awareness, and innovative thinking to the table. Together, they can create campaigns that are not only data-driven but also deeply resonant with consumers.

Examples of Synergy:

- Marketers can use AI to analyse consumer data and identify emerging trends, which can then inspire creative campaign ideas.
- AI tools can generate multiple content variations, while humans curate and refine the most compelling options.
- Emotional intelligence and cultural nuances, which are difficult for AI to fully comprehend, can be infused into campaigns by human marketers.

EMERGING TRENDS AND FUTURE PROSPECTS

As AI technology continues to evolve, its applications in brand marketing are expected to expand in scope and sophistication. Several emerging trends warrant attention:

1. Multimodal AI

Future AI systems may integrate multiple data modalities—such as text, image, audio, and video—to provide a more comprehensive understanding of consumer behaviour. This development could enhance personalized marketing and improve content creation.

2. Emotional AI

Advances in affective computing will enable AI to better understand and respond to human emotions. Emotional AI could play a pivotal role in creating empathetic and engaging customer interactions, fostering deeper connections between brands and consumers.

3. Sustainability-Focused Marketing

AI can help brands align their marketing efforts with sustainability goals by optimizing resource allocation, reducing waste, and promoting eco-friendly products. As consumers increasingly prioritize sustainability, this area of research will become increasingly relevant.

4. Ethical AI Development

The development of frameworks for ethical AI practices in marketing will be a critical area of focus. Research into bias mitigation, transparency, and accountability will guide the responsible use of AI in the industry.

5. Advanced Personalization

AI's capacity for hyper-personalization will continue to grow, allowing brands to deliver highly customized experiences at scale. However, this trend will also require careful management to avoid over-segmentation and ensure inclusivity.

CHALLENGES AND CONSIDERATIONS

Despite its potential, the integration of AI into brand marketing is not without its challenges. Some of the key issues include:

1. Data Quality and Availability:

The effectiveness of AI depends on the quality and quantity of data it processes. Incomplete or biased data can lead to inaccurate insights and suboptimal decisions.

2. Technological Complexity:

Implementing AI solutions requires technical expertise and substantial investment, which may be a barrier for smaller brands.

3. Regulatory Compliance:

Adhering to data protection regulations while leveraging AI for marketing purposes is a delicate balancing act. Brands must navigate this landscape carefully to avoid legal and reputational risks.

4. Consumer Trust:

The use of AI in marketing raises questions about transparency and authenticity. Brands must communicate their AI practices openly to build and maintain consumer trust.

Final Thoughts

The future of brand marketing lies in the integration of AI with human creativity, ethical practices, and consumer-centric strategies. As AI continues to evolve, its potential to enhance marketing effectiveness will grow, offering brands new opportunities to engage with consumers in meaningful ways. However, realizing this potential requires a commitment to responsible AI development, ongoing research, and a focus on building trust and transparency.

Ultimately, AI is not a replacement for human ingenuity but a powerful tool that can augment it. By embracing this synergy, brands can create campaigns that are not only data-driven but also emotionally resonant, ethically sound, and aligned with consumer values. In this new era of marketing innovation, the collaboration between humans and AI will shape the future of brand-consumer relationships, driving success in an increasingly digital world.

REFERENCES

Anagnostopoulos, D., Tsekouras, G., & Kotsakis, R. (2021). Emotional AI: Technology for personalized and emotional responses in brand marketing. *Journal of Consumer Psychology*, 31(3), 615–630.

Brodie, R. J., Hollenbeck, L. D., Juric, B., & Ilic, A. (2011). Customer engagement: Conceptual domain, fundamental propositions, and implications for research. *Journal of Service Research*, 14(3), 252–271. DOI: 10.1177/1094670511411703

Brown, T., Mann, B., Ryder, N., Subbiah, M., Kaplan, J. D., Dhariwal, P., & Amodei, D. (2020). Language models are few-shot learners. *arrive preprintarXiv:2005.14165.*

Brown, T., Mann, B., Ryder, N., Subbiah, M., Kaplan, J. D., Dhariwal, P., . . . Amodei, D. (2020). Language models are few-shot learners. arXiv preprint arXiv:2005.14165

Chaffey, D. (2020). Multi-channel marketing automation using AI: Current trends. *Journal of Digital Marketing*, 12(1), 34–42.

Chaffey, D., & Ellis-Chadwick, F. (2019). *Digital marketing: Strategy, implementation, and practice.* Pearson UK.

Chaffey, D., & Ellis-Chadwick, F. (2019). *Digital marketing: Strategy, implementation, and practice.* Pearson UK.

Chaffey, D., & Ellis-Chadwick, F. (2019). *Digital marketing: Strategy, implementation, and practice.* Pearson UK.

Chen, H., Chiang, R. H., & Storey, V. C. (2012). Business intelligence and analytics: From big data to big impact. *Management Information Systems Quarterly*, 36(4), 1165–1188. DOI: 10.2307/41703503

Chintalapati, M., & Pandey, R. (2022). Chatbots and virtual try-ons: AI's role in retail engagement. *Journal of Digital Marketing*. Brown, T., Mann, B., Ryder, N., Subbiah, M., Kaplan, J. D., Dhariwal, P., ... & Amodei, D. (2020). Language models are few-shot learners. *arXiv preprint arXiv:2005.14165.*

Chui, M., Manyika, J., & Miremadi, M. (2018). What AI Can and Cannot Do (Yet) for Your Business. *Harvard Business Review.* Retrieved from https://hbr.org

Chui, M., Manyika, J., & Miremadi, M. (2018). What AI Can and Can't Do (Yet) for Your Business. *Harvard Business Review.* Retrieved from https://hbr.org

Collins, S., & McCarthy, I. (2021). How AI is Revolutionizing Content Creation in Marketing. *Journal of Marketing Technology*, 12(3), 45–57.

Colson, E., & Ramsay, M. (2020). Human-AI partnerships in the workplace: Enhancing creativity and performance. *MIT Sloan Management Review*, 61(4), 58–63.

Cowley, S. J. (2020). Human-AI Interaction: An Experience-Centered Perspective on AI Implementation in Marketing. *Journal of Interactive Marketing*, 51, 36–45.

Davenport, T., & Ronanki, R. (2018). Artificial intelligence for the real world. *Harvard Business Review*, 96(1), 108–116.

Davenport, T., & Ronanki, R. (2018). Artificial intelligence for the real world. *Harvard Business Review*, 96(1), 108–116.

Davenport, T., & Ronanki, R. (2018). Artificial intelligence for the real world. *Harvard Business Review*, 96(1), 108–116.

Davenport, T., & Ronanki, R. (2018). Artificial intelligence for the real world. *Harvard Business Review*, 96(1), 108–116.

Davenport, T. H., Guha, A., Grewal, D., & Bressgott, T. (2021). How Artificial Intelligence Will Transform Marketing Strategy. *Journal of the Academy of Marketing Science*, 49(1), 24–42. DOI: 10.1007/s11747-019-00696-0

Davenport, T. H., & Ronanki, R. (2018). Artificial intelligence for the real world. *Harvard Business Review*, 96(1), 108–116.

de Reuver, M., Sørensen, C., & Basole, R. C. (2018). The digital platform: A research agenda. *Journal of Information Technology*, 33(2), 124–135. DOI: 10.1057/s41265-016-0033-3

Dwork, C., & Roth, A. (2014). The algorithmic foundations of differential privacy. *Foundations and Trends in Theoretical Computer Science*, 9(3–4), 211–407.

Firat, D., et al. (2020). Enhancing brand growth through AI. *International Journal for Multidisciplinary Research.*

Flavián, C., Ibáñez-Sánchez, S., & Orús, C. (2019). Integrating virtual reality and augmented reality in brand engagement. *Journal of Business Research*, 100, 547–560.

Georgieva, S. (2021). AI and the Future of Creativity: How Generative AI Will Transform the Role of Marketers. *Journal of Creative AI*, 6(2), 45–58.

Grewal, D., Hulland, J., Kopalle, P. K., & Karahanna, E. (2020). The Future of Technology and Marketing: A Multidisciplinary Perspective. *Journal of Marketing*, 84(1), 1–6. DOI: 10.1177/0022242919879877

Guha, S., Harrigan, P., & Soutar, G. (2018). Linking social media to customer relationship management (CRM): A qualitative study on SMEs. *Journal of Small Business and Enterprise Development*, 25(6), 844–868.

Huang, M. H., & Rust, R. T. (2018). Artificial intelligence in service. *Journal of Service Research*, 21(2), 155–172. DOI: 10.1177/1094670517752459

Huang, M. H., & Rust, R. T. (2018). Artificial intelligence in service. *Journal of Service Research*, 21(2), 155–172. DOI: 10.1177/1094670517752459

Huang, M. H., & Rust, R. T. (2018). Artificial intelligence in service. *Journal of Service Research*, 21(2), 155–172. DOI: 10.1177/1094670517752459

Huang, M. H., & Rust, R. T. (2021). A Strategic Framework for Artificial Intelligence in Marketing. *Journal of the Academy of Marketing Science*, 49(1), 30–50. DOI: 10.1007/s11747-020-00749-9

Jannach, D., & Adomavicius, G. (2016). Recommendation systems: Challenges, insights, and research opportunities. *AI Magazine*, 37(3), 73–85.

Jannach, D., & Adomavicius, G. (2016). Recommendation systems: Challenges, insights, and research opportunities. *AI Magazine*, 37(3), 73–85.

Jannach, D., & Adomavicius, G. (2016). Recommendation systems: Challenges, insights, and research opportunities. *AI Magazine*, 37(3), 73–85.

Jannach, D., & Adomavicius, G. (2016). Recommendation systems: Challenges, insights, and research opportunities. *AI Magazine*, 37(3), 73–85.

Keegan, P. (2019). The Netflix effect: AI in entertainment. *Harvard Business Review*.

Kingsnorth, S. (2019). *Digital marketing strategy: An integrated approach to online marketing*. Kogan Page Publishers.

Kingsnorth, S. (2019). *Digital marketing strategy: An integrated approach to online marketing*. Kogan Page Publishers.

Kingsnorth, S. (2019). *Digital marketing strategy: An integrated approach to online marketing*. Kogan Page Publishers.

Kingsnorth, S. (2019). *Digital marketing strategy: An integrated approach to online marketing*. Kogan Page Publishers.

Kumar, V., Rajan, B., Venkatesan, R., & Lecinski, J. (2020). Understanding the Role of Artificial Intelligence in Personalized Marketing. *Journal of Interactive Marketing*, 51, 26–42. DOI: 10.1016/j.intmar.2020.04.005

Lambrecht, A., & Tucker, C. (2019). Algorithmic bias? An empirical study of apparent gender-based discrimination in the display of STEM career ads. *Management Science*, 65(7), 2966–2981. DOI: 10.1287/mnsc.2018.3093

Lambrecht, A., & Tucker, C. (2019). Algorithmic bias? An empirical study of apparent gender-based discrimination in the display of STEM career ads. *Management Science*, 65(7), 2966–2981. DOI: 10.1287/mnsc.2018.3093

Lambrecht, A., & Tucker, C. (2019). Algorithmic bias? An empirical study of apparent gender-based discrimination in the display of STEM career ads. *Management Science*, 65(7), 2966–2981. DOI: 10.1287/mnsc.2018.3093

Lambrecht, A., & Tucker, C. (2019). Algorithmic bias? An empirical study of apparent gender-based discrimination in the display of STEM career ads. *Management Science*, 65(7), 2966–2981. DOI: 10.1287/mnsc.2018.3093

Lee, I., & Lee, K. (2015). The Internet of Things (IoT): Applications, investments, and challenges for enterprises. *Business Horizons*, 58(4), 431–440. DOI: 10.1016/j.bushor.2015.03.008

Lemon, K. N., & Verhoef, P. C. (2016). Understanding customer experience throughout the customer journey. *Journal of Marketing*, 80(6), 69–96. DOI: 10.1509/jm.15.0420

Li, F. (2021). Artificial Intelligence, Platform Economy, and the Future of Marketing. *Journal of Business Research*, 125, 410–420. DOI: 10.1016/j.jbusres.2020.12.035

Li, H., & Kannan, P. K. (2014). Attributing conversions in a multichannel online marketing environment: An empirical model and a field experiment. *JMR, Journal of Marketing Research*, 51(1), 40–56. DOI: 10.1509/jmr.13.0050

Li, H., & Kannan, P. K. (2014). Attributing conversions in a multichannel online marketing environment: An empirical model and a field experiment. *JMR, Journal of Marketing Research*, 51(1), 40–56. DOI: 10.1509/jmr.13.0050

Merrilees, B. (2019). Interactive Brand Experience Pathways to Customer-Brand Engagement and Value Co-Creation. *Journal of Product and Brand Management*, 28(4), 477–487. DOI: 10.1108/JPBM-12-2018-2166

Mittelstadt, B. D., Allo, P., Taddeo, M., Wachter, S., & Floridi, L. (2016). The ethics of algorithms: Mapping the debate. *Big Data & Society*, 3(2), 205–220. DOI: 10.1177/2053951716679679

Noble, S. U. (2018). *Algorithms of oppression: How search engines reinforce racism*. NYU Press. DOI: 10.18574/nyu/9781479833641.001.0001

Noble, S. U. (2018). *Algorithms of oppression: How search engines reinforce racism*. NYU Press. DOI: 10.18574/nyu/9781479833641.001.0001

Noble, S. U. (2018). *Algorithms of oppression: How search engines reinforce racism*. NYU Press. DOI: 10.18574/nyu/9781479833641.001.0001

Noble, S. U. (2018). *Algorithms of oppression: How search engines reinforce racism*. NYU Press. DOI: 10.18574/nyu/9781479833641.001.0001

O'Neil, C. (2016). *Weapons of Math Destruction: How Big Data Increases Inequality and Threatens Democracy*. Crown Publishing Group.

Pantano, E., & Timmermans, H. (2014). What is smart for retailing? Insights on the future of retailing in the smart city. *International Journal of Retail & Distribution Management*, 42(3), 176–188.

Parasuraman, R., Sheridan, T. B., & Wickens, C. D. (2000). A model for types and levels of human interaction with automation. *IEEE Transactions on Systems, Man, and Cybernetics. Part A, Systems and Humans*, 30(3), 286–297. DOI: 10.1109/3468.844354 PMID: 11760769

Picard, R. W. (2010). Affective computing: From laughter to IEEE. *IEEE Transactions on Affective Computing*, 1(1), 11–17. DOI: 10.1109/T-AFFC.2010.10

Poria, S., Cambria, E., Bajpai, R., & Hussain, A. (2017). A review of affective computing: From unimodal analysis to multimodal fusion. *Information Fusion*, 37, 98–125. DOI: 10.1016/j.inffus.2017.02.003

Ramanathan, R.. (2019). Customer relationship management with AI: A systematic review. *Journal of Business Research*.

Russell, S., & Norvig, P. (2016). *Artificial intelligence: A modern approach*. Pearson Education Limited.

Russell, S., & Norvig, P. (2016). *Artificial intelligence: A modern approach*. Pearson Education Limited.

Russell, S., & Norvig, P. (2016). *Artificial intelligence: A modern approach*. Pearson Education Limited.

Russell, S., & Norvig, P. (2016). *Artificial intelligence: A modern approach*. Pearson Education Limited.

Rust, R. T. (2020). The Future of Marketing. *International Journal of Research in Marketing*, 37(1), 15–26. DOI: 10.1016/j.ijresmar.2019.08.002

Rust, R. T., & Huang, M. H. (2021). The future of artificial intelligence in marketing: Looking forward to 2025. *Journal of the Academy of Marketing Science*, 49(5), 946–958.

Shankar, V. (2018). What an artificial intelligence (AI) is reshaping retailing. *Journal of Retailing*, 94(4), vi–xi. DOI: 10.1016/S0022-4359(18)30076-9

Shankar, V. (2018). How artificial intelligence (AI) is reshaping retailing. *Journal of Retailing*, 94(4), vi–xi. DOI: 10.1016/S0022-4359(18)30076-9

Shankar, V. (2018). How artificial intelligence (AI) is reshaping retailing. *Journal of Retailing*, 94(4), vi–xi. DOI: 10.1016/S0022-4359(18)30076-9

Sheth, J. N., Sisodia, R. S., & Sharma, A. (2000). The antecedents and consequences of customer-centric marketing. *Journal of the Academy of Marketing Science*, 28(1), 55–66. DOI: 10.1177/0092070300281006

Shmueli, G., & Koppius, O. R. (2011). Predictive analytics in information systems research. *Management Information Systems Quarterly*, 35(3), 553–572. DOI: 10.2307/23042796

Smith, P. R., & Chaffey, D. (2020). *Digital Marketing Excellence: Planning, Optimizing and Integrating Online Marketing*. Routledge.

Sullivan, R. (2020). *Predictive analytics in marketing: Starbucks' AI-driven recommendations*. Marketing Science Review.

Török, Á., Török, M., & Jacobsen, G. (2022). Ethical consumerism: The role of AI in promoting sustainable choices. *Journal of Marketing Ethics*, 15(2), 77–90.

Török, Á., Török, M., & Jacobsen, G. (2022). Ethical consumerism: The role of AI in promoting sustainable choices. *Journal of Marketing Ethics*, 15(2), 77–90.

Vesanen, J. (2007). What is personalization? A conceptual framework. *European Journal of Marketing*, 41(5/6), 409–418. DOI: 10.1108/03090560710737534

Westerman, G., Bonnet, D., & McAfee, A. (2014). *Leading digital: Turning technology into business transformation*. Harvard Business Review Press.

Ziakis, C., & Vlachopoulou, M. (2023). Artificial intelligence in digital marketing: Insights from a comprehensive review. *Information (Basel)*, 14(12), 664.

Zuboff, S. (2019). *The age of surveillance capitalism: The fight for a human future at the new frontier of power*. PublicAffairs.

Zuboff, S. (2019). *The age of surveillance capitalism: The fight for a human future at the new frontier of power*. PublicAffairs.

Zuboff, S. (2019). *The age of surveillance capitalism: The fight for a human future at the new frontier of power*. PublicAffairs.

Zuboff, S. (2019). *The age of surveillance capitalism: The fight for a human future at the new frontier of power*. PublicAffairs. Noble, S. U. (2018). *Algorithms of oppression: How search engines reinforce racism*. NYU Press.

Zuboff, S. (2019). *The age of surveillance capitalism: The fight for a human future at the new frontier of power*. PublicAffairs.

Compilation of References

Abdeldayem, M. M., & Aldulaimi, S. H. (2020). Trends and opportunities of artificial intelligence in human resource management: Aspirations for public sector in Bahrain. *International journal of scientific and technology research, 9*(1), 3867-3871.

Abdullah Aman Tutul, M.. (2021). Impact of AI Inaccuracy on Trust in Human-AI Collaboration for Speech-Based Fear Estimation. *IEEE Access : Practical Innovations, Open Solutions*, 9, 122272–122283.

Abdullah Aman Tutul, M.. (2024). Factors Influencing Trust in Human-AI Collaboration for Speech-Based Fear Estimation: A Multi-faceted Perspective. *IEEE Access : Practical Innovations, Open Solutions*, 12, 9050–9063.

Abhilasha, A., Sreenivasulu, A., & Manimozhi, T. KUMAR, P. S., Selvakumar, P., & Singh, P. (2022, April). The Model of Smart Sensing Device For Sensitive Nanoclusters Modification in Sensing Properties. In 2022 2nd International Conference on Advance Computing and Innovative Technologies in Engineering (ICACITE) (pp. 1043-1047). IEEE.

Abirami, S., Pethuraj, M., Uthayakumar, M., & Chitra, P. (2024). A systematic survey on big data and artificial intelligence algorithms for intelligent transportation system. *Case Studies on Transport Policy*, 17, 101247. DOI: 10.1016/j.cstp.2024.101247

Achmad Wandi, A., & Ashari, M. (2023). IoT-Based Flood Early Warning System Using Arduino and Raspberry Pi. *Journal of Physics: Conference Series*, 2289(1), 012022.

Adam, H.. (2022). The impact of AI bias on human decision-making: An experimental investigation. *Journal of Experimental Social Psychology*, 105, 104272.

Adam, P.. (2012). Real-time communication networks for emergency preparedness and response. *IEEE Communications Magazine*, 50(11), 96–103.

Ado, M., & Amitab, K. (2023, March). Landslide susceptibility mapping using support vector machine for Meghalaya, India. In *2023 4th International Conference on Computing and Communication Systems (I3CS)* (pp. 1-6). IEEE.

Ado, M., Amitab, K., Maji, A. K., Jasińska, E., Gono, R., Leonowicz, Z., & Jasiński, M. (2022). Landslide susceptibility mapping using machine learning: A literature survey. *Remote Sensing (Basel)*, 14(13), 3029. DOI: 10.3390/rs14133029

African Union. (2023). *Smart Cities Initiative*. Retrieved from African Union Asian Development Bank. (2023). *ASEAN's Role in Strengthening Regional Resilience through Smart Cities*.

Agiwal, M., Roy, A., & Saxena, N. (2016). Next generation 5G wireless networks: A comprehensive survey. *IEEE Communications Surveys and Tutorials*, 18(3), 1617–1655. DOI: 10.1109/COMST.2016.2532458

Aguas, C., Asuncion, J., Dumulot, A., Gepulle, D., Larin, J., Pingol, H., & Salas, L. (2022). Solar Panel Powered Mobile Charging Station. Academia. https://www.academia.edu/31982060/Republic_of_the_Philippines_COLLEGE_OF_TECHNOLOGY_TITLE_OF_THE_PROJECT_Solar_Panel_Powered_Mobile_Charging_Station

Ahmad, H. W., Bibi, H. A., Chandrasekaran, M., Ahmad, S., & Kyriakopoulos, G. L. (2024). Sustainable Wastewater Treatment Strategies in Effective Abatement of Emerging Pollutants. *Water (Basel)*, 16(20), 20. Advance online publication. DOI: 10.3390/w16202893

Ahmad, S.. (2023). The Impact of Surveillance Technology on Privacy Concerns: A Review. *International Journal of Advanced Research in Engineering and Technology*, 15(3), 1–10.

Ahmed, M. M., Hassanien, E., & Hassanien, A. E. (2023). IoT-based intelligent waste management system. *Neural Computing & Applications*, 35(32), 23551–23579. DOI: 10.1007/s00521-023-08970-7

Ahmed, M., Khan, S., & Ali, A. (2023). Quantum sampling techniques for improved performance in handling large datasets for flood forecasting. *Journal of Hydrology (Amsterdam)*, 605, 127–139.

Al Harbi, S.. (2023). Responsible Design Patterns for AI: A Systematic Literature Review. *IEEE Access : Practical Innovations, Open Solutions*, 11, 10668–10695.

Alahi, M. E. E., Sukkuea, A., Tina, F. W., Nag, A., Kurdthongmee, W., Suwannarat, K., & Mukhopadhyay, S. C. (2023). Integration of IoT-Enabled Technologies and Artificial Intelligence (AI) for Smart City Scenario: Recent Advancements and Future Trends. *Sensors (Basel)*, 23(11), 5206. DOI: 10.3390/s23115206 PMID: 37299934

Al-Ani, A. K., Laghari, S. U. A., Manoharan, H., Selvarajan, S., & Uddin, M. (2023). Improved Transportation Model with Internet of Things Using Artificial Intelligence Algorithm. *Computers, Materials & Continua/Computers. Computers, Materials & Continua*, 76(2), 2261–2279. DOI: 10.32604/cmc.2023.038534

Albahri, A. S., Khaleel, Y. L., Habeeb, M. A., Ismael, R. D., Hameed, Q. A., Deveci, M., Homod, R. Z., Albahri, O. S., Alamoodi, A. H., & Alzubaidi, L. (2024). A systematic review of trustworthy artificial intelligence applications in natural disasters. *Computers & Electrical Engineering*, 118, 109409. DOI: 10.1016/j.compeleceng.2024.109409

Aldrich, D. P., Bhatt, M., & Charpentier, M. (2021). *Policy challenges in deploying emerging technologies for disaster management.* Policy Review-Journal.

Alenezi, A. (2023). Teacher perspectives on AI-driven gamification: Impact on student motivation, engagement, and learning outcomes.

Alhussen, A., & Ansari, A. S. (2024). Real-Time prediction of urban traffic problems based on Artificial Intelligence-Enhanced Mobile Ad Hoc Networks (MANETS). *Computers, Materials & Continua/Computers. Computers, Materials & Continua*, 79(2), 1903–1923. DOI: 10.32604/cmc.2024.049260

Ali, Y., Raadsen, M. P., & Bliemer, M. C. (2023). Modelling speed reduction behaviour on variable speed limit-controlled highways considering surrounding traffic pressure: A random parameters duration modelling approach. *Analytic Methods in Accident Research*, 40, 100290. DOI: 10.1016/j.amar.2023.100290

Alkhunaizan, A. (2020). The Implication of Mobile Phone Charging Station. *International Journal of Computer and Network Security, VOL*, 20(1). http://paper.ijcsns.org/07_book/202001/20200117.pdf

Almlöf, E., Nybacka, M., & Pernestål, A. (2020). Will public transport be relevant in a self-driving future? A demand model simulation of four scenarios for Stockholm, Sweden. *Transportation Research Procedia*, 49, 60–69. DOI: 10.1016/j.trpro.2020.09.006

Almusaed, A., Almssad, A., Yitmen, I., & Homod, R. Z. (2023). Enhancing student engagement: Harnessing "AIED"'s power in hybrid education—A review analysis. *Education Sciences*, 13(7), 632. DOI: 10.3390/educsci13070632

Alves, A. A., Monteiro, V., Pinto, J. G., Afonso, J. L., & Afonso, J. A. (2020). Development of an Internet of things system for smart home HVAC monitoring and control. *Lecture Notes of the Institute for Computer Sciences, Social Informatics and Telecommunications Engineering*, 315, 197–208. DOI: 10.1007/978-3-030-45694-8_15

Alvi, M., & Zhao, L. (2023). *AI-powered delivery systems for disaster response: Autonomous solutions in crisis management*. Robotics and AI Journal.

Alwindawi, A. F., Uçan, O. N., Ibrahim, A. A., & Abbas, S. A. (2022). Semi-supervised learning with ensemble deep learning networks for descriptor generation. *Optik (Stuttgart)*, 170450, 170450. Advance online publication. DOI: 10.1016/j.ijleo.2022.170450

Al-Yahyai, S., Charabi, Y., & Gastli, A. (2010). Review of the use of numerical weather prediction (NWP) models for wind energy assessment. *Renewable & Sustainable Energy Reviews*, 14(9), 3192–3198. DOI: 10.1016/j.rser.2010.07.001

Amaral, R. E. C., Brito, J., Buckman, M., Drake, E., Ilatova, E., Rice, P., Sabbagh, C., Voronkin, S., & Abraham, Y. S. (2020). Waste Management and Operational Energy for Sustainable Buildings: A Review. *Sustainability (Basel)*, 12(13), 13. Advance online publication. DOI: 10.3390/su12135337

Anaba, D. C.. (2024). Impact of Artificial Intelligence, Internet of Things and Blockchain on Supply Chain Management: A Review. *International Journal of Supply Chain Management*, 13(2), 111–125.

Anagnostopoulos, D., Tsekouras, G., & Kotsakis, R. (2021). Emotional AI: Technology for personalized and emotional responses in brand marketing. *Journal of Consumer Psychology*, 31(3), 615–630.

Andrade, Y., & Yoo, S. G. (2019). A comprehensive study of the use of LoRa in the developmentof smart cities. *Applied Sciences (Basel, Switzerland)*, 9(22), 4753. DOI: 10.3390/app9224753

Angelidou, M. (2014). Smart city policies: A spatial approach. *Cities (London, England)*, 41, S3–S11. DOI: 10.1016/j.cities.2014.06.007

Annual Report 2021. (n.d.). UNEP - UN Environment Programme. https://www.unep.org/resources/annual-report-2021

Anthony, B.Jnr. (2024). Artificial intelligence of things and distributed technologies as enablers for intelligent mobility services in smart Cities-A survey. *Internet of Things : Engineering Cyber Physical Human Systems*, 101399, 101399. Advance online publication. DOI: 10.1016/j.iot.2024.101399

Anthopoulos, L. G. (2015). Understanding the smart city domain: A literature review. *Transforming city governments for successful smart cities*, 9-21.

Anthopoulos, L. G. (2017). The rise of the smart city. In *Understanding smart cities: a tool for smart government or an industrial trick?* (pp. 5–45). Springer., DOI: 10.1007/978-3-319-57015-0_2

Antora, N. A., Rahman, M. A., Mosharraf, A. A., Ibn Ehsan, M., Alve, M., & Elahi, M. M. (2022)"Design and Implementation of a Smart Bin using IOT for an Efficient Waste ManagementSystem,"25th International Conference on Computer and Information Technology(ICCIT), Cox's Bazar, Bangladesh, 2022, pp. 774-779, DOI: 10.1109/ICCIT57492.2022.10055998

Ao, X., He, R., Zhang, C., & Wang, L. (2019). A New Outdoor Energy Mobile Phone Charging Station. 34rd Youth Academic Annual Conference of Chinese Association of Automation. https://ieeexplore.ieee.org/document/8787589

Ap. (2023, August 19). San Francisco launches driverless bus service following robotaxi expansion. *The Economic Times*. https://economictimes.indiatimes.com/tech/technology/san-francisco-launches-driverless-bus-service-following-robotaxi-expansion/articleshow/102847234.cms?from=mdr

Apexon. (2024). Applying quantum computing and AI in healthcare and financial services.

Arnaud, A., Saeidi, M., & Vallee, M. (2019). A study of performance metrics for earthquake early warning systems. *Bulletin of the Seismological Society of America*, 109(4), 1492–1505.

Arshad, M. W., Murtza, I., & Arshad, M. A. (2023). Applications of quantum computing in the health sector. *Journal of Data Science and Intelligent Systems*, 1(1), 19–24. DOI: 10.47852/bonviewJDSIS3202656

Asian Development Bank. (2021). *Asian Development Outlook 2021: Financing a green and inclusive recovery*.

Asian Development Bank. (2024). *Flood Management in North Java Project*.

Awan, R. (2024). Impact of AI on personalized learning and curriculum delivery. *Multidisciplinary Journal of Emerging Needs of Curriculum*, 1(2), 1–9.

Aybar-Ruiz, A., Jiménez-Fernández, S., Cornejo-Bueno, L., Casanova-Mateo, C., Sanz-Justo, J., Salvador-González, P., & Salcedo-Sanz, S. (2016). A novel Grouping Genetic Algorithm–Extreme Learning Machine approach for global solar radiation prediction from numerical weather models inputs. *Solar Energy*, 132, 129–142. DOI: 10.1016/j.solener.2016.03.015

Ayeni, O. O., Al Hamad, N. M., Chisom, O. N., Osawaru, B., & Adewusi, O. E. (2024). AI in education: A review of personalized learning and educational technology. *GSC Advanced Research and Reviews*, 18(2), 261–271. DOI: 10.30574/gscarr.2024.18.2.0062

Azar, A. T. (2012). Overview of Type-2 Fuzzy Logic Systems. [IJFSA]. *International Journal of Fuzzy System Applications*, 2(4), 1–28. DOI: 10.4018/ijfsa.2012100101

Azar, A. T., & Serrano, F. E. (2015). Adaptive Sliding Mode Control of the Furuta Pendulum. In Azar, A. T., & Zhu, Q. (Eds.), *Advances and Applications in Sliding Mode Control systems* (pp. 1–42). Springer International Publishing., DOI: 10.1007/978-3-319-11173-5_1

Badiger, M., & Mathew, J. A. (2022, March). Retrospective Review of Activation Functions in Artificial Neural Networks. In Proceedings of Third International Conference on Communication, Computing and Electronics Systems: ICCCES 2021 (pp. 905-919). Singapore: Springer Singapore. DOI: 10.1007/978-981-16-8862-1_59

Bagrow, J. P., Benson, T., & Crane, R. (2011). Emergent communication networks and the scaling of human collective action. *Proceedings of the National Academy of Sciences of the United States of America*, 108(Supplement 3), 18214–18220.

Bakıcı, T., Almirall, E., & Wareham, J. (2013). A smart city initiative: The case of Barcelona. *Journal of the Knowledge Economy*, 4(2), 135–148. DOI: 10.1007/s13132-012-0084-9

Balasubramanian, M. (2018). Municipal Solid Waste Management in India: Status,Problems andChallenges. *International Journal of Environment and Waste Management*, 21(4), 253–265. . DOI: 10.1504/IJEWM.2018.093435

Ball State University. (2022). Smart City History. https://digitalresearch.bsu.edu/immersive-learning-showcase-spring-2022/exhibits/show/smart-city-muncie/definition-and-history-of-smar

Balzano, W., & Vitale, F. (2017, March). DiG-Park: a smart parking availability searching method using V2V/V2I and DGP-class problem. In *2017 31st International Conference on Advanced Information Networking and Applications Workshops (WAINA)* (pp. 698-703). IEEE.

Band, S. S., Ardabili, S., Sookhak, M., Chronopoulos, A. T., Elnaffar, S., Moslehpour, M., Csaba, M., Torok, B., Pai, H.-T., & Mosavi, A. (2022). When smart cities get smarter via machine learning: An in-depth literature review. *IEEE Access : Practical Innovations, Open Solutions*, 10, 60985–61015. DOI: 10.1109/ACCESS.2022.3181718

Bano, A., Ud Din, I., & Asma, A. (2020). An IoT-Based Smart Bin for Real-Time Monitoring and Management of Solid Waste. *Scientific Programming*, 2020, 6613263. DOI: 10.1155/2020/6613263

Barcelona City Council. (2023). *AI and the City: Ethical Guidelines for Smart City Projects.*

Barrera-Animas, A. Y., Oyedele, L. O., Bilal, M., Akinosho, T. D., Delgado, J. M. D., & Akanbi, L. A. (2022). Rainfall prediction: A comparative analysis of modern machine learning algorithms for time-series forecasting. *Machine Learning with Applications*, 7, 100204. DOI: 10.1016/j.mlwa.2021.100204

Barrionuevo, J. M., Berrone, P., & Ricart, J. E. (2012). Smart cities, sustainable progress. *IESE insight, 14*(14), 50-57.

Barr, P. J.. (2016). The changing burden of disease in Africa: Implications for research and policy. *Lancet*, 387(10021), 955–970.

Beba, H., & Öztürk, Z. (2024). Investigation of models predicting NOx level in the sample region and the use of intelligent transportation system. *Environmental Challenges*, 16, 100990. DOI: 10.1016/j.envc.2024.100990

Behnia, F., Schuelke-Leech, B., & Mirhassani, M. (2024). Optimizing Sustainable Urban Mobility: A Comprehensive review of electric bus scheduling strategies and future directions. *Sustainable Cities and Society*, 108, 105497. DOI: 10.1016/j.scs.2024.105497

'Benefits of Cellphone Charging Station', Veloxity, 2016, accessed 12 September 2022, https://veloxity.us/benefits-cell-phone-charging-station/

Bertilsson, L., Wiklund, K., de Moura Tebaldi, I., Rezende, O. M., Veról, A. P., & Miguez, M. G. (2019). Urban flood resilience–A multi-criteria index to integrate flood resilience into urban planning. *Journal of Hydrology (Amsterdam)*, 573, 970–982. DOI: 10.1016/j.jhydrol.2018.06.052

Bhavsar, P., Safro, I., Nidhal, B., Robi, P., Dera, D., Dutta, P., & Aminul, O. (2024). Artificial intelligence in transportation data analytics. In *Elsevier eBooks* (pp. 337–382). https://doi.org/DOI: 10.1016/B978-0-443-13878-2.00008-4

Bhoi, A., Balabantaray, R. C., Sahoo, D., Dhiman, G., Khare, M., Narducci, F., & Kaur, A. (2022). Mining social media text for disaster resource management using a feature selection based on forest optimization. *Computers & Industrial Engineering*, 169, 108280. DOI: 10.1016/j.cie.2022.108280

Bibri, S. E., & Krogstie, J. (2020). The emerging data–driven Smart City and its innovative applied solutions for sustainability: The cases of London and Barcelona. *Energy Informatics*, 3(1), 5. DOI: 10.1186/s42162-020-00108-6

Bibri, S., & Krogstie, J. (2022). Governance and the Evolution of Smart Cities: From Technology-Driven to Citizen-Centric Models. *Technological Forecasting and Social Change.*

bin Salem, I. (2024). Integrating artificial intelligence in personalized learning: A future-oriented approach to enhance student engagement and achievement. International Journal of Post Axial: Futuristic Teaching and Learning, 111-119.

Biswas, R. A., & Molla, A. (2024, August). Impact of artificial intelligence on personalized learning. In *Artificial Intelligence in Education* (p. 302).

Bjarnason, E.. (2009). Simulation-based optimization of emergency medical service resource allocation. *European Journal of Operational Research*, 193(1), 1–18.

Boer, D. (2015). *The Routledge Handbook of Disaster Risk Reduction*. Routledge.

Bogdanov, V. (2023, July 10). *How machine learning is used in autonomous vehicles*. rinf.tech. https://www.rinf.tech/how-machine-learning-is-used-in-autonomous-vehicles/

Boopathi, S., Haribalaji, V., Mageswari, M., & Asif, M. M. (2022). "Agricultural Waste Management Systems Using Artificial Intelligence Techniques". In Agricultural Waste Management Systems Using Artificial Intelligence Techniques (pp. 1-23).

Bosch, M., Conroy, C., Ortiz, B., & Bogden, P. (2020, September). Improving emergency response during hurricane season using computer vision. In *Earth Resources and Environmental Remote Sensing/GIS Applications XI* (Vol. 11534, pp. 72–79). SPIE., DOI: 10.1117/12.2574639

Boza, I., & Evgeniou, T. (2021). Towards an Ethical Framework for AI: A Regulatory Perspective. *arXiv preprint arXiv:2109.02208.*

Brester, C., Kallio-Myers, V., Lindfors, A. V., Kolehmainen, M., & Niska, H. (2023). Evaluating neural network models in site-specific solar PV forecasting using numerical weather prediction data and weather observations. *Renewable Energy*, 207, 266–274. DOI: 10.1016/j.renene.2023.02.130

Brodie, R. J., Hollenbeck, L. D., Juric, B., & Ilic, A. (2011). Customer engagement: Conceptual domain, fundamental propositions, and implications for research. *Journal of Service Research*, 14(3), 252–271. DOI: 10.1177/1094670511411703

Brown, T., Mann, B., Ryder, N., Subbiah, M., Kaplan, J. D., Dhariwal, P., . . . Amodei, D. (2020). Language models are few-shot learners. arXiv preprint arXiv:2005.14165

Brown, M., & Clifton, J. (2023). Public-private partnerships in global smart cities: A framework for collaborative innovation. *International Journal of Smart Cities*, 9(2), 45–60.

Brown, T., Mann, B., Ryder, N., Subbiah, M., Kaplan, J. D., Dhariwal, P., & Amodei, D. (2020). Language models are few-shot learners. *arrive preprintarXiv:2005.14165.*

Brynjolfsson, E., & McAfee, A. (2015). The second machine age: Work, progress, and prosperity in a time of brilliant technologies. *Choice (Chicago, Ill.)*, 52(06), 52–3201. DOI: 10.5860/CHOICE.184834

Buhl, S. (2024). Citizen Engagement in Smart Cities: The Case of Copenhagen. *Urban Studies Journal*, 61(1), 125–140.

Bui, T. A., Lee, P. J., Lum, K. Y., Loh, C., & Tan, K. (2020). Deep learning for landslide recognition in satellite architecture. *IEEE Access : Practical Innovations, Open Solutions*, 8, 143665–143678. DOI: 10.1109/ACCESS.2020.3014305

Cardullo, P., et al. (2022). *Citizen Participation and Urban Co-Creation in Smart City Development.* Journal of Urban Studies.

Catania, V., & Ventura, D. (2014, April). An approch for monitoring and smart planning of urban solid waste management using smart-M3 platform. In *Proceedings of 15th conference of open innovations association FRUCT* (pp. 24-31). IEEE.

Cavoukian, A. (2023). Privacy and Security in Smart Cities: The Role of Technology. *Smart Cities Journal*, 5(1), 10–21. DOI: 10.1016/j.scij.2023.100051

Central Weather Bureau. (2023). *Earthquake Report App.*https://www.cwa.gov.tw/V8/E/S/eservice/app/app_e.html

Cervero, R., & Murakami, J. (2010). Effects of Built Environments on Vehicle Miles Traveled: Evidence from 370 US Urbanized Areas. *Environment & Planning A*, 42(2), 400–418. DOI: 10.1068/a4236

Cha, E. (2024). AI Regulation for Emergency Response: Lessons from Nuclear Safety Law. *Science*, 384(6693), 464–466.

Chaffey, D. (2020). Multi-channel marketing automation using AI: Current trends. *Journal of Digital Marketing*, 12(1), 34–42.

Chaffey, D., & Ellis-Chadwick, F. (2019). *Digital marketing: Strategy, implementation, and practice.* Pearson UK.

Chafle, K., Faruk, B., Shrivas, S., & Sawarkar, N. (2015). Coin Based Mobile Charger on Solar Energy. International Journal of Research- Granthaalayah, 3(1):13-18. https://www.researchgate.net/publication/306103054_COIN_BASED_MOBILE_CHARGER_ON_SOLAR_ENERGY

Chaganti, P. C. V., Vasireddy, K., Reddy, E. R., Dondeti, R. R., & Syama, S. (2023, July). Predicting Landslides and Floods with Deep Learning. In 2023 4th International Conference on Electronics and Sustainable Communication Systems (ICESC) (pp. 1259-1265). IEEE. DOI: 10.1109/ICESC57686.2023.10193456

Chakrabortty, R., Pal, S. C., Ruidas, D., Roy, P., Saha, A., & Chowdhuri, I. (2023). Living with floods using state-of-the-art and geospatial techniques: Flood mitigation alternatives, management measures, and policy recommendations. *Water (Basel)*, 15(3), 558. DOI: 10.3390/w15030558

Chakraborty, T., & Datta, S. K. (2017). Home automation using edge computing and Internet of things. *2017 IEEE International Symposium on Consumer Electronics (ISCE)*. https://doi.org/DOI: 10.1109/ISCE.2017.8355544

Chakravarty, P., & Gattupalli, S. (2024). Integration of Indigenous Traditional Knowledge and AI in Hurricane Resilience and Adaptation. In *Advances in Hurricane Risk in a Changing Climate* (pp. 125–158). Springer Nature Switzerland., DOI: 10.1007/978-3-031-63186-3_6

Chamola, V., Hassija, V., Gupta, S., Goyal, A., Guizani, M., & Sikdar, B. (2021). Disaster and pandemic management using machine learning: A survey. *IEEE Internet of Things Journal*, 8(21), 16047–16071. DOI: 10.1109/JIOT.2020.3044966 PMID: 35782181

Chandrashekhar, T., Swaminaidu, G., & Rao, B. Ch. (2014). Mobile Charger Based on Coin by Using Solar Tracking System. International Journal of Innovative Reseach in Science, Engineering and Technology, Vol. 3, Issue 2, February 2014. http://www.ijirset.com/upload/2014/february/82_Mobile.pdf

Chang, F., & Das, D. (2020). Smart Nation Singapore: Developing Policies for a Citizen-Oriented Smart City Initiative. Developing National Urban Policies. *doi, 10*, 978-981. https://doi.org/DOI: 10.1007/978-981-15-3738-7_18

Chan, M. P.. (2004). Information technology in disaster medicine: A review. *Disaster Medicine and Public Health Preparedness*, 2(3), 132–140.

Chan, M., Bonhomme, S., Estève, D., & Campo, E. (2009). Individual movement trajectories in smart homes. *IFMBE Proceedings*, 1014-1018, 1014–1018. Advance online publication. DOI: 10.1007/978-3-540-92841-6_250

Chaurasia, S. H. U. B. H. A. M., Jain, S. H. U. B. H. A., Vishwkarma, H. O., & Singh, N. I. S. H. A. N. T. (2023). Conversational AI Unleashed: A Comprehensive Review of NLP-Powered Chatbot Platforms. *Iconic Research and Engineering Journals*, 7(3), 1–8.

Chavhan, R. D., & Sambare, G. B. (2024). AI-Driven Traffic Management Systems In Smart Cities: A Review. *Educational Administration: Theory and Practice*, 30(5), 105–116. DOI: 10.53555/kuey.v30i5.2780

Chen, X., Li, Y., & Zhang, Q. (2023). Advances in AI predictive models for disaster management. Journal of Machine Learning in Environmental Studies.

Chen, Y., & Zhang, T. (2022). Cybersecurity risks in IoT-based disaster management. Journal of Internet Security.

Chen, C.-C., Lin, C.-W., Huang, Y.-M., & Hsu, Y.-C. (2020). A review of earthquake early warning systems: From concepts to implementations. *Sensors (Basel)*, 20(6), 1740. DOI: 10.3390/s20061740 PMID: 32245028

Cheng, M., & Wang, R. (2023). Computer vision applications in disaster response: A case study approach. *IEEE Transactions on Image Processing*.

Chen, H., Chiang, R. H., & Storey, V. C. (2012). Business intelligence and analytics: From big data to big impact. *Management Information Systems Quarterly*, 36(4), 1165–1188. DOI: 10.2307/41703503

Chen, J.-Y., Feng, Y.-W., Teng, D., Lu, C., & Fei, C.-W. (2022). Support vector machine-based similarity selection method for structural transient reliability analysis. *Reliability Engineering & System Safety*, 223, 108513. DOI: 10.1016/j.ress.2022.108513

Chen, L., Chen, P., & Lin, Z. (2020). Artificial intelligence in education: A review. *IEEE Access : Practical Innovations, Open Solutions*, 8, 75264–75278. DOI: 10.1109/ACCESS.2020.2988510

Chen, T. (2010). Smart grids, smart cities need better networks [Editor's note]. *IEEE Network*, 24(2), 2–3. DOI: 10.1109/MNET.2010.5430136

Chen, Y., & Liu, J. (2024). Quantum reinforcement learning for adaptive flood prediction models. *Environmental Modelling & Software*, 156, 105–117.

Chettri, L., & Bera, R. (2019). A comprehensive survey on Internet of Things (IoT) toward 5G wireless systems. *IEEE Internet of Things Journal*, 7(1), 16–32. DOI: 10.1109/JIOT.2019.2948888

Chintalapati, M., & Pandey, R. (2022). Chatbots and virtual try-ons: AI's role in retail engagement. *Journal of Digital Marketing*. Brown, T., Mann, B., Ryder, N., Subbiah, M., Kaplan, J. D., Dhariwal, P., ... & Amodei, D. (2020). Language models are few-shot learners. *arXiv preprint arXiv:2005.14165.*

Chourabi, H.. (2021). *Citizen Engagement and Transparency in Smart Cities: The Role of Governance in Urban Planning*. Cities Journal.

Chui, M., Manyika, J., & Miremadi, M. (2018). What AI Can and Can't Do (Yet) for Your Business. *Harvard Business Review.* Retrieved from https://hbr.org

Chui, M., Manyika, J., & Miremadi, M. (2018). What AI Can and Cannot Do (Yet) for Your Business. *Harvard Business Review.* Retrieved from https://hbr.org

CIVITAS. (2022). Citizen Engagement in Smart City Projects. *Best Practice.*

Clements, L. M., & Kockelman, K. M. (2017). Economic effects of automated vehicles. *Transportation Research Record: Journal of the Transportation Research Board*, 2606(1), 106–114. DOI: 10.3141/2606-14

Cleveland Clinic. (2022). Enabling medical research and precision medicine with quantum computing.

Cloud Security Alliance. (2023). *AI Security and Privacy Best Practices.*

CNA. (2024). Singapore ranked 5th in Smart City Index.

Cohen, B. (2015). The 3 Generations Of Smart Cities. https://www.fastcompany.com/3047795/the-3-generations-of-smart-cities

Collini, E., Palesi, L. I., Nesi, P., Pantaleo, G., Nocentini, N., & Rosi, A. (2022). Predicting and understanding landslide events with explainable AI. *IEEE Access : Practical Innovations, Open Solutions*, 10, 31175–31189. DOI: 10.1109/ACCESS.2022.3158328

Collins, S., & McCarthy, I. (2021). How AI is Revolutionizing Content Creation in Marketing. *Journal of Marketing Technology*, 12(3), 45–57.

Colson, E., & Ramsay, M. (2020). Human-AI partnerships in the workplace: Enhancing creativity and performance. *MIT Sloan Management Review*, 61(4), 58–63.

Constantinides, G.. (2009). The role of information and communication technologies in disaster preparedness and response. *Social Science & Medicine*, 69(10), 1557–1566.

Contissa, C.. (2013). Liability in complex socio-technical systems: The challenge of distributed agency. *Safety Science*, 51, 102–114.

Cowley, S. J. (2020). Human-AI Interaction: An Experience-Centered Perspective on AI Implementation in Marketing. *Journal of Interactive Marketing*, 51, 36–45.

Cretu, L. G. (2012). Smart cities design using event-driven paradigm and semantic web. *Informações Econômicas*, 16(4), 57.

Cruz Ulloa, D. A.. (2023). Deep Learning-Based Victim Detection Using Multispectral Imagery in Outdoor Environments. *Remote Sensing*, 15(13), 3265.

da Silva, L. S. A., Lúcio, Y. L. S., Coelho, L. dos S., Mariani, V. C., & Rao, R. V. (2023). A comprehensive review on Jaya optimization algorithm. *Artificial Intelligence Review*, 56(5), 4329–4361. DOI: 10.1007/s10462-022-10234-0

Damaševičius, R.. (2023). The Role of Technology in Disaster Response: A Review. *Sustainability*, 15(12), 9490.

Dameri, R. P. (2017). *Smart city implementation. Progress in IS*. Springer., DOI: 10.1007/978-3-319-45766-6

Dandachi, I. E. (2024). AI-powered personalized learning: Toward sustainable education. In *Navigating the Intersection of Business, Sustainability and Technology* (pp. 109–118). Springer Nature Singapore.

Daneshvar, M. H., Sarmadi, H., & Yuen, K.-V. (2023). A locally unsupervised hybrid learning method for removing environmental effects under different measurement periods. *Measurement*, 208, 112465. DOI: 10.1016/j.measurement.2023.112465

Danish, M. (2022). "Chapter 1 - artificial intelligence and machine learning in water resources engineering," in Current directions in water scarcity research. Editors M. ZAKWAN, A. WAHID, M. NIAZKAR, and U. CHATTERJEE (Elsevier).

Darji, J., Lodha, P., & Tyagi, S. (2022). Assimilative capacity and water quality modeling of rivers: A review. AQUA - Water Infrastructure. *Journal of Water Supply: Research & Technology - Aqua*, 71(10), 1127–1147. DOI: 10.2166/aqua.2022.063

Dass-Brailsford, P. E., & Myrick, R. B. (2010). Integrated treatment for co-occurring disorders: A review of models and outcomes. *Clinical Psychology Review*, 30(4), 381–394.

Davenport, T. H., Guha, A., Grewal, D., & Bressgott, T. (2021). How Artificial Intelligence Will Transform Marketing Strategy. *Journal of the Academy of Marketing Science*, 49(1), 24–42. DOI: 10.1007/s11747-019-00696-0

Davenport, T., & Ronanki, R. (2018). Artificial intelligence for the real world. *Harvard Business Review*, 96(1), 108–116.

Dawdi, A.. (2020). UAV-based victim detection using thermal and visible imagery. *IEEE Access : Practical Innovations, Open Solutions*, 8, 12227–12236.

de Reuver, M., Sørensen, C., & Basole, R. C. (2018). The digital platform: A research agenda. *Journal of Information Technology*, 33(2), 124–135. DOI: 10.1057/s41265-016-0033-3

Deed - Attribution-NonCommercial-NoDerivatives 4.0 International - Creative Commons. (n.d.). https://creativecommons.org/licenses/by-nc-nd/4.0

Deep, S. (2023). Role of Artificial Intelligence in Smart Cities for Sustainable Development. *Sustainability*, 15(11), 8352.

Deloitte Team. (2024). Quantum machine learning for next-day flood predictions: Assessing prerequisites for larger-scale predictions. *Climate Resilience Journal*, 12(1), 45–58.

Deloitte. (2023). *The Future of Urban Mobility: Smart Cities and AI*.

Delzendeh, E., Wu, S., Lee, A., & Zhou, Y. (2017). The impact of occupants' behaviours on building energy analysis: A research review. *Renewable & Sustainable Energy Reviews*, 80, 1061–1071. DOI: 10.1016/j.rser.2017.05.264

Dharmarathne, G., Waduge, A. O., Bogahawaththa, M., Rathnayake, U., & Meddage, D. P. P. (2024). Adapting cities to the surge: A comprehensive review of climate-induced urban flooding. *Results in Engineering*, 102123, 102123. Advance online publication. DOI: 10.1016/j.rineng.2024.102123

Dignum, V. (2019). Responsible innovation: A framework for ethical and trustworthy AI. *AI & Society*, 33(1), 1–11.

Dignum, V. (2022). *Responsible Artificial Intelligence: Towards Principle-Centered AI*. Springer International Publishing.

Diop, P. A., Gueye, A. D., & Sall, A. (2024). Detecting road signs and deformations for road safety in Senegal. *Procedia Computer Science*, 237, 221–228. DOI: 10.1016/j.procs.2024.05.099

Doerksen, K., Gal, Y., Kalaitzis, F., Rossi, C., Petit, D., Li, S., & Dadson, S. (2023, July). Precipitation-Triggered Landslide Prediction in Nepal Using Machine Learning and Deep Learning. In IGARSS 2023-2023 IEEE International Geoscience and Remote Sensing Symposium (pp. 4962-4965). IEEE. DOI: 10.1109/IGARSS52108.2023.10283036

Dominguez, F.. (2014). Situation-aware system for enhancing evacuation plans. *Simulation Modelling Practice and Theory*, 45, 106–117.

Dominic Abuga, N. S. (2021). Raghava,(2021) Real-time smart garbage bin mechanism for solid wastemanagement in smart cities. *Sustainable Cities and Society*, 75, 103347. DOI: 10.1016/j.scs.2021.103347

dos Santos Silva, F. J., Catunda, S. Y. C., Dorea, C. E. T., van Haandel, A. C., & dos Santos, H. R. (2019). Oxygen Uptake Rate Measurement Using Kalman Filter and PWM Control in Activated Sludge Systems. IEEE Transactions on Instrumentation and Measurement, 68(11), 4493–4501. *IEEE Transactions on Instrumentation and Measurement*. Advance online publication. DOI: 10.1109/TIM.2018.2886941

Douha, J. (2024). Fluid insights: Navigating water's quantum potential in computing. *TechRxiv*.

Duhayyim, M. A., Albraikan, A. A., Al-Wesabi, F. N., Burbur, H. M., Alamgeer, M., Hilal, A. M., Hamza, M. A., & Rizwanullah, M. (2021). Modeling of Artificial Intelligence Based Traffic Flow Prediction with Weather Conditions. *Computers, Materials & Continua/Computers. Materials & Continua (Print)*, 71(2), 3953–3968. DOI: 10.32604/cmc.2022.022692

Dwork, C., & Roth, A. (2014). The algorithmic foundations of differential privacy. *Foundations and Trends in Theoretical Computer Science*, 9(3–4), 211–407.

Ehsanifar, M., Dekamini, F., Spulbar, C., Birau, R., Khazaei, M., & Bărbăcioru, I. C. (2023). A sustainable pattern of waste management and energy efficiency in smart homes using the Internet of things (IoT). *Sustainability (Basel)*, 15(6), 5081. DOI: 10.3390/su15065081

El Mestari, Y.. (2023). Privacy-Preserving Machine Learning: A Survey. *IEEE Access : Practical Innovations, Open Solutions*, 11, 12227–12246.

Elassy, M., Al-Hattab, M., Takruri, M., & Badawi, S. (2024). Intelligent transportation systems for sustainable smart cities. *Transportation Engineering*, 16, 100252. DOI: 10.1016/j.treng.2024.100252

Ellikkal, A., & Rajamohan, S. (2024). *AI-enabled personalized learning: Empowering management students for improving engagement and academic performance*. Vilakshan-XIMB Journal of Management.

Elvas, L. B., Mataloto, B. M., Martins, A. L., & Ferreira, J. C. (2021). Disaster management in smart cities. *Smart Cities*, 4(2), 819–839. DOI: 10.3390/smartcities4020042

Eremia, M., Toma, L., & Sanduleac, M. (2017). The smart city concept in the 21st century. *Procedia Engineering*, 181, 12–19. DOI: 10.1016/j.proeng.2017.02.357

Esrafilian-Najafabadi, M., & Haghighat, F. (2022). Impact of predictor variables on the performance of future occupancy prediction: Feature selection using genetic algorithms and machine learning. *Building and Environment*, 219, 109152. DOI: 10.1016/j.buildenv.2022.109152

European Commission. (2022). *EU Cybersecurity Act: Strengthening Europe's Cybersecurity*. European Commission. (2024). EU Urban Mobility Observatory. Directorate-General for Mobility and Transport.

Ezer, N., ... (2019). Building Trust in Human-AI Teams: A Framework for Trustworthy AI. *arXiv preprint arXiv:1904.09936*.

Fakhar, M. Z., Yalcin, E., & Bilge, A. (2023). A survey of smart home energy conservation techniques. *Expert Systems with Applications*, 213, 118974. DOI: 10.1016/j.eswa.2022.118974

Farooq, M. S., Khan, M., & Abid, A. (2020). A framework to make charity collection transparent and auditable using blockchain technology. *Computers & Electrical Engineering*, 83, 106588.

Firat, D., et al. (2020). Enhancing brand growth through AI. *International Journal for Multidisciplinary Research*.

Fischer, J., Wirtz, S., & Scherer, V. (2023). Random forest classifier and neural network for fraction identification of refuse-derived fuel images. *Fuel*, 341, 127712. DOI: 10.1016/j.fuel.2023.127712

Fisher, A.. (2023). The impact of AI on health and safety at work equity: A call for interdisciplinary research. *Work, Employment and Society*, 37(1), 12–33.

Flavián, C., Ibáñez-Sánchez, S., & Orús, C. (2019). Integrating virtual reality and augmented reality in brand engagement. *Journal of Business Research*, 100, 547–560.

Fletcher, J. M.. (2021). Ensuring equitable and appropriate AI in global health. *Nature Medicine*, 27(12), 1973–1980.

Fowdur, T. P., & Nassir-Ud-Diin Ibn Nazir, R. M. (2022). A real-time collaborative machine learning based weather forecasting system with multiple predictor locations. *Array (New York, N.Y.)*, 14, 100153. DOI: 10.1016/j.array.2022.100153

Freeman, S., & Rossi, M. S. (2012). Interagency collaboration in emergency management: A review of the literature. *International Journal of Mass Emergencies and Disasters*, 30(2), 117–142.

Gaber, R. M., El-Kader, A., Hussein, M., & Okba, E. M. (2022). The Resilience Performance Index, a Fuzzy Logic Approach to Assess Urban Resilience. *International Journal of Sustainable Development and Planning*, 17(4), 1225–1235. Advance online publication. DOI: 10.18280/ijsdp.170421

Gad, A. G., Mosa, D. T., Abualigah, L., & Abohany, A. A. (2022). Emerging trends in blockchain technology and applications: A review and outlook. *Journal of King Saud University. Computer and Information Sciences*, 34(9), 6719–6742.

Ganguly, K. K., Nahar, N., & Hossain, B. M. M. (2019). A machine learning-based prediction and analysis of flood affected households: A case study of floods in Bangladesh. *International Journal of Disaster Risk Reduction*, 34, 283–294. DOI: 10.1016/j.ijdrr.2018.12.002

Georgieva, S. (2021). AI and the Future of Creativity: How Generative AI Will Transform the Role of Marketers. *Journal of Creative AI*, 6(2), 45–58.

Gernaey, K. V., Flores-Alsina, X., Rosen, C., Benedetti, L., & Jeppsson, U. (2011). Dynamic influent pollutant disturbance scenario generation using a phenomenological modelling approach. *Environmental Modelling & Software*, 26(11), 1255–1267. DOI: 10.1016/j.envsoft.2011.06.001

Gevaert, T.. (2021). Unpacking algorithmic risk assessment: Exploring hidden geographic biases and accountability gaps. *Policing and Society*, 31(6), 659–677.

Ghai, G., Khanna, A., & Nirmal Kumar, S. J. (2024). "Confront Acknowledgment and OTP Based Security Bolt Framework, " Procedures of the 6th Universal Conference on Communications and Cyber Physical Designing, vol. 096, pp. 729.

Ghasemi, A., & Rahman, S. (2021). Natural language processing for crisis management. Journal of Applied AI.

Ghazal, T. M., Hasan, M. K., Alshurideh, M. T., Alzoubi, H. M., Ahmad, M., Akbar, S. S., Al Kurdi, B., & Akour, I. A. (2021). IoT for smart cities: Machine learning approaches in smart healthcare—A review. *Future Internet*, 13(8), 218. DOI: 10.3390/fi13080218

Ghebleh Goydaragh, M., Taghizadeh-Mehrjardi, R., Golchin, A., Asghar Jafarzadeh, A., & Lado, M. (2021). Predicting weathering indices in soils using FTIR spectra and random forest models. *Catena*, 204, 105437. DOI: 10.1016/j.catena.2021.105437

Ghoushchi, S. J., Haghshenas, S. S., Vahabzadeh, S., Guido, G., & Geem, Z. W. (2024). An integrated MCDM approach for enhancing efficiency in connected autonomous vehicles through augmented intelligence and IoT integration. *Results in Engineering*, 23, 102626. DOI: 10.1016/j.rineng.2024.102626

Giesen, D., & Kristen, G. (2014). Accountability for robots: A new challenge for legal systems. *Ethics and Information Technology*, 16(4), 359–374.

Giffinger, R., Fertner, C., Kramar, H., Kalasek, R., Pichler-Milanović, N., & Meijers, E. (2007). Smart cities ranking of European medium-sized cities. https://www.smart-cities.eu/download/smart_cities_final_report.pdf

Given, D. D., Allen, R. M., Hauksson, E., Crowell, B. W., Hellweg, M., & Heaton, T. H. (2018). The ShakeAlert earthquake early warning system in California: Status as of March 2018. *Seismological Research Letters*, 89(1), 171–179. DOI: 10.1785/0220170140

Gohar, A., & Nencioni, G. (2021). The role of 5G technologies in a smart city: The case for intelligent transportation system. *Sustainability (Basel)*, 13(9), 5188. DOI: 10.3390/su13095188

Goldbach, C., Sickmann, J., Pitz, T., & Zimasa, T. (2021). Towards autonomous public transportation: Attitudes and intentions of the local population. *Transportation Research Interdisciplinary Perspectives*, 13, 100504. DOI: 10.1016/j.trip.2021.100504

Gomez Llerena, J. A. (2023). *Investigation of the Integrity of Tailings Dam Surfaces Using Aerial Imaging and Computer Vision* (Doctoral dissertation).

Gomez, H., Ramírez, L., & Yang, J. (2024). *Drone and robotics swarms for disaster response applications*. Robotics and Disaster Management.

Gort, M. (2024). *European and Asian cities dominate latest smart city rankings*. Workplace Insight.

Gössling, S., & Choi, A. S. (2015). Transport transitions in Copenhagen: Comparing the cost of cars and bicycles. *Ecological Economics*, 113, 106–113. DOI: 10.1016/j.ecolecon.2015.03.006

GovInsider. (2024). *ASEAN Smart Cities and Cybersecurity*. GovInsider Insights.

Graham, S., & Marvin, S. (1996). *Telecommunications and the city: Electronic Spaces, Urban Places*. Psychology Press.

Grant, M. J., & Booth, A. (2009). A typology of reviews: An analysis of 14 review types and associated methodologies. *Health Information and Libraries Journal*, 26(2), 91–108. DOI: 10.1111/j.1471-1842.2009.00848.x PMID: 19490148

Greeff, M.. (2017). Human-robot collaboration in disaster response: Experiences from large-scale field experiments. *Journal of Field Robotics*, 34(3), 439–460.

Green, T. R.. (2015). The future of human-computer interaction. *Interaction*, 22(1), 16–21.

Grewal, D., Hulland, J., Kopalle, P. K., & Karahanna, E. (2020). The Future of Technology and Marketing: A Multidisciplinary Perspective. *Journal of Marketing*, 84(1), 1–6. DOI: 10.1177/0022242919879877

Grunow, M.. (2011). Logistics: A key enabler for sustainable supply chains. *Procedia Engineering*, 9, 124–131.

Gubbi, J., Buyya, R., Marusic, S., & Pavithran, P. (2023). Internet of Things (IoT) in Smart Cities: Challenges and Opportunities. *Journal of Urban Technology*, 30(1), 1–18. DOI: 10.1080/10630732.2023.2157513

Gu, D., Andreev, K., & Dupre, M. E. (2021). Major Trends in Population Growth Around the World. *China CDC Weekly*, 3(28), 604–613. DOI: 10.46234/ccdcw2021.160 PMID: 34594946

Guest, S. D.. (2013). Real-time spatial analytics for building evacuation: A decision support system for first responders. *Computers, Environment and Urban Systems*, 39, 1–13.

Guha, S., Harrigan, P., & Soutar, G. (2018). Linking social media to customer relationship management (CRM): A qualitative study on SMEs. *Journal of Small Business and Enterprise Development*, 25(6), 844–868.

Gupta, A., & Jha, R. K. (2015). A survey of 5G network: Architecture and emerging technologies. *IEEE Access : Practical Innovations, Open Solutions*, 3, 1206–1232. DOI: 10.1109/ACCESS.2015.2461602

Habib, M., Habib, A., Albzaie, M., & Farghal, A. (2024). Sustainability benefits of AI-based engineering solutions for infrastructure resilience in arid regions against extreme rainfall events. *Discover Sustainability*, 5(1), 278. DOI: 10.1007/s43621-024-00500-2

Haidar, N., Tamani, N., Nienaber, F., Wesseling, M. T., Bouju, A., & Ghamri-Doudane, Y. (2019). Data collection period and sensor selection method for smart building occupancy prediction. *2019 IEEE 89th Vehicular Technology Conference (VTC2019-Spring)*, 1-6. https://doi.org/DOI: 10.1109/VTCSpring.2019.8746447

Hammons, T. M., & Myers, S. L. (2019). Smart cities: A review of enabling technologies, challenges, and opportunities. *International Journal of Information Management*, 49, 101819.

Han, H., Wu, X., & Qiao, J. (2019). A Self-Organizing Sliding-Mode Controller for Wastewater Treatment Processes. IEEE Transactions on Control Systems Technology, 27(4), 1480–1491. *IEEE Transactions on Control Systems Technology*. Advance online publication. DOI: 10.1109/TCST.2018.2836358

Han, S., Park, Y., Noh, N., Kim, J.-H., Kim, J.-J., Kim, B.-M., & Choi, W. (2023). Spatiotemporal variability of the PM2.5 distribution and weather anomalies during severe pollution events: Observations from 462 air quality monitoring stations across South Korea. *Atmospheric Pollution Research*, 14(3), 101676. DOI: 10.1016/j.apr.2023.101676

Haque, E. U., Shah, A., Iqbal, J., Ullah, S. S., Alroobaea, R., & Hussain, S. (2024). A scalable blockchain based framework for efficient IoT data management using lightweight consensus. *Scientific Reports*, 14(1), 7841. DOI: 10.1038/s41598-024-58578-7 PMID: 38570648

Harbud, N. (2022). Process automation systems. In Perlmutter, B. A. (Ed.), *Integration and Optimization of Unit Operations* (pp. 191–218). Elsevier., DOI: 10.1016/B978-0-12-823502-7.00021-9

Harith, M. Z. M. Z., Hossain, M. A., Ahmedy, I., Idris, M. Y. I., Soon, T. K., & Noor, R. M. (2020). "Prototype development of IoT based smart waste management system for smart city". IOPConf. *IOP Conference Series. Materials Science and Engineering*, 884(1), 012051. DOI: 10.1088/1757-899X/884/1/012051

Hassan, A. (2024). Cloud Security with AI: A Comprehensive Review. *International Journal of Advanced Research in Engineering and Technology*, 15(4), 1–11.

Haver, S. M., Gedamke, J., Hatch, L. T., Dziak, R. P., Van Parijs, S., Mckenna, M. F., Barlow, J., Berchok, C., DiDonato, E., Hanson, B., Haxel, J., Holt, M., Lipski, D., Matsumoto, H., Meinig, C., Mellinger, D. K., Moore, S. E., Oleson, E. M., Soldevilla, M. S., & Klinck, H. (2018). Monitoring long-term soundscape trends in U.S. Waters: The NOAA/NPS ocean noise reference station network. *Marine Policy*, 90, 6–13. DOI: 10.1016/j.marpol.2018.01.023

Hawkins, T. R., Singh, B., Majeau-Bettez, G., & Strømman, A. H. (2012). Comparative Environmental Life cycle assessment of conventional and electric vehicles. *Journal of Industrial Ecology*, 17(1), 53–64. DOI: 10.1111/j.1530-9290.2012.00532.x

Heitmeyer, C.. (2013). High-assurance systems engineering: A perspective. *IEEE Software*, 30(4), 66–73.

Hellwig, D., Wendt, K., Babich, V., & Huchzermeier, A. Playing with DISASTER: A Blockchain-enabled Supply Chain Simulation Platform for Studying Shortages and the Competition for Scarce Resources (February 26, 2022). In: Lee, H., Ernst, R., Huchzermeier, A., & Cui, S. (Eds.): Creating Values with Operations & Analytics, Springer Series in Supply Chain Management, Forthcoming, Georgetown McDonough School of Business Research Paper No. 4044385, Available at SSRN: https://ssrn.com/abstract=4044385 DOI: 10.1007/978-3-031-08871-1_9

Henaien, A., Ben Elhadj, H., & Fourati, L. C. (2024). A sustainable smart IoT-basedsolid waste management system. *Future Generation Computer Systems*, 157, 587–602. DOI: 10.1016/j.future.2024.03.056

Hendawy, M., & Ghoz, L. (2024). A starting framework for urban AI applications. *Ain Shams Engineering Journal*, 102987(11), 102987. Advance online publication. DOI: 10.1016/j.asej.2024.102987

Henkey, G. (2017). *Urban emergency management: A handbook for the 21st century*. Routledge.

Henry, J., & Maxwell, T. (2024). Harnessing AI for personalized learning: Innovations in EdTech and pedagogy.

Herbon, A., & Hadas, Y. (2014). Determining optimal frequency and vehicle capacity for public transit routes: A generalized newsvendor model. *Transportation Research Part B: Methodological*, 71, 85–99. DOI: 10.1016/j.trb.2014.10.007

Hidalgo, D., & Huizenga, C. (2012). Implementation of sustainable urban transport in Latin America. *Research in Transportation Economics*, 40(1), 66–77. DOI: 10.1016/j.retrec.2012.06.034

Ho, C. Q., & Tirachini, A. (2023). Mobility-as-a-Service and the role of multimodality in the sustainability of urban mobility in developing and developed countries. *Transport Policy*, 145, 161–176. DOI: 10.1016/j.tranpol.2023.10.013

Hofmann, S. Z. (2021). 100 Resilient Cities program and the role of the Sendai framework and disaster risk reduction for resilient cities. *Progress in Disaster Science*, 11, 100189. DOI: 10.1016/j.pdisas.2021.100189

Home. (2024, October 29). https://www.who.int/

Hoshiba, M., Aoki, S., & Horiuchi, S. (2016). Earthquake Early Warning in Japan: Warning the general public and its evaluation. *Annual Review of Earth and Planetary Sciences*, 44(1), 435–468. DOI: 10.1146/annurev-earth-063016-015546

Hou, M.. (2024). IMPACTS: A Homeostasis-Based Trust Management System for Human-AI Collaboration. *IEEE Transactions on Human-Machine Systems*, 54(1), 17–28.

How will autonomous vehicles shape future urban mobility? (2024, October 16). World Economic Forum. https://www.weforum.org/agenda/2024/10/how-will-autonomous-vehicles-shape-urban-mobility/

Hsu, Y.-C., Chen, C.-C., Lee, C.-H., Huang, Y.-M., Chen, Y.-L., & Lai, Y.-H. (2021). The feasibility of using smartphones with embedded micro-electro-mechanical systems accelerometers for structural health monitoring. *Sensors (Basel)*, 21(8), 2896. DOI: 10.3390/s21082896

Huang, M. H., & Rust, R. T. (2018). Artificial intelligence in service. *Journal of Service Research*, 21(2), 155–172. DOI: 10.1177/1094670517752459

Huang, M. H., & Rust, R. T. (2021). A Strategic Framework for Artificial Intelligence in Marketing. *Journal of the Academy of Marketing Science*, 49(1), 30–50. DOI: 10.1007/s11747-020-00749-9

Huang, Q., Huang, S., Qin, H., Wang, J., & Chen, J. (2019). Earthquake risk assessment based on GIS and crowdsourcing data: A case study of the 2017 Jiuzhaigou earthquake, China. *International Journal of Disaster Risk Reduction*, 39, 101225.

Huang, Y., & Wang, X. (2023). Quantum computing applications in hydrology: A review. *Journal of Hydrology (Amsterdam)*, 615, 128–145.

Humayun, M.. (2020). Smart Governance for Sustainable Development: A Conceptual Framework. *Sustainability*, 12(11), 4371.

Hunt, K., & Zhuang, J. (2022). Blockchain for Disaster Management. In Emrouznejad, A., & Charles, V. (Eds.), *Big Data and Blockchain for Service Operations Management. Studies in Big Data* (Vol. 98). Springer., DOI: 10.1007/978-3-030-87304-2_10

Hussain, F. (2023). Cybersecurity in Smart Cities: Challenges and Solutions. *Cybersecurity Journal*, 6(1), 45–60. DOI: 10.1016/j.cyber.2023.100082

IBM. (2023). *AI for Smart Cities: Bogotá's Transportation Optimization.*

IDB. (2022). *Digital Cities: Transforming Urban Services in Latin America.*

IEA – International Energy Agency. (n.d.). IEA. https://www.iea.org/

IEEE Communications Magazine | IEEE Communications Society. (2024, October 1). https://www.comsoc.org/publications/magazines/ieee-communications-magazine

IEEE. (2021). *IEEE.* Transparency in Autonomous Systems.

IEEE. (2023). *Global Initiative on Ethics of Autonomous and Intelligent Systems.*

Ifraheem, S., Rasheed, M., & Siddiqui, A. (2024). Transforming education through artificial intelligence: Personalization, engagement and predictive analytics. *Journal of Asian Development Studies*, 13(2), 250–266. DOI: 10.62345/jads.2024.13.2.22

Illahi, U., & Mir, M. S. (2021). Impact of COVID-19 on Supply Chain Management: A Review. *International Journal of Supply Chain Management*, 10(4), 213–221.

IMD. (2024). *Smart City Index 2024.* IMD World Competitiveness Center.

Inam, M. (2024, July 10). *Example of a Smart City: A Case Study into Barcelona.* Minnovation Technologies. https://minnovation.com.au/smart-cities-2/example-of-a-smart-city-a-case-study-into-barcelona/

Intergovernmental Panel on Climate Change. (2021). [*The physical science basis.*]. *Climatic Change*, •••, 2021.

International Banker. (2024). Why Singapore is Consistently Regarded as Asia's Smartest City. Kohli, S., & Vala, P. (2023). *Aligning AI standards: A cross-border approach to data privacy and ethics.AI and Society Journal*, 17(4), 202–225.

Iqbal, M. (2023). AI in education: Personalized learning and adaptive assessment. *Cosmic Bulletin of Business Management*, 2(1), 280–297.

Irandoust, M., & Benaskeur, A. (2020). Human-autonomy teaming in command and control: A review. *IEEE Transactions on Systems, Man, and Cybernetics. Systems*, 50(11), 4007–4021.

Islam, M. R., Shahin, M. I., & Ahmed, S. (2023). Flood prediction using machine learning techniques: A case study of Bangladesh. *Applied Water Science*, 13, 8. DOI: 10.1007/s13201-022-01638-x

Issac, R., & Akshai, M. (2013, August). SVASTHA: An effective solid waste management system for Thiruvalla Municipality in Android OS. In *2013 IEEE Global Humanitarian Technology Conference: South Asia Satellite (GHTC-SAS)* (pp. 254-259). IEEE.

ITU. (2022). *Global Connectivity Report 2022*. International Telecommunication Union.

Iyer, L. S. (2021). AI enabled applications towards intelligent transportation. *Transportation Engineering*, 5, 100083. DOI: 10.1016/j.treng.2021.100083

Jadhav, S. P., Suryavanshi, J. R., Samant, R.s C., & Archana, , S. (2024). Design and Development of a Smart IoT Enabled Garbage Bin Mechanism for Optimized Volume Utilization. *African Journal of Biological Sciences*, 6(14), 10806–10818.

Jain, P., Chaudhary, T., & Gajjar, S. (2023) "Design and Development ofSmartWasteManagement System," *International Conference on Communication System, Computing and IT Applications (CSCITA)*, Mumbai, India, 2023, pp. 199-203, DOI: 10.1109/CSCITA55725.2023.10104960

Jameel, S. M., Kumar, S., Mohsin, M., & Islam, M. S. (2022). Assessment of advanced machine learning algorithms for the prediction of global solar radiation. *Journal of Cleaner Production*, 341, 130922. DOI: 10.1016/j.jclepro.2021.130922

Jannach, D., & Adomavicius, G. (2016). Recommendation systems: Challenges, insights, and research opportunities. *AI Magazine*, 37(3), 73–85.

Jatobá, A., de Castro Nunes, P., & de Carvalho, P. V. (2023). A framework to assess potential health system resilience using fuzzy logic. *Revista Panamericana de Salud Pública = Pan American Journal of Public Health*, 47, e73. DOI: 10.26633/RPSP.2023.73 PMID: 37123641

Javvaji, V. (2023). Surveillance Technology and Its Impact on Privacy: A Review. *International Journal of Advanced Research in Engineering and Technology*, 15(3), 1–10.

Jeon, H., Lee, J., & Sohn, K. (2017). Artificial intelligence for traffic signal control based solely on video images. *Journal of Intelligent Transport Systems*, 22(5), 433–445. DOI: 10.1080/15472450.2017.1394192

Jiang, Y., & Liu, H. (2021). A comprehensive review on big data analytics and artificial intelligence for disaster risk reduction: Trends, challenges, and future directions. *Future Generation Computer Systems*, 120, 445–467. DOI: 10.1016/j.future.2020.10.034

Johnson, L., & Smith, R. (2023). Quantum-enhanced flood forecasting: Techniques and applications. *Environmental Modelling & Software*, 150, 105–120.

Jolly, P., & Raj, A. (2020). Ethical and Legal Considerations for Artificial Intelligence in Emergency Response. *International Journal of Advanced Research in Engineering and Technology*, 13(2), 111–125.

Joseph, S. B., Dada, E. G., Abidemi, A., Oyewola, D. O., & Khammas, B. M. (2022). Metaheuristic algorithms for PID controller parameters tuning: Review, approaches and open problems. *Heliyon*, 8(5), e09399. DOI: 10.1016/j.heliyon.2022.e09399 PMID: 35600459

Joshi, A., Kanungo, D. P., & Panigrahi, R. K. (2023, January). Development of landslide forecasting system using deep learning. In *2023 IEEE Applied Sensing Conference (APSCON)* (pp. 1-3). IEEE. DOI: 10.1109/APSCON56343.2023.10101223

Jung, D., Tuan, V. T., Tran, D. Q., Park, M., & Park, S. (2020). Conceptual Framework of an Intelligent Decision Support System for Smart City Disaster Management. *Applied Sciences (Basel, Switzerland)*, 10(2), 666. DOI: 10.3390/app10020666

Kaftan, V., Kandalov, W., Molodtsov, I., Sherstobitova, A., & Strielkowski, W. (2023). Socio-economic stability and sustainable development in the Post-COVID era: Lessons for the business and economic leaders. *Sustainability (Basel)*, 15(4), 2876. DOI: 10.3390/su15042876

Kalinsky, M., Sykora, M., Markova, J., & Maňas, P. (2024). Nonlinear dynamic finite element analysis of vehicle impacts into road restraint systems. *Results in Engineering*, 23, 102726. DOI: 10.1016/j.rineng.2024.102726

Kalkman, L.. (2019). Ethical and legal challenges of secondary use of health data for research. *The Lancet. Digital Health*, 1(1), e12–e19.

Kamat, A., Kulkarni, A., Akshay, K., Kasturi, R., & Balur, N. (2017). Coin Insert Mobile Charger. Communication on Applied Electronics, Volume 6 No. 7. https://www.caeaccess.org/archives/volume6/number7/708-2017652515

Kamble, P., Khot, S., Solkar, S., & Patil, S. (2022). Mobile Charging Vending Machine. International Research Journal of Modernization in Engineering Technology and Science, Volume: 04/Issue: 07/July-2022. https://www.irjmets.com/uploadedfiles/paper//issue_7_july_2022/27570/final/fin_irjmets1657227298.pdf

Kanipriya, M., Sugumaran, D., Pandiyarajan, V., Yogananda, B. S., Selvakumar, P., & Mohan, S. (2024, January). Analysis of Electromyography Signals for Control of Mechanical Prosthesis using Machine Learning Techniques. In 2024 2nd International Conference on Intelligent Data Communication Technologies and Internet of Things (IDCIoT) (pp. 1059-1067). IEEE.

Kantarcioglu, M., & Shaon, S. (2019). Privacy-preserving data mining: Challenges and opportunities. *Wiley Interdisciplinary Reviews. Data Mining and Knowledge Discovery*, 9(5), e1329.

Kapse, A., Kamble, T., Lohar, A., Chaudhari, S., & Puri, D. (2022). "Confront acknowledgment Participation framework utilizing Hoard and CNN calculation," ITM Web of Conferences, vol. 44, pp. 03028.

Kar, A.. (2017). A Review of Smart City Initiatives: A Global Perspective. *Renewable & Sustainable Energy Reviews*, 70, 1181–1199.

Karballaeezadeh, N., Zaremotekhases, F., Shamshirband, S., Mosavi, A., Nabipour, N., Csiba, P., & Várkonyi-Kóczy, A. R. (2020). Intelligent road inspection with advanced machine learning; hybrid prediction models for smart mobility and transportation maintenance systems. *Energies*, 13(7), 1718. DOI: 10.3390/en13071718

Karimzadeh, S.. (2019). Real-time communication networks for disaster management: A review. *Sensors (Basel)*, 19(10), 2297. PMID: 31109028

Karrar, A. E. (2022). The Effect of Using Data Pre-Processing by Imputations in Handling Missing Values. [IJEEI]. *Indonesian Journal of Electrical Engineering and Informatics*, 10(2), 2. Advance online publication. DOI: 10.52549/ijeei.v10i2.3730

Karunakar, M., Sai, C. A., Chandra, K., & Kumar, K. A. (2020). Shrewd Participation Checking Framework (SAMS): A Confront Acknowledgment Based Participation Framework for Classroom Environment. *Worldwide Diary for Later Improvements in Science and Innovation*, 4(5), 194–201.

Kaswan, K. S., Dhatterwal, J. S., & Ojha, R. P. (2024). AI in personalized learning. In *Advances in Technological Innovations in Higher Education* (pp. 103–117). CRC Press. DOI: 10.1201/9781003376699-9

Kathiresan, R., Ashwin, V., & Raj, M. (2016). Coin Based Mobile Charger. International Journal of Advanced Research in Electrical, Electronics and Instrumentation Engineering, Volume 5, Special Issue 1, March 2016. https://www.ijareeie.com/upload/2016/ncrte/29_EIE29.pdf

Katiyar, N., Awasthi, M. V. K., Pratap, R., Mishra, M. K., Shukla, M. N., & Tiwari, M. (2024). AI-driven personalized learning systems: Enhancing educational effectiveness. *Educational Administration: Theory and Practice*, 30(5), 11514–11524.

Kaur, P., Namboori, M., Pandey, A., & Tyagi, K. (2023). "Profound Learning Based Shrewd Participation Framework," Procedures of World Conference on Manufactured Insights: Progresses and Applications, pp. 485,.

Kaur, D., & Singh, A. (2023). *Addressing bias in AI disaster prediction models: Toward inclusive risk management*. AI Ethics Journal.

Kaur, S.. (2024). Smart Governance: A Framework for Sustainable Development in Smart Cities. *International Journal of Advanced Research in Engineering and Technology*, 15(3), 1–10.

Keegan, P. (2019). The Netflix effect: AI in entertainment. *Harvard Business Review*.

Kelly, S. D., Suryadevara, N. K., & Mukhopadhyay, S. C. (2013). Towards the implementation of IoT for environmental condition monitoring in homes. *IEEE Sensors Journal*, 13(10), 3846–3853. DOI: 10.1109/JSEN.2013.2263379

Khan, M. A., Sajjad, I. A., Tahir, M., & Haseeb, A. (2022). IOT application for energy management in smart homes. *The 7th International Electrical Engineering Conference*, 43. https://doi.org/DOI: 10.3390/engproc2022020043

Khare, S., Prakash, V., Talwandi, N. S., Soni, P., & Kumar, S. "Revolutionizing Cyber Security Incident Response with Smart Contracts," *2024 International Conference on Computational Intelligence and Computing Applications (ICCICA)*, Samalkha, India, 2024, pp. 86-90, DOI: 10.1109/ICCICA60014.2024.10584955

Khosravi, A., & Zare, A. (2023). Quantum-enhanced decision support systems for flood management: A review of methodologies and applications. *Environmental Modelling & Software*, •••, 156.

Khutade, N.. (2024). Narrative Review: A Comprehensive Guide. *International Journal of Advanced Research in Engineering and Technology, 15*(4), 1-11.Luckett, D. J. (2023). Artificial Intelligence and the Law: Challenges and Opportunities. *Annual Review of Law and Social Science*, 19, 1–24.

Kikuchi, S. (2009). Artificial intelligence in transportation analysis: Approaches, methods, and applications. *Transportation Research Part C, Emerging Technologies*, 17(5), 455. DOI: 10.1016/j.trc.2009.04.002

Kim, S., & Choi, Y. (2021). Blockchain technology in disaster risk management: Transparency and accountability. Journal of Emergency Management.

Kim, J., Lee, S., & Park, M. (2023). *5G-enabled emergency response systems in South Korea*. Telecommunications Journal.

Kim, S., & Lee, J. (2023). Quantum machine learning for hydrological predictions. *Water Resources Research*, 59(4), 1–15.

Kim, Y., & Park, J. (2024). *Future directions in 5G and 6G for DRM applications*. Telecommunications Journal.

Kingsnorth, S. (2019). *Digital marketing strategy: An integrated approach to online marketing*. Kogan Page Publishers.

Kirci, P., Arslan, D., & Dincer, S. F. (2023). A communication, management, and tracking mobile application for enhancing earthquake preparedness and situational awareness in the event of an earthquake. *Sustainability (Basel)*, 15(970), 970. Advance online publication. DOI: 10.3390/su15020970

Knoppers, B. M. (2014). The Global Alliance for Genomics and Health: A framework for responsible data sharing. *Genome Medicine*, 6(1), 1–2. PMID: 24433494

Kong, Q., Zhou, C., Yan, J., & Chen, X. (2020). Earthquake early-warning with deep learning based on multiple seismic observations. *Geophysical Research Letters, 47*(14), e2020GL088148.

Koulu, H. (2020). Human agency and algorithmic decision-making: Towards a framework for understanding human discretion in automated systems. *Artificial Intelligence and Law*, 28(1), 1–26.

Kourtit, K., & Nijkamp, P. (2012). Smart cities in the innovation age. *Innovation (Abingdon)*, 25(2), 93–95. DOI: 10.1080/13511610.2012.660331

Kowshika, S.. (2024). A Review on Drone Technology for Disaster Management. *Journal of Physics: Conference Series*, 2289(1), 012024.

Kruijff-Korbayová, I.. (2015). TRADR: Towards Robust and Dependable Human-Robot Collaboration in Disaster Response. *IEEE Robotics & Automation Magazine*, 22(3), 72–84.

Kruijff, W. J. M.. (2014). Human-robot collaboration in disaster response. *Robotics and Autonomous Systems*, 62(12), 1729–1744.

Kumar, A., Kumar, A., & Singh, A. (2017). Energy efficient and low cost air quality sensor for smart buildings. *2017 3rd International Conference on Computational Intelligence & Communication Technology (CICT)*, 1-4. https://doi.org/DOI: 10.1109/CIACT.2017.7977310

Kumar, R., & Chen, H. (2023). Smart city infrastructure and disaster resilience: A case for IoT-enabled urban systems. Journal of Urban Resilience.

Kumar, A., Bag, A., Anand, A., Saha, S., Mohapatra, H., & Kolhar, M. (2024). Examining healthcare services utilizing cloud technology in intelligent urban environments. *Advances in Healthcare Information Systems and Administration*, •••, 77–98. DOI: 10.4018/979-8-3693-7225-8.ch004

Kumar, A., Gupta, R., & Singh, P. (2023). Cultural Sensitivity in AI Deployment: Designing for Diverse Societies. *AI & Society*, 38(2), 345–359. DOI: 10.1007/s00146-023-00388-9

Kumar, A., & Kumari, S. (2022). *Internet of Things in disaster management: Enhancing resilience.* Springer.

Kumara, V., & Ganesan, E. (2024). A Novel Approach to Wastewater Treatment Control: A Self-Organizing Fuzzy Sliding Mode Controller. *International Journal of Fuzzy Logic and Intelligent Systems*, 24(4), 440–450.

Kumar, V., Rajan, B., Venkatesan, R., & Lecinski, J. (2020). Understanding the Role of Artificial Intelligence in Personalized Marketing. *Journal of Interactive Marketing*, 51, 26–42. DOI: 10.1016/j.intmar.2020.04.005

Kung, P.. (2015). Privacy in context: Technology, policy, and the integrity of social life. *Science*, 347(6221), 510–513.

Kurama, V. (2020)"A audit of well known profound learning models: AlexNet, VGG16, and GoogleNet,".

Kutty, A. A., Kucukvar, M., Onat, N. C., Ayvaz, B., & Abdella, G. M. (2023). Measuring sustainability, resilience and livability performance of European smart cities: A novel fuzzy expert-based multi-criteria decision support model. *Cities (London, England)*, 137, 104293. DOI: 10.1016/j.cities.2023.104293

Kyriakopoulos, G. L., Kalavrouziotis, I. K., & Zamparas, M. G. (2023). Public health risks associated with the reuse of wastewater for irrigation. In Zamparas, M. G., & Kyriakopoulos, G. L. (Eds.), *Water Management and Circular Economy* (pp. 409–424). Elsevier., DOI: 10.1016/B978-0-323-95280-4.00009-6

Lakshmi, K., Sadik Khan, D. P., Wagh, V., & Vasanti, G. (2024). AI-powered learning analytics: Transforming educational outcomes through ICT integration. *Library Progress International*, 44(3), 17910–17918.

Lambrecht, A., & Tucker, C. (2019). Algorithmic bias? An empirical study of apparent gender-based discrimination in the display of STEM career ads. *Management Science*, 65(7), 2966–2981. DOI: 10.1287/mnsc.2018.3093

Lazaroiu, G. C., & Roscia, M. (2012). Definition methodology for the smart cities model. *Energy*, 47(1), 326–332. DOI: 10.1016/j.energy.2012.09.028

Lee, D., Yang, R., & Kim, S. (2022). Artificial intelligence and drone technology in wildfire management. *Journal of Environmental Protection.*

Lee, I., & Lee, K. (2015). The Internet of Things (IoT): Applications, investments, and challenges for enterprises. *Business Horizons*, 58(4), 431–440. DOI: 10.1016/j.bushor.2015.03.008

Lee, S., Tan, H., & Wen, Q. (2023). *Blockchain for cross-border data sharing in disaster management.* Blockchain Review.

Lee, T., & Zhang, H. (2022). Quantum variational circuits for capturing environmental dynamics in flood prediction models. *Computers & Geosciences*, 164, 104–115.

Lemon, K. N., & Verhoef, P. C. (2016). Understanding customer experience throughout the customer journey. *Journal of Marketing*, 80(6), 69–96. DOI: 10.1509/jm.15.0420

Li, J., Liu, X., Liang, W., & Zhu, C. "Research on network resource sharing based on distributed ledger technology," 2023 IEEE 3rd International Conference on Information Technology, Big Data and Artificial Intelligence (ICIBA), Chongqing, China, 2023, pp. 1509-1513, DOI: 10.1109/ICIBA56860.2023.10165046

Liakot Ali, Md., Mahbubul Alam and Md. Abu Innovative Research in Computer andNayeem Redwanur Rahaman. (2012). RFID based [-Dr. Santosh Pandurang Jadhav/Afr.J.Bio.Sc.]. *Communication Engineering*, 4, 1267.

Li, C., Jia, W., Tao, Q., & Sun, M. (2011, April).. . *Solar Cellphone Charger Performance in Indoor Environment*, 01-03. Advance online publication. DOI: 10.1109/NEBC.2011.5778623

Lichocik, G., & Sadowski, A. (2013). The impact of information technology on supply chain management. *Procedia Engineering*, 57, 131–138.

Li, F. (2021). Artificial Intelligence, Platform Economy, and the Future of Marketing. *Journal of Business Research*, 125, 410–420. DOI: 10.1016/j.jbusres.2020.12.035

Li, F., & Zhang, H. (2023). *5G networks and disaster management*. Telecommunications Journal.

Li, H., & Kannan, P. K. (2014). Attributing conversions in a multichannel online marketing environment: An empirical model and a field experiment. *JMR, Journal of Marketing Research*, 51(1), 40–56. DOI: 10.1509/jmr.13.0050

Li, H., & Zhao, Q. (2023). Integrating quantum computing with hydrological models for improved flood prediction. *Hydrological Processes*, 37(2), 345–360.

Lijesh, L., & Saroja, G. A. S. (2023, January). Analysis based on Different Optimization Algorithms for Landslide Detection. In 2023 3rd International Conference on Intelligent Communication and Computational Techniques (ICCT) (pp. 1-6). IEEE.

Li, M., Gu, W., Chen, W., He, Y., Wu, Y., & Zhang, Y. (2018). Smart home: Architecture, technologies and systems. *Procedia Computer Science*, 131, 393–400. DOI: 10.1016/j.procs.2018.04.219

Lim, Y.-C., Chen, C.-F., Tsai, M.-L., Wu, C.-H., Lin, Y.-L., Wang, M.-H., Albarico, F. P. J. B., Chen, C.-W., & Dong, C.-D. (2022). Impacts of Fishing Vessels on the Heavy Metal Contamination in Sediments: A Case Study of Qianzhen Fishing Port in Southern Taiwan. *Water (Basel)*, 14(7), 7. Advance online publication. DOI: 10.3390/w14071174

Linardos, V., Drakaki, M., Tzionas, P., & Karnavas, Y. L. (2022). Machine learning in disaster management: Recent developments in methods and applications. *Machine Learning and Knowledge Extraction*, 4(2), 446–473. Advance online publication. DOI: 10.3390/make4020020

Lin, C.-W., Huang, Y.-M., Chen, C.-C., Hsu, Y.-C., Tu, Y.-T., & Lee, C.-H. (2020). Development of a smartphone-based earthquake early warning system with an onsite warning determination algorithm. *Sensors (Basel)*, 20(23), 6512. DOI: 10.3390/s20236512

Li, S., Da Xu, L., & Zhao, S. (2018). 5G Internet of Things: A survey. *Journal of Industrial Information Integration*, 10, 1–9. DOI: 10.1016/j.jii.2018.01.005

Liu, X., & Di, X. (2020). TanhExp: A Smooth Activation Function with High Convergence Speed for Lightweight Neural Networks (arXiv:2003.09855). arXiv. https://doi.org//arXiv.2003.09855DOI: 10.48550

Liu, H., & Pei, L. (2022). Bias in AI algorithms for disaster prediction: A case study on demographic disparities. *AI & Society*.

Liu, S., Zhao, J., & Wang, G. (2022). Earthquake early warning system based on the Internet of Things and machine learning. *IEEE Internet of Things Journal*, 9(10), 7742–7752.

Liu, T., Chen, T., Niu, R., & Plaza, A. (2021). Landslide detection mapping employing CNN, ResNet, and DenseNet in the three gorges reservoir, China. *IEEE Journal of Selected Topics in Applied Earth Observations and Remote Sensing*, 14, 11417–11428. DOI: 10.1109/JSTARS.2021.3117975

Liu, Y., Feng, Y., Wang, J., & Gao, L. (2021). Statistical learning-based drought forecasting: A review of machine learning techniques and their applications. *Water (Basel)*, 13, 21. DOI: 10.3390/w13010021

Li, W., Chai, Y., Khan, F., Jan, S. R. U., Verma, S., Menon, V. G., Kavita, , & Li, X. (2021). A comprehensive survey on machine learning-based big data analytics for IoT-enabled smart healthcare system. *Mobile Networks and Applications*, 26(1), 234–252. DOI: 10.1007/s11036-020-01700-6

Li, W., Fu, Y., Fan, S., Xin, M., & Bai, H. (2023). DCI-PGCN: Dual-channel interaction portable graph convolutional network for landslide detection. *IEEE Transactions on Geoscience and Remote Sensing*, 61, 1–16. DOI: 10.1109/TGRS.2023.3273623

Li, X., & Chen, Z. (2022). Pooling resources for AI-enabled cities: The economic value of collaboration. *Global Innovation Journal*, 6(3), 150–172.

Li, Z., Hu, X., Xu, Q., Chen, W., & Liu, L. (2021). Integrating remote sensing and machine learning for urban land cover classification: A case study of Beijing. *ISPRS Journal of Photogrammetry and Remote Sensing*, 171, 113–128. DOI: 10.1016/j.isprsjprs.2020.11.005

Li, Z., Meier, P., Hauksson, E., Zhan, Z., & Andrews, J. (2018). Machine learning predicts earthquake stress drop. *Geophysical Research Letters*, 45(10), 4773–4779. DOI: 10.1029/2018GL077870

Li, Z., Yu, X., & Chen, T. (2023). *Blockchain and 5G: Transformative technologies in crisis management*. Elsevier.

Lombardi, P., Giordano, S., Farouh, H., & Yousef, W. (2012). Modelling the smart city performance. *Innovation (Abingdon)*, 25(2), 137–149. DOI: 10.1080/13511610.2012.660325

Lopes, V., & Alexandre, L. A. (2018). Blockchain for Robotics: A Survey. *IEEE Access : Practical Innovations, Open Solutions*, 6, 15505–15520.

Lozano, J. M., & Tien, I. (2023). Data collection tools for post-disaster damage assessment of building and lifeline infrastructure systems. *International Journal of Disaster Risk Reduction*, 94, 103819. DOI: 10.1016/j.ijdrr.2023.103819

Lundin, A. C., & Özkil, A. G., & Schuldt-Jensen, J. (2017). Smart cities: casestudywastemonitoring and management. In Proceedings of the 50th Hawaii International Conference onSystem Sciences (HICSS 2017) (pp. 1392-1401) https://doi.org/ Nations Human Settlement Program (UN-HABITAT). Solid Waste Managementin the World, Cities. London, UK: Gutenberg Press; 2010DOI: 10.24251/HICSS.2017.167United

Luque-Ayala, A., & Marvin, S. (2023). The Future of Smart Cities: Co-Creation, Sustainability, and Citizen Empowerment. *Journal of Urban Planning and Development.*

Lv, L., Chen, T., & Li, J. (2023, July). An Interpretation Study on the ML Models for Landslide Susceptibility Mapping. In IGARSS 2023-2023 IEEE International Geoscience and Remote Sensing Symposium (pp. 7241-7244). IEEE. DOI: 10.1109/IGARSS52108.2023.10282171

Mahadev Madgule, N. (2023). Vinayaka, Yeshwant M. Sonkhaskar, Dhiren Ramanbhai Patel, R.Karthikeyan, P. Selvakumar,(2023). Mechanical properties and microstructure of activated TIG welded similar joints of Inconel alloys by desirability approaches. *Materials Today: Proceedings*, 77, 528–533. DOI: 10.1016/j.matpr.2022.12.250

Mahmood, M. (2016). Enhancing Citizens' Trust and Confidence in Government through Digital Transformation. *International Journal of Electronic Government Research*, 12(1), 99–110. DOI: 10.4018/IJEGR.2016010105

Mahmood, O. T., Wan Hasan, W. Z., Ismail, L. I., Ramli, H. R. H., Azis, N., & Norsahperi, N. M. H. (2024). Sliding Mode Controller Optimization-Based Three-Phase Rectifier: Review Study. *IEEE Access : Practical Innovations, Open Solutions*, 12, 101457–101483. DOI: 10.1109/ACCESS.2024.3423776

Mahmoud, C. F., & Sørensen, J. T. (2024). Artificial intelligence in personalized learning with a focus on current developments and future prospects. *Research and Advances in Education*, 3(8), 25–31. DOI: 10.56397/RAE.2024.08.04

Mailley, J., Whitehead, S., & Farrell, G. (2006). Progress and prospects in the prevention of mobile phone theft. Loughborough University, Volume 170, Issue 3, June 2006. https://www.academia.edu/29412701/Progress_and_prospects_in_the_prevention_of_mobile_phone_theft

Makinde, A. I., Adeleye, S. A., Oronti, A. O., & Jimoh, I. T. (2024). Revolutionizing education. *Artificial Intelligence for Wireless Communication Systems: Technology and Applications*, 103.

Manoharan, G., Razak, A., Rao, B. S., Singh, R., Ashtikar, S. P., & Nivedha, M. (2024). Navigating the Crescendo of Challenges in Harnessing Artificial Intelligence for Disaster Management. In *Predicting Natural Disasters With AI and Machine Learning* (pp. 64–94). IGI Global., DOI: 10.4018/979-8-3693-2280-2.ch003

Manzolli, J. A., Trovão, J. P., & Antunes, C. H. (2022). A review of electric bus vehicles research topics – Methods and trends. *Renewable & Sustainable Energy Reviews*, 159, 112211. DOI: 10.1016/j.rser.2022.112211

Mapari, N., Khadija, A., Taiba, A., & Azima, S. (2019, August). Smart Charging Vending Machine. *IIEE Xplore*, 29(d), 414–417. Advance online publication. DOI: 10.1109/ICCMC.2019.8819821

Markets, R. A. (2020, October 5). Smart Cities Market Report 2020 - Global Forecast to 2025: Market Size is Expected to Grow from $410.8 Billion in 2020 to $820.7 Billion. *GlobeNewswire News Room*. https://www.globenewswire.com/news-release/2020/10/05/2103315/0/en/Smart-Cities-Market-Report-2020-Global-Forecast-to-2025-Market-Size-is-Expected-to-Grow-from-410-8-Billion-in-2020-to-820-7-Billion.html

Maroma, N. (2014, December). Solar Powered Cellphone Charging Station. *OAlib*, 1(9), 1–7. Advance online publication. DOI: 10.4236/oalib.1101156

Marsal-Llacuna, M. L., Colomer-Llinàs, J., & Meléndez-Frigola, J. (2015). Lessons in urban monitoring taken from sustainable and livable cities to better address the Smart Cities initiative. *Technological Forecasting and Social Change*, 90, 611–622. DOI: 10.1016/j.techfore.2014.01.012

Martinez, A., & Garcia, M. (2023). Quantum algorithms for flood risk management. *Natural Hazards*, 110(3), 1234–1250.

Martin, P., & Lewis, Q. (2024). Validating quantum-enhanced statistical techniques in flood prediction models: A comparative study with classical approaches. *Environmental Science & Policy*, 132, 25–36.

Mayer, H.. (2019). Drones in disast er response: A review. *International Journal of Disaster Risk Reduction*, 32, 372–386.

Mazumdar, A.. (2016). Citizen-centric emergency response system using social media and mobile technology. *IEEE Internet of Things Journal*, 3(1), 137–146.

McCarroll, C., & Cugurullo, F. (2022). No city on the horizon: Autonomous cars, artificial intelligence, and the absence of urbanism. *Frontiers in Sustainable Cities*, 4, 937933. Advance online publication. DOI: 10.3389/frsc.2022.937933

Meduri, Y., Singh, R., & Manoharan, G. (2024). Key networks to create disaster resilient Smart Cities Mission: A case for remodeling India's Smart Cities Mission to include disaster resilience. *Resilient Cities and Structures*, 3(4), 52–62. DOI: 10.1016/j.rcns.2024.09.001

Meena, M. D., Dotaniya, M. L., Meena, B. L., Rai, P. K., Antil, R. S., Meena, H. S., & Meena, R. B. (2023). Municipal solid waste: Opportunities, challenges and management policies in India: A review. *Waste Management Bulletin*, 1(1), 4–18.

Meghanadh, D., Maurya, V. K., Kumar, M., & Dwivedi, R. (2021, July). Automatic detection of landslides based on machine learning framework. In 2021 IEEE International Geoscience and Remote Sensing Symposium IGARSS (pp. 8460-8463). IEEE. DOI: 10.1109/IGARSS47720.2021.9553341

Meier, P. (2018). *Earthquake early warning systems: A technological and social challenge*. Springer.

Merrilees, B. (2019). Interactive Brand Experience Pathways to Customer-Brand Engagement and Value Co-Creation. *Journal of Product and Brand Management*, 28(4), 477–487. DOI: 10.1108/JPBM-12-2018-2166

Miller, P., De Barros, A. G., Kattan, L., & Wirasinghe, S. C. (2016). Public transportation and sustainability: A review. *KSCE Journal of Civil Engineering*, 20(3), 1076–1083. DOI: 10.1007/s12205-016-0705-0

Ministry of Digital Development and Information. (2023). Singapore and Shenzhen Drive Smart City Collaboration with 14 New Projects. https://www.mddi.gov.sg/media-centre/press-releases/singapore-and-shenzhen-drive-smart-city-collaboration/

Ministry of Housing and Urban Affairs. (2022). *Smart Cities Mission: Annual Report 2022*.

Mishra, V., Mishra, R., Rani, R., & Khan, I. (2022). Participation Administration Framework Utilizing Confront Acknowledgment. *Progresses in Computational Insights and Communication Innovation*, 399, 47.

Mitchell, M. T., & Egudo, R. R. (2003). Narrative review: New wine in old bottles? *Qualitative Health Research*, 13(10), 1350–1361.

Mittelstadt, B. D., Allo, P., Taddeo, M., Wachter, S., & Floridi, L. (2016). The ethics of algorithms: Mapping the debate. *Big Data & Society*, 3(2), 205–220. DOI: 10.1177/2053951716679679

Mohammad El-Basioni, B. M., Moustafa, A. I., Abd El-Kader, S. M., & Konber, H. A. (2017). Designing a channel access mechanism for wireless sensor network. *Wireless Communications and Mobile Computing*, 2017, 1–31. DOI: 10.1155/2017/7493269

Mohanty, K., Mishra, D., & Ray, P. K. (2016). Smart cities: A perspective. *IEEE Internet of Things Journal*, 3(6), 1211–1227.

Mohapatra, H. (2021). Socio-technical challenges in the implementation of smart city. *2021 International Conference on Innovation and Intelligence for Informatics, Computing, and Technologies (3ICT)*, 57-62. https://doi.org/DOI: 10.1109/3ICT53449.2021.9581905

Mohapatra, H., & Rath, A. K. (2021). IoE based framework for smart agriculture. *Journal of Ambient Intelligence and Humanized Computing*, 13(1), 407–424. DOI: 10.1007/s12652-021-02908-4

Mohd Ali, N. A.. (2023). A Review on Drone Technology for Search and Rescue Operations. *Journal of Physics: Conference Series*, 2289(1), 012023.

Moon, M. J. (2023). Searching for inclusive artificial intelligence for social good: Participatory governance and policy recommendations for making AI more inclusive and benign for society. *Public Administration Review*, 83(6), 1496–1505. DOI: 10.1111/puar.13648

Mora, C.. (2020). Technology-Driven Smart Cities: Evolution, Challenges, and Future Prospects. *Journal of Urban Technology*.

Morales, J., Saito, T., & Chen, L. (2023). *Drones and robotics in disaster response*. Robotics Review.

Moreira, M. A.. (2018). IoT-Based Early Warning System for Accidents in Port Regions. *IEEE Internet of Things Journal*, 5(6), 4730–4741.

Moreno, M., Úbeda, B., Skarmeta, A., & Zamora, M. (2014). How can we tackle energy efficiency in IoT BasedSmart buildings? *Sensors (Basel)*, 14(6), 9582–9614. DOI: 10.3390/s140609582 PMID: 24887040

Morimoto, A. (2024). History and prospects of traffic enforcement for traffic safety in Japan. *IATSS Research*, 48(2), 129–135. DOI: 10.1016/j.iatssr.2024.02.004

Mouha, R. A. R. A. (2021). Internet of things (IoT). *Journal of Data Analysis and Information Processing*, 9(02), 77–101. DOI: 10.4236/jdaip.2021.92006

Moura, F., & de Abreu e Silva, J. (2021). Smart cities: definitions, evolution of the concept, and examples of initiatives. In Industry, innovation and infrastructure (pp. 989-997). Cham: Springer International Publishing.

Muhadi, N. A., Abdullah, A. F., Bejo, S. K., Mahadi, M. R., & Mijic, A. (2021). Deep learning semantic segmentation for water level estimation using surveillance camera. *Applied Sciences (Basel, Switzerland)*, 11(20), 9691. DOI: 10.3390/app11209691

Muhammad, M., & Safdar, G. A. (2021). 5G-based V2V broadcast communications: A security perspective. *Array (New York, N.Y.)*, 11, 100084. DOI: 10.1016/j.array.2021.100084

Müller, V., & Van der Meer, T. (2023). Ethical AI in Urban Planning: A Framework for Accountability. *Urban Studies Journal*, 60(2), 345–360. DOI: 10.1177/00420980211053754

Muratore, G., Vannelli, A., & Micheli, D. (2023). Rome vehicle accidents analysis and city riskiness prediction by Artificial Intelligence. *Transportation Engineering*, 12, 100172. DOI: 10.1016/j.treng.2023.100172

Murni, S., & Alhadi, W. (2019). Enhancing Collaboration in Disaster Management: A Review of Literature. *International Journal of Disaster Risk Reduction*, 32, 360–371.

Murtagh, C. E.. (2018). Ethical considerations for data sharing in health research: A framework for responsible data stewardship. *Journal of Medical Ethics*, 44(1), 16–21.

Muthusamy, V., Karthik, R., & Kumar, S. (2022). Review of machine learning approaches in drought forecasting: A focus on various meteorological and hydrological parameters. *Geoscience Frontiers*, 13, 101152. DOI: 10.1016/j.gsf.2021.11.004

Myeong, S., & Shahzad, K. (2021). Integrating data-based strategies and advanced technologies with efficient air pollution management in smart cities. *Sustainability (Basel)*, 13(13), 7168. DOI: 10.3390/su13137168

Naik, S. R., & Sahu, N. K. (2021). Machine learning-based approaches for forecasting rainfall: A review. *Meteorology and Atmospheric Physics*, 133, 253–271. DOI: 10.1007/s00703-020-00712-2

Nastjuk, S.. (2022). Smart Cities: A Review of the Concept and Its Implementation. *Sustainability*, 14(15), 9274.

National Science and Technology Center for Disaster Reduction. (2023). *National Disaster Prevention and Protection Information Website*.https://www.nfa.gov.tw/cht/index.php

Nazer, B. B.. (2023). Addressing Bias in Healthcare AI: A Review. *IEEE Journal of Biomedical and Health Informatics*, 27(1), 1–14.

Nearing, G. S., et al. (2023). Artificial Intelligence for Flood Forecasting in Ungauged Basins: A Review. *Water Resources Research, 59*(6), e2022WR034352.

Nethravathi, P., Aithal, P., Soans, S., & Yadav, N. (2021). Coin-Based Mobile Charger Using Solar Tracker. *International Journal of Applied Engineering and Management Letter*, 5(2). Advance online publication. DOI: 10.47992/IJAEML.2581.7000.0104

Newton, J.. (2022). Addressing disparities in access to emergency services: A framework and case study. *Journal of Urban Health*, 99(4), 704–716.

Nguyen, T., & Tran, H. (2023). Hybrid ML models with QML integration: Showed significant improvements in operational efficiency during disaster management scenarios. *Journal of Hydrology (Amsterdam)*, 605, 140–152.

Niittymäki, J., & Pursula, M. (2001). The 11th Mini-EURO conference: Artificial intelligence on transportation systems and science. *European Journal of Operational Research*, 131(2), 229–231. DOI: 10.1016/S0377-2217(00)00122-3

Nikitas, A., Michalakopoulou, K., Njoya, E. T., & Karampatzakis, D. (2020). Artificial intelligence, transport and the smart City: Definitions and dimensions of a new mobility era. *Sustainability (Basel)*, 12(7), 2789. DOI: 10.3390/su12072789

Nishant, R., Kennedy, M., & Corbett, J. (2020). Artificial intelligence for sustainability: Challenges, opportunities, and a research agenda. *International Journal of Information Management*, 53, 102104. DOI: 10.1016/j.ijinfomgt.2020.102104

Nithya, S., Revathi, M., Sathiya Sree, A., Sivapriya, T., & Vaishnavi, P. (2022). Guaranteeing the Nearness of a Individual Amid Virtual Classes Utilizing Histogram of Arranged Angles (Hoard) Calculation. *Troublesome Advances for Enormous Information and Cloud Applications*, 905, 287.

Niu, H., Feng, Z., Xiao, Q., & Zhang, Y. (2020). A PID control method based on optimal control strategy. Numerical Algebra. *Control and Optimization*, 11(1), 117–126. DOI: 10.3934/naco.2020019

Noble, S. U. (2018). *Algorithms of oppression: How search engines reinforce racism*. NYU Press. DOI: 10.18574/nyu/9781479833641.001.0001

Nursalim, A., Judijanto, L., & Asfahani, A. (2022). Educational revolution through the application of AI in the digital era. *Journal of Artificial Intelligence and Development*, 1(1), 31–40.

Nyika, J. MegersaOlumanaDinka(2024), "The role of solid waste management in boosting sustainable development goals"chapter 8, In the book "Exploring waste management in sustainable development", IGI global publications, 2024.

O'Brien, J., Smith, R., & Johnson, L. (2022). Real-time flood forecasting with QML: Provided timely alerts and decision support for emergency services during floods. *Scientific Reports*, 12(1), 12345. PMID: 35853970

O'Neil, C. (2022). *Weapons of Math Destruction: How Big Data Increases Inequality and Threatens Democracy*. Crown Publishing Group.

Obcnauf, M.. (2019). The Potential of Connected and Autonomous Vehicles for Emergency Response. *International Journal of Disaster Risk Reduction*, 32, 350–359.

Ochieng, J. (2023). Enhancing Road Safety through AI in Kenya. *East African Journal of Science and Technology*, 2(1), 45–60. DOI: 10.1007/s00500-022-06371-8

OECD. (2021). *OECD Principles on AI.*

OECD. (2023). Digital. *Economic Outlook*, •••, 2023.

Oerlemans, S.. (2023). Uncertainty in healthcare: A call for a more holistic understanding. *Journal of Evaluation in Clinical Practice*, 29(7), 1946–1953.

Oh, J., & Lee, J. (2020). A Review of Emergency Management Research: Trends and Future Directions. *Sustainability*, 12(13), 5458.

Olayode, O., Tartibu, L., & Okwu, M. (2020). Application of artificial intelligence in traffic control system of non-autonomous vehicles at signalized road intersection. *Procedia CIRP*, 91, 194–200. DOI: 10.1016/j.procir.2020.02.167

Omar, Z. F., Mior Harun, M. H., Mohd Ishar, N. I., Mustapha, N. A., & Ismail, Z. (2024). Enhancing professional development and training through AI for personalized learning: A framework to engaging learners. *International Journal of e-Learning and Higher Education (IJELHE), 19*(3), 115-138.

OpenGov Asia. (2024). *RescueAI: Smart City Disaster Management System in Malaysia.* OpenGov Asia.

Organisation for Economic Co-operation and Development. (2019). OECD AI principles overview (May). https://oecd.ai/en/ai-principles

Oxendine, L. B.. (2014). Using non-authoritative data sources to mitigate data deficiencies in emergency scenarios. *International Journal of Critical Infrastructure Protection*, 7, 18–27.

Özkaya Marangoz, A. (2023). The Impact of Artificial Intelligence on Language Services: A Review. *International Journal of Emerging Technologies in Learning*, 18(1), 1–17.

Pamucar, D., Deveci, M., Gokasar, I., Tavana, M., & Köppen, M. (2022). A metaverse assessment model for sustainable transportation using ordinal priority approach and Aczel-Alsina norms. *Technological Forecasting and Social Change*, 182, 121778. DOI: 10.1016/j.techfore.2022.121778

Pan, C., Hu, H., Xu, C., & Zhang, X. (2023). Predicting soil moisture using machine learning methods: A case study in the Loess Plateau of China. *Catena*, 207, 105792. DOI: 10.1016/j.catena.2021.1057

Panda, A. K., Lenka, A. A., Mohapatra, A., Rath, B. K., Parida, A. A., & Mohapatra, H. (2024). Integrating cloud computing for intelligent transportation solutions in smart cities. *Advances in Civil and Industrial Engineering*, •••, 121–142. DOI: 10.4018/979-8-3693-6695-0.ch005

Pandey, H. (2016). *Jaya a Novel Optimization Algorithm: What.* How and Why., DOI: 10.1109/CONFLUENCE.2016.7508215

Panduman, Y. Y. F., Funabiki, N., Fajrianti, E. D., Fang, S., & Sukaridhoto, S. (2024). A Survey of AI Techniques in IoT Applications with Use Case Investigations in the Smart Environmental Monitoring and Analytics in Real-Time IoT Platform. *Information (Basel)*, 15(3), 153. DOI: 10.3390/info15030153

Pang, Y., & Pang, G. (2022). Artificial Intelligence for Disaster Management: A Review. *Sustainability*, 14(18), 11200.

Pantano, E., & Timmermans, H. (2014). What is smart for retailing? Insights on the future of retailing in the smart city. *International Journal of Retail & Distribution Management*, 42(3), 176–188.

Panwar, A., Singh, A., Kumawat, R., Jaidka, S., & Garg, K. (2017). Eyrie smart home automation using Internet of things. *2017 Computing Conference*, 1368-1370. https://doi.org/DOI: 10.1109/SAI.2017.8252269

Papageorgiou, G., Tsappi, E., & Wang, T. (2024). Smart Urban Systems Planning for active mobility and sustainability. *IFAC-PapersOnLine*, 58(10), 261–266. DOI: 10.1016/j.ifacol.2024.07.350

Parasuraman, R., Sheridan, T. B., & Wickens, C. D. (2000). A model for types and levels of human interaction with automation. *IEEE Transactions on Systems, Man, and Cybernetics. Part A, Systems and Humans*, 30(3), 286–297. DOI: 10.1109/3468.844354 PMID: 11760769

Park, Y., Li, X., & Chen, Y. (2023). IoT for real-time disaster management applications. International Journal of IoT.

Parker, D. J. (2020). Disaster resilience–a challenged science. *Environmental Hazards*, 19(1), 1–9. DOI: 10.1080/17477891.2019.1694857

Parkes, M. (2022). Control of non-point source pollution Summary. DOI: 10.13140/RG.2.2.26413.56801

Parmar, A.. (2007). Enhancing coordination in disaster relief: The role of donors. *Disasters*, 31(1), 111–132.

Parrado, N., & Donoso, Y. (2015). Congestion based mechanism for route discovery in a V2I-V2V system applying smart devices and IoT. *Sensors (Basel)*, 15(4), 7768–7806. DOI: 10.3390/s150407768 PMID: 25835185

Parthasarathy, S., & Katzman, D. (2024). Designing AI for Equity: A Framework for Collaboration with Disadvantaged Communities. *AI & Society*, 38(1), 11–28.

Patekar, A. S., Daniel, E., Seetha, S., & Manimekalai, M. A. P. (2024, March). Applying U-Net CNN Approach for Landslide Susceptibility Mapping. In *2024 Third International Conference on Intelligent Techniques in Control, Optimization and Signal Processing (INCOS)* (pp. 1-6). IEEE. DOI: 10.1109/INCOS59338.2024.10527582

Patel, A., & Verma, S. (2024). Quantum neural networks with temporal analysis: Capturing complex patterns in flooding data effectively. *Hydrological Processes*, 38(5), e14356.

Pawar, A. (2015). Coin Based Solar Mobile Charger. International Journal of Engineering and Technical Research, Volume-3, Issue-5, May 2015. https://www.erpublication.org/published_paper/IJETR032148.pdf

Pera, A. (2017). Piso Automated Cellphone Charging Utility Model. Slideshare. https://www.slideshare.net/angelitopera/chapter1-3-74917597

Pérez, P., & Rodríguez, M. (2023). Smart Water Management in Barcelona: An AI Approach. *Journal of Urban Technology*, 30(3), 25–40. DOI: 10.1080/10630732.2023.2157514

Perol, T., Gharbi, M., & Denolle, M. (2018). Convolutional neural network for earthquake detection and location. *Science Advances*, 4(2), e1700578. DOI: 10.1126/sciadv.1700578 PMID: 29487899

Perwej, Y. (2015). An Assessment of Profound Learning Smaller than expected Concerning in Delicate Computing [IJARCCE]. *Universal Diary of Progressed Investigate in Computer and Communication Building*, 4(2), 10–16.

Pham, N., Gye-An Lee, & De Flaviis, F. (2004). Minimized dual-band coupled line meander antenna for system-in-a-package applications. *IEEE Antennas and Propagation Society Symposium, 2004*, 1451-1454 Vol.2. https://doi.org/DOI: 10.1109/APS.2004.1330461

Philip, A., Islam, S. N., Phillips, N., & Anwar, A. (2022). Optimum energy management for air conditioners in IoT-enabled smart home. *Sensors (Basel)*, 22(19), 7102. DOI: 10.3390/s22197102 PMID: 36236199

Picard, R. W. (2010). Affective computing: From laughter to IEEE. *IEEE Transactions on Affective Computing*, 1(1), 11–17. DOI: 10.1109/T-AFFC.2010.10

PMC. (2023). *Global, regional and national trends and impacts of natural floods from 1990 to 2022*. PMC.

Poria, S., Cambria, E., Bajpai, R., & Hussain, A. (2017). A review of affective computing: From unimodal analysis to multimodal fusion. *Information Fusion*, 37, 98–125. DOI: 10.1016/j.inffus.2017.02.003

Priyadarshi, S., Subudhi, S., Kumar, S., Bhardwaj, D., & Mohapatra, H. (2024). Analysis on enhancing urban mobility with IoT-integrated parking solutions. *Advances in Civil and Industrial Engineering*, •••, 143–172. DOI: 10.4018/979-8-3693-6695-0.ch006

Priyanka, G., Anisha, S., & Padma Shri, P. (2018). Coin Based Mobile Charger. *International Journal of Pure and Applied Mathematics*, 119(12), 13695–13701. http://www.ijpam.eu

Pucoe, G., & Obagbuwa, I. C. (2024, April). Deep Learning Approach with a Convolutional Neural Network for Landslide Detection Using Remotely Sensed Images. In 2024 International Conference on Science, Engineering and Business for Driving Sustainable Development Goals (SEB4SDG) (pp. 1-9). IEEE. DOI: 10.1109/SEB4SDG60871.2024.10629839

Qiu, D.. (2022). Digital Transformation for Urban Resilience: A Systematic Literature Review. *Sustainability*, 14(18), 11449.

Rabah, K. (2018). The Impact of Industry 4.0 on the Economy. *International Journal of Engineering & Technology, 7*(3.3), 246-250.

Radu, L. D. (2020). Disruptive technologies in smart cities: A survey on current trends and challenges. *Smart Cities*, 3(3), 1022–1038. DOI: 10.3390/smartcities3030051

Raghavendra, R., Chand, R., Bhadani, A. J., & Singh, A. (2023). Deep learning for flood prediction: A systematic review and future directions. *Journal of Hydrology (Amsterdam)*, 610, 127713. DOI: 10.1016/j.jhydrol.2022.127713

Raghavendra, S. (2022). AI-Powered Cybersecurity: A Comprehensive Review. *Materials Today: Proceedings*, 48, 1616–1621.

Rahimi, H., & Verma, R. (2022). *Autonomous sensing in disaster zones: IoT and AI applications*. Smart Systems Review.

Raj, A. (2023). A Review on Facial Recognition Technology and Its Applications. *International Journal of Advanced Research in Engineering and Technology*, 15(3), 1–10.

Rajesh, P., Shajin, F. H., & Kannayeram, G. (2022). A novel intelligent technique for energy management in smart home using Internet of things. *Applied Soft Computing*, 128, 109442. DOI: 10.1016/j.asoc.2022.109442

Ramakrishnan, T., Mohan Gift, M. D., Chitradevi, S., Jegan, R., & Subha Hency Jose, P. Nagaraja, Rajneesh Sharma H.N., Selvakumar P., Sintayehu Mekuria Hailegiorgis, (2022). "Study of Numerous Resins Used in Polymer Matrix Composite Materials" Advances in Materials Science and Engineering, vol. 2022, Article ID 1088926, 8 pages, 2022. DOI: 10.1155/2022/1088926

Ramanathan, R.. (2019). Customer relationship management with AI: A systematic review. *Journal of Business Research.*

Ramirez Lopez, L. J., & Grijalba Castro, A. I. (2020). Sustainability and resilience in smart city planning: A review. *Sustainability (Basel)*, 13(1), 181. DOI: 10.3390/su13010181

Ramos, L., Casas, E., Bendek, E., Romero, C., & Rivas-Echeverría, F. (2024). Computer vision for wildfire detection: A critical brief review. *Multimedia Tools and Applications*, 83(35), 1–44. DOI: 10.1007/s11042-024-18685-z

Rane, N., Choudhary, S., & Rane, J. (2023). Education 4.0 and 5.0: Integrating artificial intelligence (AI) for personalized and adaptive learning. Available at *SSRN* 4638365. DOI: 10.2139/ssrn.4638365

Rane, N., Choudhary, S., & Rane, J. (2024). Artificial intelligence for enhancing resilience. *Journal of Applied Artificial Intelligence*, 5(2), 1–33. DOI: 10.48185/jaai.v5i2.1053

Rasheed, F., Yau, K. A., Noor, R. M., Wu, C., & Low, Y. (2020). Deep Reinforcement Learning for Traffic Signal Control: A review. *IEEE Access : Practical Innovations, Open Solutions*, 8, 208016–208044. DOI: 10.1109/ACCESS.2020.3034141

Rasheed, Z., Ghwanmeh, S., & Abualkishik, A. Z. (2023). Harnessing artificial intelligence for personalized learning: A systematic review. *Data and Metadata*, 2, 146–146. DOI: 10.56294/dm2023146

Rauf, A., Guru, A., & Mohapatra, H. (2024). Exploring the social model of smart home devices. *Advances in Human and Social Aspects of Technology*, •••, 233–256. DOI: 10.4018/979-8-3693-9235-5.ch012

Rawat, P. S., Yadav, A. K., Barthwal, V., & Gaur, S. (2023, October). Rainfall Induced Landslide Detection in Study Area using Machine Learning Model. In 2023 7th International Conference on I-SMAC (IoT in Social, Mobile, Analytics and Cloud)(I-SMAC) (pp. 734-739). IEEE. DOI: 10.1109/I-SMAC58438.2023.10290177

Raymond, C.. (2016). Ethical considerations for data sharing in humanitarian settings. *The Lancet. Global Health*, 4(11), e744–e750.

Ray, S., & Sikdar, D. P. (2024). AI-driven flipped classroom: Revolutionizing education through digital pedagogy. *Psychology (Irvine, Calif.)*, 7(2), 169–179.

Reddy, S., Kumar, P., & Singh, J. (2023). *Economic challenges in implementing disaster management technologies*. Economics & Environment.

Reed, C.. (2016). The legal implications of machine learning. *Science*, 353(6294), 284–285.

Ren, S., & Li, Y. (2023). A review of high performance computing applications in high-speed rail systems. *High-speed Railway*, 1(2), 92–96. DOI: 10.1016/j.hspr.2023.05.001

Restackio Research Team. (2023). Hybrid classical and quantum machine learning models for improving flood prediction accuracy and reducing computational time: An interdisciplinary approach. *Journal of Environmental Management*, 305, 113–125.

Rizk, M.. (2021). Real-time person recognition with UAVs using deep learning and coordinated communication. *IEEE Internet of Things Journal*, 8(18), 14286–14299.

Rizvi, S. (2023). Revolutionizing student engagement: Artificial intelligence's impact on specialized learning motivation. *International Journal of Advanced Engineering Research and Science*, 10(09), 27–31. DOI: 10.22161/ijaers.109.4

Roberts, T., & Nguyen, P. (2023). Quantum-driven hydrological models: A new frontier in flood prediction. *Advances in Water Resources*, 145, 103–118.

Robinson, C., & Clark, T. (2022). Ensemble learning with quantum algorithms: Achieved robust predictions by combining multiple quantum models. *Journal of Environmental Management*, 305, 113–125.

Rodrigue, J. (2024). *The geography of transport systems*. https://doi.org/DOI: 10.4324/9781003343196

Rodríguez, M., & Vargas, E. (2020). *Blockchain for transparent disaster relief in Puerto Rico*. Disaster Response Journal.

Roger, S., Martín-Sacristán, D., Garcia-Roger, D., Monserrat, J. F., Kousaridas, A., Spapis, P., & Ayaz, S. (2020). 5G V2V communication with antenna selection based on context awareness: Signaling and performance study. *IEEE Transactions on Intelligent Transportation Systems*, 23(2), 1044–1057. DOI: 10.1109/TITS.2020.3019530

Roselli, T.. (2019). Addressing bias in artificial intelligence for healthcare. *Nature Medicine*, 25(11), 1753–1758. PMID: 31700174

Rossi, P., & Singh, M. (2024). Multi-hazard sensing networks for disaster management. International Journal of IoT Applications.

Rossi, G., & Petruzzi, L. (2022). Challenges in regulatory harmonization for international AI collaborations. *Policy & Data Ethics Review*, 11(1), 89–105.

Roy, A. C., & Islam, M. M. (2019, September). Predicting the probability of landslide using artificial neural network. In 2019 5th International Conference on Advances in Electrical Engineering (ICAEE) (pp. 874-879). IEEE. DOI: 10.1109/ICAEE48663.2019.8975696

Roy, S., Jana, D. K., & Mishra, A. (2024). Linguistic interval type 2 fuzzy logic-based Exigency Vehicle routing: IoT system development for smart city applications with soft computing-based optimization. *Franklin Open*, 6, 100057. DOI: 10.1016/j.fraope.2023.100057

Rudewicz, P. (2019). Smart cities: A review of enabling technologies, challenges, and opportunities. *Future Internet*, 11(1), 1–26.

Rudolph, S., Schmid, K., & Bauer, W. (2023). The Role of Standards in Smart Cities. *Journal of Urban Technology*, 30(2), 125–140. DOI: 10.1080/10630732.2023.2147532

Russell, S., & Norvig, P. (2016). *Artificial intelligence: A modern approach*. Pearson Education Limited.

Rust, R. T. (2020). The Future of Marketing. *International Journal of Research in Marketing*, 37(1), 15–26. DOI: 10.1016/j.ijresmar.2019.08.002

Rust, R. T., & Huang, M. H. (2021). The future of artificial intelligence in marketing: Looking forward to 2025. *Journal of the Academy of Marketing Science*, 49(5), 946–958.

Saina, M. (2021). The impact of artificial intelligence on language learning and teaching. *International Journal of Emerging Technologies in Learning*, 16(4), 1–15.

Saisubramanian, S.. (2015). Risk-based optimization of emergency medical service vehicle deployment. *Computers & Operations Research*, 61, 122–132.

Salcedo, J. D., Vargas, G. M., Rojas, J. C., & Chavarria, D. (2023). Integrating remote sensing and machine learning for flood risk management: A systematic review. *Water (Basel)*, 15, 245. DOI: 10.3390/w15020245

Salman, L., Salman, S., Jahangirian, S., Abraham, M., German, F., Blair, C., & Krenz, P. (2016). Energy efficient IoT-based smart home. *2016 IEEE 3rd World Forum on Internet of Things (WF-IoT)*, 526-529. https://doi.org/DOI: 10.1109/WF-IoT.2016.7845449

Sanchez Gracias, M. J.. (2023)... *Sensors (Basel)*, 23(10), 4784. DOI: 10.3390/s23104784 PMID: 37430698

Sangeeta, R.. (2023). A Review on Drone Technology in Search and Rescue Operations. *Materials Today: Proceedings*, 60, 1420–1427.

Santos, F., Araújo, L., & Ribeiro, C. (2023). AI and Urban Mobility in Brazil: Challenges and Opportunities. *Urban Planning Review*, 12(2), 78–93. DOI: 10.5555/1989-4924-12-2-78

Sato, K., Ishikawa, T., & Nakamura, Y. (2022). AI and seismic prediction: Japan's earthquake early warning system. *Journal of Seismology*.

Scantlebury, A., Lopez, T., Chen, Y., & Morales, D. (2022). Knowledge sharing and innovation in cross-border AI collaborations. *Journal of Urban Technology Innovation*, 14(3), 112–134.

Shackelford, S. J.. (2022). Cybersecurity and Smart Cities: A Global Perspective. *Journal of Cyber Policy*, 7(1), 95–113. DOI: 10.1080/23738871.2022.2074173

Shafique, K., Khawaja, B. A., Sabir, F., Qazi, S., & Mustaqim, M. (2020). Internet of things (IoT) for next-generation smart systems: A review of current challenges, future trends, and prospects for emerging 5G-IoT scenarios. *IEEE Access : Practical Innovations, Open Solutions*, 8, 23022–23040. DOI: 10.1109/ACCESS.2020.2970118

Shahid, A., & Adnan, A. (2023). AI for Sustainable Urban Development: A Review of Current Trends. *Sustainability Journal*, 15(12), 4567. DOI: 10.3390/su15124567

Shaiken, H. (1986). *Work transformed: Automation and labor in computer and electronics industries.* Holt, Rinehart and Winston.

Shankar, V. (2018). What an artificial intelligence (AI) is reshaping retailing. *Journal of Retailing*, 94(4), vi–xi. DOI: 10.1016/S0022-4359(18)30076-9

Sharifi, A., Khavarian-Garmsir, A. R., & Kummitha, R. K. R. (2021). Contributions of Smart City Solutions and Technologies to Resilience against the COVID-19 Pandemic: A Literature Review. *Sustainability (Basel)*, 13(14), 8018. DOI: 10.3390/su13148018

Sharma, A., N., Mukherjee, S., Kumar, S., Kumar, S., & Mohapatra, H. (2025). Analyzing territorial variations in the implementation of IoT-based smart homes: A comprehensive review. Utilizing Technology to Manage Territories, 305–334.

Sharma, A., Anand, N., Mukherjee, S., Kumar, S., Kumar, S., & Mohapatra, H. (2024). Analyzing territorial variations in the implementation of IoT-based smart homes. *Advances in Electronic Government, Digital Divide, and Regional Development*, •••, 305–334. DOI: 10.4018/979-8-3693-6854-1.ch010

Sharma, P., Jindal, R., & Borah, M. D. (2023). A review of smart contract-based platforms, applications, and challenges. *Cluster Computing*, 26(1), 395–421. DOI: 10.1007/s10586-021-03491-1

Sharma, S. (2020). Augmented reality instructional components for improving spatial awareness and cognitive mapping in multilevel structures. *Applied Ergonomics*, 87, 103145.

Sharma, S., & Kumari, D. (2022). Quantum computing for healthcare applications. *International Journal of Developmental Research*, 12(10), 28041–28045.

Shawon, R. E. R., Miah, M. N. I., & Islam, M. Z. (2023). Enhancing US education systems with AI: Personalized learning and academic performance prediction. *International Journal of Advanced Engineering Technologies and Innovations*, 1(01), 518–540.

Shaygan, M., Meese, C., Li, W., Zhao, X., & Nejad, M. (2022). Traffic prediction using artificial intelligence: Review of recent advances and emerging opportunities. *Transportation Research Part C, Emerging Technologies*, 145, 103921. DOI: 10.1016/j.trc.2022.103921

Sheth, J. N., Sisodia, R. S., & Sharma, A. (2000). The antecedents and consequences of customer-centric marketing. *Journal of the Academy of Marketing Science*, 28(1), 55–66. DOI: 10.1177/0092070300281006

Shi, X., Yang, Q., & Zhao, P. (2022). *Machine learning applications in environmental disaster prediction*. Springer.

Shmueli, G., & Koppius, O. R. (2011). Predictive analytics in information systems research. *Management Information Systems Quarterly*, 35(3), 553–572. DOI: 10.2307/23042796

Shojaei, S., & Shojaei, S. (2021). Optimization of Process Conditions in Wastewater Degradation Process. In Karri, R. R., Ravindran, G., & Dehghani, M. H. (Eds.), *Soft Computing Techniques in Solid Waste and Wastewater Management* (pp. 381–392). Elsevier., DOI: 10.1016/B978-0-12-824463-0.00010-0

Shukla, M. K. (2024). Ethical Considerations for Artificial Intelligence: A Review. *International Journal of Advanced Research in Engineering and Technology*, 15(3), 1–10.

Shyam, R., Mishra, A., Kumar, A., Chowdhary, A., & Kr, A. (2024). Srivastava, "Recording of Course Participation Utilizing DL-Based Confront Acknowledgment Strategy,". *Information Science and Applications*, 818, 249.

Siddique, M. N. I.. (2023). IoT-Based Early Warning System for Flood Disaster Management: A Review. *Sensors (Basel)*, 23(10), 4784. PMID: 37430698

Sin, C. (2024). Singapore is world's 5th smartest city, top in Asia: Global index. The Straits Times. https://www.straitstimes.com/singapore/singapore-is-5th-smartest-city-in-the-world-top-in-asia-global-index

Singh, C. R., & Manoharan, G. (2024). Strengthening Resilience: AI and Machine Learning in Emergency Decision-Making for Natural Disasters. In *Internet of Things and AI for Natural Disaster Management and Prediction* (pp. 249-278). IGI Global. https://doi.org/DOI: 10.4018/979-8-3693-4284-8.ch012

Singh, R., Manoharan, G., & Singh, S. (2024). Strategic Deployment of AI and Drones Enhancing Disaster Management in Natural Disasters. *AI and IoT for Proactive Disaster Management*, 172-203. https://doi.org/DOI: 10.4018/979-8-3693-3896-4.ch009

Singh, B. J., Chakraborty, A., & Sehgal, R. (2023). A systematic review of industrial wastewater management: Evaluating challenges and enablers. *Journal of Environmental Management*, 348, 119230. DOI: 10.1016/j.jenvman.2023.119230 PMID: 37832302

Singh, M., & Gupta, N. (2023). *Standards for interoperability in disaster risk management technologies*. International Standards Review.

Singh, T. M., Reddy, C. K. K., Murthy, B. R., Nag, A., & Doss, S. (2025). AI and education: Bridging the gap to personalized, efficient, and accessible learning. In *Internet of Behavior-Based Computational Intelligence for Smart Education Systems* (pp. 131–160). IGI Global.

Sit, M. A., Koylu, C., & Demir, I. (2020). Identifying disaster-related tweets and their semantic, spatial and temporal context using deep learning, natural language processing and spatial analysis: a case study of Hurricane Irma. In *Social Sensing and Big Data Computing for Disaster Management* (pp. 8–32). Routledge. DOI: 10.4324/9781003106494-2

Smart Nation Singapore. (2023). *AI Ethics Advisory Council: Guidelines and Recommendations*.

Smith, J., & Johnson, K. (2021). Quantum support vector machines: Enhanced classification of flood risk zones using QSV models. *Journal of Hydrology (Amsterdam)*, 598, 123–135.

Smith, P. R., & Chaffey, D. (2020). *Digital Marketing Excellence: Planning, Optimizing and Integrating Online Marketing*. Routledge.

Sohag, M. U., & Podder, A. K. (2020). Smart garbage menagement system for a sustainableurban life: An IoT based application. *Internet of Things : Engineering Cyber Physical Human Systems*, 11, 100–255. DOI: 10.1016/j.iot.2020.100255

Sriskanthan, N., Das, A., Loh, P., & Leong, A. H. (1994). An adaptive switching architecture for multiprocessor networks. *Microprocessors and Microsystems*, 18(6), 307–314. DOI: 10.1016/0141-9331(94)90045-0

Srivastava, R. K. (2020). Urbanization-Led Neo Risks and Vulnerabilities: A New Challenge. In *Disaster Studies and Management* (pp. 251–296). https://doi.org/DOI: 10.1007/978-981-15-2410-3_6

State of Green. (2023). Fostering cross-border collaboration in the Capital of Architecture 2023. https://stateofgreen.com/en/news/fostering-collaboration-and-inspiration-in-the-capital-of-architecture-2023/

Su, C., & Chuang, H. (2011). Toward mass customized product deployment in e-Commerce: The Modularization function and postponement strategy. *Journal of Organizational Computing and Electronic Commerce*, 21(1), 24–49. DOI: 10.1080/10919392.2011.541000

Sudha, S. (2020, October). Prediction of earthquake induced landslide using deep learning models. In 2020 5th International Conference on Computing, Communication and Security (ICCCS) (pp. 1-6). IEEE.

Sufi, F. K., & Khalil, I. (2022). Automated disaster monitoring from social media posts using AI-based location intelligence and sentiment analysis. *IEEE Transactions on Computational Social Systems*. Advance online publication. DOI: 10.1109/TCSS.2022.3157142

Sufi, M. S. (2022). AI-Social Disaster: A novel framework for real-time natural disaster detection and information dissemination using social media. *International Journal of Disaster Risk Reduction*, 66, 102835.

Sukhera, J. (2022). Narrative Review: A Comprehensive Guide. *International Journal of Advanced Research in Engineering and Technology*, 14(3), 1–10.

Suliman, M. A.. (2013). Flood forecasting using artificial neural networks and support vector machines. *Journal of Hydrology (Amsterdam)*, 499, 145–157.

Sullivan, R. (2020). *Predictive analytics in marketing: Starbucks' AI-driven recommendations*. Marketing Science Review.

Sumathy, V., & Navamani, G. (2024). AI-driven personalized learning: Enhancing student success through adaptive technologies. *Library Progress International*, 44(3), 16235–16242.

Sungheetha, S., & Sharma, R. S. (2020). IoT Based Early Warning System for Forest Fire Detection Using Drone and Wireless Sensor Network. *Materials Today: Proceedings*, 27, 1580–1584.

Supporting the global shift to electric mobility. (n.d.). UNEP - UN Environment Programme. https://www.unep.org/topics/transport/electric-mobility/supporting-global-shift-electric-mobility

Suvarnamma, A., & Pradeepkiran, J. A. (2021). SmartBin system with waste tracking and sorting mechanism using IoT. *Cleaner Engineering and Technology*, 5, 100348. DOI: 10.1016/j.clet.2021.100348

Swargiary, K. (2024). The impact of AI-driven personalized learning and intelligent tutoring systems on student engagement and academic achievement: Ethical implications and the digital divide. Available at *SSRN* 4897241. DOI: 10.2139/ssrn.4897241

Syam, S. K., & Sharma, S. (2023). A machine learning framework for predicting flood events in a catchment area. *Journal of Hydrology (Amsterdam)*, 611, 128008. DOI: 10.1016/j.jhydrol.2022.128008

Szpilko, D., de la Torre Gallegos, A., Jimenez Naharro, F., Rzepka, A., & Remiszewska, A. (2023). Waste Management in the Smart City: Current Practices and Future Directions. *Resources*, 12(115), 1–25. DOI: 10.3390/resources12100115

Talebkhah, M., Sali, A., Marjani, M., Gordan, M., Hashim, S. J., & Rokhani, F. Z. (2020, September). Edge computing: architecture, applications and future perspectives. In *2020 IEEE 2nd International Conference on Artificial Intelligence in Engineering and Technology (IICAIET)* (pp. 1-6). IEEE. https://doi.org/DOI: 10.1109/IICAIET49801.2020.9257824

Talebkhah, M., Sali, A., Marjani, M., Gordan, M., Hashim, S. J., & Rokhani, F. Z. (2021). IoT and big data applications in smart cities: Recent advances, challenges, and critical issues. *IEEE Access : Practical Innovations, Open Solutions*, 9, 55465–55484. DOI: 10.1109/ACCESS.2021.3070905

Tanaka, Y., & Martinez, R. (2022). Blockchain for transparent disaster aid distribution. Journal of Humanitarian Aid and Technology.

Tang, R., De Donato, L., Bešinović, N., Flammini, F., Goverde, R. M. P., Lin, Z., Liu, R., Tang, T., Vittorini, V., & Wang, Z. (2022). A literature review of Artificial Intelligence applications in railway systems. *Transportation Research Part C, Emerging Technologies*, 140, 103679. DOI: 10.1016/j.trc.2022.103679

Tarek, M. M., & Ouf, S. A. (2021). Sustainable Smart Cities: A Review. *Sustainable Cities and Society*, 69, 102958.

Tariq A, Muhammad I, Abdullah SA and Adam G (2019). IoT-Based Smart Waste BinMonitoring and Municipal Solid Waste Management System for Smart Cities 2020Computer Engineering and Computer Science

Tariq, M. U. (2024). Generative AI in Curriculum Development in Higher Education. In Fields, Z. (Ed.), *Impacts of Generative AI on Creativity in Higher Education* (pp. 227–258). IGI Global., DOI: 10.4018/979-8-3693-2418-9.ch009

Tariq, M. U. (2024). Leveraging AI for Entrepreneurial Innovation in Healthcare. In Özsungur, F. (Ed.), *Generating Entrepreneurial Ideas With AI* (pp. 192–216). IGI Global., DOI: 10.4018/979-8-3693-3498-0.ch009

Tariq, M. U. (2025). *AI-Driven Learning and Engagement in Higher Education*. IGI Global., DOI: 10.4018/979-8-3693-4074-5

Tariq, M. U. (2025). Harnessing Generative AI for Enhanced Web Application Security. In Shah, I., & Jhanjhi, N. (Eds.), *Generative AI for Web Engineering Models* (pp. 161–192). IGI Global., DOI: 10.4018/979-8-3693-3703-5.ch008

Tariq, M. U. (2025). Navigating Bias and Fairness in Digital AI Systems. In Bhattacharya, P., Hassan, A., Liu, H., & Bhushan, B. (Eds.), *Ethical Dimensions of AI Development* (pp. 127–156). IGI Global., DOI: 10.4018/979-8-3693-4147-6.ch006

Thales. (2023). Singapore: The World's Smartest City. https://www.thalesgroup.com/en/worldwide-digital-identity-and-security/iot/magazine/singapore-worlds-smartest-city

Thomas, P., & Narayan, A. (2022). *Cross-border disaster management and policy harmonization*. Disaster Policy Journal.

Thompson, M., Lee, T., & Zhang, H. (2023). Quantum phase estimation: Improved estimation of flood risk factors. *Water Science and Technology*, 88(4), 789–802.

Tiedemann, J., & Thottingal, S. (2020). OPUS-MT: Open-source machine translation. *arXiv preprint arXiv:2007.01775*.

Titko, M., & Ristvej, J. (2020). Assessing importance of disaster preparedness factors for sustainable disaster risk management: The case of the Slovak Republic. *Sustainability (Basel)*, 12(21), 9121. DOI: 10.3390/su12219121

Tiwari, P. (2023). The machine learning framework for traffic management in smart cities. *Management of Environmental Quality*, 35(2), 445–462. DOI: 10.1108/MEQ-08-2022-0242

Tiwari, S., & Dahiya, A. (2023). A review on the applications of machine learning in disaster management. *Journal of Ambient Intelligence and Humanized Computing*. Advance online publication. DOI: 10.1007/s12652-023-04661-x

Tokody, D., Ady, L., Hudasi, L. F., Varga, P. J., & Hell, P. (2020). Collaborative Robotics Research: Subiko Project. *Procedia Manufacturing*, 46, 467–474. DOI: 10.1016/j.promfg.2020.03.068

Török, Á., Török, M., & Jacobsen, G. (2022). Ethical consumerism: The role of AI in promoting sustainable choices. *Journal of Marketing Ethics*, 15(2), 77–90.

Townsend, A. (2021). *Smart Cities: Big Data, Civic Hackers, and the Quest for a New Utopia*. W.W. Norton & Company.

Treiblmaier, H., & Rejeb, A. (2023). Exploring blockchain for disaster prevention and relief: A comprehensive framework based on industry case studies. *Journal of Business Logistics*, 44(4), 550–582. DOI: 10.1111/jbl.12345

Tripathi, G., Ahad, M. A., & Casalino, G. (2023). A comprehensive review of blockchain technology: Underlying principles and historical background with future challenges. Decision Analytics Journal, 100344.

Trivedi, N. K.. (2023). Role of Emerging Technologies in Industry 4.0: A Review. *Materials Today: Proceedings*, 60, 1428–1435.

Tsai, H.-C., Chiu, H.-C., & Lee, T.-W. (2022). Predicting extreme rainfall using machine learning techniques. *Atmosphere*, 13, 160. DOI: 10.3390/atmos13020160

Tuan, T. A., Pha, P. D., Tam, T. T., & Bui, D. T. (2023). A new approach based on Balancing Composite Motion Optimization and Deep Neural Networks for spatial prediction of landslides at tropical cyclone areas. *IEEE Access : Practical Innovations, Open Solutions*, 11, 69495–69511. DOI: 10.1109/ACCESS.2023.3291411

Tullio, R. (2024). AI in Urban Mobility: How Copenhagen is Leading the Charge. *Journal of Smart City Innovation*, 4(2), 45–60.

Tunau, J., & Wanda, J. (2020). Inter-organizational collaboration in disaster management: A review. *Sustainability*, 12(1), 326.

Twigg, J., & Mosel, I. (2017). Emergent groups and spontaneous volunteers in urban disaster response. *Environment and Urbanization*, 29(2), 443–445. DOI: 10.1177/0956247817721413

Ullo, S. L., Langenkamp, M. S., Oikarinen, T. P., Del Rosso, M. P., Sebastianelli, A., Piccirillo, F., & Sica, S. (2019, July). Landslide geohazard assessment with convolutional neural networks using sentinel-2 imagery data. In IGARSS 2019-2019 IEEE International Geoscience and Remote Sensing Symposium (pp. 9646-9649). IEEE. DOI: 10.1109/IGARSS.2019.8898632

Umlauft, J.. (2016). A multi-sensor solution for near real-time situational awareness in disaster response. *Sensors (Basel)*, 16(1), 132. PMID: 26805842

UNESCO. (2023). DDS: De Digitale Stad / The Digital City. https://www.unesco.org/en/memory-world/dds-de-digitale-stad/digital-city?hub=80094#:~:text=On%20January%2015%2C%201994%2C%20De,unknown%20world%20of%20the%20Internet

UN-Habitat. (2022). *Global review of smart city governance practices.*

UN-Habitat. (2023). *Promoting Smart Cities for Sustainable Development.*

United Nations Economic and Social Commission for Asia and the Pacific. (2023). *Urban Cybersecurity Challenges in ASEAN Smart Cities.*

United Nations Office for Disaster Risk Reduction [UNDRR]. (2020). Global assessment report on disaster risk reduction. UNDER.

United Nations. (2017). World Population Prospects The 2017 Revision. https://population.un.org/Wpp/Publications/Files/Wpp2017_Keyfindings.Pdf

United Nations. (2022). *AI for Good Global Summit: Advancing AI to Benefit Humanity.*

Urban development. (n.d.). World Bank. https://www.worldbank.org/en/topic/urbandevelopment

Urban Mobility, E. I. T. (2023). CIVITAS Initiative: Transforming Urban Mobility Across Europe. EIT Urban Mobility. Retrieved from EIT Urban Mobility European Commission. (n.d.). Smart Cities. https://commission.europa.eu/eu-regional-and-urban-development/topics/cities-and-urban-development/city-initiatives/smart-cities_en#related-policies

Ushakov, D., Dudukalov, E., Shmatko, L., & Shatila, K. (2022a). Artificial Intelligence as a factor of public transportations system development. *Transportation Research Procedia*, 63, 2401–2408. DOI: 10.1016/j.trpro.2022.06.276

Vallor, S. (2023). Ethics in AI Development: Transparency and Public Trust. *AI & Society*, 38(3), 497–511. DOI: 10.1007/s00146-022-00492-5

Van Dijk, B., & Roos, L. (2021). *IoT and flood management in the Netherlands: Lessons from practice.* Water Management Review.

Varsha, V. R., Naganandini, S., & Hariharan, C. (2024). Utilizing AI and machine learning for natural disaster management: predicting natural disasters with AI and machine learning. In *Internet of Things and AI for Natural Disaster Management and Prediction* (pp. 279–304). IGI Global., DOI: 10.4018/979-8-3693-4284-8.ch013

Vashishth, T. K., Sharma, V., Sharma, K. K., Kumar, B., Panwar, R., & Chaudhary, S. (2024). AI-driven learning analytics for personalized feedback and assessment in higher education. In *Using traditional design methods to enhance AI-driven decision making* (pp. 206–230). IGI Global. DOI: 10.4018/979-8-3693-0639-0.ch009

Vasylchenko, A., Schols, Y., De Raedt, W., & Vandenbosch, G. (2009). Quality assessment of computational techniques and software tools for planar-antenna analysis. *IEEE Antennas & Propagation Magazine*, 51(1), 23–38. DOI: 10.1109/MAP.2009.4939017

Venkadesh, P., Divya, S. V., Marymariyal, P., & Keerthana, S. (2024). Predicting Natural Disasters With AI and Machine Learning. In *Utilizing AI and Machine Learning for Natural Disaster Management* (pp. 39–64). IGI Global., DOI: 10.4018/979-8-3693-3362-4.ch003

Verma, S., & Shah, A. (2021). Artificial intelligence and machine learning applications for flood forecasting: A systematic review. *Journal of Hydrology (Amsterdam)*, 603, 126740. DOI: 10.1016/j.jhydrol.2021.126740

Vermiglio, G.. (2020). Smart Cities for Disaster Resilience: A Review. *Sustainable Cities and Society*, 55, 102004.

Vesanen, J. (2007). What is personalization? A conceptual framework. *European Journal of Marketing*, 41(5/6), 409–418. DOI: 10.1108/03090560710737534

Visave, J. (2024). AI in Emergency Management: Ethical Considerations and Challenges. *Journal of Emergency Management and Disaster Communications*, 4(3), 1–19. DOI: 10.1142/S268998092450009X

Vishal, & V. Reddy, (2021). A Review on Internet of Things, Big Data, and Artificial Intelligence for Smart Cities. *Materials Today: Proceedings, 48*, 1610-1615.

Vishnu, S., Ramson, S. R. J., Senith, S., Anagnostopoulos, T., Abu-Mahfouz, A. M., Fan, Z., Srinivasan, S., & Kirubaraj, A. A. (2021). IoT-enabled solid waste management in smart cities. *Smart Cities*, 4(3), 1004–1017. DOI: 10.3390/smartcities4030053

Wagner, A., & Bojic, M. (2021). Privacy and data protection in IoT-enabled disaster response. *Journal of Information Ethics.*

Wang, M. (2021). *Active learning for improved damage detection and disaster response* (Doctoral dissertation).

Wang, Q., Chang, R., Xu, Q., Liu, X., Jian, I. Y., Ma, Y., & Wang, Y. (2021). The impact of personality traits on household energy conservation behavioral intentions – An empirical study based on theory of planned behavior in Xi'an. *Sustainable Energy Technologies and Assessments*, 43, 100949. DOI: 10.1016/j.seta.2020.100949

Wang, Q., & Liu, Y. (2023). Enhancing Collaborative Emergency Management: Leveraging Blockchain Distributed Ledger for Inter-Agency Data Sharing. *Proceedings of the 8th ACM SIGSPATIAL International Workshop on Security Response using GIS*. DOI: 10.1145/3615884.3629423

Wang, S., & Nishio, M. (2024). Review for vision-based structural damage evaluation in disasters focusing on nonlinearity. *Smart Structures and Systems*, 33(4), 263. DOI: 10.12989/sss.2024.33.4.263

Wang, Y., & Chen, H. (2022). Blockchain: A potential technology to improve the performance of collaborative emergency management with multi-agent participation. *International Journal of Disaster Risk Reduction*, 72, 102867.

Wang, Y.-H., Chen, Y.-L., Lai, Y.-H., Huang, Y.-M., Lin, C.-W., & Hsu, Y.-C. (2021). Development of an earthquake early warning system for the Taipei Rapid Transit System. *Sensors (Basel)*, 21(17), 5806. DOI: 10.3390/s21175806 PMID: 34502697

Wang, Y., Li, X., & Zhao, J. (2023). Hybrid quantum machine learning with satellite data to enhance flood forecasting capabilities: A case study of the Yangtze River Basin. *Water Science and Technology*, 88(4), 789–802.

Waqas Arshad, M., Murtza, I., & Abdullah Arshad, M. (2023). Applications of quantum computing in health sector: A review of current trends and future directions.

Weisberg, E. M., Chu, L. C., & Fishman, E. K. (2020). The first use of artificial intelligence (AI) in the ER: Triage not diagnosis. *Emergency Radiology*, 27(4), 361–366. DOI: 10.1007/s10140-020-01773-6 PMID: 32643069

Westerman, G., Bonnet, D., & McAfee, A. (2014). *Leading digital: Turning technology into business transformation*. Harvard Business Review Press.

White Paper on Artificial Intelligence: a European approach to excellence and trust | European Commission. (n.d.). European Commission. https://commission.europa.eu/document/d2ec4039-c5be-423a-81ef-b9e44e79825b_en

Whitehead, S., Mailley, J., Storer, I., McCardle, J., Torrens, G., & Farrell, G. (2008). In safe hands: A review of mobile phone anti-theft designs. springer science + business media, Vol. 14, Iss. 1, (2008): 39-60. https://doi.org/DOI: 10.1007/s10610-007-9040-9

Williams, S., & Singh, R. (2023). Economic synergies in AI development: Financing the future of smart cities. *Urban Economics Review*, 15(5), 300–320.

Wogi, L., Ayana, T., Morawiec, M., & Jąderko, A. (2022). A Comparative Study of Fuzzy SMC with Adaptive Fuzzy PID for Sensorless Speed Control of Six-Phase Induction Motor. *Energies*, 15(21), 21. Advance online publication. DOI: 10.3390/en15218183

World Economic Forum. (2020). *Governing smart cities: Policy benchmarks for ethical and responsible smart city development.*

WorldBank. (n.d.). *World Bank Group - International Development, Poverty and Sustainability*. https://www.worldbank.org/ext/en/home

Wright, D. (2023). Data Protection and Privacy in Cross-Border Contexts. *International Data Privacy Law*, 13(1), 1–12. DOI: 10.1093/idpl/ipac015

Wu, L., Liu, J., Vazquez, S., & Mazumder, S. K. (2021). Sliding mode control in power converters and drives: A review. IEEE/CAA Journal of Automatica Sinica, 9(3), 392-406.

Wu, L., Liu, R., Li, G., Gou, J., & Lei, Y. (2022, August). Landslide detection methods based on deep learning in remote sensing images. In 2022 29th International Conference on Geoinformatics (pp. 1-4). IEEE. DOI: 10.1109/Geoinformatics57846.2022.9963885

Wu, J., Wang, X., Dang, Y., & Lv, Z. (2022). Digital twins and artificial intelligence in transportation infrastructure: Classification, application, and future research directions. *Computers & Electrical Engineering*, 101, 107983. DOI: 10.1016/j.compeleceng.2022.107983

Wu, Y., Fu, G., Wu, Z., Wang, Y., Xie, X., Han, M., & Lyu, Q. (2022). A popular systemic accident model in China: Theory and applications of 24Model. *Safety Science*, 159, 106013. DOI: 10.1016/j.ssci.2022.106013

Wu, Y.-M., Chen, C.-C., & Lin, C.-W. (2021). Earthquake early warning system for high-speed rail in Taiwan. *Sensors (Basel)*, 21(11), 3874. DOI: 10.3390/s21113874 PMID: 34199735

Wu, Y.-M., Teng, T.-L., & Shin, T.-C. (2020). Earthquake early warning system in Taiwan: Current status and future challenges. *Territorial, Urban, and Environmental Planning*, 1(1), 1–15.

Xie, X., Fu, G., Xue, Y., Zhao, Z., Chen, P., Lu, B., & Jiang, S. (2018). Risk prediction and factors risk analysis based on IFOA-GRNN and apriori algorithms: Application of artificial intelligence in accident prevention. *Process Safety and Environmental Protection*, 122, 169–184. DOI: 10.1016/j.psep.2018.11.019

Yadav, R., Kumar, S., & Singh, P. (2021). Quantum machine learning with historical flood event analysis: Leveraging past data for better future predictions in hydrology systems. *Journal of Hydrology (Amsterdam)*, 598, 123–135.

Yang, B. (2023). Research on vehicle detection and recognition technology based on artificial intelligence. *Microprocessors and Microsystems*, 104937, 104937. Advance online publication. DOI: 10.1016/j.micpro.2023.104937

Yang, J., Liang, W., Wang, X., Li, S., Jiang, X., Mu, Y., & Zeng, S. (2024). DarkMor: A framework for darknet traffic detection that integrates local and spatial features. *Neurocomputing*, 607, 128377. DOI: 10.1016/j.neucom.2024.128377

Yigitcanlar, T., Kamruzzaman, M., Foth, M., Sabatini-Marques, J., Da Costa, E., & Ioppolo, G. (2019). Can cities become smart without being sustainable? A systematic review of the literature. *Sustainable Cities and Society*, 45, 348–365. DOI: 10.1016/j.scs.2018.11.033

Yuryeva, O., Kovaleva, N., & Shukhova, O. (2023). Increasing economic losses from natural disasters as a last decade trend. In *E3S Web of Conferences* (Vol. 458, p. 05005). EDP Sciences., DOI: 10.1051/e3sconf/202345805005

Zaheeruddin, M.. (2021). Optimization of ambulance deployment using genetic algorithm: A review. *Journal of Ambient Intelligence and Humanized Computing*, 12(10), 4735–4751.

Zanella, A., Bui, N., & Castellani, A. (2023). Smart Cities: An Overview of Smart City Development Worldwide. *Smart Cities*, 6(4), 577–593. DOI: 10.3390/smartcities6040039

Zang, X., Wang, T., Zhang, X., Gong, J., Gao, P., & Zhang, G. (2024). Encrypted malicious traffic detection based on natural language processing and deep learning. *Computer Networks*, 250, 110598. DOI: 10.1016/j.comnet.2024.110598

Zarour, M., Alenezi, M., Ansari, M. T. J., Pandey, A. K., Ahmad, M., Agrawal, A., Kumar, R., & Khan, R. A. (2021, April 16). Ensuring data integrity of healthcare information in the era of digital health. *Healthcare Technology Letters*, 8(3), 66–77. DOI: 10.1049/htl2.12008 PMID: 34035927

Zeb, A., Ali, Q., Saleem, M. Q., Awan, K. M., Alowayr, A. S., Uddin, J., & Bashir, F. (2019). 'A proposed IoT-enabled smart waste bin management system and efficient route selection. *Journal of Computer Networks and Communications*, 2019, 1–9. DOI: 10.1155/2019/7043674

Zeng, W., & Wu, L. (2023). A multi-source data-driven approach for drought forecasting using machine learning: A case study in the Yellow River Basin, China. *Journal of Hydrology (Amsterdam)*, 614, 128530. DOI: 10.1016/j.jhydrol.2022.128530

Zhang, L., & Chen, Q. (2022). Significant improvements in flood forecasting accuracy using quantum neural networks compared to classical methods: An empirical evaluation study on hydrological data sets from Asia-Pacific regions. Water Resources Research, 58(7), e2021WR030123.

Zhang, D., Yang, J., Li, F., Han, S., Qin, L., & Li, Q. (2022). Landslide risk prediction model using an attention-based temporal convolutional network connected to a recurrent neural network. *IEEE Access : Practical Innovations, Open Solutions*, 10, 37635–37645. DOI: 10.1109/ACCESS.2022.3165051

Zhang, Y., Zhao, Y., & Zhang, L. (2023). Machine learning for flood forecasting: A comprehensive review. *Journal of Hydrology (Amsterdam)*, 619, 126837. DOI: 10.1016/j.jhydrol.2023.126837

Zhao, D., Qin, H., Song, B., Zhang, Y., Du, X., & Guizani, M. (2020). A reinforcement learning method for joint mode selection and power adaptation in the V2V communication network in 5G. *IEEE Transactions on Cognitive Communications and Networking*, 6(2), 452–463. DOI: 10.1109/TCCN.2020.2983170

Zharmukhanbetov, S., & Singh, C. P. (2023, November). Enhancing flipped classroom engagement and personalized learning through AI-powered adaptive content delivery. In *2023 3rd International Conference on Technological Advancements in Computational Sciences (ICTACS)* (pp. 1411-1416). IEEE. DOI: 10.1109/ICTACS59847.2023.10389841

Zheng, X.. (2004). Human-robot interaction for search and rescue. *Autonomous Robots*, 16(1), 97–113.

Zhou, J., Gao, G., & Luo, X. (2023, September). Landslide Displacement Prediction Based on EEMD and CNN-LSTM. In *2023 IEEE 6th International Conference on Information Systems and Computer Aided Education (ICISCAE)* (pp. 618-622). IEEE. DOI: 10.1109/ICISCAE59047.2023.10391866

Zhou, L., Wang, R., Zhu, Y., Li, J., & Luo, X. (2020, September). Landslide displacement prediction based on integrated neural network. In *2020 8th International Conference on Digital Home (ICDH)* (pp. 57-62). IEEE. DOI: 10.1109/ICDH51081.2020.00018

Zhou, D., & Liu, J. (2024). Quantum computing applications in hydrology: Transforming flood prediction through advanced algorithms and data integration techniques. *Hydrological Processes*, 38(5), e14345.

Zhou, H., Dai, K., Pirasteh, S., Li, R., Xiang, J., & Li, Z. (2023). InSAR spatial-heterogeneity tropospheric delay correction in steep mountainous areas based on deep learning for landslides monitoring. *IEEE Transactions on Geoscience and Remote Sensing*, 61, 1–14. DOI: 10.1109/TGRS.2023.3307477

Zhou, X., & Chen, Y. (2014). A GIS-based earthquake emergency response system for lifeline facilities. *Natural Hazards*, 74(2), 1105–1123.

Zhou, Z., Liu, Y., & Kim, P. (2022). *Smart infrastructure and IoT for disaster resilience*. Urban Resilience Journal.

Zhu, S., Li, D., & Feng, H. (2019). Is smart city resilient? Evidence from China. *Sustainable Cities and Society*, 50, 101636. DOI: 10.1016/j.scs.2019.101636

Zhu, S., Li, D., Feng, H., Gu, T., Hewage, K., & Sadiq, R. (2020). Smart city and resilient city: Differences and connections. *Wiley Interdisciplinary Reviews. Data Mining and Knowledge Discovery*, 10(6), e1388. DOI: 10.1002/widm.1388

Ziakis, C., & Vlachopoulou, M. (2023). Artificial intelligence in digital marketing: Insights from a comprehensive review. *Information (Basel)*, 14(12), 664.

Zohuri, B., & Mossavar-Rahmani, F. (2024). Revolutionizing education: The dynamic synergy of personalized learning and artificial intelligence. *International Journal of Advanced Engineering and Management Research*, 9(1), 143–153. DOI: 10.51505/ijaemr.2024.9111

Zuboff, S. (2019). *The age of surveillance capitalism: The fight for a human future at the new frontier of power*. PublicAffairs. Noble, S. U. (2018). *Algorithms of oppression: How search engines reinforce racism*. NYU Press.

Zuboff, S. (2019). *The age of surveillance capitalism: The fight for a human future at the new frontier of power*. PublicAffairs.

Zunder, T. H., Carrillo, A., & Robinson, M. (2023). Consensus and dissent for rail visions of 2050: Insights from a Delphi study. *Transportation Research Procedia*, 72, 844–851. DOI: 10.1016/j.trpro.2023.11.476

Zuo, Y., & Qi, Z. (2022). A Blockchain-Based IoT Framework for Oil Field Remote Monitoring and Control. *IEEE Access : Practical Innovations, Open Solutions*, 10, 2497–2514. DOI: 10.1109/ACCESS.2021.3139582

Zygiaris, S. (2013). Smart city reference model: Assisting planners to conceptualize the building of smart city innovation ecosystems. *Journal of the Knowledge Economy*, 4(2), 217–231. DOI: 10.1007/s13132-012-0089-4

About the Contributors

Froilan Mobo He is a Doctor of Public Administration graduate from the Urdaneta City University Class of 2016 and a graduate of the 2nd Doctorate Degree (Ph.D.) in Development Education program at the Central Luzon State University, Nueva Ecija, Philippines, Class of 2022. On March 11, 2024, Dr. Mobo was accredited and reclassified by the Commission on Higher Education (CHED) to the position of Professor II in the Philippine Merchant Marine Academy (PMMA), and this allowed him to work with different international research institutions, such as the Director and Research Consultant of the IKSAD Research Institute, Turkey. At present, he is in the process of finishing his 3rd master's degree, leading to social studies education at Bicol University. Recently, Dr. Mobo passed Batch 3—Certified Research Professional—and ranked in the top 5 in the National Examination and also passed the Certified Human Resource Associate (CHRA). He was appointed Editor-in-Chief of the International Journal of Multidisciplinary: Applied Business and Education Research, Malang, Indonesia, and appointed as a technical research evaluator by the Department of Science and Technology and Book Chapter Editor in IGI Global Publisher based in USA. He has published 111 research articles with 217 citations indexed in the Web of Science, Scopus, and ASEAN Citation Index

***

John Mark R. Asio currently works and designated as Research Director of the Research Development and Community Extension Services at Gordon College, Olongapo City. He does research in Higher Education, Educational Policy, Educational Management, Pedagogy, Social Sciences, Public Administration, and Organizational Management. He is also a Registered Nurse, Registered Midwife, Licensed Professional Teacher, and Civil Service Eligible (Professional).

Manjunatha Badiger is an Assistant Professor in the Department of VLSI Design & Technology at NMAM Institute of Technology, Nitte. He has over 12 years of experience in academics, research, and administration. Dr. Badiger holds a Ph.D. from Visvesvaraya Technological University, an M.Tech in VLSI & Embedded System Design from Reva Institute of Technology and Management, and a B.E. in Electronics & Communication Engineering from NMAM Institute of Technology, Nitte. His research interests span Machine Learning, Deep Learning, Image Processing, and VLSI. Dr. Badiger has made significant contributions to national and international journals, book chapters, and has completed various NPTEL courses. Previously, he served as an Assistant Professor in the Department of Electronics and Communication Engineering at Sahyadri College of Engineering & Management and Shree Devi Institute of Technology. Dr. Badiger is accessible via email at manjunatha.badiger@nitte.edu.in or badiger_manju@yahoo.com.

Sohini Banerjee is an Assistant Professor and head of the department of Computer Science and Engineering department at Abacus Institute of Engineering and Management.

Sumanta Bhattacharya from Salkia, Howrah, West Bengal is an Expert commissioner member IUCN Environmental, economics and social policy, Individual Policy Analyst (ADB). He is among the top 2% world scientist (AD Index), Special civilian member Ministry of Defence, (USI) Member at Ministry of Culture (AS). Member of Ministry of Defence { [Defence and strategic analysis] IDSA }, Member of World Economic Forum, littérateur at National Economic Research Bureau, Member of Ministry of Personnel, Pension and Public grievance (IIPA) He completed his B.Tech, M.Tech in Textile Technology along with that he has MA in Development studies, M.Sc in Environmental Science, LLB, MA in security and defence law from NALSAR, MA in Economics, MBA, Post Graduate Diploma in Environment and Sustainable Development, DIA&D, DS&DS, MPI(Oxford University) . Being a research scholar he has 216 research papers published in International and Scopus index journal, Wiley, Springer, Hindawi in various sectors like law, economic, Development studies, Textiles, Climate Change, SDGs, Public policy, Nanotechnology, Biotechnology, cancer, environment studies, 57 Book chapters published, Edited 2 books, wrote 1 book, 80 patents International and National, 4 copyrights, attended 93 International and National conference and presented papers and won 25 awards for his excellent in education, policy making and innovation which includes Best Water Management Solution Award 2022, Young Researcher Award 2020, Bravo International Book of Records, Mahatma Gandhi Ratna Award 2020, Sustainable Technology Award 2022, Global Change Maker Award 2020, Best Water Management Solution Award at the Rural & Urban Development Summit & Awards 2022, New Delhi, The Progress Global Awards 2023 for Excellence in Research, GIS -Person of the year – Water Conservation, Green India award, 2023, Global Legend Award 2023 by GSIL Educational Society . He is also a member of various International and National organization which includes IEI, IICHE, Asiatic Society, IIPA, IDSA, YOUNGO, IUCN, USI and many more .

Balaji Chandrasekaran is working as an Assistant Professor in the Department of Computer Science and Engineering, Anna University, Chennai

Kavitha Dhanushkodi is currently working as an Assistant Professor in the School of Computer Science and Engineering(SCOPE) at Vellore Institute of Technology, Chennai Campus, Chennai, Tamil Nadu, India. She has completed Ph.D and Master of Engineering in Computer Science and Engineering from Anna University, Chennai, Her research interests are mainly focused on Software Security, Internet of Things and Cyber Security. She has an overall teaching experience of 14 years in various academic institutions. She has published more than 40 Research papers to her credit in reputed international journals and conferences with high impact factors.

Sudeshna Das is an Assistant Professor of the department of Computer Science and Engineering department at Abacus Institute of Engineering and Management.

Pranoy Debnath have done my B.Tech and M.Tech both from NIT Agartala. I have also completed my Ph.D. from IIT (ISM) Dhanbad. Now I am pursuing my Post Doctoral position in the Civil Engineering Department of IIT Bombay. My research interests are Structural dynamics, Earthquake engineering, URM, Finite-element method, Strengthening, Vulnerability, Pushover analysis.

G. Sowmya is an Assistant Professor in the Department of Computer Science and Engineering at MLRIT, Hyderabad, with over a decade of teaching experience. Currently pursuing a Ph.D. in CSE from JNTUH, her research focuses on the formal verification of smart contracts. Sowmya holds an M.Tech and B.Tech in CSE and has received the prestigious Prathibha Award for academic excellence. She has contributed to numerous international journals and conferences, authored several patents, and actively engages in various FDPs and certifications, particularly in Blockchain, AI, and IoT. She has also conducted workshops and webinars on cutting-edge technologies such as blockchain and machine learning

Smrithy G S received her B. Tech degree in Information Technology from the University of Kerala, India, in 2011, her M. Tech degree in Engineering Statistics from the Cochin University of Science and Technology, India, in 2014 and PhD from National Institute of Technology, Tiruchirappalli, India in 2021. Currently, she is working as an Assistant Professor in the School of Computer Science and Engineering at VIT Chennai. Her research interests include statistical computing, anomaly detection, cloud computing, and the internet of things.

G. Sangeetha is an Associate Professor at R.M.K. College of Engineering and Technology, Puduvoyal, with nearly 20 years of teaching experience. She earned her B.E. in Computer Science and Engineering in 2003 and her M.E. in Computer Science and Engineering in 2005. She also holds a Ph.D. in Information and Communication Engineering from Anna University, Chennai. In 2006, she received the "Best Teacher Award." She has published her innovative ideas for patents and many papers in international journals. Actively engaged in research, she has participated in various consultancy works. Her areas of interest include Distributed Computing, Cloud Computing, Data Security, Data Science, and Machine Learning. She has contributed numerous book chapters to IGI Global publications and has delivered guest lectures and seminars on recent computing trends and technologies at various college. Scopus Id:57226863957, Orcid Id: 0000-0001-7347-2162, Google Scholar Id: VGEoJc0AAAAJ, Web of Science Researcher ID: HOH-6833-2023.es.

Vishal Jain is presently working as Professor at the Department of Computer Science and Engineering, School of Engineering and Technology, Sharda University, Greater Noida, India. Before that, he worked for several years as an Associate Professor at Bharati Vidyapeeth's Institute of Computer Applications and Management (BVICAM), New Delhi. He has more than 18 years of experience in the academics. He has earned degrees: Ph.D (CSE), M.Tech (CSE), MBA (HR), MCA, MCP, and CCNA. He has more than 2200 research citations with Google Scholar (h-index score 24 and i-10 index 55) and has authored more than 150 research papers in professional journals and conferences. He has authored and edited more than 60 books (most of them are indexed at the Scopus) with various reputed publishers, including Elsevier, Springer, IET, Apple Academic Press, CRC, Taylor and Francis Group, Scrivener, Wiley, Emerald, NOVA Science, River Publishers, IGI-Global and Bentham Science. He has more than 200 Scopus publications. His research areas include machine learning, information retrieval, semantic web, ontology engineering, data mining, ad hoc networks, sensor networks and network security.

S.V Annlin Jeba, professor in computer science and Engineering department, Sree Buddha college of Engineering, Pattoor, Area of Interest: Network Security, Machine learning

Varuna Kumara a Research Scholar in the department of Electronics Engineering at JAIN (Deemed to be University), Bengaluru, India. He also received his B. E and M. Tech from Visvesvaraya Technological University, Belagavi, India in 2009 and 2012 respectively. He is currently Assistant Professor at Electronics and Communication Engineering in Moodlakatte Institute of Technology, Kundapura, India. His research interests are in Artificial Intelligence, Signal Processing, Control Systems.

Jeni Precilla M is currently an M.Tech student in the School of Computer Science and Engineering, VIT Chennai.

T.C. Manjunath is a distinguished professor in the Department of Computer Science & Engineering at Rajarajeswari College of Engineering, located in Bangalore, Karnataka. He specializes in IoT Cybersecurity and Blockchain Technology and currently serves as the Dean of Research & Development at the institution. With extensive academic and research experience, Dr. Manjunath has contributed significantly to the advancement of technology in his fields of expertise. His work and leadership in research have made him a prominent figure in the academic community. The college is located on Mysore Road, Ramohalli Cross, Kumbalgodu Post, in Bangalore, with the postal code 560074.

Geetha Manoharan is currently working in Telangana as an assistant professor at SR University. She is the university-level PhD program coordinator and has also been given the additional responsibility of In Charge Director of Publications and Patents under the Research Division at SR University. Under her tutelage, students are inspired to reach their full potential in all areas of their education and beyond through experiential learning. It creates an atmosphere conducive to the growth of students into independent thinkers and avid readers. She has more than ten years of experience across the board in the business world, academia,and the academy. She has a keen interest in the study of organizational behavior and management. More than forty articles and books have been published in scholarly venues such as UGC-refereed, SCOPUS, Web of Science, and Springer. Over the past six-plus years, she has participated in varied research and student exchange programs at both the national and international levels. A total of five of her collaborative innovations in this area have already been published and patented. Emotional intelligence, selfefficacy, and work-life balance are among her specialties. She organizes programs for academic organizations. She belongs to several professional organizations, including the CMA and the CPC. The TIPS GLOBAL Institute of Coimbatore has recognized her twice (in 2017 and 2018) for her outstanding academic performance.

Wes Harven Guillemer Maravilla has been an esteemed EFL/ESL language expert in Asia since 2019, bringing his expertise to diverse learners ranging from primary school students to adults. An advocate for the Communicative Language Approach and Project-Based Approach, Mr. Maravilla has dedicated his career to enhancing English language education. He holds a Master of Arts in Education, majoring in English Language Teaching from Chiang Kai Shek College in Manila, Philippines. Mr. Maravilla's research interests are vast and encompass technology and innovation, artificial intelligence, EFL/ESL methodologies, sustainable development goals, and specialized areas of English, including reading, oral expression, composition writing, and effective listening. His scholarly contributions include notable publications such as “Content-Based Learning,” “Communicative Language Approach,” “Assessment of Effectiveness of Zoom Platform on English Language Learning,” and studies on “Thai and Taiwanese University Students' English Language Learning Experiences.” Driven by a passion for improving language education, Mr. Maravilla constantly seeks new teaching methods and technologies to create effective and inclusive classrooms. He is also committed to using innovative strategies and technology to make everyday life easier for the community. By embracing modern educational tools, he strives to make learning more accessible and convenient, enhancing the well-being and growth of the society he serves.

Archan Mitra is an Assistant Professor at School of Media Studies (SOMS) at Presidency University, Bangalore. He is the author of two book "Cases for Classroom Media and Entertainment Business" and "Multiverse and Media", he also has other several edited books to his credit. He has done his doctorate from Visva-Bharati Santiniketan, West Bengal in the field of "environmental informatics and communication for sustainability". In addition to that he is a certified Science Communicator and Journalism from Indian Science Communication Society (ISCOS), certified Corporate Trainer with Amity Institute of Training and Development, Certified Social Media Network Analyst. He has a strong interest in environmental communication. He was awarded certificate of merit by PRSI, Kolkata Chapter and Medal of Honor by Journalistic Club of Kolkata. He was working as a research assistant with the World Bank's "Environmental Capacity Building in Southeast Asia" project at IIM Kashipur. He was instrumental in launching the World Bank's Green MBA MOOC, he has also assisted in the research project on Uttarakhand disaster mitigation by ICSSR, the leading research on Uttarakhand disaster.

Hitesh Mohapatra received the B.E. degree in Information Technology from Biju Patnaik University of Technology (BPUT), Odisha in 2006, and the MTech. Degree in CSE from Odisha University of Technology and Research (OUTR) University, Odisha in 2009. He received his Ph.D. in Computer Science & Engineering in 2021 from Veer Surendra Sai University of Technology (VSSUT), Burla, India. He has contributed 30 SCI and Scopus indexed research papers, 21 international/national conferences and books on Software Engineering and C Programming respectively. He has 15 years of teaching experience both in industry and academia. He has served the research community with various capacities like session chair, technical chair, keynote speaker etc. His research interests include wireless sensor network, smart city, smart grid and smart water. Currently he is working as Associate professor at School of Computer Engineering, KIIT DU, Bhubaneswar, and Odisha.

Anshit Mukherjee is presently pursuing Bachelor of Technology (Hons.) in Computer Science and Engineering from the Abacus Institute of Engineering and Management, which is affiliated with Maulana Abul Kalam Azad University of Technology, and is presently in final year. He was the zonal topper of the AMBIKA KALNA zone in I.C.S.E. 2019 (10 TH STANDARD) and the second zonal topper in I.S.C. 2021 (12 TH STANDARD). He is also awarded with NPTEL Discipline Star in July–December 2023 and Motivated Learner in January-June 2024.

P. Selvakumar is currently an Associate Professor in the Department of Science and Humanities at Nehru Institute of Technology, Coimbatore, Tamil Nadu, India. Since 2023, he has served as the Department IQAC Coordinator at the institute. Prior to joining Nehru Institute of Technology, he held positions as an Assistant Professor in the Department of Science and Humanities various engineering colleges in Coimbatore, Tamil Nadu, India, spanning from 2011 to 2023. Dr.P. Selvakumar's academic and research interests lie primarily in the field of phytochemistry. He has made significant contributions to his field, with an impressive record that includes 30 journal publications, 12 patents, and 20 conference papers. Additionally, he has organized 5 seminars and conferences, further demonstrating his commitment to academic and scientific exchange. Throughout his career, Dr. P.Selvakumar has been recognized for his dedication and excellence in teaching, having been honored with several awards, including the Best Teacher Awards. He holds a prominent position in the academic community, contributing actively to research.

Poonam Rani is a seasoned data analyst with 8.5 years of experience in data analysis, interpretation and full-cycle data processing. Holds an MSc in Statistics from Panjab University, Chandigarh. Technical expertise includes Power BI, SQL, SPSS, R and Python, which has been utilized to develop dashboards, optimize workflows and derive actionable insights. Contributed to several Scopus-indexed journals, focusing on healthcare research and statistical analysis. Currently, she serves as an Assistant Professor at Chandigarh Group of Colleges, bringing her expertise to both academic and practical applications.

Thangam S, BE, ME, Ph.D., Dr. Thangam S. currently serves as an Assistant Professor (SG) at the Amrita School of Computing, Bengaluru.,Amrita Vishwa Vidyapeetham, India. Her area of interest in research includes Service Oriented Architecture, Networks, IoT, Artificial Intelligence, Data Structures and Cloud Computing. She completed her Ph. D. in Computer Science and Engineering from Anna University, Chennai. India. She has 26 years of teaching experience. She has published her research works in 60 international journals and conferences. She is a member of ISTE and IEEE

Col Ravinder Singh has served in the military for 35 years. An alumnus of National Defence Academy and Indian Military Academy, he has a vast military experience in various terrains. He holds a Master of Science in (Technology) Weapon Systems from Pune University and also qualified in Disaster Management from Gujarat Institute of Disaster Management, Ahmedabad. He also holds a Post Graduate Diploma in Mass Communication and Journalism from National Institute of Mass Communication and Journalism, Ahmedabad He conducts lectures on Disaster Management at Gujarat Institute of Disaster Management, Colleges, schools and for Associate NCC officers and NCC cadets. He is also holding a degree in MBA from ICFAI University. He has also completed Post Graduate Diploma in Disaster Management from IGNOU and currently he is research scholar with SR University.

Muhammad Usman Tariq has more than 16+ year's experience in industry and academia. He has authored more than 200+ research articles, 110+ case studies, 115+ book chapters and several books other than 4 patents. He is founder and CEO of The Case HQ, a unique repository for courses, narrative and video case studies. He has been working as a consultant and trainer for industries representing six sigma, quality, health and safety, environmental systems, project management, and information security standards. His work has encompassed sectors in aviation, manufacturing, food, hospitality, education, finance, research, software and transportation. He has diverse and significant experience working with accreditation agencies of ABET, ACBSP, AACSB, WASC, CAA, EFQM and NCEAC. Additionally, Dr. Tariq has operational experience in incubators, research labs, government research projects, private sector startups, program creation and management at various industrial and academic levels. He is Certified Higher Education Teacher from Harvard University, USA, Certified Online Educator from HMBSU, Certified Six Sigma Master Black Belt and has been awarded PFHEA, SFSEDA, SMIEEE, and CMBE.

Ajay Upadhyaya is presently working as a Professor in the Department of Computer Engineering at SAL Engineering & Technical Institute, SAL Education, Ahmedabad, Gujarat. He has completed his PhD in Computer Engineering and his M.Tech. in Computer Science and Engineering. He has 18 years of teaching experience. His areas of specialisation include Mobile and Vehicular Ad-hoc Networks, Security in Networks, Machine Learning, Data Science, and IOT. In addition to his academic endeavours, he participates actively in faculty development programs, short-term initiatives, and seminars. He conducted more than 120 workshops and training programs on various technical topics at different institutes for students and faculty members. He has more than 22 publications in various refereed national and international conferences and journals.

Index

Symbols

A

B

C

D

E

F

G

H

I

J

L

M

N

O

P

Q

R

S

T

U

V

W

www.ingramcontent.com/pod-product-compliance
Lightning Source LLC
LaVergne TN
LVHW062105300926
849161LV00010B/2375